W0099531

BIRKHÄUSER

Frontiers in Mathematics

Advisory Editorial Board

Leonid Bunimovich (Georgia Institute of Technology, Atlanta, USA)
Benoît Perthame (Ecole Normale Supérieure, Paris, France)
Laurent Saloff-Coste (Cornell University, Rhodes Hall, USA)
Igor Shparlinski (Macquarie University, New South Wales, Australia)
Wolfgang Sprössig (TU Bergakademie, Freiberg, Germany)
Cédric Villani (Ecole Normale Supérieure, Lyon, France)

Victor D. Didenko
Bernd Silbermann

Approximation of Additive Convolution-Like Operators

Real C*-Algebra Approach

Birkhäuser Verlag
Basel · Boston · Berlin

Authors:

Victor D. Didenko
Department of Mathematics
University of Brunei Darussalam
Gadong BE 1410
Brunei
e-mail: diviol@gmail.com
victor@fos.ubd.edu.bn

Bernd Silbermann
Fakultät für Mathematik
TU Chemnitz
09107 Chemnitz
Germany
e-mail: bernd.silbermann@mathematik.tu-chemnitz.de

2000 Mathematical Subject Classification: 31-xx, 35-xx,45-xx, 46-xx, 47-xx, 65-xx, 74-xx, 76-xx, 76-xx. In particular: 31A10, 31A30, 35Q15, 35Q30, 45A05, 45B05, 45Exx, 45L05, 45P05, 46N20, 46N40, 47B35, 47Gxx, 47N40, 65E05, 65Jxx, 65R20, 74B10, 74G15, 74Kxx, 74S15, 76D05

Library of Congress Control Number: 2008925217

Bibliographic information published by Die Deutsche Bibliothek
Die Deutsche Bibliothek lists this publication in the Deutsche Nationalbibliografie; detailed bibliographic data is available in the Internet at <http://dnb.ddb.de>.

ISBN 978-3-7643-8750-1 Birkhäuser Verlag AG, Basel · Boston · Berlin

This work is subject to copyright. All rights are reserved, whether the whole or part of the material is concerned, specifically the rights of translation, reprinting, re-use of illustrations, recitation, broadcasting, reproduction on microfilms or in other ways, and storage in data banks. For any kind of use permission of the copyright owner must be obtained.

© 2008 Birkhäuser Verlag AG
Basel · Boston · Berlin
P.O. Box 133, CH-4010 Basel, Switzerland
Part of Springer Science+Business Media
Printed on acid-free paper produced from chlorine-free pulp. TCF ∞
Printed in Germany

ISBN 978-3-7643-8750-1 e-ISBN 978-3-7643-8751-8

9 8 7 6 5 4 3 2 1 www.birkhauser.ch

Contents

Preface

Various aspects of numerical analysis for equations arising in boundary integral equation methods have been the subject of several books published in the last 15 years [95, 102, 183, 196, 198]. Prominent examples include various classes of one-dimensional singular integral equations or equations related to single and double layer potentials. Usually, a mathematically rigorous foundation and error analysis for the approximate solution of such equations is by no means an easy task. One reason is the fact that boundary integral operators generally are neither integral operators of the form identity plus compact operator nor identity plus an operator with a small norm. Consequently, existing standard theories for the numerical analysis of Fredholm integral equations of the second kind are not applicable. In the last 15 years it became clear that the Banach algebra technique is a powerful tool to analyze the stability problem for relevant approximation methods [102, 103, 183, 189]. The starting point for this approach is the observation that the stability problem is an invertibility problem in a certain Banach or C^*-algebra. As a rule, this algebra is very complicated – and one has to find relevant subalgebras to use such tools as local principles and representation theory.

However, in various applications there often arise continuous operators acting on complex Banach spaces that are not linear but only additive – i.e.,

$$A(x + y) = Ax + Ay$$

for all x, y from a given Banach space. It is easily seen that additive operators are $\mathbb{R}$-linear provided they are continuous[1]. As an example, let us mention the one-dimensional singular integral operators with conjugation often arising in mechanics. It is known that the study of such operators can be reduced to $\mathbb{C}$-linear operators, but with matrix-valued coefficients. In passing note that this observation is one of a number of motivations to study singular integral operators with matrix-valued coefficients.

The present book is devoted to numerical analysis for certain classes of additive operators and related equations, including singular integral operators with conjugation, the Riemann-Hilbert problem, Mellin operators with conjugation and the famous Muskhelishvili equation. Until now, most relevant material is only

[1] Here and subsequently, $\mathbb{R}$ and $\mathbb{C}$ denote the fields of real and complex numbers, respectively.

found in journal papers, and there is no book offering a systematic study of this topic. Banach algebras play an important role in this book. However, the algebras that arise are not complex but real, and are not as familiar as complex algebras. Therefore, here we present certain results on real algebras and demonstrate their use in stability problems. In particular, we obtain stability conditions for various approximation methods, including spline Galerkin, collocation, qualocation and quadrature methods for equations with additive operators for both smooth and non-smooth data. Error analysis and convergence rates are present only occasionally, since rather more standard.

This book is addressed to a wide audience. We hope that it can be useful for both mathematicians working in theoretical fields of numerical analysis and engineers wishing to have practically realizable concepts for computations. Let us give a short overview of the content of this book.

Chapter 1 contains theoretical background. Here we have collected facts of functional analysis, necessary for understanding the approach proposed. Since real C^*-algebras play an important role in our investigations, elementary properties of such algebras are discussed. Moreover, a method to obtain real C^*-algebras by extending complex C^*-algebras (by adding a special element m) is described. Features of this procedure have previously been used in the study of one-dimensional singular integral equations with conjugation. As already mentioned, the stability problem for operator sequences can be interpreted as an invertibility problem in suitable real or complex Banach algebras. Thus we are accustomed to studying invertibility in Banach algebras or, more specifically, in C^*-algebras. Over the last 40 years certain concepts known as local principles were worked out. We present related results with special attention paid to the case of real algebras. The concluding part of Chapter 1 is devoted to the theory of singular integral operators and to Mellin operators. It is notable that all operators in this book have the property that *locally* they are Mellin operators.

Chapter 2 deals with polynomial and spline approximation methods for the Cauchy singular integral equation

$$(A\varphi)(t) = a(t)\varphi(t) + \frac{b(t)}{\pi i}\int_{\Gamma_0} \frac{\varphi(\tau)}{\tau - t}d\tau + \overline{c(t)\varphi(t)} + \overline{\frac{d(t)}{\pi i}\int_{\Gamma_0} \frac{\varphi(\tau)}{\tau - t}d\tau} = f(t)\,,$$

in the space $L^2(\Gamma_0)$, where Γ_0 is the unit circle with center at the origin and the functions a, b, c, d are continuous or piecewise continuous. For operators A without conjugation (i.e., $c = d \equiv 0$), there is a vast literature concerning the approximation methods under consideration (see e.g., [102, 183] and comments and remarks for the related chapters of the present book). Thus various complex C^*-algebras generated by approximation sequences for singular integral operators are completely described. Such algebras can be extended to real C^*-algebras that contain operator sequences associated with approximation methods for singular integral equations with conjugation. In particular, a real C^*-algebra generated by paired circulants and by the operator of complex conjugation is studied. This algebra contains a variety of approximation sequences, including spline Galerkin and spline collocation methods sequences, and others arising in quadrature and qualocation methods. The stability result is that a sequence from this algebra is stable if and only if a family of associated operators consists of invertible elements only. In the case of a simple closed Lyapunov contour Γ, the study of approximation methods for the operators mentioned can be reduced to the case of Γ_0.

Chapter 3 presents approximation methods for the following Riemann-Hilbert problem: Given an $(m\times m)$-matrix function G and a real vector-function f on Γ_0, find a vector-function φ that is analytic in the unit disc $\mathbf{D} := \{z \in \mathbb{C}, |z| < 1\}$, and such that $\operatorname{Im}\varphi(0) = 0$, and

$$\frac{1}{2}\left(G\varphi + \overline{G\varphi}\right) = f$$

on Γ_0. The new aspect is that the operator corresponding to this problem acts in a pair of spaces, so the algebraic methods used to study the stability of related approximation sequences have to be modified. This is done by using para-algebras. The same concept is employed to study approximation sequences associated with the generalized Riemann-Hilbert-Poincaré problem.

Chapter 4 is again concerned with approximation methods for the Cauchy singular integral equations with conjugation, but more general conditions are imposed on the curve Γ. Thus we now assume that Γ is a simple open or closed piecewise smooth curve in the complex plane $\mathbb{C}$. It is notable that the double layer potential operator is contained in the aforementioned class of operators. Given smooth boundaries, the stability of the corresponding projection methods for the double layer potential operator can be studied without great effort, since this operator is compact. However, if Γ is piecewise smooth, the algebras of Mellin operators with conjugation have to be invoked. Using this approach, we study various approximation sequences. As before, the stability of these sequences relies on the invertibility of the members of families of associated operators. The invertibility of the occurring operators is extremely difficult to check, especially if they are connected with corner points, so approximation methods based on cut-off techniques are also studied. This approach allows us to simplify conditions of the applicability of the corresponding methods.

Chapter 5 is devoted to the famous Muskelishvilli equation

$$(R\varphi)(t) = -k\overline{\varphi(t)} - \frac{k}{2\pi i}\int_\Gamma \overline{\varphi(\tau)}\,d\log\frac{\overline{\tau}-\overline{t}}{\tau - t} - \frac{1}{2\pi}\int_\Gamma \varphi(\tau)\,d\frac{\overline{\tau}-\overline{t}}{\tau - t} = f_0(t)\,, \qquad (1)$$

and its approximate solution. The Muskhelishvili equation arose in investigation and solution of various biharmonic problems, especially in elasticity theory and hydrodynamics. Notwithstanding its exceptional importance, approximation methods for this equation have not been developed, mainly due to the fact that the operator R is not invertible in the functional spaces under interest. Fortunately,

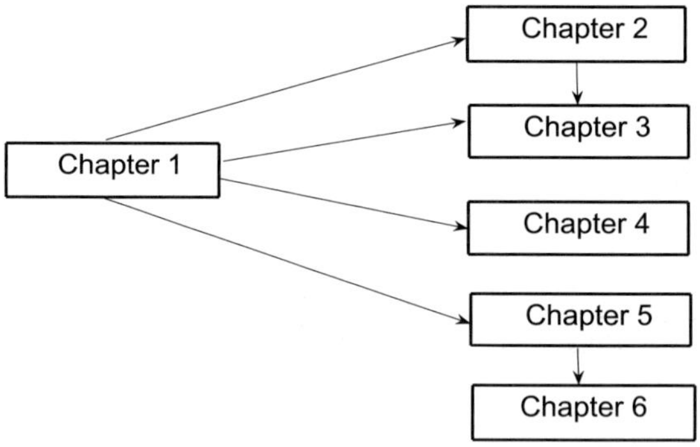

Figure 1: Book structure

this can be corrected by introducing a new operator that possesses all the necessary properties in order to find approximate solutions of the Muskhelishvili equation.

The idea of such correction is due to D.I. Sherman, although he only studied the solvability of the equation – not the invertibility of the associated operator, which is very important for the stability of approximation methods. Here we present all the results needed to construct and study projection methods for equation (1) in spaces L_p with weight. Let us note that we again use the fact that *locally* the operator R is a Mellin convolution operator.

Finally, Chapter 6 presents a few numerical results showing that the proposed approximation methods behave fairly well.

How to read this book? Probably, the best way is to single out a topic of interest and immediately read the related chapter. (If necessary, the reader can consult Chapter 1 for some background.) Of course, the later chapters of the book contain some material from chapters other, then the first but all connections can be easily traced. In particular, Chapter 3 also uses results from Chapter 2. The connection between different parts of the book is shown in Figure 1.

Although the attitudes and approaches of this book are solely the responsibility of the authors, we are indebted to our colleagues, friends, and collaborators for useful suggestions and ideas. It is a pleasure to mention here Roger Hosking and Steffen Roch, who read the early drafts of the manuscript and saved us from a number of embarrassing solecisms and ambiguities with detailed criticisms and generous advises. Ezio Venturino provided numerical examples and graphs presented in Chapter 6. We are grateful to Wolfgang Sprössig who has significantly facilitated our work. Finally our gratitude is due to an anonymous reviewer who contributed substantial improvement of this book.

Chapter 1

Complex and Real Algebras

A complex algebra is an associative ring $\mathcal{A}$ which is also a complex vector space. It is assumed that vector space addition and ring addition coincide and that the operations of multiplication and multiplication by scalars satisfy the relation

$$\lambda(xy) = (\lambda x)y = x(\lambda y) \tag{1.1}$$

for all $x, y \in \mathcal{A}$ and for all complex numbers λ.

Correspondingly, a real algebra is an associative ring $\mathcal{A}$ which is a real vector space with the ring and vector space additions coinciding, and with relation (1.1) satisfied for all $x, y \in \mathcal{A}$ and for all *real* numbers λ.

Obviously, every complex algebra is a real algebra but the reverse inclusion does not always hold. One reason for this is the absence of the operation of multiplication by complex scalars in $\mathcal{A}$. However, many algebras emerging from applications do have this operation but fail to satisfy relation (1.1). As a result, standard tools used in the complex situation either are not available or have to be properly redesigned.

1.1 Complex and Real C^*-Algebras

Definition 1.1.1. *A complex algebra $\mathcal{A}$ is called an algebra with involution or a *-algebra if there exists a map $^* : \mathcal{A} \to \mathcal{A}$, called involution, such that $(a+b)^* = a^* + b^*$, $(ab)^* = b^*a^*$, $(\lambda a)^* = \overline{\lambda} a^*$ and $(a^*)^* = a$ for all $a, b \in \mathcal{A}$ and for all $\lambda \in \mathbb{C}$.*

Definition 1.1.2. *A complex Banach algebra is a complex algebra $\mathcal{A}$ equipped with a Banach space norm $||\cdot||$ such that*

$$||xy|| \leq ||x||\,||y||$$

for all $x, y \in \mathcal{A}$.

Let $\mathcal{A}$ be a real or complex Banach algebra. A subset $\mathcal{I}$ is called left (right) ideal of $\mathcal{A}$ if:

1. $\mathcal{I}$ is a linear subspace of $\mathcal{A}$;
2. $ja \in \mathcal{I}$ ($aj \in \mathcal{I}$) for all $j \in \mathcal{I}$ and for all $a \in \mathcal{A}$.

$\mathcal{I}$ is referred to as an ideal or two-sided ideal of $\mathcal{A}$ if it is both a left and a right ideal simultaneously. An ideal $\mathcal{I} \subset \mathcal{A}$ is called proper if $\mathcal{I} \neq \{0\}$ and $\mathcal{I} \neq \mathcal{A}$. Finally, a proper ideal $\mathcal{I} \subset \mathcal{A}$ is said to be maximal if, whenever $\mathcal{J} \subset \mathcal{A}$ is an ideal of $\mathcal{A}$ such that $\mathcal{I} \subset \mathcal{J}$, either $\mathcal{I} = \mathcal{J}$ or $\mathcal{J} = \mathcal{A}$. Note that any maximal ideal $\mathcal{I}$ is closed – i.e., the subspace $\mathcal{I}$ is closed.

A Banach algebra $\mathcal{A}$ is called unital if it possesses an identity element e. Note that any Banach algebra $\mathcal{A}$ is contained in a unital Banach algebra $\tilde{\mathcal{A}}$, as an ideal of codimension 1 [96]. Therefore, without loss of generality, we may assume that all algebras under consideration possess an identity element e unless otherwise specified.

An element $a \in \mathcal{A}$ is called left (right) invertible if there exists $b \in \mathcal{A}$ ($c \in \mathcal{A}$) such that $ba = e$ ($ac = e$). If a is left (right) invertible, then the corresponding element b or c is called a left (right) inverse for a and is denoted by a_l^{-1} (respectively a_r^{-1}). If $a \in \mathcal{A}$ is both left and right invertible, then it is called invertible. In this case the right and left inverses coincide and their common value is denoted by a^{-1}.

To each Banach algebra $\mathcal{A}$ and each closed two-sided ideal $\mathcal{I}$ of $\mathcal{A}$ one can assign another Banach algebra $\mathcal{A}/\mathcal{I}$ which is called the quotient or factor algebra. The elements of $\mathcal{A}/\mathcal{I}$ are the cosets $a + \mathcal{I}, a \in \mathcal{A}$. For instance, it is not hard to see that the product $(a+\mathcal{I})(b+\mathcal{I}) := ab + \mathcal{I}$ is correctly defined. The norm on $\mathcal{A}/\mathcal{I}$ is given by

$$||a + \mathcal{I}|| := \inf_{j \in \mathcal{I}} ||a + j||.$$

Throughout this work we use various notions of homomorphisms defined on appropriate algebras. Thus a complex algebra homomorphism is a mapping between complex algebras which preserves addition, multiplication, scalar multiplication and the identity element, whereas a complex *-algebra homomorphism has to preserve the operation of involution as well. Corresponding real algebra homomorphisms are defined similarly.

A complex C^*-algebra is a complex Banach *-algebra such that $||a^*a|| = ||a||^2$ for all $a \in \mathcal{A}$.

Now let us turn to real C^*-algebras.

Definition 1.1.3. *A real algebra $\mathcal{A}$ is called a *-algebra or algebra with involution if there exists a map $^* : \mathcal{A} \to \mathcal{A}$ such that $(a+b)^* = a^* + b^*$, $(ab)^* = b^*a^*$, $(\lambda a)^* = \lambda a^*$ and $(a^*)^* = a$ for all $a, b \in \mathcal{A}$ and for all $\lambda \in \mathbb{R}$.*

The definition of a real Banach algebra literally repeats the corresponding definition for the complex case. However, unlike the complex case, the definition of a real C^*-algebra contains an additional condition.

Definition 1.1.4. *A real C^*-algebra $\mathcal{A}$ is a real Banach *-algebra such that $||a^*a|| = ||a||^2$ for all $a \in \mathcal{A}$, and such that the element $e + a^*a$ is invertible in $\mathcal{A}$ for any $a \in \mathcal{A}$.*

The invertibility of the elements $e + a^*a, a \in \mathcal{A}$ plays an important role in different aspects of the theory of C^*-algebras, so initially this condition was included in the definition of any C^*-algebra, whether real or complex. However, as was shown later, in the complex case other axioms already provide invertibility of the element $e + a^*a$ for any $a \in \mathcal{A}$. As far as real C^*-algebras are concerned, this condition does not follow from other axioms and has to be postulated. Indeed, consider the set of the complex numbers $\mathbb{C}$ as a real Banach *-algebra with the usual absolute value norm $||a|| := |a|$ and with the involution $a^* := a$. Such an algebra satisfies all but one axiom of real C^*-algebras. Thus if $a := i$, then $b := 1 + i^*i = 0$ and this element b is not invertible.

It turns out that a closed two-sided ideal $\mathcal{I}$ of a C^*-algebra $\mathcal{A}$ is a *-ideal, that is $a \in \mathcal{I}$ implies $a^* \in \mathcal{I}$. Moreover, $\mathcal{A}/\mathcal{I}$ is a C^*-algebra [66, 96].

An element a belonging to a real or complex C^*-algebra $\mathcal{A}$ is called self-adjoint if $a = a^*$, and normal if $aa^* = a^*a$.

Let $\mathcal{A}$ and $\mathcal{D}$ be real or complex C^*-algebras. A homomorphism $\varphi : \mathcal{A} \mapsto \mathcal{D}$ is called *-homomorphism if $\varphi(a^*) = \varphi(a)^*$ for all $a \in \mathcal{A}$. Notice that any *-homomorphism φ is automatically continuous, the image set $\operatorname{im}\varphi$ is again a C^*-algebra, and an injective *-homomorphism is isometric [66, 96].

Example 1.1.5. Let $\mathcal{A}$ be a real or complex Hilbert space, and let $\mathcal{L}(\mathcal{H})$ denote the collection of all linear bounded operators acting on $\mathcal{H}$. Then $\mathcal{L}(\mathcal{H})$ forms respectively a real or complex C^*-algebra, where the involution $A \mapsto A^*$ is the Hilbert adjunction. The collection $\mathcal{K}(\mathcal{H}) \subset \mathcal{L}(\mathcal{H})$ of all compact operators forms a closed two-sided ideal in $\mathcal{L}(\mathcal{H})$.

We introduce now a class of operators $A \in \mathcal{L}(\mathcal{H})$ which are close to invertible operators.

Let $\mathcal{H}$ be a real or complex Hilbert space. An operator $A \in \mathcal{L}(\mathcal{H})$ is called Fredholm if $\dim\ker A < \infty$, $\dim\ker A^* < \infty$ and the image space $\operatorname{im} A$ is closed, that is $\operatorname{im} A = \overline{\operatorname{im} A}$.

If A is a Fredholm operator, then the number

$$\operatorname{ind} A := \dim\ker A - \dim\ker A^*$$

is called the index of the operator A.

Let us list some properties of Fredholm operators.

(a) A is Fredholm if and only if there are compact operators K_1, K_2 and an operator $B \in \mathcal{L}(\mathcal{H})$ such that

$$AB = I + K_1, \quad BA = I + K_2.$$

The operator B is also called a regularizer of A.

(b) A is Fredholm if and only if the coset $A + \mathcal{K}(\mathcal{H})$ is invertible in the Calkin algebra $\mathcal{L}(\mathcal{H})/\mathcal{K}(\mathcal{H})$.

(c) The set of all Fredholm operators is open in $\mathcal{L}(\mathcal{H})$.

(d) If $A \in \mathcal{L}(\mathcal{H})$ is Fredholm and $K \in \mathcal{K}(\mathcal{H})$, then $A + K$ is Fredholm and

$$\operatorname{ind}(A + K) = \operatorname{ind} A.$$

(e) If $A, B \in \mathcal{L}(\mathcal{H})$ are Fredholm operators, then AB is a Fredholm operator and

$$\operatorname{ind}(AB) = \operatorname{ind} A + \operatorname{ind} B.$$

If $\mathcal{H}$ is a complex Hilbert space and $A \in \mathcal{L}(\mathcal{H})$, then the essential spectrum of A is the set

$$\operatorname{ess\,sp} A := \{\lambda \in \mathbb{C} : A - \lambda I \quad \text{is not Fredholm}\}.$$

It will follow from the discussion in Section 1.4 that ess sp$A$ is a non-empty compact subset of $\mathbb{C}$.

1.2 Real Extensions of Complex *-Algebras

In many applications the real C^*-algebras arising have special structures, so that the algebra and properties of its elements can be studied more effectively. One such construction will be considered here.

Let $\mathcal{R}$ be a real algebra. Assume that this algebra contains a complex C^*-algebra $\mathcal{A}$ with identity $e \neq 0$, and an element m that does not belong to $\mathcal{A}$ and satisfies the following assumptions:

(A_1) For each $a \in \mathcal{A}$ the element mam belongs to the C^*-algebra $\mathcal{A}$ as well.

(A_2) For each $a \in \mathcal{A}$ and for each $\lambda \in \mathbb{C}$ the relation $m(\lambda a) = \overline{\lambda} ma$ holds.

(A_3) $m^2 = e$ and $me = m$.

(A_4) The null element 0 of the C^*-algebra $\mathcal{A}$ is also that of the algebra $\mathcal{R}$.

(A_5) For each $a \in \mathcal{A}$, $(mam)^* = ma^*m$, where "$*$" means the given involution on $\mathcal{A}$.

Note that the above product $\overline{\lambda} ma$, $\lambda \in \mathbb{C}$ is well defined. Indeed, $\overline{\lambda} ma = (\overline{\lambda}(mam))m$ and by (A_1) the element mam belongs to the complex C^*-algebra $\mathcal{A}$.

Let $\mathcal{A}m$ stand for the set of those elements of $\mathcal{R}$ which have the form am with $a \in \mathcal{A}$. The following lemma is an immediate consequence of the assumptions (A_1) – (A_4) and the fact that $\mathcal{R}$ is an algebra.

Lemma 1.2.1. *Let the assumptions* (A_1) – (A_4) *be satisfied. Then*

1. *For any* $b \in \mathcal{A}$ *the element* mb *belongs to* $\mathcal{A}m$.
2. $em = m$.
3. *If* $a \in \mathcal{A}$ *and* $am = 0$, *then* $a = 0$.

Corollary 1.2.2. *Under the assumptions of Lemma* 1.2.1 *one has:*

1. $mem = e$.
2. *For any* $a, b \in \mathcal{A}$ *the element* amb *belongs to* $\mathcal{A}m$.

The proofs of the above results follow from simple computations.

Lemma 1.2.3. *The intersection of the sets* $\mathcal{A}$ *and* $\mathcal{A}m$ *consists of the element* 0 *only.*

Proof. Let us suppose that there is an element $b \in \mathcal{R}$ such that $b \in \mathcal{A} \cap \mathcal{A}m$ and $b \neq 0$. Then b has the form $b = cm$, where $c \in \mathcal{A}$. Multiplying the last equality by i on the left side and by $(-ie)$ on the right, we obtain $b = -cm$. Comparing these two expressions for b and using Lemma 1.2.1, we see that $b = 0$ and $c = 0$. □

We consider now the subset $\tilde{\mathcal{A}} \subseteq \mathcal{R}$ which consists of all elements $\tilde{a} \in \mathcal{R}$ having the form

$$\tilde{a} = b + cm, \quad b, c \in \mathcal{A}. \tag{1.2}$$

Lemma 1.2.3 yields the uniqueness of this representation for the elements from $\tilde{\mathcal{A}}$.

Lemma 1.2.4. *If an element* $\tilde{a} \in \tilde{\mathcal{A}}$ *has two representations* $\tilde{a} = b_1 + c_1 m$ *and* $\tilde{a} = b_2 + c_2 m$, *then* $b_1 = b_2$ *and* $c_1 = c_2$.

Indeed, $b_1 + c_1 m = b_2 + c_2 m$, then $b_1 - b_2 = (c_2 - c_1)m$, hence $b_1 = b_2$ and $c_1 = c_2$.

It should be noted that the set $\tilde{\mathcal{A}}$ possesses operations of addition and multiplication which are inherited from the algebra $\mathcal{R}$. For instance, if $\tilde{a}_1, \tilde{a}_2 \in \tilde{\mathcal{A}}$, $\tilde{a}_1 = b_1 + c_1 m$, $\tilde{a}_2 = b_2 + c_2 m$, then we can write

$$\tilde{a}_1 \cdot \tilde{a}_2 = [b_1 b_2 + c_1(m c_2 m)] + [b_1 c_2 + c_1(m b_2 m)]m.$$

Additionally, a scalar multiplication can be introduced in the following way: For each $\tilde{a} \in \tilde{\mathcal{A}}$, $\tilde{a} = b + cm$ and for each $\lambda \in \mathbb{C}$ we put

$$\lambda \tilde{a} = \lambda b + \lambda cm.$$

Due to the assumptions (A_1) – (A_3), the result of each of the above operations is in $\tilde{\mathcal{A}}$ again. Besides, it is easily seen that the set $\tilde{\mathcal{A}}$ with these operations is a linear space, and the multiplication is distributive with respect to addition and also associative. However, $\tilde{\mathcal{A}}$ does not become a complex algebra, because in general

$$\tilde{a}(\lambda \tilde{b}) \neq \lambda(\tilde{a}\tilde{b}), \quad \tilde{a}, \tilde{b} \in \tilde{\mathcal{A}}, \quad \lambda \in \mathbb{C}.$$

But it does form a real algebra. Moreover this set can be made into a real *-algebra by introducing an involution $\star : \tilde{\mathcal{A}} \to \tilde{\mathcal{A}}$ by

$$\tilde{a}^\star := b^* + mc^* \tag{1.3}$$

where $\tilde{a} = b+cm$. Such definition is correct because the elements of $\tilde{\mathcal{A}}$ have unique representation (1.2). Moreover it is easily seen that the operation '$\star$' possesses most of the basic properties for involutions on complex algebras. For instance:

1. For any $\tilde{a} \in \tilde{\mathcal{A}}$ the element $\tilde{a}^\star$ belongs to the algebra $\tilde{\mathcal{A}}$ as well.
2. $(\tilde{a}^\star)^\star = \tilde{a}$ for any $\tilde{a} \in \tilde{\mathcal{A}}$.
3. For any $\tilde{a}, \tilde{b} \in \tilde{\mathcal{A}}$ and for any $\alpha, \beta \in \mathbb{R}$,

$$(\tilde{a} + \tilde{b})^\star = \tilde{a}^\star + \tilde{b}^\star,$$

$$(\tilde{a} \cdot \tilde{b})^\star = \tilde{b}^\star \cdot \tilde{a}^\star,$$

$$(\alpha\tilde{a} + \beta\tilde{b})^\star = \alpha\tilde{a}^\star + \beta\tilde{b}^\star.$$

Thus the set $\tilde{\mathcal{A}}$ equipped with the above operations of addition, multiplication, scalar multiplication and the involution '$\star$' becomes a real *-algebra.

In addition, one can see that $m^\star = m$, and $a^\star = a^*$ for any $a \in \mathcal{A}$, therefore it will cause no confusion if we use the same symbol '*' to denote the involution on the set $\tilde{\mathcal{A}}$. We emphasize that the above involution on the set $\tilde{\mathcal{A}}$ depends on the element m. In the sequel, we study a number of algebras $\tilde{\mathcal{A}}$ which can be defined by different elements m. In these cases a special notation is used to make it clear which involution is meant. Thus the corresponding involution is called m-involution and the corresponding extension is denoted by $\tilde{\mathcal{A}}_m$. We study the uniqueness of this involution and establish conditions when different elements m define the same involution on the algebra under consideration. Meanwhile let us denote by $\mathcal{M}(\mathcal{A})$ the set of all elements m of $\tilde{\mathcal{A}}$ which satisfy axioms $(A_1)-(A_3)$ and axiom (A_5).

A simple example of a pair $(\mathcal{A}, m)$ which satisfies conditions $(A_1)-(A_5)$ is given below. Let $X = \mathbb{C}$, and let $\mathcal{A}$ be the C^*-algebra of all continuous linear operators acting in X. Evidently, $\mathcal{A}$ consists of the multiplication operators by complex numbers only. We define the element m by

$$mx = \overline{x}, \quad x \in \mathbb{C},$$

where the over-bar denotes the operation of complex conjugation. It is easily seen that the pair $(\mathcal{A}, m)$ satisfies conditions $(A_1)-(A_5)$ if the set of all additive continuous operators in $\mathbb{C}$ is taken as the corresponding real algebra $\mathcal{R}$.

1.3 Uniqueness of Involution in Real Extensions of Complex *-Algebras

Suppose we have two different elements m_1 and m_2 of $\mathcal{R}$ possessing the properties $(A_1)-(A_5)$ and producing just the same algebra $\tilde{\mathcal{A}}$. If $\tilde{a} \in \tilde{\mathcal{A}}$, then it is not known whether the corresponding involutions

$$\begin{aligned} (\tilde{a})^*_{m_1} &= (b_1 + c_1 m_1)^*_{m_1} = b_1^* + m_1 c_1^* \qquad (b_1, c_1 \in \mathcal{A}), \\ (\tilde{a})^*_{m_2} &= (b_2 + c_2 m_2)^*_{m_2} = b_2^* + m_2 c_2^* \qquad (b_2, c_2 \in \mathcal{A}) \end{aligned}$$

for the element $\tilde{a}$ coincide. In this section we give some conditions when the equality $(\tilde{a})^*_{m_1} = (\tilde{a})^*_{m_2}$ is fulfilled for each $\tilde{a}$ of $\tilde{\mathcal{A}}$ provided that $\tilde{\mathcal{A}}_{m_1} = \tilde{\mathcal{A}}_{m_2}$ $(= \tilde{\mathcal{A}})$.

Theorem 1.3.1. *If $\mathcal{A}$ is a complex C^*-algebra, and $m_1, m_2 \in \mathcal{M}(\mathcal{A})$, then the following assertions are equivalent:*

1. $\tilde{\mathcal{A}}_{m_1} = \tilde{\mathcal{A}}_{m_2}$ $(= \tilde{\mathcal{A}})$ *and, for each* $\tilde{a} \in \tilde{\mathcal{A}}$,
$$(\tilde{a})^*_{m_1} = (\tilde{a})^*_{m_2}.$$
2. *The element $m_1 m_2$ belongs to the algebra $\mathcal{A}$ and satisfies the relation*
$$(m_1 m_2)^* = m_2 m_1. \tag{1.4}$$

Proof. Necessity. Let the algebra $\tilde{\mathcal{A}}_{m_1}$ coincide with $\tilde{\mathcal{A}}_{m_2}$, and let them both have just the same involution. Then there exist $f, g \in \mathcal{A}$ such that $m_1 = f + g m_2$. Following the proof of Lemma 1.2.3 we get $m_1 = -f + g m_2$. Hence, $m_1 = g m_2$, or

$$m_1 m_2 = g, \tag{1.5}$$

i.e., $m_1 m_2 \in \mathcal{A}$.

To establish relation (1.4), let us compute each of the involutions for the element m_1. Thus we have

$$(m_1)^*_{m_1} = m_1 \tag{1.6}$$

and

$$(m_1)^*_{m_2} = (g m_2)^*_{m_2} = m_2 g^*. \tag{1.7}$$

Comparing (1.6) with (1.7) we obtain $m_1 = m_2 g^*$. Multiplying this equality by m_2 from the left and recalling (1.5) completes the proof.

Sufficiency. Assume that $m_1 m_2$ is an element from the algebra $\tilde{\mathcal{A}}$ and equality (1.4) is satisfied. Let us show that algebras $\tilde{\mathcal{A}}_{m_1}$ and $\tilde{\mathcal{A}}_{m_2}$ coincide. If $\tilde{a} \in \tilde{\mathcal{A}}_{m_1}$, then there are elements b and c in the algebra $\mathcal{A}$ such that $\tilde{a} = b + c m_1$. Hence

$$\tilde{a} = b + c m_1 = b + c m_1 \cdot m_2^2 = b + c \cdot (m_1 m_2) m_2 = b + c_1 m_2,$$

Example 1.3.6. Consider the Hilbert space $L_2(\Gamma_0)$ again, and take a continuously differentiable function $\alpha = \alpha(t), t \in \Gamma_0$ such that

$$\alpha^2(t) = \alpha(\alpha(t)) = t \qquad \text{for all} \quad t \in \Gamma_0. \tag{1.10}$$

Suppose that

$$|\alpha'(t)| = 1 \qquad \text{for all} \quad t \in \Gamma_0. \tag{1.11}$$

An example of the function which satisfies equations (1.10), (1.11) is $\alpha(t) = 1/t,\ t \in \Gamma_0$. By W_α we denote the operator of the Carleman shift

$$(W_\alpha\varphi)(t) = \varphi(\alpha(t)), \qquad \varphi \in L_2(\Gamma_0), \quad t \in \Gamma_0$$

and by B_a the operator of multiplication by a continuous function a, i.e.,

$$(B_a\varphi)(t) = a(t)\varphi(t), \quad \varphi \in L_2(\Gamma_0), \quad t \in \Gamma_0.$$

Let $\mathcal{A}$ be the smallest closed C^*-subalgebra of the C^*-algebra $\mathcal{L}(L_2(\Gamma_0))$ containing the operator W_α and all operators B_a with $a \in \mathbf{C}(\Gamma_0)$. We consider the extensions of this algebra by the elements M and $M_1 = MW_\alpha$.

It is easily seen that $M \in \mathcal{M}(\mathcal{L}(L_2(\Gamma_0)))$. Therefore, we only have to check the conditions of Corollary 1.3.3 for the operator W_α. Indeed, it follows from (1.10) and (1.11) that $W_\alpha^2 = I$ and $W_\alpha^* = B_{|\alpha|}W_\alpha = W_\alpha$. In addition, a straightforward computation gives

$$MW_\alpha M = W_\alpha = W_\alpha^*,$$

i.e., all conditions of Corollary 1.3.3 are satisfied. Thus the elements M and M_1 define the same algebra $\tilde{\mathcal{A}}$ and the same involution on this algebra.

Let $\mathcal{H}$ be a complex Hilbert space and let $\mathcal{L}_{add}(\mathcal{H})$ be the set of all additive continuous operators acting in $\mathcal{H}$. By $\mathfrak{M}(\mathcal{H})$ we denote the set of continuous additive operators M which act on the space $\mathcal{H}$ and satisfy the conditions:

1. $M^2 = I$, where I is the identical operator of $\mathcal{L}(\mathcal{H})$.
2. $(M\varphi, \psi) = \overline{(\varphi, M\psi)}$ for all $\varphi, \psi \in \mathcal{H}$.

Lemma 1.3.7. *Let $M \in \mathfrak{M}(\mathcal{H})$. Then:*

1. *M is an antilinear operator, i.e.,*

$$M(\lambda\varphi) = \overline{\lambda}M\varphi \tag{1.12}$$

for all $\lambda \in \mathbb{C}$ and for all $\varphi \in \mathcal{H}$.

2. *For each $A \in \mathcal{L}(\mathcal{H})$, the operator MAM belongs to $\mathcal{L}(\mathcal{H})$ again and*

$$(MAM)^* = MA^*M.$$

3. *For each* $A \in \mathcal{L}(\mathcal{H})$,

$$||MAM|| = ||A||.$$

Proof. First we show that each operator M satisfying the above conditions is antilinear. Indeed, for any $\varphi, \psi \in \mathcal{H}$ one has

$$(M(\lambda\varphi), \psi) = \overline{\lambda}\ \overline{(\varphi, M\psi)} = \overline{\lambda}(M\varphi, \psi) = (\overline{\lambda}M\varphi, \psi)$$

which implies (1.12). The second assertion can be verified analogously. For the third, one assumes that $A \in \mathcal{L}(\mathcal{H})$ and φ is an element of $\mathcal{H}$. Then

$$\begin{aligned} ||MAM\varphi||^2 &= (MAM\varphi, MAM\varphi) = \overline{(AM\varphi, M^2AM\varphi)} \\ &= (AM\varphi, AM\varphi) = ||AM\varphi||^2 \leq ||A||^2||M\varphi||^2 \\ &= ||A||^2(M\varphi, M\varphi) = ||A||^2\overline{(\varphi, M^2\varphi)} = ||A||^2||\varphi||^2, \end{aligned}$$

i.e.,

$$||MAM|| \leq ||A|| \quad \text{for all} \quad A \in \mathcal{L}(\mathcal{H}).$$

Conversely, set $B = MAM$. Then

$$||A|| = ||MBM|| \leq ||B||,$$

and the proof is complete. □

Lemma 1.3.7 shows that every operator $M \in \mathfrak{M}(\mathcal{H})$ necessarily belongs to $\mathcal{M}(\mathcal{L}(\mathcal{H}))$. Given such an element M, one can construct the real extension $\tilde{\mathcal{L}}(\mathcal{H})$ of the algebra of linear continuous operators $\mathcal{L}(\mathcal{H})$. Thus it is important to know whether for a given Hilbert space $\mathcal{H}$, the set $\mathfrak{M}(\mathcal{H})$ contains at least one element.

Theorem 1.3.8. *For any Hilbert space* $\mathcal{H}$*, the set* $\mathfrak{M}(\mathcal{H})$ *is not empty.*

Proof. Let Φ be an orthonormal basis for $\mathcal{H}$. Then any element $h \in \mathcal{H}$ can be represented in the form [77, pp. 252–253]

$$h = \sum_{\phi \in \Phi} (h, \phi)\phi, \tag{1.13}$$

and

$$||h||^2 = \sum_{\phi \in \Phi} |(h, \phi)|^2. \tag{1.14}$$

Let us now define an operator M_Φ by

$$M_\Phi h := \sum_{\phi \in \Phi} \overline{(h, \phi)}\phi, \quad h \in \mathcal{H}. \tag{1.15}$$

It is easily seen that M_Φ is an additive operator and

$$M_\Phi^2 = I.$$

Moreover, using (1.14) one obtains

$$||M_\Phi h|| = ||h||, \quad h \in \mathcal{H},$$

hence $M \in \mathcal{L}_{add}(\mathcal{H})$. It remains to show the validity of the equation

$$(M_\Phi h, g) = \overline{(h, M_\Phi g)}, \quad h, g \in \mathcal{H}.$$

Let $h, g \in \mathcal{H}$, the element h have representation (1.13), and let

$$g = \sum_{\phi \in \Phi} (g, \phi)\phi. \tag{1.16}$$

Note that there are at most a countable number of non-zero scalar products $(h, \phi), (g, \phi)$ in the series (1.13), (1.16). Then

$$\begin{aligned}(M_\Phi h, g) &= \left(\sum_{\phi \in \Phi} \overline{(h, \phi)}\phi, \sum_{\phi \in \Phi} (g, \phi)\phi \right) \\ &= \sum_{\phi \in \Phi} \overline{(h, \phi)}\ \overline{(g, \phi)} = \overline{(h, M_\Phi g)}.\end{aligned}$$

Thus $M_\Phi \in \mathfrak{M}(\mathcal{H})$ and the proof is complete. □

Example 1.3.9. Consider the Hilbert space $L_2(\Gamma_0)$ again. The system $\Phi := \{t^k, k \in \mathbb{Z}\}$ is an orthonormal basis for $L_2(\Gamma_0)$. Then the operator M_Φ defined by (1.15) admits the representation

$$(M_\Phi h)(t) = \overline{h(1/t)}.$$

On the other hand, choosing in $L_2(\Gamma_0)$ an orthonormal basis Ψ that consists of only real-valued elements, one obtains the operator of complex conjugation

$$(M_\Psi h)(t) = \overline{h(t)}.$$

Analogously to (1.9), each additive operator $\tilde{A} \in \mathcal{L}_{add}(\mathcal{H})$ can be represented in the form

$$\tilde{A} = A_1 + A_2 M, \tag{1.17}$$

where A_1 and $A_2 = A_a M$ are linear continuous operators on the space $\mathcal{H}$. Hence any operator $M \in \mathfrak{M}(\mathcal{H})$ produces an extension of $\mathcal{L}(\mathcal{H})$ by M that coincides with $\mathcal{L}_{add}(\mathcal{H})$, and furthermore an involution , viz., for each operator $\tilde{A} \in \mathcal{L}_{add}(\mathcal{H})$ one can define $\tilde{A}^*$ by

$$\tilde{A}^* = A_1^* + M A_2^*.$$

Therefore the real algebra $\mathcal{L}_{add}(\mathcal{H})$ can be considered as an extension of the complex C^*-algebra $\mathcal{L}(\mathcal{H})$ by any operator $M \in \mathfrak{M}(\mathcal{H})$.

Now we are in a position to show that all such operators produce the same involution on $\mathcal{L}_{add}(\mathcal{H})$. Indeed, by (1.12) the operator M_1M_2 belongs to $\mathcal{L}(\mathcal{H})$ for any $M_1, M_2 \in \mathfrak{M}(\mathcal{H})$. Moreover, for any $\varphi, \psi \in \mathcal{H}$ one has

$$(M_1M_2\varphi, \psi) = \overline{(M_2\varphi, M_1\psi)} = (\varphi, M_2M_1\psi).$$

Thus $(M_1M_2)^* = M_2M_1$, and Theorem 1.3.1 finishes the proof.

We see that a wide class of antilinear operators produces just the same involution on $\mathcal{L}_{add}(\mathcal{H})$ (in the sense of definition (1.3)). Moreover, it turns out that in the case $\mathcal{L}_{add}(\mathcal{H})$ all possible involutions (1.3) coincide! The proof of this result is given below.

Let $\mathcal{H}_\mathbb{R}$ be the same Hilbert space $\mathcal{H}$ considered as a real space. As a scalar product on $\mathcal{H}_\mathbb{R}$ we will use the form

$$\langle \varphi, \psi \rangle = \langle \varphi, \psi \rangle_{\mathcal{H}_\mathbb{R}} = \operatorname{Re}(\varphi, \psi), \qquad \varphi, \psi \in \mathcal{H}_\mathbb{R}. \tag{1.18}$$

Note that some features of this real space $\mathcal{H}_\mathbb{R}$ will be discussed in Section 1.6.

For each $A \in \mathcal{L}_{add}(\mathcal{H})$, let $A^*_\mathbb{R}$ refer to an operator $B \in \mathcal{L}_{add}(\mathcal{H})$ which satisfies the relation

$$\langle A\varphi, \psi \rangle = \langle \varphi, B\psi \rangle \qquad \text{for all} \quad \varphi, \psi \in \mathcal{H}.$$

Let us list elementary properties of these operators:

1. For each $A \in \mathcal{L}_{add}(\mathcal{H})$, the operator $A^*_\mathbb{R}$ exists and is uniquely determined;
2. For any $A, B \in \mathcal{L}_{add}(\mathcal{H})$, $(A+B)^*_\mathbb{R} = A^*_\mathbb{R} + B^*_\mathbb{R}$ and $(AB)^*_\mathbb{R} = B^*_\mathbb{R}A^*_\mathbb{R}$;
3. For each $A \in \mathcal{L}_{add}(\mathcal{H})$, $(A^*_\mathbb{R})^*_\mathbb{R} = A$;
4. For each $A \in \mathcal{L}(\mathcal{H})$, $A^*_\mathbb{R} = A^*$;
5. For each $M \in \mathcal{M}(\mathcal{L}(\mathcal{H}))$, the operator $MM^*_\mathbb{R}$ belongs to $\mathcal{L}(\mathcal{H})$.

Theorem 1.3.10. *Let $\mathcal{H}$ be a complex Hilbert space. Then any two operators M_1 and M_2 from the set $\mathcal{M}(\mathcal{L}(\mathcal{H}))$ produce just the same involution on $\mathcal{L}_{add}(\mathcal{H})$.*

Proof. First of all, we suppose that $\dim \mathcal{H} > 1$ and consider an arbitrary operator $M \in \mathcal{M}(\mathcal{L}(\mathcal{H}))$. Let $A \in \mathcal{L}(\mathcal{H})$. On account of properties 1) – 4) one can write

$$(MA^*M)^*_\mathbb{R} = M^*_\mathbb{R}AM^*_\mathbb{R}.$$

On the other hand,

$$(MA^*M)^*_\mathbb{R} = (MA^*M)^* = MAM.$$

Therefore, according to condition (A_3) we get

$$(MM^*_\mathbb{R})A = A(MM^*_\mathbb{R}) \qquad \text{for all } A \in \mathcal{L}(\mathcal{H}).$$

Since $\dim \mathcal{H} > 1$ the algebra $\mathcal{L}(\mathcal{H})$ is irreducible, and $MM_{\mathbb{R}}^* \in \mathcal{L}(\mathcal{H})$ is an operator of multiplication by a scalar [66], i.e.,

$$MM_{\mathbb{R}}^* = \lambda I \tag{1.19}$$

for a $\lambda \in \mathbb{C}$. However, the operator $MM_{\mathbb{R}}^*$ is self-adjoint. This implies $\lambda \in \mathbb{R}$, and applying axiom (A_3) once again we obtain $M_{\mathbb{R}}^* M = \lambda I$, so multiplying the last expression by (1.19) one gets $\lambda^2 = 1$ or $\lambda = \pm 1$. Thus, there may exist only two situations, $M_{\mathbb{R}}^* = M$ or $M_{\mathbb{R}}^* = -M$.

Let us now show that the second case is impossible. Indeed, if we suppose that $M_{\mathbb{R}}^* = -M$, then $I + MM_{\mathbb{R}}^* = 0$. Therefore for any $\varphi \in \mathcal{H}$ we have

$$\begin{aligned}
||\varphi||^2 &\leq (\varphi, \varphi) + (M_{\mathbb{R}}^*\varphi, M_{\mathbb{R}}^*\varphi) \\
&= \langle \varphi, \varphi \rangle + \langle MM_{\mathbb{R}}^*\varphi, \varphi \rangle \\
&= \langle (I + MM_{\mathbb{R}}^*)\varphi, \varphi \rangle \\
&\leq ||(I + MM_{\mathbb{R}}^*)\varphi||\, ||\varphi||
\end{aligned}$$

using the Cauchy-Schwarz inequality, whence

$$||\varphi|| \leq ||(I + MM_{\mathbb{R}}^*)\varphi|| = 0 \qquad \text{for all} \quad \varphi \in \mathcal{H}$$

which is not possible. Hence, for each $M \in \mathcal{M}(\mathcal{L}(\mathcal{H}))$ we obtain $M_{\mathbb{R}}^* = M$.

It remains to use the necessary and sufficient condition (1.4). Thus for any $M_1, M_2 \in \mathcal{M}(\mathcal{L}(\mathcal{H}))$ we can write

$$(M_1M_2)^* = (M_1M_2)_{\mathbb{R}}^* = (M_2)_{\mathbb{R}}^*(M_1)_{\mathbb{R}}^* = M_2M_1.$$

Hence if $\dim \mathcal{H} > 1$, all the operators of $\mathcal{M}(\mathcal{L}(\mathcal{H}))$ produce just the same involution on $\mathcal{L}_{add}(\mathcal{H})$. Let us now consider the situation $\dim \mathcal{H} = 1$. In this case the algebra $\mathcal{L}(\mathcal{H})$ consists of the operators of multiplication by complex scalars only, and it is a simple matter to check that

$$\mathcal{M}(\mathcal{L}(\mathcal{H})) = \left\{ e^{i\varphi} M : \varphi \in [0, 2\pi) \right\}$$

where M is the operator of complex conjugation. However, all such elements produce the same involution on $\mathcal{L}_{add}(\mathcal{H})$. □

Remark 1.3.11. Theorem 1.3.10 can be generalized on further real C^*-algebras $\tilde{\mathcal{A}}$. Axioms 1) - 5) show what assumptions have to be made to guarantee the uniqueness of involution in such algebras.

All the previous considerations have dealt with cases where different elements of $\mathcal{M}(\mathcal{A})$ produced just the same algebra $\tilde{\mathcal{A}}$. However, there are examples where different elements m_1 and m_2 generate different real algebras $\tilde{\mathcal{A}}_{m_1}$ and $\tilde{\mathcal{A}}_{m_1}$.

Example 1.3.12. Assume that a continuous function α satisfies relation (1.10), and let B_a, W_α, M and M_1 be the operators considered in Example 1.3.6. By $\mathcal{A}$ we now denote the smallest closed C^*-subalgebra of $\mathcal{L}(L_2(\Gamma))$ containing all operators B_a, with function a being in $\mathbf{C}(\Gamma)$. It is easily seen that

$$M_1 B_a M_1 = B_{\overline{a \circ \alpha}} \qquad \text{and} \qquad (M_1 B_a M_1)^* = B_{a \circ \alpha} = M_1 B_a^* M_1,$$

therefore the element M_1 generates a real algebra $\tilde{\mathcal{A}}_{M_1}$. However, this real extension $\tilde{\mathcal{A}}_{M_1}$ of the complex C^*-algebra $\mathcal{A}$ does not coincide with the extension $\tilde{\mathcal{A}}_M$.

1.4 Real and Complex Spectrum. Inverse Closedness

Let $\mathcal{A}$ be a unital complex C^*-algebra, and let $b \in \mathcal{A}$.

Definition 1.4.1. *The spectrum of b in $\mathcal{A}$, denoted by $\mathrm{sp}_{\mathcal{A}} b$, is the set of all complex numbers λ such that the element $b - \lambda e$ is not invertible in $\mathcal{A}$.*

Let us recall some known properties of unital complex C^*-algebras. If $\mathcal{A}$ is a unital complex C^*-algebra and $b \in \mathcal{A}$, then

(a) The spectrum $\mathrm{sp}_{\mathcal{A}} b$ is a non-empty compact subset of $\mathbb{C}$.

(b) If $r(b)$ denotes the spectral radius of $b \in \mathcal{A}$, i.e.,

$$r(b) := \max_{\lambda \in \mathrm{sp}_{\mathcal{A}} b} |\lambda|,$$

then

$$r(b) = \inf_{n \in \mathbb{N}} ||b^n||^{1/n} = \lim_{n \to \infty} ||b^n||^{1/n}.$$

Moreover, if the element b is self-adjoint or normal, then

$$r(b) = ||b||.$$

(c) If $\mathcal{D} \subset \mathcal{A}$ is a C^*-subalgebra containing the unit element of $\mathcal{A}$, then $\mathcal{D}$ is inverse closed in $\mathcal{A}$, that is if $b \in \mathcal{D}$ is invertible in $\mathcal{A}$, then b is also invertible in $\mathcal{D}$.

(d) If b is a self-adjoint element of $\mathcal{A}$, then $\mathrm{sp}_{\mathcal{A}} b \subset \mathbb{R}$.

Assume now that b is an element of a real C^*-algebra. Then for $\lambda \in \mathbb{C}$ the element λb may not be defined, so in the above definition of the spectrum the set of complex numbers $\mathbb{C}$ has to be replaced by the set of real numbers $\mathbb{R}$. However, as already mentioned, real algebras often possess an operation of multiplication by complex scalars. For such algebras one can distinguish between real and complex spectra. More precisely, we define the real spectrum of an element as follows.

Definition 1.4.2. *Let $\mathcal{A}$ be a real C^*-algebra and let $b \in \mathcal{A}$. The real spectrum of b in $\mathcal{A}$, denoted by $\mathrm{sp}^r_{\mathcal{A}} b$, is the set of all numbers $\lambda \in \mathbb{R}$ such that the element $b - \lambda e$ is not invertible in $\mathcal{A}$.*

If a real C^*-algebra $\mathcal{A}$ has an operation of multiplication by complex scalars, we will also consider the complex spectrum of elements in $\mathcal{A}$. It is defined analogously to the real spectrum, but the set of real numbers $\mathbb{R}$ in Definition 1.4.2 is replaced by the set of complex numbers $\mathbb{C}$. For the sake of simplicity, for this spectrum we use the same notation as for a spectrum in complex algebras, viz., $\mathrm{sp}_{\mathcal{A}} b$.

It can happen that the real spectrum of an element from a unital real C^*-algebra is empty. An obvious example of such a situation is provided by the imaginary unit $i \in \mathbb{C}$ when the set of the complex numbers $\mathbb{C}$ is considered as a real C^*-algebra.

However, there is a nice substitution for the real spectrum of an element b from real C^*- or Banach algebra $\mathcal{A}$, viz., the complexified spectrum $\mathrm{sp}^c_{\mathcal{A}} b$ defined by

$$\mathrm{sp}^c_{\mathcal{A}} b := \{\alpha + i\beta : \alpha, \beta \in \mathbb{R} \quad \text{and} \quad (\alpha - b)^2 + \beta^2 \quad \text{is not invertible in} \quad \mathcal{A}\}.$$

It is clear that if $\alpha + i\beta \in \mathrm{sp}^c_{\mathcal{A}} b$, then $\alpha - i\beta \in \mathrm{sp}^c_{\mathcal{A}} b$ as well.

There is another definition of the complexified spectrum as the familiar (complex) spectrum in the complexification of the algebra $\mathcal{A}$. We are not going to discuss this construction but mention that it immediately implies that the complexified spectrum of any element from any real algebra is a non-empty compact subset of $\mathbb{C}$. Moreover, the complexified spectral radius

$$r^c_{\mathcal{A}}(b) := \max_{\lambda \in \mathrm{sp}^c_{\mathcal{A}} b} |\lambda|$$

is connected with the norm of the element b by

$$r^c_{\mathcal{A}}(b) = \inf_{n \in \mathbb{N}} ||b^n||^{1/n} = \lim_{n \to \infty} ||b^n||^{1/n}.$$

If b is a self-adjoint element in a unital real C^*-algebra $\mathcal{A}$, then $\mathrm{sp}^c_{\mathcal{A}} b \in \mathbb{R}$. This leads to the identity

$$\mathrm{sp}^c_{\mathcal{A}} b = \mathrm{sp}^r_{\mathcal{A}} b,$$

so that the spectrum $\mathrm{sp}^r_{\mathcal{A}} b$ is not empty. Moreover

$$r^c_{\mathcal{A}}(b) = r_{\mathcal{A}}(b) = \inf_{n \in \mathbb{N}} ||b^n||^{1/n} = \lim_{n \to \infty} ||b^n||^{1/n}.$$

Finally, if $\mathcal{A}$ is a unital complex C^*-algebra and $\mathcal{A}_{\mathbb{R}}$ denotes $\mathcal{A}$ considered as a real algebra, then

$$\mathrm{sp}^c_{\mathcal{A}_{\mathbb{R}}} b = \mathrm{sp}_{\mathcal{A}} b \cup \{\overline{\lambda} : \lambda \in \mathrm{sp}_{\mathcal{A}} b\}.$$

Note that further information and proofs can be found in [96, 131].

We need some known results concerning real algebras.

Theorem 1.4.3 ([96], Chapter 11 or [135], Appendix 1). *Let b be a self-adjoint element of a unital (real or complex) C^*-algebra $\mathcal{A}$, and let $\mathcal{R}(b)$ be the smallest closed real C^*-subalgebra containing the elements b and e. Then $\mathcal{R}(b)$ is isometrically isomorphic to the algebra $\mathbf{C}(\mathrm{sp}^r_{\mathcal{R}(b)}b)$ of all continuous real functions on $\mathrm{sp}^r_{\mathcal{R}(b)}b$.*

Lemma 1.4.4. *Let a be a self-adjoint element of the complex C^*-algebra $\mathcal{U}$. Then*

$$\mathrm{sp}^r_{\mathcal{R}(a)}a = \mathrm{sp}_{\mathcal{U}}a \, (= \mathrm{sp}^r_{\mathcal{U}}a)$$

where $\mathcal{R}(a)$ is the smallest closed real C^*-algebra of $\mathcal{U}$ generated by the elements a and e.

Proof. Since for each self-adjoint element $a \in \mathcal{U}$,

$$\mathrm{sp}_{\mathcal{U}}a \subset \mathbb{R},$$

we only have to prove that each self-adjoint invertible element a of $\mathcal{U}$ is also invertible in $\mathcal{R}(a)$. To do this we use the fact that by Theorem 1.4.3 the algebra $\mathcal{R}(a)$ is isometrically isomorphic to the algebra $\mathbf{C}(\mathrm{sp}^r_{\mathcal{R}(a)}a)$ of all real continuous functions on the spectrum $\mathrm{sp}^r_{\mathcal{R}(a)}a$. Let us now suppose that the element a is invertible in $\mathcal{U}$ but not in $\mathcal{R}(a)$. Then the point 0 belongs to the $\mathrm{sp}^r_{\mathcal{R}(a)}a$. We define the function f_n on $\mathrm{sp}^r_{\mathcal{R}(a)}a$ in the following way:

$$f_n(t) = \begin{cases} n, & \text{if } t \in \mathrm{sp}^r_{\mathcal{R}(a)}a \cap [-1/n, 1/n], \\ 1/|t|, & \text{if } t \in (\mathbb{R} \cap \mathrm{sp}^r_{\mathcal{R}(a)}a) \backslash [-1/n, 1/n]. \end{cases}$$

The function $f_n \in \mathbf{C}(\mathrm{sp}^r_{\mathcal{R}(a)}a)$ and $|tf_n(t)| \leq 1$ for all $t \in \mathrm{sp}^r_{\mathcal{R}(a)}a$. Hence

$$||af_n(a)|| \leq 1.$$

If the element a is invertible in $\mathcal{U}$, then there exists $b \in \mathcal{U}$ such that $ba = e$, so

$$||f_n(a)|| = ||baf_n(a)|| \leq ||b||.$$

We have obtained a contradiction, because the norm of $f_n(a)$ can be made sufficiently large. □

Corollary 1.4.5. *Let $\mathcal{B}$ be a real C^*-subalgebra of the complex C^*-algebra $\mathcal{A}$. Then $\mathcal{B}$ is inverse closed in the algebra $\mathcal{A}$, i.e., if an element $b \in \mathcal{B}$ is invertible in $\mathcal{A}$, then it is also invertible in $\mathcal{B}$.*

Proof. Let the element $b \in \mathcal{B}$ be invertible in $\mathcal{A}$. Then b^* is also invertible in $\mathcal{A}$. Since b^*b is a self-adjoint element, by the previous lemma the element $(b^*b)^{-1}$ belongs to the algebra $\mathcal{B}$. Hence the element $c = (b^*b)^{-1}b^*$ belongs to the real C^*-subalgebra $\mathcal{B}$, and a simple computation shows that $c = b_l^{-1}$. The right invertibility of b in the subalgebra $\mathcal{B}$ can be proved analogously. □

Below we describe relations between the complex spectrum of elements from the real extensions $\tilde{\mathcal{A}}$ of complex algebras $\mathcal{A}$ and the real spectra of auxiliary elements of a complex algebra.

Let $\mathcal{A}^{2\times 2}$ be the complex C^*-algebra of all 2×2 matrices with entries from $\mathcal{A}$. We consider a subset $E_{\tilde{\mathcal{A}}}^{2\times 2}$ of $\mathcal{A}^{2\times 2}$, which by definition consists of all matrices A having the form

$$A = \begin{pmatrix} b & c \\ mcm & mbm \end{pmatrix}, \quad b, c \in \mathcal{A}, \tag{1.20}$$

where the element m satisfies axioms $(A_1) - (A_5)$ of Section 1.2.

Provided with natural operations of multiplication, addition, involution and with the norm of $\mathcal{A}^{2\times 2}$, this set becomes a *real* C^*-subalgebra of $\mathcal{A}^{2\times 2}$. Let us now define a mapping $\Psi : \tilde{\mathcal{A}} \to E_{\tilde{\mathcal{A}}}^{2\times 2}$ by

$$\Psi(\tilde{a}) = \Psi(b + cm) := \begin{pmatrix} b & c \\ mcm & mbm \end{pmatrix}. \tag{1.21}$$

Lemma 1.4.6. *If $\tilde{a}_1, \tilde{a}_2 \in \tilde{\mathcal{A}}$, then*

1) $\Psi(\lambda\tilde{a}) = \lambda\Psi(\tilde{a})$ *for all $\tilde{a} \in \tilde{\mathcal{A}}$ and for all $\lambda \in \mathbb{R}$.*
2) $\Psi(\tilde{a}_1 + \tilde{a}_2) = \Psi(\tilde{a}_1) + \Psi(\tilde{a}_2)$.
3) $\Psi(\tilde{a}_1 \cdot \tilde{a}_2) = \Psi(\tilde{a}_1) \cdot \Psi(\tilde{a}_2)$.
4) $\Psi(\tilde{a}_1^*) = \Psi(\tilde{a}_1)^*$.
5) *The mapping $\Psi : \tilde{\mathcal{A}} \to E_{\tilde{\mathcal{A}}}^{2\times 2}$ is continuously invertible.*

Proof. Properties 1) – 4) immediately follow from the definition of the operations of addition, multiplication and involution on $\tilde{\mathcal{A}}$. It is also clear that $\Psi : \tilde{\mathcal{A}} \to E_{\tilde{\mathcal{A}}}^{2\times 2}$ is an isomorphism, and it is closed. By the closed graph theorem of Banach, both the operators Ψ and Ψ^{-1} are continuous. □

An immediate consequence of Lemma 1.4.6 and Corollary 1.4.5 is the following result.

Corollary 1.4.7. *Let $\tilde{a} \in \tilde{\mathcal{A}}$. The element $\tilde{a}$ is invertible in $\tilde{\mathcal{A}}$ if and only if the element $\Psi(\tilde{a})$ is invertible in $\mathcal{A}^{2\times 2}$.*

Remark 1.4.8. From Lemma 1.4.6 the set $\tilde{\mathcal{A}}$ can be provided with the norm

$$||\tilde{a}||_{\tilde{\mathcal{A}}} := ||\Psi(\tilde{a})||_{\mathcal{A}^{2\times 2}}, \tag{1.22}$$

so $\tilde{\mathcal{A}}$ becomes a real C^*-algebra. Moreover, Lemma 1.4.6 and inclusion $\Psi(e+\tilde{a}\tilde{a}^*) \in \mathcal{A}^{2\times 2}$ show that the element $e + \tilde{a}\tilde{a}^*$ is invertible for any $\tilde{a} \in \tilde{\mathcal{A}}$, since $\mathcal{A}^{2\times 2}$ is a complex C^*-algebra. Hence the mapping $\Psi : \tilde{\mathcal{A}} \to E_{\tilde{\mathcal{A}}}^{2\times 2}$ is a real C^*-algebra isomorphism. Thus the norm (1.22) is the most natural norm for $\tilde{\mathcal{A}}$, although in some cases it is more convenient to use an equivalent norm

$$||\tilde{a}|| := ||b||_{\mathcal{A}} + ||c||_{\mathcal{A}}.$$

Let $\mathcal{A}$ and m be as before.

Definition 1.4.9. *A complex C^*-subalgebra $\mathcal{C}$ of a real C^*-algebra $\mathcal{A}$ is called m-closed if*

$$m\mathcal{C}m \subseteq \mathcal{C},$$

and if the axioms (A_2) – (A_5) with respect to $\mathcal{C}$ and $\mathcal{A}$ are satisfied.

Note that for any m-closed C^*-subalgebra $\mathcal{C}$ one has

$$m\mathcal{C}m = \mathcal{C}.$$

Corollary 1.4.10. *If $\mathcal{C}$ is an m-closed C^*-subalgebra of an m-closed complex C^*-algebra $\mathcal{A}$, then the real C^*-subalgebra $\tilde{\mathcal{C}}$ is inverse closed in the real algebra $\tilde{\mathcal{A}}$.*

This follows by combining Corollary 1.4.5 and Lemma 1.4.6. □

Corollary 1.4.11. *Let $M \in \mathfrak{M}(\mathcal{H})$ and let $\mathcal{A}$ be an M-closed C^*-subalgebra of $\mathcal{L}(\mathcal{H})$. Then $\widetilde{\mathcal{A}}$ is inverse closed in $\mathcal{L}_{add}(\mathcal{H})$.*

Now, we are in position to give a complete description of the complex spectrum for elements from the real algebra $\tilde{\mathcal{A}}$ in terms of real components of the spectra of elements from the complex algebra $\mathcal{A}^{2\times 2}$.

Theorem 1.4.12. *Let $\tilde{a} = b + cm \in \tilde{\mathcal{A}}$. Then*

$$\mathrm{sp}_{\tilde{\mathcal{A}}}\tilde{a} = \bigcup_{\varphi\in[0,2\pi)} \{\mathrm{sp}^+_{\mathcal{A}^{2\times 2}}\Psi(\tilde{a}_\varphi)\}\, e^{i\varphi}$$

where

$$\tilde{a}_\varphi = e^{-i\varphi}b + e^{i\varphi}cm \qquad (\varphi \in [0, 2\pi))$$

and

$$\mathrm{sp}^+_{\mathcal{B}} b := \{\mathrm{sp}_{\mathcal{B}} b\} \cap \{\mathbb{R}^+ \cup 0\}.$$

Proof. For each $\lambda = |\lambda|e^{i\varphi}$ we have

$$\tilde{a} - \lambda\mathbf{e} = (\tilde{a}_\varphi - |\lambda|\mathbf{e})e^{i\varphi},$$

so the element $\tilde{a} - \lambda\mathbf{e}$ is invertible in $\tilde{\mathcal{A}}$ if and only if all elements $\tilde{a}_\varphi - |\lambda|\mathbf{e}$ for $\varphi \in [0, 2\pi)$ are. We consider the subset $E^{2\times 2}_{\tilde{\mathcal{A}}}$ of $\mathcal{A}^{2\times 2}$ which consists of all matrices of $\mathcal{A}^{2\times 2}$ having the form (1.20). By Lemma 1.4.6 an element $\tilde{a}$ is invertible in $\tilde{\mathcal{A}}$ if and only if $\Psi(\tilde{a})$ is invertible in $E^{2\times 2}_{\tilde{\mathcal{A}}}$. Taking into account the relation

$$\Psi(\tilde{a}_\varphi - |\lambda|\mathbf{e}) = \Psi(\tilde{a}_\varphi) - |\lambda|\mathcal{E}$$

where

$$\mathcal{E} = \begin{pmatrix} \mathbf{e} & 0 \\ 0 & \mathbf{e} \end{pmatrix},$$

we obtain

$$\mathrm{sp}_{\tilde{\mathcal{A}}}\tilde{a} = \bigcup_{\varphi\in[0,2\pi)} \left\{\mathrm{sp}^+_{E^{2\times 2}_{\tilde{\mathcal{A}}}}\Psi(\tilde{a}_\varphi)\right\}e^{i\varphi}.$$

Reference to Corollary 1.4.5 completes the proof. □

Corollary 1.4.13. *Let $\tilde{a} \in \tilde{\mathcal{A}}$. Then*

$$\mathrm{sp}^r_{\tilde{\mathcal{A}}} \tilde{a} = \mathrm{sp}_{\mathcal{A}^{2\times 2}} \Psi(\tilde{a}).$$

It should be noted that such an equality is not always true for the complex spectrum $\mathrm{sp}_{\tilde{\mathcal{A}}} \tilde{a}$ of $\tilde{a}$. Consider the following example.

Example 1.4.14. Let $\mathcal{H} = \mathbb{C}$, $\mathcal{A}$ be the algebra of all operators of multiplications by $a \in \mathbb{C}$, with the involution $a^* = \overline{a}$, and let $M(x) = \overline{x}$ for all $x \in \mathbb{C}$. Then

$$\tilde{\mathcal{A}} = \{b + cM : b, c \in \mathcal{A}\}.$$

We consider a self-adjoint element $\tilde{a}$ of $\tilde{\mathcal{A}}$. It is clear that $\tilde{a} = b + cM$ is self-adjoint if and only if $\overline{b} = b$. If we exploit transformation (1.21) again, the complex spectrum $\mathrm{sp}_{\tilde{\mathcal{A}}} \tilde{a}$ of the element $\tilde{a}$ is the circle of radius $|c|$ with center at the point b, i.e., if $c \neq 0$, then the complex spectrum of $\tilde{a}$ contains complex points.

Remark 1.4.15. Let $\mathcal{B}$ be a complex Banach algebra with identity $e \neq 0$ included into a real algebra $\mathcal{R}$. Assume that the last algebra contains an element m, which does not belong to $\mathcal{B}$ and which satisfies the following assumptions:

($\widehat{A}_1$) For each $a \in \mathcal{B}$ the element mam belongs to the Banach algebra $\mathcal{B}$ as well.

($\widehat{A}_2$) For each $a \in \mathcal{B}$ and for each $\lambda \in \mathbb{C}$ the relation $m(\lambda a) = \overline{\lambda} ma$ holds.

($\widehat{A}_3$) $m^2 = e$ and $me = m$.

($\widehat{A}_4$) The null element 0 of the Banach algebra $\mathcal{B}$ is also that of the algebra $\mathcal{R}$.

Let $\mathcal{B}m$ stand for the set of those elements of $\mathcal{R}$ which have the form am with $a \in \mathcal{B}$. Define a real algebra $\widetilde{\mathcal{B}}$ by $\widetilde{\mathcal{B}} := \mathcal{B} + \mathcal{B}m$, and consider the mapping $\Psi : \widetilde{\mathcal{B}} \mapsto \mathcal{B}^{2\times 2}$ defined by (1.21).

Lemma 1.4.16. *Let $\tilde{a}_1, \tilde{a}_2 \in \tilde{\mathcal{B}}$. Then*

1) $\Psi(\lambda\tilde{a}) = \lambda\Psi(\tilde{a})$ *for all* $\tilde{a} \in \tilde{\mathcal{B}}$ *and for all* $\lambda \in \mathbb{R}$.
2) $\Psi(\tilde{a}_1 + \tilde{a}_2) = \Psi(\tilde{a}_1) + \Psi(\tilde{a}_2)$.
3) $\Psi(\tilde{a}_1 \cdot \tilde{a}_2) = \Psi(\tilde{a}_1) \cdot \Psi(\tilde{a}_2)$.
4) *The mapping* $\Psi : \tilde{\mathcal{B}} \to E^{2\times 2}_{\tilde{\mathcal{B}}}$ *is continuously invertible.*

Note that the last result is useful in situations when one considers operators in Banach spaces.

1.5 Moore-Penrose Invertibility in Algebra $\tilde{\mathcal{A}}$

After introducing operations of multiplication and involution on the algebra $\tilde{\mathcal{A}}$ we may also consider the Moore-Penrose invertibility in this algebra.

Definition 1.5.1. *An element $\tilde{a} \in \tilde{\mathcal{A}}$ is said to be m-Moore-Penrose invertible in the algebra $\tilde{\mathcal{A}}$ if there exists an element $\tilde{b} \in \tilde{\mathcal{A}}$ such that the relations*

$$\tilde{a}\tilde{b}\tilde{a} = \tilde{a}, \qquad \tilde{b}\tilde{a}\tilde{b} = \tilde{b}, \qquad (\tilde{a}\tilde{b})_m^* = \tilde{a}\tilde{b}, \qquad (\tilde{b}\tilde{a})_m^* = \tilde{b}\tilde{a}$$

hold.

If such an element $\tilde{b}$ exists, then it is called an m-Moore-Penrose inverse of $\tilde{a}$ and denoted by $\tilde{a}_m^+$.

Lemma 1.5.2. *Let the elements m_1 and m_2 produce just the same algebra $\tilde{\mathcal{A}}$. If $(m_1m_2)^* = m_2m_1$, then each element $\tilde{a} \in \tilde{\mathcal{A}}$ is m_1-Moore-Penrose invertible if and only if it is m_2-Moore-Penrose invertible and $\tilde{a}_{m_1}^+ = \tilde{a}_{m_2}^+$. In particular, if $\tilde{a}$ is m-Moore-Penrose invertible, then its m-Moore-Penrose inverse is unique.*

Proof. If $(m_1m_2)^* = m_2m_1$, then the m_1-involution on $\tilde{\mathcal{A}}$ coincides with the m_2-involution. It implies the validity of the first assertion of Lemma 1.5.2. The calculation

$$\begin{aligned}
\tilde{a}_{m_1}^+ &= \tilde{a}_{m_1}^+ \tilde{a}\tilde{a}_{m_1}^+ = \tilde{a}_{m_1}^+ (\tilde{a}\tilde{a}_{m_1}^+)_{m_1}^* = \tilde{a}_{m_1}^+ (\tilde{a}_{m_1}^+)_{m_1}^* \tilde{a}_{m_2}^* \\
&= \tilde{a}_{m_1}^+ (\tilde{a}_{m_1}^+)_{m_1}^* \tilde{a}_{m_2}^* (\tilde{a}_{m_2}^+)_{m_2}^* \tilde{a}_{m_2}^* = \tilde{a}_{m_1}^+ ((\tilde{a}_{m_1}^+)_{m_1}^* \tilde{a}_{m_1}^*)(\tilde{a}_{m_2}^+)_{m_2}^* \tilde{a}_{m_2}^* \\
&= \tilde{a}_{m_1}^+ \tilde{a}\tilde{a}_{m_1}^+ (\tilde{a}\tilde{a}_{m_2}^+)_{m_2}^* = \tilde{a}_{m_1}^+ \tilde{a}\tilde{a}_{m_2}^+ = (\tilde{a}_{m_1}^+ \tilde{a})_{m_1}^* \tilde{a}_{m_2}^+ = \tilde{a}_{m_1}^* (\tilde{a}_{m_1}^+)_{m_1}^* \tilde{a}_{m_2}^+ \\
&= \tilde{a}_{m_2}^* (\tilde{a}_{m_2}^+)_{m_2}^* \tilde{a}_{m_2}^* (\tilde{a}_{m_1}^+)_{m_1}^* \tilde{a}_{m_2}^+ = (\tilde{a}_{m_2}^+ \tilde{a})_{m_2}^* (\tilde{a}_{m_1}^+ \tilde{a})_{m_1}^* \tilde{a}_{m_2}^+ \\
&= \tilde{a}_{m_2}^+ (\tilde{a}\tilde{a}_{m_1}^+ \tilde{a})\tilde{a}_{m_2}^+ = \tilde{a}_{m_2}^+ \tilde{a}\tilde{a}_{m_2}^+ = \tilde{a}_{m_2}^+
\end{aligned}$$

implies the second assertion. Setting $m = m_1 = m_2$ one obtains the uniqueness of the m-Moore-Penrose inverse if it exists. □

Corollary 1.5.3. *Let $\mathcal{H}$ be a complex Hilbert space, and let $M_1, M_2 \in \mathcal{M}(\mathcal{L}(\mathcal{H}))$. Then an operator $A \in \mathcal{L}_{add}(\mathcal{H})$ is M_1-Moore-Penrose invertible if and only if it is M_2-Moore-Penrose invertible and $A_{M_1}^+ = A_{M_2}^+$.*

The proof of this corollary immediately follows from Theorem 1.3.10.

Thus in the case of the real C^*-algebra $\mathcal{L}_{add}(\mathcal{H})$, Moore-Penrose inverses do not depend on the element M. However, extensions of arbitrary complex C^*-algebras may not possess this property. Therefore, in the following we always assume that the element m is fixed or that any element m of $\mathcal{M}(\mathcal{A})$ produces the same involution, so we can write $\tilde{a}^*$ and $\tilde{a}^+$ instead of $\tilde{a}_m^*$ and of $\tilde{a}_m^+$, respectively.

Let us recall that the Moore-Penrose invertibility in complex C^*-algebras can be studied via spectral characteristics of suitably chosen self-adjoint elements [103]. However, as Example 1.4.14 shows, self-adjoint elements in real algebras do not always have the properties that are customary for complex algebras. It means that the respective "complex" results cannot be immediately used to study the Moore-Penrose invertibility in general real algebras with involution. Nevertheless, if we confine our study to the above extensions of complex C^*-algebras, the results obtained are quite complete. Thus using the mapping (1.21), one can translate the problem of Moore-Penrose invertibility for $\tilde{\mathcal{A}}$ into a similar problem for a real subalgebra of the complex algebra $\mathcal{A}^{2\times 2}$.

Proposition 1.5.4. *The element $\tilde{a} \in \tilde{\mathcal{A}}$ is Moore-Penrose invertible if and only if the element $\Psi(\tilde{a})$ is Moore-Penrose invertible in $E^{2\times 2}_{\tilde{\mathcal{A}}}$ and*

$$\Psi(\tilde{a})^+ = \Psi(\tilde{a}^+), \tag{1.23}$$

$$\Psi^{-1}(A)^+ = \Psi^{-1}(A^+). \tag{1.24}$$

Relations (1.23), (1.24) show that Moore-Penrose invertibility in $\tilde{\mathcal{A}}$ is equivalent to Moore-Penrose invertibility in $E^{2\times 2}_{\tilde{\mathcal{A}}}$. However, the latter is a real subalgebra of the complex algebra $\mathcal{A}^{2\times 2}$ that makes our task simpler.

Proposition 1.5.5. *An element $\tilde{a} = b + cm \in \tilde{\mathcal{A}}$ is Moore-Penrose invertible in $\tilde{\mathcal{A}}$ if and only if the element $\hat{A} = \Psi(\tilde{a}^*\tilde{a})$,*

$$\hat{A} = \begin{pmatrix} b^*b + mc^*cm & b^*c + mc^*bm \\ c^*b + mb^*cm & c^*c + mb^*bm \end{pmatrix},$$

is invertible in the real algebra $E^{2\times 2}_{\tilde{\mathcal{A}}}$ or if 0 is an isolated point of the real spectrum of $\hat{A}$ in $E^{2\times 2}_{\tilde{\mathcal{A}}}$.

Proof. Necessity. Suppose that the element $\tilde{a}$ is Moore-Penrose invertible in $\tilde{\mathcal{A}}$, and let $\tilde{a}^+$ be the Moore-Penrose inverse for $\tilde{a}$. By B we denote the element $\Psi(\tilde{a}^+)$. Evidently, B belongs to $E^{2\times 2}_{\tilde{\mathcal{A}}}$, so for all real λ such that $0 < |\lambda| < ||BB^*||^{-1}$, the element $\mathcal{E} - \lambda BB^*$ is invertible in $E^{2\times 2}_{\tilde{\mathcal{A}}}$. Since λ is a real number, the element

$$(\mathcal{E} - \lambda BB^*)^{-1}BB^* - \frac{1}{\lambda}(\mathcal{E} - BB^*) \tag{1.25}$$

also belongs to $E^{2\times 2}_{\tilde{\mathcal{A}}}$. However, as noted in [193], the element (1.25) is the inverse for $\hat{A} - \lambda\mathcal{E}$, i.e., $\hat{A}$ is invertible or 0 is an isolated point of the spectrum of the element $\hat{A}$ in $E^{2\times 2}_{\tilde{\mathcal{A}}}$. (Note that $\mathcal{E}$ denotes the identity element of $\mathcal{A}^{2\times 2}$.)

Sufficiency. If $\hat{A}$ is invertible in $E^{2\times 2}_{\tilde{\mathcal{A}}}$, then $B = (\hat{A})^{-1}A^*$ belongs to $E^{2\times 2}_{\tilde{\mathcal{A}}}$ and is the Moore-Penrose inverse for $A = \Psi(\tilde{a})$. Assume that 0 is an isolated point of $\mathrm{sp}_{E^{2\times 2}_{\tilde{\mathcal{A}}}}\hat{A}$ and consider the real C^*-algebra $\mathcal{R}(\hat{A})$ generated by the self-adjoint element $\hat{A}$ and by the unit $\mathcal{E}$. It is clear that $\mathcal{R}(\hat{A}) \subseteq E^{2\times 2}_{\tilde{\mathcal{A}}}$ and from Theorem 1.4.3 the algebra $\mathcal{R}(\hat{A})$ is isometrically isomorphic to the algebra of all real continuous functions on $\mathrm{sp}_{\mathcal{R}(\hat{A})}\hat{A}$. The rest of the proof runs analogously to [192], Proposition 3.2. □

Corollary 1.5.6. *The element $\tilde{a} \in \tilde{\mathcal{A}}$ is Moore-Penrose invertible if and only if the element $\Psi(\tilde{a}^*\tilde{a})$ is invertible in $\mathcal{A}^{2\times 2}$ or 0 is an isolated point of the spectrum of $\Psi(\tilde{a}^*\tilde{a})$ in $\mathcal{A}^{2\times 2}$.*

Proof. It is clear that

$$\mathrm{sp}_{\mathcal{A}^{2\times 2}}\Psi(\tilde{a}^*\tilde{a}) \subseteq \mathrm{sp}_{E^{2\times 2}_{\tilde{\mathcal{A}}}}\Psi(\tilde{a}^*\tilde{a}) \subseteq \mathrm{sp}_{\mathcal{R}(\Psi(\tilde{a}^*\tilde{a}))}\Psi(\tilde{a}^*\tilde{a}).$$

Since $\Psi(\tilde{a}^*\tilde{a})$ is self-adjoint, we have

$$\mathrm{sp}_{\mathcal{A}^{2\times 2}}\Psi(\tilde{a}^*\tilde{a}) = \mathrm{sp}_{\mathcal{R}(\Psi(\tilde{a}^*\tilde{a}))}\Psi(\tilde{a}^*\tilde{a}) \tag{1.26}$$

from Lemma 1.4.4, and Proposition 1.5.5 completes the proof. □

An important property of complex C^*-subalgebras of complex C^*-algebras is that they are inverse closed with respect to the Moore-Penrose invertibility. In the case of real extensions, we have to impose an additional condition on the corresponding C^*-subalgebra. More precisely, the following corollary is true.

Corollary 1.5.7. *Let $\mathcal{C}$ be an m-closed C^*-algebra with identity of a C^*-algebra $\mathcal{A}$. Then $\tilde{\mathcal{C}}$ is inverse closed with respect to Moore-Penrose invertibility, i.e., if an element $\tilde{a} \in \tilde{\mathcal{C}}$ has a Moore-Penrose inverse in $\tilde{\mathcal{A}}$, then this one also belongs to $\tilde{\mathcal{C}}$.*

Corollary 1.5.8. *Let $\mathcal{U}$ be a complex C^*-algebra of a C^*-algebra $\mathcal{A}$, which contains a real C^*-algebra $\mathcal{B}_{\mathbb{R}}$. The element $b \in \mathcal{B}_{\mathbb{R}}$ is Moore-Penrose invertible in $\mathcal{B}_{\mathbb{R}}$ if and only if it is Moore-Penrose invertible in $\mathcal{U}$.*

The proof of these two results follows from relation (1.26).

As usual, an element $\tilde{p} \in \tilde{\mathcal{A}}$ is said to be a projection if $\tilde{p}^2 = \tilde{p}$ and $\tilde{p}^* = \tilde{p}$. The following results as well as their proofs literally repeat the corresponding formulations for complex C^*-algebras. It should be noted that in order to prove the uniqueness of the below arising projection $\tilde{p}$ we have to use the connection between the algebras $\tilde{\mathcal{A}}$ and $E^{2\times 2}_{\tilde{\mathcal{A}}}$.

Proposition 1.5.9. *Let $\tilde{\mathcal{A}}$ be the real extension of a unital complex C^*-algebra $\mathcal{A}$ by an element m. If $\tilde{a} \in \tilde{\mathcal{A}}$, then the following assertions are equivalent:*

1) *$\tilde{a}$ is Moore-Penrose invertible.*
2) *$\tilde{a}^*\tilde{a}$ is invertible or 0 is an isolated point of the real spectrum of $\tilde{a}^*\tilde{a}$.*
3) *There exists a projection $\tilde{p}$ in $\mathcal{R}_{\tilde{\mathcal{A}}}(\tilde{a}^*\tilde{a})$ such that $\tilde{a}^*\tilde{a}\tilde{p} = 0$ and $\tilde{a}^*\tilde{a} + \tilde{p}$ is invertible.*
4) *There exists a projection $\tilde{q}$ in $\tilde{\mathcal{A}}$ such that $\tilde{a}\tilde{q} = 0$ and $\tilde{a}^*\tilde{a} + \tilde{q}$ is invertible.*

If one of these conditions is satisfied, then $\tilde{q}$ is uniquely determined and $\tilde{a}^+ = (\tilde{a}^*\tilde{a} + \tilde{q})^{-1}\tilde{a}^*$.

In connection with Proposition 1.5.9, let us list some elementary properties of projection elements in the algebra $\tilde{\mathcal{A}}$.

Proposition 1.5.10. *Let $f, g \in \mathcal{A}$. The element $\tilde{p} = f + gm \in \tilde{\mathcal{A}}$ is a projection if and only if the elements f and g satisfy the following four relations:*

i) $f^* = f$;

ii) $g^* = mgm$;

iii) $f = f^2 + fg^*$;

iv) $(f - g)g^* = 0$.

For the proof, one can exploit the transformation (1.21).

Proposition 1.5.10 allows one to obtain the following description of projection elements in the algebra $\tilde{\mathcal{A}}$.

Corollary 1.5.11. *Let $\tilde{p} = f + gm$ be a projection in $\tilde{\mathcal{A}}$. Then one and only one of the following assertions holds:*

a) *The element g is not invertible from the left side.*

b) $\tilde{p} = \frac{1}{2}(e + em)$, *where e is the identity element in $\mathcal{A}$.*

Proof. If the element g is invertible from the left, then relation *iv)* implies $f = g$. Using the equality $g = g^*$, one obtains $g = e/2$. □

1.6 Operator Sequences: Stability

Let X, Y be real or complex Banach spaces, and let $(P_n^X), (P_n^Y)$ be sequences of projection operators such that these sequences and the sequences $((P_n^X)^*), ((P_n^Y)^*)$ converge strongly to the identity operators in the corresponding Banach spaces.

Consider the operator equation

$$Ax = y, \quad x \in X, \quad y \in Y, \quad A \in \mathcal{L}(X, Y). \tag{1.27}$$

To solve equation (1.27) approximately, let us construct the sequence of equations

$$A_n P_n^X x_n = P_n^Y y, \quad x_n \in \operatorname{im} P_n^X, n = 1, 2, \ldots, \tag{1.28}$$

where A_n is a linear and bounded operator defined on the space $\operatorname{im} P_n^X$ and taking its values in $\operatorname{im} P_n^Y$. We say that the approximation method (1.28) applies to the operator A and write $A \in \Pi\{A_n, P_n^Y\}$ if there is an $n_0 \in \mathbb{N}$ such that for all $n \geq n_0$ equations (1.28) are solvable and for any right-hand side $y \in Y$, solutions of (1.28) converge to the solution of equation (1.27).

Let us assume that the operators A_n are connected with A in the following sense:

The sequences $(A_n P_n^X)$, $(A_n^*(P_n^Y)^*)$ converge strongly to A and A^*, respectively.

Note that for the strong limit of a sequence (A_n) we usually use the notation $s - \lim A_n$.

Assume that the operator A is invertible, the operators A_n are invertible for all $n \geq n_0$ and the inverses $A_n^{-1}, n \geq n_0$ are uniformly bounded. Then the sequence $(A_n^{-1}P_n^Y)_{n\geq n_0}$ converges strongly to A^{-1}. Indeed, for any $y \in Y$ we have

$$||A^{-1}y - A_n^{-1}P_n^Y y|| \leq ||A_n^{-1}P_n^Y|| \, ||A_n P_n^X A^{-1}y - P_n^Y y||,$$

so the claim follows from the boundedness of the sequence $(||A_n^{-1}P_n^Y||)_{n\geq n_0}$. Thus the (unique) solution of equation (1.27) can be approximated as close as desired by the elements $A_n^{-1}P_n^Y y$. Moreover, it is not hard to estimate the speed of the convergence, viz., if $Ax = y$ and $A_n x_n = P_n^Y y$, then for any $n \geq n_0$ one has

$$\begin{aligned} ||x - x_n|| &\leq \inf_{v\in \operatorname{im} P_n^X} (||x - v|| + ||v - x_n||) \\ &\leq \inf_{v\in \operatorname{im} P_n^X} (||x - v|| + ||A_n^{-1}P_n^Y|| \, ||A_n v - P_n^Y y||) \\ &\leq \inf_{v\in \operatorname{im} P_n^X} (||x - v|| + ||A_n^{-1}P_n^Y||(||A_n v - Ax|| + ||y - P_n^Y y||)). \quad (1.29) \end{aligned}$$

In particular, for the projection operators $A_n = P_n^Y A P_n^X$, inequality (1.29) leads to the estimate

$$||x - x_n|| \leq \inf_{v\in \operatorname{im} P_n^X} (||x - v|| + ||A_n^{-1}P_n^Y|| \, ||P_n^Y A|| \, ||x - v||);$$

that is

$$||x - x_n|| \leq C \inf_{v\in \operatorname{im} P_n^X} ||x - v||, \quad (1.30)$$

so the sequence $x_n = A_n^{-1}P_n^Y y$ converges quasi-optimally to the solution $x = A^{-1}y$.

The previous considerations show that the uniform boundedness of the sequence (A_n^{-1}) plays a very important role in the study of approximation methods. This motivates us to introduce the following characteristic of approximating sequences.

Definition 1.6.1. *A sequence (A_n), $A_n : \operatorname{im} P_n^X \mapsto \operatorname{im} P_n^Y$ of linear and bounded operators is called stable if for n large enough (say for $n \geq n_0$) the operators A_n are continuously invertible and*

$$\sup_{n\geq n_0} ||A_n^{-1}|| < \infty.$$

Remark 1.6.2. If (A_n) is stable, then

$$||A_n P_n^X x|| \geq C||P_n^X x|| \quad (1.31)$$

for n large enough, where the constant $C > 0$ does not depend on n and x. If $A_n P_n^X$ converges strongly to A, then (1.31) implies the inequality

$$||Ax|| \geq C||x||,$$

that is $\ker A = \{0\}$ and $\operatorname{im} A$ is closed. If $A_n^*(P_n^Y)^*$ also tends strongly to A^*, then the same argumentation shows that $\ker A^* = \{0\}$ and $\operatorname{im} A^*$ is closed. Hence A is necessarily invertible.

Let $\mathbf{Z}$ be any unbounded subset of the set of non-negative integers. By $S = \mathcal{S}(\mathcal{P}^X, \mathcal{P}^Y)$, $\mathcal{P}^X := (P_n^X)$, $\mathcal{P}^Y := (P_n^Y)$ we denote the collection of all sequences $(A_n)_{n\in\mathbf{Z}}$ of linear bounded operators $A_n : \operatorname{im} P_n^X \mapsto \operatorname{im} P_n^Y$ such that $\sup_n ||A_n P_n|| < \infty$; on defining $(A_n) + (B_n) := (A_n + B_n)$, $\lambda(A_n) := (\lambda A_n)$ and the norm

$$||(A_n)||_{\mathcal{S}} = \sup ||A_n P_n^X||$$

the set $\mathcal{S}$ becomes a (real or complex) Banach space. Further, let $\mathcal{S}_C \subset \mathcal{S}(\mathcal{P}^X, \mathcal{P}^Y)$ be the closed subspace of S consisting of all sequences (A_n) such that $(A_n P_n^X)$ and $(A_n^*(P_n^Y)^*)$ possess strong limits in $\mathcal{L}(X, Y)$ and $\mathcal{L}(Y^*, X^*)$, respectively. The Banach spaces $\mathcal{S}(\mathcal{P}^Y, \mathcal{P}^X), \mathcal{S}_C(\mathcal{P}^Y, \mathcal{P}^X)$, $\mathcal{S}(\mathcal{P}^X) := \mathcal{S}(\mathcal{P}^X, \mathcal{P}^X)$, $\mathcal{S}_C(\mathcal{P}^X)$ and $\mathcal{S}(\mathcal{P}^Y)$, $\mathcal{S}_C(\mathcal{P}^Y)$ are defined analogously. Note that for elements $(A_n) \in \mathcal{S}(\mathcal{P}^X, \mathcal{P}^Y)$, $(B_n) \in \mathcal{S}(\mathcal{P}^Y, \mathcal{P}^X)$ the products $(B_n)(A_n) := (A_n B_n) \in \mathcal{S}(\mathcal{P}^X)$ and $(A_n)(B_n) := (A_n B_n) \in \mathcal{S}(\mathcal{P}^Y)$ are well defined and

$$||(A_n)(B_n)|| \leq ||(A_n)|| \, ||(B_n)||,$$
$$||(B_n)(A_n)|| \leq ||(B_n)|| \, ||(A_n)||.$$

Let us also note that $\mathcal{S}(\mathcal{P}^X)$ and $\mathcal{S}(\mathcal{P}^Y)$ actually form Banach algebras with the unit elements (P_n^X) and (P_n^Y), respectively. Clearly, everything that was said concerning the spaces $\mathcal{S}$ remains true for the spaces $\mathcal{S}_C$.

Let $\mathcal{G} \subset \mathcal{S}(\mathcal{P}^X, \mathcal{P}^Y)$ be the collection of all sequences (A_n) such that the norms $||A_n P_n^X||$ tend to 0 as n tends to ∞. For the remaining spaces $\mathcal{S}(\mathcal{P}^X), \mathcal{S}(\mathcal{P}^Y), \mathcal{S}(\mathcal{P}^Y, \mathcal{P}^X)$ the set $\mathcal{G}$ is defined analogously. Moreover, the symbol $\mathcal{G}$ is again used for every set $\mathcal{G}(\mathcal{P}^X), \mathcal{G}(\mathcal{P}^Y)$, and $\mathcal{G}(\mathcal{P}^Y, \mathcal{P}^X)$. Note that $\mathcal{G}$ is a closed subspace in each space $\mathcal{S}$ and $\mathcal{S}_C$.

For cosets $(A_n) + \mathcal{G}$ and $(B_n) + \mathcal{G}$ that respectively belong to the quotient spaces $\mathcal{S}(\mathcal{P}^X, \mathcal{P}^Y)/\mathcal{G}$ and $\mathcal{S}(\mathcal{P}^Y, \mathcal{P}^X)/\mathcal{G}$, their products are defined by

$$((A_n) + \mathcal{G})\,((B_n) + \mathcal{G}) := (A_n)(B_n) + \mathcal{G} \in \mathcal{S}(\mathcal{P}^Y)/\mathcal{G},$$
$$((B_n) + \mathcal{G})\,((A_n) + \mathcal{G}) := (B_n)(A_n) + \mathcal{G} \in \mathcal{S}(\mathcal{P}^X)/\mathcal{G}.$$

It is easily seen that these products are correctly defined, and the related sets $\mathcal{G}$ form closed two-sided ideals in Banach algebras $\mathcal{S}(\mathcal{P}^X)$ and $\mathcal{S}(\mathcal{P}^Y)$. Of course, the spaces $\mathcal{S}$ can again be replaced by the spaces $\mathcal{S}_C$, so all the previous statements hold. According to the language used in operator theory, an element $(A_n) \in \mathcal{S}(\mathcal{P}^X, \mathcal{P}^Y)$ is called invertible, if there is an element $(B_n) \in \mathcal{S}(\mathcal{P}^Y, \mathcal{P}^X)$ such that $(B_n)(A_n) = \mathcal{P}^X$ and $(A_n)(B_n) = \mathcal{P}^Y$. By $\mathbf{G}\mathcal{S}(\mathcal{P}^X, \mathcal{P}^Y)$ we denote the set of all invertible elements from $\mathcal{S}(\mathcal{P}^X, \mathcal{P}^Y)$. Similar definitions and notations are used for the spaces $\mathcal{S}_C$, as well as for the associated quotient spaces $\mathcal{S}/\mathcal{G}$ and $\mathcal{S}_C/\mathcal{G}$.

Let π and π_C denote the canonical homomorphisms $\pi : \mathcal{S} \mapsto \mathcal{S}/\mathcal{G}$ and $\pi_C : \mathcal{S}_C \mapsto \mathcal{S}_C/\mathcal{G}$, respectively.

Proposition 1.6.3. *Suppose* $P_n^X \to I$, $P_n^Y \to I$, $(P_n^X)^* \to I$ *and* $(P_n^Y)^* \to I$ *strongly as* $n \to \infty$, *the sequence* (A_n) *belongs to* $\mathcal{S}_C$, *and let* A *denote the strong limit of* (A_n). *Then the following statements are equivalent:*

(a) *The sequence* (A_n) *is stable.*

(b) A *is continuously invertible and* $\pi_C(A_n) \in \mathbf{G}(\mathcal{S}_C/\mathcal{G})$.

(c) A *is continuously invertible and* $\pi(A_n) \in \mathbf{G}(\mathcal{S}/\mathcal{G})$.

Proof. (a) $\Rightarrow$ (b): This was proved earlier.

(b) $\Rightarrow$ (c): Obvious.

(c) $\Rightarrow$ (a): Let for instance $\mathcal{S} = \mathcal{S}(\mathcal{P}^X, \mathcal{P}^Y)$. Suppose there is a sequence $(B_n) \in \mathcal{S}(\mathcal{P}^Y, \mathcal{P}^X)$ such that

$$B_n A_n = P_n^X + C_n, \quad A_n B_n = P_n^Y + D_n$$

where $(C_n) \in \mathcal{G}(\mathcal{P}^X), (D_n) \in \mathcal{G}(\mathcal{P}^Y)$. Then there is an n_0 such that $||C_n P_n^X|| < 1/2$ and $||D_n P_n^Y|| < 1/2$ for all $n \geq n_0$. Thus by the Neumann series theorem, the operators $P_n^X + C_n$ and $P_n^Y + D_n$, regarded as acting on $\operatorname{im} P_n^X$ and $\operatorname{im} P_n^Y$ respectively, are invertible for $n \geq n_0$ and the norms of their inverses are uniformly bounded. This proves the claim. □

The importance of this result is that the stability problem for $X = Y$ is indeed an invertibility problem in a Banach algebra, viz. in algebra $\mathcal{S}/\mathcal{G}$ or in $\mathcal{S}_C/\mathcal{G}$, hence all results on invertibility in Banach algebras can now be used to study the stability of approximation sequences. If $X = Y$ but $\mathcal{P}^X \neq \mathcal{P}^Y$, or if $X \neq Y$, the set $\mathcal{S}$ does not necessarily form an algebra. Nevertheless, some kind of algebraization is still possible. Thus, the concept of para-algebra introduced in [184], gives us all tools to handle this situation.

From now on and to the end of this section, we assume that $X = Y$ and $\mathcal{P}^X = \mathcal{P}^Y$. We write P_n for P_n^X and $\mathcal{P}$ for $\mathcal{P}^X$.

Proposition 1.6.4. *Let* $\mathcal{K}(X)$ *denote the ideal of all compact operators in* $\mathcal{L}(X)$. *Then the collection of sequences*

$$\mathcal{J}_1 := \{(A_n) \in \mathcal{S}_C : A_n = P_n T P_n + C_n, \quad T \in \mathcal{K}(X) \quad \textit{and} \quad (C_n) \in \mathcal{G}\}$$

forms a closed two-sided ideal in $\mathcal{S}_C$. *Moreover, the mapping* $W : \mathcal{S}_C \mapsto \mathcal{L}(X)$, $(A_n) \mapsto s\text{-}\lim A_n P_n$, *is a Banach algebra homomorphism and the sequence* $(A_n) \in \mathcal{S}_C$ *is stable if and only if* $W(A_n)$ *is invertible in* $\mathcal{L}(X)$ *and the coset* $(A_n) + \mathcal{J}_1$ *is invertible in* $\mathcal{S}_C/\mathcal{J}_1$.

Proof. The fact that $\mathcal{J}_1$ forms an ideal is a consequence of the following well-known result:

If sequences (A_n) and (B_n^*) converge strongly to operators A and B^*, respectively, and if K is a compact operator, then

$$\lim_{n\to\infty} ||A_n K B_n - AKB|| = 0. \tag{1.32}$$

The closedness of $\mathcal{J}_1$ can be shown immediately. Let us consider the invertibility statement. The necessity is obvious. The sufficiency can be proved as follows. Since $(A_n) + \mathcal{J}_1$ is invertible in $\mathcal{S}_C/\mathcal{J}_1$ there is a sequence $(B_n) \in \mathcal{S}_C$ such that

$$A_n B_n = P_n + P_n T_1 P_n + C_n,$$
$$B_n A_n = P_n + P_n T_2 P_n + D_n,$$

where $T_1, T_2 \in \mathcal{K}(X)$ and $(C_n), (D_n) \in \mathcal{G}$. Then

$$W(A_n B_n) = W(A_n)W(B_n) = I + T_1$$
$$W(B_n A_n) = W(B_n)W(A_n) = I + T_2.$$

Therefore $W(A_n)$, and also $W(B_n)$, are Fredholm operators. The invertibility of $W(A_n)$ allows us to consider the operators

$$B_n' := B_n - P_n W(A_n)^{-1} T_1 P_n,$$

so

$$A_n B_n' = P_n + C_n', \quad (C_n') \in \mathcal{G}.$$

Analogously, there is a sequence $(B_n'') \in \mathcal{S}_C$ such that

$$B_n'' A_n = P_n + C_n'', \quad (C_n'') \in \mathcal{G},$$

hence $(A_n) + \mathcal{G}$ is invertible in $\mathcal{S}_C/\mathcal{G}$ and Proposition 1.6.3 yields the stability of the sequence (A_n). □

The idea of this proof also applies in more general situations. Thus if $\mathcal{H}$ is a separable Hilbert space and $\{\ldots, e_{-1}, e_0, e_1, \ldots\}$ is a complete orthonormal basis in $\mathcal{H}$, then by P_n we denote the orthogonal projection onto $\operatorname{span}\{e_{-n}, e_{-n+1}, \ldots, e_0, e_1, \ldots, e_{n-1}, e_n\}$. Consider the related algebras $\mathcal{S}$ and $\mathcal{S}_C$, which proved to be C^*-algebras. Let $a = \sum_{k=-n}^{n} a_k e_k$, where $a_i \in \mathbb{R}$ or $a_i \in \mathbb{C}$ in dependence on whether $\mathcal{H}$ is a real or complex space. We define the operators $W_n : \operatorname{im} P_n \mapsto \operatorname{im} P_n$ by

$$W_n(a) = a_{-1}e_{-n} + \ldots + a_{-n}e_{-1} + a_n e_0 + a_{n-1}e_1 + \ldots + a_0 e_n.$$

Then $W_n^2 = P_n$, and W_n tends weakly to 0. The last assertion is a consequence of the fact that for any $\varphi \in \mathcal{H}$ the scalar products $< e_k, \varphi > \to 0$ as $k \to +\infty$ or $k \to -\infty$.

Introduce now the collection $\mathcal{A}$ of all sequences $(A_n) \subset \mathcal{S}_C$, such that in addition to the previous requirements, the sequences $W_n A_n W_n \to \widetilde{A} := \widetilde{W}(A_n)$ and

$W_n A_n^* W_n \to \widetilde{A}^*$ strongly. The collection $\mathcal{A}$ actually forms a unital C^*-subalgebra of $\mathcal{S}_C$ and the mappings

$$W : \mathcal{A} \mapsto \mathcal{L}(\mathcal{H}), \quad \widetilde{W} : \mathcal{A} \mapsto \mathcal{L}(\mathcal{H})$$

are *-homomorphisms. Note that the involution in $\mathcal{A}$ is defined by

$$(A_n)^* := (A_n^*).$$

In addition to the above-defined ideal $\mathcal{J}_1$, there is another ideal $\mathcal{J}_2$ inside of the algebra $\mathcal{A}$, viz.,

$$\mathcal{J}_2 := \{(A_n) \in \mathcal{A} : A_n = W_n T W_n + C_n, \quad T \in \mathcal{K}(\mathcal{H}), \quad (C_n) \in \mathcal{G}\}.$$

To see that $\mathcal{J}_2$ is really an ideal in $\mathcal{A}$, one has to use the fact that compact operators transform the weakly convergent sequences into strongly convergent.

Note that such algebras play an important role in the study of approximation methods for different classes of operators. For example, the following proposition holds.

Proposition 1.6.5. *Let $\mathcal{J}$ be the smallest closed two-sided ideal containing the ideals $\mathcal{J}_1$ and $\mathcal{J}_2$. A sequence $(A_n) \in \mathcal{A}$ is stable if and only if the operators $W(A_n)$, $\widetilde{W}(A_n)$, and the coset $(A_n) + \mathcal{J} \in \mathcal{A}/\mathcal{J}$ are invertible.*

The proof of Proposition 1.6.5 runs parallel to the proof of Proposition 1.6.4 and is omitted here.

Let us also consider a modification of this result. This modification will be used in the first two sections of Chapter 2. If $\mathcal{H}$ is a complex Hilbert space and the operators P_n and W_n are defined as above, then the underlying algebra $\mathcal{A}$ is a complex C^*-algebra. On the other hand, the space $\mathcal{H}$ can also be viewed as a real Hilbert space. Recall that such a real Hilbert space $\mathcal{H}_\mathbb{R}$ is provided with the scalar product $< h_1, h_2 >_{\mathcal{H}_\mathbb{R}} := \operatorname{Re}(h_1, h_2)$, cf. (1.18). If $\{e_k\}_{k\in\mathbb{Z}}$ is a complete orthogonal basis in $\mathcal{H}$, then $\{e_k, ie_k\}_{k\in\mathbb{Z}}$ forms a complete orthogonal basis in $\mathcal{H}_\mathbb{R}$. Indeed, since

$$\begin{aligned} ||h||^2_{\mathcal{H}} &= \sum_{k\in\mathbb{Z}} |(h, e_k)|^2 = \sum_{k\in\mathbb{Z}} \left((\operatorname{Re}(h, e_k))^2 + (\operatorname{Im}(h, e_k))^2\right) \\ &= \sum_{k\in\mathbb{Z}} \left((< h, e_k >_{\mathcal{H}_\mathbb{R}})^2 + (< h, ie_k >_{\mathcal{H}_\mathbb{R}})^2\right) = ||h||^2_{\mathcal{H}_\mathbb{R}}, \end{aligned}$$

the completeness of $\mathcal{H}_\mathbb{R}$ follows.

On real Hilbert space $\mathcal{H}_\mathbb{R}$ we consider modified operators W_n and P_n, viz., if $h \in \mathcal{H}_\mathbb{R}$, then $h = \sum_{k\in\mathbb{Z}} (a_k e_k + ib_k e_k)$, so we set

$$\begin{aligned} W_n h := {} & (a_{-1}e_{-n} + ib_{-1}e_{-n}) + \ldots + (a_{-n}e_{-1} + ib_{-n}e_{-1}) \\ & + (a_n e_0 + ib_n e_0) + \ldots + (a_0 e_n + ib_0 e_n) \end{aligned}$$

and $P_n := (W_n)^2$. It is clear that $(P_n)^2 = P_n$, and $P_n, n \in \mathbb{N}$ are orthogonal projections such that on the space $\mathcal{H}_{\mathbb{R}}$ the sequence (P_n) strongly converges to the identity operator. Moreover, the sequence (W_n) weakly converges to zero. Notice that, as mappings, the operators P_n, W_n in $\mathcal{H}_{\mathbb{R}}$ coincide with the operators P_n, W_n in $\mathcal{H}$. Of course, Proposition 1.6.5 remains valid in this situation. Consider now the bounded and linear operator $M : \mathcal{H}_{\mathbb{R}} \mapsto \mathcal{H}_{\mathbb{R}}$ which is defined on the basis elements e_k, ie_k by

$$M(e_k) = e_{-k}, \quad M(ie_k) := -ie_{-k}, \quad k \in \mathbb{Z}.$$

Now let $\mathcal{A}_{\mathbb{R}}$ be the algebra associated with the operators $P_n, W_n, n \in \mathbb{N}$ considered on the real Hilbert space $\mathcal{H}_{\mathbb{R}}$. It is easily seen that the previously defined algebra $\mathcal{A}$ is a complex C^*-subalgebra of the real algebra $\mathcal{A}_{\mathbb{R}}$. Let us further notice that $MP_n = P_nM$ for all $n \in \mathbb{N}$ and that the sequence (W_nMW_n) strongly converges to the operator $\widehat{M} = M\mathcal{U}$, where $\mathcal{U}$ is the unitary operator defined on the Hilbert space $\mathcal{H}$ by

$$\mathcal{U}e_k := e_{k+1}, \quad k \in \mathbb{Z}.$$

Because of the relation $(W_nMW_n)^* = W_nMW_n$, the sequence of the adjoint operators $((W_nMW_n)^*)$ strongly converges to the same operator $\widehat{M}$. This also leads to the equality $\mathcal{U}M = M\mathcal{U}(= \widehat{M})$. The sequence (P_nMP_n) does not belong to the C^*-algebra $\mathcal{A}$, however $(P_nMP_n) \in \mathcal{A}_{\mathbb{R}}$. In addition, the element $m := (P_nMP_n)$ satisfies the axioms $(A_1) - (A_5)$ from Section 1.2 with respect to the algebras $\mathcal{A}$ and $\mathcal{A}_{\mathbb{R}}$. Let $\tilde{\mathcal{A}}$ denote the extension of the algebra $\mathcal{A}$ by the element m. Using ideas from Section 1.3.5 it is not difficult to prove that $\tilde{\mathcal{A}} = \mathcal{A}_{\mathbb{R}}$.

Next we consider another generalization of Proposition 1.6.5, which plays an extremely important role in applications considered below. Let $\mathcal{H}$ be an infinite-dimensional Hilbert space, and let $\mathcal{H}_n$ be a sequence of closed subspaces such that the orthogonal projections P_n from $\mathcal{H}$ onto $\mathcal{H}_n$ converge strongly to the identity operator I on $\mathcal{H}$. Further, let T be a possibly infinite index set, and suppose that for every $t \in T$ we are given an infinite-dimensional Hilbert space $\mathcal{H}^t$ with the identity operator I^t, as well as a sequence (E_n^t) of partial isometries $E_n^t : \mathcal{H}^t \mapsto \mathcal{H}$ such that:

- the initial projections P_n^t of E_n^t converge strongly to I^t as $n \to \infty$,
- the range projection of E_n^t is P_n,
- the separation condition

  $$(E_n^s)^* E_n^t \to 0 \quad \text{weakly as} \quad n \to \infty,$$

  holds for every $s, t \in T$ with $s \neq t$.

Recall that an operator $E : \mathcal{H}' \mapsto \mathcal{H}''$ is a partial isometry if $EE^*E = E$. Then E^*E and EE^* are orthogonal projections, which are respectively called the initial and range projections. For the sake of brevity, we will write E_{-n}^t instead of $(E_n^t)^*$.

Let $\mathcal{S}^T$ stand for the set of all sequences $(A_n) \in \mathcal{S}$ such that for every $t \in T$, the strong limits

$$s - \lim E^t_{-n} A_n E^t_n \quad \text{and} \quad s - \lim (E^t_{-n} A_n E^t_n)^*$$

exist, so one can define the mappings $W^t : \mathcal{S}^T \mapsto \mathcal{L}(\mathcal{H}^t)$ by

$$W^t(A_n) := s - \lim E^t_{-n} A_n E^t_n.$$

It is easy to check that $\mathcal{S}^T$ is a C^*-subalgebra of $\mathcal{S}$ and the mappings W^t are *-homomorphisms, i.e., $W^t(A_n)^* = (W^t(A_n))^*$ for each $(A_n) \in \mathcal{S}^T$. Moreover, the separation condition ensures that, for every $t \in T$ and for every compact operator $K^t \in \mathcal{K}(\mathcal{H}^t)$, the sequence $(E^t_n K^t E^t_{-n})$ belongs to the algebra $\mathcal{S}^T$, and

$$W^s(E^t_n K^t E^t_{-n}) = \begin{cases} K^t & \text{if} \quad s = t, \\ 0 & \text{if} \quad s \neq t \end{cases} \tag{1.33}$$

for all $s \in T$. Introduce the smallest closed ideal $\mathcal{J}^T$ of $\mathcal{S}^T$ which contains all sequences $(E^t_n K^t E^t_{-n})$ with $t \in T$ and $K^t \in \mathcal{K}(\mathcal{H}^t)$, as well as all sequences $(C_n) \in \mathcal{G}$.

Theorem 1.6.6. (a) *A sequence $(A_n) \in \mathcal{S}^T$ is stable if and only if the operators $W^t(A_n)$ are invertible in $\mathcal{L}(\mathcal{H}^t)$ for every $t \in T$ and the coset $(A_n) + \mathcal{J}^T$ is invertible in the quotient algebra $\mathcal{S}^T/\mathcal{J}^T$.*

(b) *If $(A_n) \in \mathcal{S}^T$ is a sequence with invertible coset $(A_n) + \mathcal{J}^T$, then all operators $W^t(A_n)$ are Fredholm on $\mathcal{H}^t$, and among the operators $W^t(A_n), t \in T$ there is only a finite number of operators that are non-invertible.*

The proof of this result is similar to the proof of Proposition 1.6.4 (cf. also [102]).

Usually, it is not easy to show that for a given operator A the corresponding approximation sequence (A_n) belongs to an algebra of the type $\mathcal{S}^T$. However, if it is shown, the above theorem applies and one is concerned with the invertibility of the operators $W^t(A_n)$ and with the invertibility of the coset $(A_n) + \mathcal{J}^T$. Note that local principles often prove to be very efficient in studying the invertibility of $(A_n) + \mathcal{J}^T$.

1.7 Asymptotic and Weak Asymptotic Moore-Penrose Invertibility of Additive Operators

Let $\mathcal{H}$ be a complex Hilbert space, and let $\mathcal{L}_{add}(\mathcal{H})$ be the set of all continuous additive operators on $\mathcal{H}$. We suppose that $\tilde{A} \in \mathcal{L}_{add}(\mathcal{H})$ and unless otherwise stated (until further notice) we assume that the sequence $(\tilde{A}_n), \tilde{A}_n \in \mathcal{L}_{add}(\mathcal{H})$ converges strongly to the operator $\tilde{A}$, i.e.,

$$\lim_{n\to\infty} \tilde{A}_n x = \tilde{A} x \quad \text{for all } x \in \mathcal{H}.$$

Suppose we are given an involution on $\mathcal{L}_{add}(\mathcal{H})$. The sequence $(\tilde{A}_n)$ is called Moore-Penrose stable if for all sufficiently large n, say for $n \geq n_0$, the operators $\tilde{A}_n$ are Moore-Penrose invertible and

$$\sup_{n \geq n_0} \left|\left| \tilde{A}_n^+ \right|\right| < +\infty.$$

The operator $\tilde{A}$ is said to be asymptotically Moore-Penrose invertible by the sequence $(\tilde{A}_n)$ if the operators $\tilde{A}_n$ are Moore-Penrose invertible for all $n > n_0$ and the sequence of their Moore-Penrose inverses $(\tilde{A}_n^+)$ converges strongly.

Now we will show how to use the previous results to study the Moore-Penrose invertibility of additive operators. Recall that for any Hilbert space $\mathcal{H}$ the set $\mathcal{L}_{add}(\mathcal{H})$ can be considered as an extension of the complex C^*-algebra $\mathcal{L}(\mathcal{H})$ by an antilinear operator $M = M_\Phi$ and, correspondingly, any additive continuous operator $\tilde{A}$ admits the representation

$$\tilde{A} = A_1 + A_2 M$$

where A_1 and $A_2 = A_a M$ are linear continuous operators on the space $\mathcal{H}$, see (1.17). Therefore, $\mathcal{L}_{add}(\mathcal{H})$ possesses a natural operation of involution, viz., for every operator $\tilde{A} \in \mathcal{L}_{add}(\mathcal{H})$ one sets

$$\tilde{A}^* := A_1^* + M A_2^*,$$

so one can use the results of Section 1.5 to study Moore-Penrose invertibility in $\mathcal{L}_{add}(\mathcal{H})$.

Proposition 1.7.1. *Let $\tilde{A} \in \mathcal{L}_{add}(\mathcal{H})$, and let the sequences $(\tilde{A}_n)$ and $(\tilde{A}_n^*)$ converge strongly to $\tilde{A}$ and $\tilde{A}^*$ respectively, as $n \to \infty$. The operator $\tilde{A}$ is asymptotically Moore-Penrose invertible by the sequence $(\tilde{A}_n)$ if and only if the operator $\tilde{A}$ is Moore-Penrose invertible and the sequence $(\tilde{A}_n)$ is Moore-Penrose stable. In this case, $(\tilde{A}_n^+)$ strongly converges to $\tilde{A}^+$ as $n \to \infty$.*

Proof. Necessity is completely analogous to the corresponding linear case, cf. [192, Theorem 1]. Let the operator $\tilde{A}$ be Moore-Penrose invertible and the sequence $(\tilde{A}_n)$ be Moore-Penrose stable. For simplicity, we suppose that $n_0 = 0$ and show that the sequence $(\tilde{A}_n^+)$ converges strongly as $n \to \infty$. Let $\mathcal{F}$ denote the set of all bounded sequences of bounded linear operators acting on the Hilbert space $\mathcal{H}$. Given natural operations of addition, multiplication, the norm

$$||(A_n)|| = \sup_n ||A_n|| , \tag{1.34}$$

and also with the involution $(A_n)^* = (A_n^*)$, the set $\mathcal{F}$ becomes a C^*-algebra with identity. We consider the subalgebra $\mathcal{B}$ of $\mathcal{F}$ consisting of all sequences (A_n) such that the sequences $\{A_n\}$ and $\{A_n^*\}$ converge strongly as $n \to \infty$. It follows from Lemma 1.3.7 that $\mathcal{B}$ is an M-closed C^*-subalgebra of the C^*-algebra $\mathcal{F}$.

Furthermore, by Corollary 1.5.7 the subalgebra $\tilde{\mathcal{B}}$ is inverse closed with respect to Moore-Penrose invertibility, i.e., if $(\tilde{A}_n) \in \tilde{\mathcal{B}}$ is Moore-Penrose stable then $(\tilde{A}_n)$ is Moore-Penrose invertible in $\tilde{\mathcal{F}}$ and so in $\tilde{\mathcal{B}}$. Hence, the sequences $(\tilde{A}_n^+)$ and $(\tilde{A}_n^+)^*$ converge strongly as $n \to \infty$. □

The notion of asymptotic Moore-Penrose invertibility is very restrictive. In our opinion, the notion of weak asymptotic Moore-Penrose invertibility defined below is more suitable for numerical analysis, because it not only contains the asymptotic Moore-Penrose invertibility but is also stable under some perturbations (cf. Remark 1.7.5).

Let us proceed to study the weak asymptotic Moore-Penrose invertibility for additive continuous operators. Let $\Pi = (P_n)$ be a sequence of linear orthogonal projection operators on the complex Hilbert space $\mathcal{H}$ which converges strongly to the identity operator I as $n \to \infty$. By $\mathcal{F}^\Pi$ we denote the set of all bounded sequences (A_n) of bounded linear operators A_n acting in $\operatorname{im} P_n$, and according to the above considerations we provide $\mathcal{F}^\Pi$ with the operations of addition, multiplication, involution, multiplication by scalars and with the norm (1.34). Consequently, the set $\mathcal{F}^\Pi$ becomes a complex C^*-algebra with identity.

Let us consider an M-closed subalgebra of $\mathcal{F}^\Pi$. Assume we are given a set T and a family of sequences of unitary operators (E_n^t), $E_n^t : \operatorname{im} P_n \to \operatorname{im} P_n$, $(E_n^t)^{-1} = (E_n^t)^* =: E_{-n}^t$, $t \in T$, such that for any $s, t \in T, s \neq t$ the sequence $E_{-n}^s E_{-n}^t P_n$ weakly converges to zero as $n \to \infty$. In addition, suppose also that

$$M(\operatorname{im} P_n) \subset \operatorname{im} P_n \quad \text{for all } n \in \mathbb{N}.$$

Let $\mathcal{F}_T$ be the set of all sequences $(A_n) \in \mathcal{F}^\Pi$ such that there exist the strong limits

$$s - \lim E_{-n}^t A_n E_n^t P_n, \quad s - \lim E_{-n}^t A_n^* E_n^t P_n,$$

$$s - \lim E_{-n}^t M A_n M E_n^t P_n, \quad s - \lim E_{-n}^t M A_n^* M E_n^t P_n,$$

as $n \to \infty$.

It is easily seen that

1) $\mathcal{F}_T$ is an M-closed C^*-subalgebra of the C^*-algebra $\mathcal{F}^\Pi$.
2) The mappings $W_t : \mathcal{F}_T \to \mathcal{L}(\mathcal{H})$, $W_t(A_n) = s - \lim E_{-n}^t A_n E_n^t P_n$ are *-homomorphisms.

Now we introduce the notion of weak asymptotic Moore-Penrose invertibility in $\tilde{\mathcal{F}}_T$. Let $\mathcal{G}$ be the set of all operator sequences $(G_n) \in \mathcal{F}$ such that $||G_n|| \to 0$ as $n \to \infty$. We say that the sequence $(\tilde{A}_n) \in \tilde{\mathcal{F}}_T$ is weakly asymptotically Moore-Penrose invertible if there exists a sequence $(B_n) \in \tilde{\mathcal{F}}_T$ such that the four sequences

$$(\tilde{A}_n \tilde{B}_n \tilde{A}_n - \tilde{A}_n), \quad (\tilde{B}_n \tilde{A}_n \tilde{B}_n - \tilde{B}_n),$$

$$((\tilde{A}_n \tilde{B}_n)^* - \tilde{A}_n \tilde{B}_n), \quad ((\tilde{B}_n \tilde{A}_n)^* - \tilde{B}_n \tilde{A}_n),$$

belong to the ideal $\tilde{\mathcal{G}}$ or, in other words, if the coset $(\tilde{\mathcal{A}}_n) + \tilde{\mathcal{G}}$ is Moore-Penrose invertible in the quotient algebra $\tilde{\mathcal{F}}/\tilde{\mathcal{G}}$ [192, 193]. We still need a special ideal of the C^*-algebra $\mathcal{F}_T$, generated by those sequences $(J_n) \in \mathcal{F}_T$ which have the form

$$J_n := \sum_{t\in T} E_n^t P_n K_t E_{-n}^t + G_n, \quad (G_n) \in \mathcal{G}$$

where K_t are compact operators in $\mathcal{L}(\mathcal{H})$ that are not equal to zero for a finite number of indices. Let us denote this ideal by $\mathcal{J}_T$. We also consider the homomorphisms $\mathbf{W}_t : \mathcal{F}_T^{2\times 2} \to \mathcal{L}^{2\times 2}(\mathcal{H})$, $t \in T$ defined by

$$\mathbf{W}_t := \begin{pmatrix} W_t & 0 \\ 0 & W_t \end{pmatrix}.$$

Theorem 1.7.2. *Let $\mathcal{B}$ be an M-closed C^*-subalgebra of the C^*-algebra $\mathcal{F}_T$, and let the homomorphisms $(\mathbf{W}_t)_{t\in T}$ possess the property*

(S) *For each sequence $(B_n) \in \mathcal{B}^{2\times 2}$ the coset $(B_n)^\circ := (B_n) + \mathcal{J}_T^{2\times 2}$ is invertible in the quotient-algebra $(\mathcal{B}^{2\times 2} + \mathcal{J}_T^{2\times 2})/\mathcal{J}_T^{2\times 2}$ whenever all operators $\mathbf{W}_t(B_n), t \in T$, are Fredholm.*

Then for the sequence $(\tilde{A}_n) \in \tilde{\mathcal{B}}$ the following assertions are equivalent:

1) *The sequence $(\tilde{A}_n)$ is weakly asymptotically Moore-Penrose invertible.*
2) *The operators $\mathbf{W}_t(\Psi(\tilde{A}_n))$ are normally solvable for all $t \in T$, and the norms of these Moore-Penrose inverses are uniformly bounded.*

Proof. 1) $\to$ 2). Let the sequence $(\tilde{A}_n)$ be weakly asymptotically Moore-Penrose invertible. Then there is a sequence $(\tilde{B}_n)\in\tilde{\mathcal{B}}$ such that $(\tilde{A}_n\tilde{B}_n\tilde{A}_n - \tilde{A}_n) \in \tilde{\mathcal{G}}$, so

$$\Psi(\tilde{A}_n\tilde{B}_n\tilde{A}_n - \tilde{A}_n) \in \Psi(\tilde{\mathcal{G}}).$$

Since Ψ and $\mathbf{W}_t$ are homomorphisms and $\Psi(\tilde{\mathcal{G}}) \subset \ker \mathbf{W}_t$ for each $t \in T$, we obtain

$$\mathbf{W}_t(\Psi(\tilde{A}_n))\mathbf{W}_t(\Psi(\tilde{B}_n))\mathbf{W}_t(\Psi(\tilde{A}_n)) = \mathbf{W}_t(\Psi(\tilde{A}_n)).$$

The second condition in the definition of Moore-Penrose invertibility can be established analogously. Thus the operators $\mathbf{W}_t(\Psi(\tilde{A}_n))$ are generalized invertible, so by [91] they are normally solvable. In addition, the norms of $\mathbf{W}_t(\Psi(\tilde{A}_n))$ are uniformly bounded because W_t are $*$–homomorphisms and [66]

$$||W_t(A_n)|| \le ||(A_n)|| \quad \text{for all } t \in T \text{ and for all } (A_n) \in \mathcal{F}_T,$$

so by Lemma 1.4.6 the homomorphism Ψ is bounded.

2) $\to$ 1). Let $\mathcal{C}(T)$ stand for the C^*-algebra of all bounded functions on T with values in $\mathcal{L}^{2\times 2}(\mathcal{H})$, and let $S(T)$ be the smallest closed subalgebra of $\mathcal{C}(T)$ which contains all functions

$$t \to \mathbf{W}_t(B_n), \quad (B_n) \in (\mathcal{B}^{2\times 2} + \mathcal{J}_T^{2\times 2}). \tag{1.35}$$

The function (1.35) is usually called the symbol of (B_n) and denoted by $\mathrm{smb}(B_n)$. Then, by [193, Lifting theorem, Part 3], condition (**S**) implies that the quotient-algebra $(\mathcal{B}^{2\times 2} + \mathcal{J}_T^{2\times 2})/\mathcal{G}^{2\times 2}$ is isometrically $*$–isomorphic to $S(T)$. Now let all the operators $\mathbf{W}_t(\tilde{A}_n)$ be normally solvable, and the norms of their Moore-Penrose inverses be uniformly bounded. Then the function $\mathrm{smb}(\tilde{A}_n)$ is Moore-Penrose invertible in $\mathcal{C}(T)$, and consequently also in $S(T)$, so the coset $\Psi(\tilde{A}_n) + \mathcal{G}^{2\times 2}$ is Moore-Penrose invertible in $(\mathcal{B}^{2\times 2} + \mathcal{J}_T^{2\times 2})/\mathcal{G}^{2\times 2}$. Since $\mathcal{B}$ is an M-closed subalgebra of $\mathcal{F}_T$, we can state that

$$(E_{\tilde{\mathcal{B}}}^{2\times 2} + \mathcal{G}^{2\times 2})/\mathcal{G}^{2\times 2} \subset (\mathcal{B}^{2\times 2} + \mathcal{J}_T^{2\times 2})/\mathcal{G}^{2\times 2}.$$

Recalling that the real algebra $E_{\tilde{\mathcal{B}}}^{2\times 2}/(E_{\tilde{\mathcal{B}}}^{2\times 2} \cap \mathcal{G}^{2\times 2})$ is isometrically isomorphic to $(E_{\tilde{\mathcal{B}}}^{2\times 2} + \mathcal{G}^{2\times 2})/\mathcal{G}^{2\times 2}$ [66] and using Corollary 1.5.8, we obtain that the coset $\Psi(\tilde{A}_n)^\circ$ is Moore-Penrose invertible in $E_{\tilde{\mathcal{B}}}^{2\times 2}/(E_{\tilde{\mathcal{B}}}^{2\times 2} \cap \mathcal{G}^{2\times 2})$, i.e., $\Psi(\tilde{A}_n)$ is weakly asymptotically Moore-Penrose invertible in $E_{\tilde{\mathcal{B}}}^{2\times 2}$. To complete the proof we use Proposition 1.5.4 once more. □

Remark 1.7.3. There are examples which show that the condition of the uniform boundedness cannot be dropped. However, if all the operators $\mathbf{W}_t(\Psi(\tilde{A}_n))$ are Fredholm, then this condition is automatically satisfied [193].

Remark 1.7.4. If all the operators $\mathbf{W}_t(\Psi(\tilde{A}_n))$ are invertible, then the sequence $(\tilde{A}_n)$ is stable.

Remark 1.7.5. The assertion 2) of Theorem 1.7.2 and the theory of normally solvable operators (e.g. [91]) shows that weakly asymptotically Moore-Penrose invertible sequences are stable under different kinds of perturbation. Specifically, if all the operators $\mathbf{W}_t(\Psi(\tilde{A}_n))$ are semi-Fredholm, then the operator sequence $(\tilde{A}_n)$ remains weakly asymptotically Moore-Penrose invertible even if it is perturbed by sequences from the ideal $\tilde{\mathcal{J}}_T$ or by sequences of $\mathcal{B}$ which have small norms.

Let us now discuss another description of weak asymptotic Moore-Penrose invertibility in the algebra $\tilde{\mathcal{F}}$. It was shown in [194] that whenever a sequence $(A_n) \in \mathcal{F}$ is weakly asymptotically Moore-Penrose invertible, the spectrum $\mathrm{sp}(A_n^* A_n)$ of $A_n^* A_n$ can be split in two parts, one of which is bounded from zero by a positive constant (independent of n), while the other part tends to zero if n tends to infinity. We are going to show that this property is also valid for elements of the real algebra $\tilde{\mathcal{F}}$. However, in our case the spectrum $\mathrm{sp}(\tilde{A}_n^* \tilde{A}_n)$, which does coincide with the real spectrum of $\tilde{A}_n^* \tilde{A}_n$ for complex C^*-algebras, should be replaced by the real spectrum $\mathrm{sp}^r(\tilde{A}_n^* \tilde{A}_n)$ of $\tilde{A}_n^* \tilde{A}_n$.

Theorem 1.7.6. *A sequence $(\tilde{A}_n)$ is weakly asymptotically Moore-Penrose invertible in $\tilde{\mathcal{F}}$ if and only if there exist non-negative numbers d and r_n with $d > 0$ and $r_n \to 0$ as $n \to \infty$ such that*

$$\mathrm{sp}^r(\tilde{A}_n^* \tilde{A}_n) \subseteq [0, r_n] \cup [d, \infty] \tag{1.36}$$

for all sufficiently large n.

To prove this assertion one can exploit the scheme of the proof of the corresponding result for complex C^*-algebras [194]. Let us comment on auxiliary results that we need.

In the theory of complex C^*-algebras an "almost projection" lemma is well known (cf. [224, Lemma 5.1.6]). A careful consideration of its proof shows that this lemma is also true for real C^*-algebras. More precisely, we have the following lemma.

Lemma 1.7.7. *Let $\mathcal{B}$ be a real C^*-algebra with identity, and let $b \in \mathcal{B}$ be a self-adjoint element with $||b^2 - b|| < 1/4$. Then there exists a self-adjoint element $g \in \mathcal{B}$ such that $b + g$ is a projection and $||g|| \le 2||b^2 - b||$.*

To verify this result, one can exploit the isometrical isomorphism between the real algebras $\mathcal{R}(b)$ and $C_{\mathbb{R}}(\mathrm{sp}^r \mathcal{R}(b))$ (cf. Theorem 1.4.3) and follow the corresponding proof for complex algebras, since this proof uses only real-valued functions.

Let $(\tilde{A}_n)$ be in $\tilde{\mathcal{F}}$. By $(\tilde{A}_n)^\circ$ we denote the coset of $\tilde{\mathcal{F}}/\tilde{\mathcal{G}}$ which contains the sequence $(\tilde{A}_n)$, i.e., $(\tilde{A}_n)^\circ := (\tilde{A}_n) + \tilde{\mathcal{G}}$. Suppose that $(\tilde{A}_n)$ is weakly asymptotically Moore-Penrose invertible. Then there is a sequence $(\tilde{P}_n) \in \tilde{\mathcal{F}}$ such that the element $q = (\tilde{P}_n)^\circ$ satisfies the assertion 4) of Proposition 1.5.9. In particular, $((\tilde{P}_n)^\circ)^2 = (\tilde{P}_n)^\circ$ and $((\tilde{P}_n)^\circ)^* = (\tilde{P}_n)^\circ$, so it is possible to find a sequence $(\tilde{P}_n) \in (\tilde{P}_n)^\circ$ such that $\tilde{P}_n^* = \tilde{P}_n$ and $||\tilde{P}_n^2 - \tilde{P}_n|| \to 0$ as $n \to \infty$. Together with Lemma 1.7.7, this implies that one can choose a sequence $(\tilde{\Pi}_n) \in (\tilde{P}_n)^\circ$ such that each member of this sequence is a projection, i.e., $\tilde{\Pi}_n^2 = \tilde{\Pi}_n$ and $\tilde{\Pi}_n^* = \tilde{\Pi}_n$ for all $n \in \mathbb{N}$.

Applying Proposition 1.5.9 once more, we obtain the following lemma.

Lemma 1.7.8. *A sequence $(\tilde{A}_n) \in \tilde{\mathcal{F}}$ is weakly asymptotically Moore-Penrose invertible if and only if there exists a sequence $(\tilde{\Pi}_n)$ of projections on $\mathcal{H}$ such that the sequence $(\tilde{A}_n^* \tilde{A}_n + \tilde{\Pi}_n)$ is stable and $||\tilde{A}_n \tilde{\Pi}_n|| \to 0$ as $n \to \infty$. The sequence $(\tilde{\Pi}_n)$ is unique modulo $\tilde{\mathcal{G}}$ and $((\tilde{A}_n^* \tilde{A}_n + \tilde{\Pi}_n)^{-1} A_n^*)^\circ = ((\tilde{A}_n)^+)^\circ$.*

It should be noted that in place of $\tilde{q}$ in assertion 4) of Proposition 1.5.9 the element $\tilde{a}^\sqcap = e - \tilde{a}^+ \tilde{a}$ can be chosen. According to this designation, we denote by $\mathcal{P}(\tilde{A}_n)$ the set of all projections in $((\tilde{A}_n)^\circ)^\sqcap$.

Lemma 1.7.9. *Let $(\tilde{A}_n) \in \tilde{\mathcal{F}}$ be weakly asymptotically Moore-Penrose invertible. Then there is a sequence $(\tilde{\Pi}_n) \in \mathcal{P}(\tilde{A}_n)$ such that $\tilde{\Pi}_n \in \mathcal{R}_{\mathcal{L}_{add}(\mathcal{H})}(\tilde{A}_n^* \tilde{A}_n)$ for all $n \in \mathbb{N}$. The sequence $(\tilde{\Pi}_n)$ is unique in the following sense: If $(\tilde{\Pi}_n^{(1)}), (\tilde{\Pi}_n^{(2)}) \in \mathcal{R}_{\mathcal{L}_{add}(\mathcal{H})}(\tilde{A}_n^* \tilde{A}_n)$ for all $n > n_0$, then $(\tilde{\Pi}_n^{(1)}) = (\tilde{\Pi}_n^{(2)})$ for all sufficiently large n.*

The proof of this lemma also follows the corresponding proof in [194, Theorem 3], and the only additional result needed here is the Gelfand-Naimark theorem for commutative real C^*-algebras [96].

Proof of Theorem 1.7.6. After proving Lemmas 1.7.8 and 1.7.9, we can use Theorem 1.4.3 to obtain relation (1.36). However, as shown before, in real algebras the spectral properties of self-adjoint elements can be different from those in complex algebras. Therefore we first show that the real spectrum of $\tilde{A}_n^* \tilde{A}_n$ does not contain negative points. Indeed, from Corollary 1.4.13 we have

$$\mathrm{sp}^r_{\mathcal{L}_{add}(\mathcal{H})}(\tilde{A}_n^* \tilde{A}_n) = \mathrm{sp}_{\mathcal{L}^{2\times 2}(\mathcal{H})}(\Psi(\tilde{A}_n)^* \Psi(\tilde{A}_n)).$$

However, the self-adjoint operator $\Psi(\tilde{A}_n)^* \Psi(\tilde{A}_n)$ does not have any negative points in its spectrum since it belongs to the complex C^*-algebra $\mathcal{L}^{2\times 2}(\mathcal{H})$, so

$$\mathrm{sp}_{\mathcal{L}_{add}(\mathcal{H})}(\tilde{A}_n^* \tilde{A}_n) \subset [0, +\infty).$$

Let $(\tilde{A}_n)$ be weakly asymptotically Moore-Penrose invertible. From Lemmas 1.7.8 and 1.7.9, there is a sequence of projections $(\tilde{\Pi}_n)$ such that $(\tilde{A}_n) \in \mathcal{R}(\tilde{A}_n^* \tilde{A}_n)$, $n \in \mathbb{N}$, the sequence $(\tilde{A}_n^* \tilde{A}_n + \tilde{\Pi}_n)$ is stable, and $||\tilde{A}_n^* \tilde{A}_n \tilde{\Pi}_n|| \to 0$ as $n \to \infty$. Then there exists a number $d > 0$ such that

$$||((\tilde{A}_n^* \tilde{A}_n + \tilde{\Pi}_n)^\circ)^{-1}||_{\tilde{\mathcal{F}}/\tilde{\mathcal{G}}} < \frac{1}{d},$$

and we put

$$r_n := ||\tilde{A}_n^* \tilde{A}_n \tilde{\Pi}_n||. \tag{1.37}$$

Using the definition of the norm in $\tilde{\mathcal{F}}/\tilde{\mathcal{G}}$, we can choose a number n_0 such that

$$||(\tilde{A}_n^* \tilde{A}_n + \tilde{\Pi}_n)^{-1}||_{\mathcal{L}_{add}(\mathcal{H})} < \frac{1}{d} \qquad (n > n_0). \tag{1.38}$$

We fix now an $n > n_0$. Due to the isometrical isomorphism between $\mathcal{R} := \mathcal{R}(\tilde{A}_n^* \tilde{A}_n)$ and $\mathbf{C}(\mathrm{sp}_{\mathcal{R}}(\tilde{A}_n^* \tilde{A}_n))$ (cf. Theorem 1.4.3), the element $\tilde{A}_n^* \tilde{A}_n$ can be identified with the function $x \to x$, and the projection $\tilde{\Pi}_n$ with the function $x \to p_n(x)$, where p_n takes the values 0 and 1 only. Then it follows from (1.37) and (1.38) that

$$x + p_n(x) > d \qquad \text{and} \qquad x p_n(x) < r_n \qquad (x \in \mathrm{sp}_{\mathcal{R}}(\tilde{A}_n^* \tilde{A}_n)),$$

completing the proof. □

Now we consider the notion of asymptotical Moore-Penrose invertibility for sequences of $\tilde{\mathcal{F}}$. We recall that a sequence $(\tilde{A}_n) \in \tilde{\mathcal{F}}$ is said to be asymptotically Moore-Penrose invertible if there is an n_0 such that the operators $\tilde{A}_n$ are Moore-Penrose invertible for all $n > n_0$ and if $\sup_{n>n_0} ||\tilde{A}_n^+|| < +\infty$.

Theorem 1.7.10. *A sequence $(\tilde{A}_n) \in \tilde{\mathcal{F}}$ is weakly asymptotically Moore-Penrose invertible if and only if it can be represented as a sum of an asymptotically Moore-Penrose invertible sequence and a sequence of $\tilde{\mathcal{G}}$.*

Proof. Necessity: Let $(\tilde{A}_n) \in \tilde{\mathcal{F}}$ be weakly asymptotically Moore-Penrose invertible, and let $(\tilde{\Pi}_n)$ be the projection sequence defined in the proof of Theorem 1.7.6. By $(\tilde{B}_n)$ we denote the operator

$$\tilde{B}_n = \tilde{A}_n(I - \tilde{\Pi}_n). \tag{1.39}$$

From the proof of Theorem 1.7.6, the real spectrum of $\tilde{B}_n^*\tilde{B}_n + \tilde{\Pi}_n$ is contained in the set $(1) \cup [d, +\infty)$, so the operator $\tilde{B}_n$ is Moore-Penrose invertible, $\tilde{B}_n^+ = (\tilde{B}_n^*\tilde{B}_n + \tilde{\Pi}_n)^{-1}\tilde{B}_n^*$ and $\|\tilde{B}_n^+\| \leq 1/\sqrt{d}$. Henceforth the sequence $(\tilde{B}_n)$ is asymptotically Moore-Penrose invertible. In addition, we have $\|\tilde{A}_n\tilde{\Pi}_n\| \to 0$ as $n \to \infty$ (cf. Lemma 1.7.8). Finally, from (1.39) we obtain

$$\tilde{A}_n = \tilde{B}_n + \tilde{G}_n \qquad \text{with } \|\tilde{G}_n\| \to 0 \text{ as } n \to \infty.$$

The proof of the sufficiency is evident. □

1.8 Approximation Methods in Para-Algebras

Let X and Y be Banach spaces. If $X \neq Y$, then the space $\mathcal{L}_{add}(X, Y)$ does not have the operation of multiplication that causes additional problems in investigation of approximation methods for operators from $\mathcal{L}_{add}(X, Y)$. However, one can use objects which are immediate generalizations of Banach algebras. Let us recall some necessary notions.

Consider a system $\mathfrak{M}$ of four abelian groups $\mathfrak{N}_1, \mathfrak{S}_1, \mathfrak{S}_2, \mathfrak{N}_2$. Each group operation is called addition and is denoted by '+'. The system $\mathfrak{M}$ is called a para-group and denoted by

$$\mathfrak{M} = \left(\begin{array}{ccc} & \mathfrak{S}_1 & \\ \mathfrak{N}_1 & & \mathfrak{N}_2 \\ & \mathfrak{S}_2 & \end{array} \right)$$

or by $\mathfrak{M} = (\mathfrak{N}_1, \mathfrak{S}_1, \mathfrak{S}_2, \mathfrak{N}_2)$ if it possesses a second operation called multiplication, and satisfies the following conditions:

1. If $n_1, n_1' \in \mathfrak{N}_1$, $n_2, n_2' \in \mathfrak{N}_2$, $s_1 \in \mathfrak{S}_1$, $s_2 \in \mathfrak{S}_2$, then the following products exist and belong to the designated groups of the system $\mathfrak{M}$:

$$\begin{array}{llll} n_1 n_1' \in \mathfrak{N}_1, & n_2 n_2' \in \mathfrak{N}_2, & s_1 s_2 \in \mathfrak{N}_2, & s_2 s_1 \in \mathfrak{N}_1, \\ s_1 n_1 \in \mathfrak{S}_1, & n_1 s_2 \in \mathfrak{S}_2, & n_2 s_1 \in \mathfrak{S}_1, & s_2 n_2 \in \mathfrak{S}_2. \end{array}$$

2. If multiplication is feasible, then it is associative and distributive with respect to the addition.

An element x is said to belong to a para-group $\mathfrak{M} = (\mathfrak{N}_1, \mathfrak{S}_1, \mathfrak{S}_2, \mathfrak{N}_2)$ if $x \in \mathfrak{N}_1 \cup \mathfrak{S}_1 \cup \mathfrak{S}_2 \cup \mathfrak{N}_2$. If each of the groups $\mathfrak{N}_1, \mathfrak{S}_1, \mathfrak{S}_2, \mathfrak{N}_2$ is a Banach space, and for any $x, y \in \mathfrak{M}$ such that the product xy is defined the inequality

$$||xy|| \leq ||x|| ||y||$$

holds, then $\mathfrak{M}$ is called para-algebra. Note that here and in the following all norms in a para-algebra are defined by the same symbol.

Example 1.8.1. If X and Y are Banach spaces, then the system

$$\mathfrak{M} = \begin{pmatrix} & \mathcal{L}_{add}(X,Y) & \\ \mathcal{L}_{add}(X) & & \mathcal{L}_{add}(Y) \\ & \mathcal{L}_{add}(Y,X) & \end{pmatrix}$$

constitutes a para-algebra with the usual operations of addition and multiplication and with operator norms.

A para-algebra

$$\mathfrak{M}' = \begin{pmatrix} & \mathfrak{S}'_1 & \\ \mathfrak{N}'_1 & & \mathfrak{N}'_2 \\ & \mathfrak{S}'_2 & \end{pmatrix}$$

is referred to as a subpara-algebra of a para-algebra $\mathfrak{M}$ if $\mathfrak{N}'_j \subset \mathfrak{N}_j$, $\mathfrak{S}'_j \subset \mathfrak{S}_j$, $j = 1, 2$ and the operations in $\mathfrak{M}'$ and $\mathfrak{M}$ coincide.

A subpara-algebra $\mathfrak{J} \subset \mathfrak{M}$ is called a two-sided ideal of the para-algebra $\mathfrak{M}$ if for any $x \in \mathfrak{J}$ and for any $y, z \in \mathfrak{M}$ such that the products xy and zx are defined, the elements xy and zx belong to the sub-para-algebra $\mathfrak{J}$.

A para-algebra $\mathfrak{M} = (\mathfrak{N}_1, \mathfrak{S}_1, \mathfrak{S}_2, \mathfrak{N}_2)$ is called unital if $\mathfrak{N}_1$ and $\mathfrak{N}_2$ are unital algebras. The unit elements in the algebras $\mathfrak{N}_1$ and $\mathfrak{N}_2$ are denoted by e_1 and e_2, respectively. (The symbol e means that unity which is relevant for the situation under consideration.) For unital para-algebras one can introduce the invertibility notion. Thus an element $x \in \mathfrak{M}$ is called right (left) invertible in $\mathfrak{M}$ if there is $y \in \mathfrak{M}$ such that $xy = e$ $(yx = e)$. An element $x \in \mathfrak{M}$ is invertible if it is right and left invertible.

Let $\mathfrak{M} = (\mathfrak{N}_1, \mathfrak{S}_1, \mathfrak{S}_2, \mathfrak{N}_2)$ be a unital para-algebra, and let $\mathfrak{J} = (\mathfrak{J}_{\mathfrak{N}_1}, \mathfrak{J}_{\mathfrak{S}_1}, \mathfrak{J}_{\mathfrak{S}_2}, \mathfrak{J}_{\mathfrak{N}_2})$ be a two-sided ideal of $\mathfrak{M}$. Consider the quotient groups $[\mathfrak{N}_j] := \mathfrak{N}_j/\mathfrak{J}_{\mathfrak{N}_j}$, $[\mathfrak{S}_j] := \mathfrak{S}_j/\mathfrak{J}_{\mathfrak{S}_j}$, $j = 1, 2$ defined by the corresponding ideals of $\mathfrak{J}$. Then the system $\mathfrak{M}/\mathfrak{J} := ([\mathfrak{N}_1], [\mathfrak{S}_1], [\mathfrak{S}_2], [\mathfrak{N}_2])$ is a para-algebra with respect to the operations induced by the operations from $\mathfrak{M}$. This para-algebra is usually called a factor or quotient para-algebra. An element $x \in \mathfrak{M}$ is said to be $\mathfrak{J}$-invertible if the coset $[x] := x + \mathfrak{J} \in \mathfrak{M}/\mathfrak{J}$ that contains the element x, is invertible in the quotient para-algebra $\mathfrak{M}/\mathfrak{J}$.

Let X and Y be Banach spaces. Consider the operator equation

$$Ax = y, \quad x \in X, \quad y \in Y, \quad A \in \mathcal{L}_{add}(X, Y), \tag{1.40}$$

and in spaces $\mathcal{L}_{add}(X)$ and $\mathcal{L}_{add}(Y)$ consider sequences of projection operators (P_n^X) and (P_n^Y), respectively. It is assumed that these sequences, and the sequences $((P_n^X)^*)$ and $((P_n^Y)^*)$ that consist of the corresponding adjoint operators, converge strongly to the identity operators in the corresponding Banach spaces.

To solve approximately equation (1.40) let us consider the sequence of equations

$$A_n P_n^X x_n = P_n^Y y, \quad x_n \in \operatorname{im} P_n^X, \quad n = 1, 2, \ldots.$$

where A_n is an additive continuous operator defined on the space $\operatorname{im} P_n^X$ and taking its values in $\operatorname{im} P_n^Y$. In the spaces $\mathcal{L}_{add}(X)$ and $\mathcal{L}_{add}(Y)$ we, respectively, consider operator sequences (W_n^X) and (W_n^Y) which satisfy the relations:

1. $W_n^X P_n^X = W_n^X, \quad (W_n^X)^2 = P_n^X, \quad n = 1, 2, \ldots$.
2. $W_n^Y P_n^Y = W_n^Y, \quad (W_n^Y)^2 = P_n^Y, \quad n = 1, 2, \ldots$.
3. The sequences $(W_n^X), (W_n^Y), ((W_n^X)^*), ((W_n^Y)^*)$ weakly converge to zero.

If such sequences exist, then $\mathcal{A}^{XY}$ denotes the set of all sequences $(\tilde{A}_n), A_n : \operatorname{im} P_n^X \mapsto \operatorname{im} P_n^Y$ such that there are operators $A, \tilde{A} \in \mathcal{L}_{add}(X, Y)$ satisfying the limit relations:

$$
\begin{aligned}
s - \lim_{n\to\infty} A_n P_n^X &= A, & s - \lim_{n\to\infty} W_n^Y A_n W_n^X &= \tilde{A}, \\
s - \lim_{n\to\infty} A_n^* (P_n^Y)^* &= A^*, & s - \lim_{n\to\infty} (W_n^Y A_n W_n^X)^* (P_n^Y)^* &= \tilde{A}^*.
\end{aligned}
$$

The set $\mathcal{A}^{XY}$ can be provided with the operations of addition, scalar multiplication and with a norm, similarly to the set $\mathcal{A}^X$. For any two sequences $(A_n) \in \mathcal{A}^{XY}$ and $(B_n) \in \mathcal{A}^{YX}$ one can also define an operation of multiplication by

$$
(A_n)(B_n) := (A_n B_n), \quad (B_n)(A_n) := (B_n A_n).
$$

Lemma 1.8.2. *The system $\mathcal{A} := (\mathcal{A}^X, \mathcal{A}^{XY}, \mathcal{A}^{YX}, \mathcal{A}^Y)$ is a unital para-algebra with respect to the defined operations and norm.*

For the proof one has to check all the definitions involved. Note that the sequences (P_n^X) and (P_n^Y) serve as the identity elements in $\mathcal{A}^X$ and $\mathcal{A}^Y$, respectively.

Let $\mathcal{J}^{XY}$ stand for the set of sequences (J_n) for which there exist compact additive operators $T_1, T_2 \in \mathcal{K}_{add}(X, Y)$ such that the operators $J_n \in \mathcal{L}_{add}(\operatorname{im} P_n^X, \operatorname{im} P_n^Y)$ can be represented in the form

$$
J_n = P_n^Y T_1 P_n^X + W_n^Y T_2 W_n^X + C_n, \quad n = 1, 2, \ldots
$$

where $||C_n||$ tends to zero as $n \to \infty$.

Lemma 1.8.3. *The above defined sets $\mathcal{J}^{XY}$ and $\mathcal{J}^{YX}$ are closed and the system $\mathcal{J} := (\mathcal{J}^X, \mathcal{J}^{XY}, \mathcal{J}^{YX}, \mathcal{J}^Y)$ is an ideal of the para-algebra $\mathcal{A}$.*

Theorem 1.8.4. *Let the $A : X \to Y$ be an additive continuous operator, the sequence $(P_n^Y A_n P_n^X) \in \mathcal{A}^{XY}$ and $P_n^Y A_n P_n^X \to A$ as $n \to \infty$. Then $A \in \Pi(A_n, P_n^Y y)$ if and only if the operators $A, \tilde{A} \in \mathbf{G}(X, Y)$ and the element $(P_n^Y A_n P_n^X)$ is $\mathcal{J}$-invertible.*

The proofs of Lemmas 1.8.2, 1.8.3 and Theorem 1.8.4 mainly follow the proofs of Section 1.6, (cf. also [102, 103, 183, 207, 208]) and are omitted here.

1.9 Local Principles

At present local principles are among the most convenient and powerful tools in operator theory and in numerical analysis to study Fredholmness or stability. The method of freezing coefficients, widely used in partial differential equations, can also be interpreted as a local principle. The essence of this method is to assign a variety of possibly simpler problems to the problem under consideration and then to study them and glue the results together to obtain information concerning the initial problem. This section presents some general relevant ideas that fit the theoretical background required later.

1.9.1 Gohberg-Krupnik Local Principle

Definition 1.9.1. *Let $\mathcal{A}$ be a real or complex Banach algebra with identity e. A subset $M \subset \mathcal{A}$ is called a localizing class if it satisfies the following two conditions:*

(L_1) $0 \notin M$.

(L_2) For any $f_1, f_2 \in M$ there exists a third element $f \in M$ such that

$$f_j f = f f_j = f, \quad j = 1, 2.$$

Two elements $a, b \in \mathcal{A}$ are said to be M-equivalent from the left (right) if

$$\inf_{f \in M} \|(a-b)f\| = 0 \qquad \left(\inf_{f \in M} \|f(a-b)\| = 0 \right).$$

An element $a \in \mathcal{A}$ is called M-invertible from the left (right) if there are elements $b \in \mathcal{A}$ and $f \in M$ such that

$$baf = f, \quad (fab = f).$$

A system $\{M_\tau\}_{\tau \in T}$ of localizing classes is said to be covering if for each set $\{f_\tau\}_{\tau \in T}$, $f_\tau \in M_\tau$ there are a finite number of elements $f_{\tau_1}, \ldots, f_{\tau_m}$ whose sum is invertible in $\mathcal{A}$.

Now suppose that T is a topological space. Then a system $\{M_\tau\}_{\tau \in T}$ of localizing classes is said to be overlapping if

(L_3) Each M_τ is a bounded subset of $\mathcal{A}$;

(L_4) $f \in M_{\tau_0}$ ($\tau_0 \in T$) implies that $f \in M_\tau$, for all τ from an open neighbourhood of τ_0;

(L_5) The elements of $\mathcal{F} = \bigcup_{\tau \in T} M_\tau$, commute pairwise.

Let $\{M_\tau\}_{\tau \in T}$ be an overlapping system of localizing classes. The commutant of $\mathcal{F}$ is the set $\mathsf{Com}\,\mathcal{F} := \{a \in \mathcal{A} : af = fa \quad \text{for all} \quad f \in \mathcal{F}\}$. It is clear that $\mathsf{Com}\,\mathcal{F}$

is a closed subalgebra of $\mathcal{A}$. For $\tau \in T$, let $\mathcal{Z}_\tau$ denote the set of all elements in $\mathsf{Com}\,\mathcal{F}$ which are M_τ-equivalent to zero, both from the left and from the right. By virtue of (L_3), $\mathcal{Z}_\tau$ is a closed two-sided ideal of $\mathsf{Com}\,\mathcal{F}$ which does not contain the identity e. If $e \in \mathcal{Z}_\tau$, then there are $f_n \in M$, such that $\|f_n\| \to 0$ as $n \to \infty$; and since there exists a $g_n \neq 0$ in M such that $f_n g_n = g_n$, it follows that $\|f_n\| \geq 1$, which is a contradiction. For $a \in \mathsf{Com}\,\mathcal{F}$, let a^τ denote the coset $a + \mathcal{Z}_\tau$ from the quotient algebra $\mathsf{Com}\,\mathcal{F}/\mathcal{Z}_\tau$.

Let us also recall that a function $f : Y \to \mathbb{R}$ given on a topological space Y is called upper semi-continuous at $y_0 \in Y$, if for each $\varepsilon > 0$ there is a neighbourhood $U_\varepsilon \subset Y$ of y_0 such that $f(y) < f(y_0) + \varepsilon$ whenever $y \in U_\varepsilon$. The function f is said to be upper semi-continuous on Y if it is upper semi-continuous at each $y \in Y$. Equivalently, f is upper semi-continuous on Y if and only if $\{y \in Y : f(y) < \alpha\}$ is an open subset of Y for every $\alpha \in \mathbb{R}$. Notice that if Y is a compact Hausdorff space and $f : Y \to \mathbb{R}$ is a bounded upper semi-continuous function on Y, then there is a $y_0 \in Y$ such that $f(y_0) = \sup\limits_{y \in Y} f(y)$.

Lemma 1.9.2. (a) *Let M be a localizing class, let $a, a_0 \in \mathcal{A}$, and suppose a and a_0 are M-equivalent from the left (right). Then a is M-invertible from the left (right) if and only if a_0 is so.*

(b) *Let $\{M_\tau\}_{\tau \in T}$ be a system of localizing classes having property (L_3), let $\tau \in T$ and $a \in \mathsf{Com}\,\mathcal{F}$. Then a is M_τ-invertible in $\mathsf{Com}\,\mathcal{F}$ from the left (right) if and only if a^τ is invertible in $\mathsf{Com}\,\mathcal{F}/\mathcal{Z}_\tau$ from the left (right).*

Proof. (a) Let a be M-invertible from the left. Then there are elements $b \in \mathcal{A}$ and $f \in M$ such that $baf = f$. Since a and a_0 are M-equivalent from the left, there is a $g \in M$ such that $\|(a - a_0)g\| < 1/\|b\|$. Choose $h \in M$ so that $fh = gh = h$. Then

$$ba_0 h = bah - b(a - a_0)h = bafh - b(a - a_0)gh = h - uh = (e - u)h\,,$$

where $u := b(a - a_0)g$. Note that $e - u$ is invertible in $\mathcal{A}$ because $\|u\| < 1$. Thus, if one sets $v := (e - u)^{-1}b$, then $va_0 h = h$, i.e., the element a_0 is M-invertible from the left.

(b) Let a^τ be left invertible in $\mathsf{Com}\,\mathcal{F}/\mathcal{Z}$. Then there is an element $b \in \mathsf{Com}\,\mathcal{F}$ such that $ba - e \in \mathcal{Z}_\tau$. This implies that ba is M_τ-equivalent to e from the left, and from part (a) we deduce that ba and hence a is M_τ-invertible from the left. Analogously, if a^τ is right invertible, then a is M_τ-invertible from the right. Conversely, if there are $b \in \mathsf{Com}\,\mathcal{F}$ and $f \in M$, such that $baf = f$, then $(ba - e)f = 0$, hence $ba - e \in \mathcal{Z}_\tau$, and thus $b^\tau a^\tau = e$. It can be shown similarly that a^τ is right invertible in case a is M_τ-invertible from the right. □

The following theorem is the local principle of Gohberg and Krupnik.

Theorem 1.9.3. *Let A be a Banach algebra with identity, let $\{M_\tau\}_{\tau \in T}$ be a covering system of localizing classes, and let $a \in \mathsf{Com}\,\mathcal{F}$.*

(a) *Suppose that for each $\tau \in T$, a is M_τ-equivalent from the left (right) to $a_\tau \in \mathcal{A}$. Then a is left (right) invertible in $\mathcal{A}$ if and only if for every $\tau \in T$ the element a_τ is M_τ-invertible from the left (right).*

(b) *Suppose the system $\{M_\tau\}_{\tau\in T}$ has property (L_3). The element a is invertible in $\mathcal{A}$ if and only if the elements a^τ are invertible in $\mathsf{Com}\,\mathcal{F}/\mathcal{Z}_\tau$ for all $\tau \in T$.*

(c) *If the system $\{M_\tau\}_{\tau\in T}$ is overlapping, then the mapping*

$$T \to \mathbb{R}^+, \quad \tau \mapsto \|a^\tau\|$$

is upper semi-continuous.

Proof. (a) If a is left-invertible, then a is M_τ-invertible from the left for all $\tau \in T$ and hence, by Lemma 1.9.2 the element a_τ is M_τ-invertible from the left for all $\tau \in T$.

Conversely, suppose a_τ is M_τ-invertible from the left for all $\tau \in T$. It again follows from Lemma 1.9.2 that the element a is M_τ-invertible from the left for all $\tau \in T$. Thus there are elements $b_\tau \in \mathcal{A}$ and $f_\tau \in M_\tau$ such that $b_\tau a f_\tau = f_\tau$. Since $\{M_\tau\}_{\tau\in T}$ is a covering system one can find elements $f_{\tau_1}, \dots, f_{\tau_m}$ so that $\sum_{i=1}^{m} f_{\tau_i}$ is invertible in $\mathcal{A}$. Set

$$s := \sum_{i=1}^{m} b_{\tau_i} f_{\tau_i} .$$

Then

$$sa = \sum_{i=1}^{m} b_{\tau_i} f_{\tau_i} a = \sum_{i=1}^{m} b_{\tau_i} a f_{\tau_i} = \sum_{i=1}^{m} f_{\tau_i},$$

which results in the element $(\sum_{i=1}^{m} f_{\tau_i})^{-1} s$ being a left-inverse for a.

(b) If a^τ is invertible from the left for all $\tau \in T$, then by virtue of part (b) of Lemma 1.9.2 and part (a) of the present theorem the element a is invertible from the left in $\mathsf{Com}\,\mathcal{F}$ and $\mathcal{A}$. On the other hand, if a is invertible from the left and one of its left inverses belongs to $\mathsf{Com}\,\mathcal{F}$, then it is obvious that a^τ is invertible from the left. If a is invertible in $\mathcal{A}$, then clearly a is invertible in $\mathsf{Com}\,\mathcal{F}$.

(c) Let $\tau_0 \in T$ and $\varepsilon > 0$. Choose an element $z \in \mathcal{Z}_{\tau_0}$ such that $\|a + z\| < \|a^{\tau_0}\| + \varepsilon/2$. Since z is M_{τ_0}-equivalent to zero from the left there is an element $f \in M$, such that $\|zf\| < \varepsilon/2$. From (L_5) we deduce that $f \in M_\tau$ for all τ in some open neighbourhood $U(\tau_0)$ of τ_0. Put $y = z - zf$. If $\tau \in U(\tau_0)$, then by (L_2) there exists an element $g \in M_\tau$ such that $fg = g$. Consequently, $yg = zg - zfg = zg - zg = 0$, and since $y \in \mathrm{Com}\mathcal{F}$ (cf. (L_5), it follows that $y \in \mathcal{Z}_\tau$, for all $\tau \in U(\tau_0)$. Hence, $\|a^\tau\| \le \|a + y\|$ for $\tau \in U(\tau_0)$. Thus, if $\tau \in U(\tau_0)$, then

$$\|a^\tau\| - \|a^{\tau_0}\| < \|a + y\| - \|a + z\| + \varepsilon/2 \le \|y - z\| + \varepsilon/2 = \|zf\| + \varepsilon/2 < \varepsilon ,$$

which proves the upper semi-continuity of $\tau \mapsto \|a^\tau\|$ at τ_0. □

1.9.2 Allan's Local Principle

Now we are going to prove a variant of Allan's local principle [1]. The proof is mainly based on Theorem 1.9.3.

Definition 1.9.4. *Let $\mathcal{A}$ be a real or complex Banach algebra with the identity e. The center* $\mathsf{Cen}\,\mathcal{A}$ *of $\mathcal{A}$ is the set of all elements $z \in \mathcal{A}$ with the property that $za = az$ for all $a \in \mathcal{A}$.*

Clearly, $\mathsf{Cen}\,\mathcal{A}$ is a closed commutative subalgebra of $\mathcal{A}$. Let $\mathcal{B}$ be a closed subalgebra of $\mathsf{Cen}\,\mathcal{A}$ containing e. Thus $\mathcal{B}$ is also commutative. If $\mathcal{N} \subset \mathcal{B}$ is a maximal ideal, then $\mathcal{J}_\mathcal{N}$ denotes the smallest closed two-sided ideal of $\mathcal{A}$ containing $\mathcal{A}$, i.e.,

$$\mathcal{J}_\mathcal{N} := \operatorname{clos}\left\{\sum_{k=1}^{m} x_k a_k : x_k \in \mathcal{N},\, a_k \in \mathcal{A},\, m \in \mathbb{N}\right\}.$$

Suppose that $\mathcal{B}$ is a complex C^*-algebra if $\mathcal{A}$ is a complex Banach algebra, and that $\mathcal{B}$ is a strictly real C^*-algebra if $\mathcal{A}$ is a real Banach algebra. The latter means that the involution in $\mathcal{B}$ is the identity operator. This makes sense because $\mathcal{B}$ is commutative. Equivalently, $\mathcal{B}$ is a commutative real Banach algebra such that

$$||b^2|| = ||b||^2 \quad \text{and} \quad e + b^2 \quad \text{is invertible for all} \quad b \in \mathcal{B}.$$

Notice that in both cases there is a Hausdorff compact $\mathfrak{M}$ such that $\mathcal{B}$ is isometrically isomorphic to the algebra of all continuous complex- or real-valued functions on $\mathfrak{M}$. For the complex case, this is the well-known Gelfand-Naimark Theorem [66]. For the real case the reader can consult [96, Chapter 11]

In what follows we identify algebra $\mathcal{B}$ with the related function algebra. Recall that the maximal ideals of $\mathcal{B}$ are precisely the sets $\mathcal{N} = \mathcal{N}_\tau$, $\tau \in \mathfrak{M}$ where

$$\mathcal{N}_\tau := \{f \in \mathcal{B} : f(\tau) = 0\}, \quad \tau \in \mathfrak{M}.$$

Theorem 1.9.5. *Let $\mathcal{A}$ be a real or complex Banach algebra with identity e, and let $\mathcal{B} \subset \mathsf{Cen}\,\mathcal{A}$ be a subalgebra of the specified type containing e.*

(a) *If $a \in \mathcal{A}$, then a is right (left, respectively, two-sided) invertible in $\mathcal{A}$ if and only if for each $\mathcal{N} \in \mathfrak{M}$ the coset $a_\mathcal{N} := a + \mathcal{N}$ is left (right, respectively, two-sided) invertible in the quotient algebra $\mathcal{A}_\mathcal{N} := \mathcal{A}/\mathcal{J}_\mathcal{N}$.*

(b) *The mapping*

$$\mathfrak{M} \mapsto \mathbb{R}^+, \quad \mathcal{N} \mapsto ||a_\mathcal{N}||$$

is upper semi-continuous.

(c) *If, in addition, $\mathcal{A}$ is a C^*-algebra, then*

$$\bigcap_{\mathcal{N} \in \mathfrak{M}} \mathcal{J}_\mathcal{N} = \{0\}.$$

(d) $$||a|| = \max_{\mathcal{N}\in\mathfrak{M}} ||a_{\mathcal{N}}||.$$

Proof. (a) and (b): For $\mathcal{N}_0 \in \mathfrak{M}$ let $L_{\mathcal{N}_0}$ be the collection of all elements b of $\mathcal{B}$ such that $b(\mathcal{N}) = 1$ for all $\mathcal{N}$ in some neighbourhood of $\mathcal{N}_0$ and b vanishes outside another neighbourhood of $\mathcal{N}_0$. It is clear that the system $\{L_{\mathcal{N}}\}_{\mathcal{N}\in\mathfrak{M}}$ forms a covering and overlapping system of localizing classes and Theorem 1.9.3 applies to any element of $\mathcal{A}$. Obviously, the assertions (a) and (b) will be proved if we show that the ideals $\mathcal{Z}_{\mathcal{N}}$ defined in Section 1.9.1 coincide with the ideals $\mathcal{J}_{\mathcal{N}}$. Let c belong to $\mathcal{Z}_{\mathcal{N}}$. Then there exists a sequence $(f_n) \subset L_{\mathcal{N}}$ such that $||f_n c|| \to 0$ as $n \to \infty$. Write $c = f_n c + (e - f_n)c$. Since $(e - f_n)c$ belongs to $\mathcal{J}_{\mathcal{N}}$ the element c can be approximated in the norm by elements from $\mathcal{J}_{\mathcal{N}}$. Hence, $\mathcal{Z}_{\mathcal{N}} \subset \mathcal{J}_{\mathcal{N}}$. Conversely, let $c \in \mathcal{J}_{\mathcal{N}}$. Then c can be approximated in the norm by elements of the form $\sum_{i=1}^{m} x_i a_i$, where $x_i \in \mathcal{B}$, $a_i \in \mathcal{A}$ and every x_i, $i = 1, 2, \dots, m$ vanishes on a certain neighbourhood of $\mathcal{N}$ (Recall that each closed ideal in $\mathbf{C}(\mathfrak{M})$ is of the form $\{f \in \mathbf{C}(\mathfrak{M}) : f(x) = 0 \quad \text{for all } x \text{ belonging to a closed subset } F \text{ of } \mathfrak{M}\}$). Then we can find an element $f \in \mathbf{C}(\mathfrak{M})$ such that

$$f\left(\sum_{i=1}^{m} x_i a_i\right) = \sum_{i=1}^{m} (f x_i) a_i = 0.$$

This proves that $\sum_{i=1}^{m} x_i a_i \in \mathcal{Z}_{\mathcal{N}}$. Since $\mathcal{Z}_{\mathcal{N}}$ is closed we get $c \in \mathcal{Z}_{\mathcal{N}}$ and $\mathcal{J}_{\mathcal{N}} \subset \mathcal{Z}_{\mathcal{N}}$. Moreover, $e \notin \mathcal{J}_{\mathcal{N}}$ because $e \notin \mathcal{Z}_{\mathcal{N}}$.

(c) and (d): Since $\mathcal{A}$ is a C^*-algebra, by part (a) of the present theorem, assertion (d) is valid for any self-adjoint element. In particular, assertion (d) holds for the elements aa^* and a^*a which implies its validity for any $a \in \mathcal{A}$.

Now let us assume that $b \in \cap_{\mathcal{N}\in\mathfrak{M}} \mathcal{J}_{\mathcal{N}}$. Then $b_{\mathcal{N}} = 0$ for all $\mathcal{N} \in \mathfrak{M}$. By part (d) we get $b = 0$, and the proof is complete. □

1.9.3 Local Principle for Para-algebras

In many cases the question concerning $\mathcal{J}$-invertibility can be resolved by using a localization technique. In this section a generalization of the Gohberg-Krupnik localization method [91] is given in the case where Banach algebras are replaced by para-algebras.

Let $\mathfrak{X} = (\mathfrak{X}_1, \mathfrak{X}_2, \mathfrak{X}_3, \mathfrak{X}_4,)$ be a unital para-algebra. A set $M^l \subset \mathfrak{X}_1$ is called a left-localizing class of the para-algebra $\mathfrak{X}$ if M^l does not contain the zero element and for any two elements $a_1, a_2 \in M^l$ there is $a \in M^l$ such that

$$a_j a = a a_j = a, \quad j = 1, 2.$$

Elements $x, y \in \mathfrak{X}_2$ are said to be M^l-equivalent if

$$\inf_{a\in M^l} ||(x - y)a|| = 0.$$

Analogously, a set $M^p \subset \mathfrak{X}_4$ is called a right-localizing class of the para-algebra $\mathfrak{X}$ if M^p does not contain the zero element and for any two elements $b_1, b_2 \in M^p$ there is $b \in M^p$ such that

$$b_j b = b b_j = b, \quad j = 1, 2.$$

Elements $x, y \in \mathfrak{X}_2$ are said to be M^p-equivalent if

$$\inf_{a \in M^p} ||b(x - y)|| = 0.$$

Elements $x, y \in \mathfrak{X}_2$ are called (M^l, M^p)-equivalent if they are simultaneously M^l- and M^p-equivalent. An element $x \in \mathfrak{X}_2$ is said to be M^l-invertible (M^p-invertible) if there exist elements $z \in \mathfrak{X}_3$ and $a \in M^l$ ($z \in \mathfrak{X}_3$ and $a \in M^p$) such that

$$zxa = a \quad (bxz = b).$$

If an element $x \in \mathfrak{X}_2$ is simultaneously M^l- and M^p-invertible, then it is called (M^l, M^p)-invertible.

The following lemma can be proved analogously to the corresponding result of [91].

Lemma 1.9.6. *Let M^l be a left-localizing class of the para-algebra $\mathfrak{X}$, and let elements $x, y \in \mathfrak{X}_2$ be M^l-equivalent. Then the element x is M^l-invertible if and only if the element y is M^l-invertible.*

Note that a similar statement holds for M^p-invertible elements.

Let A be an index set. Following [91], the system $\{M_\alpha\}_{\alpha \in A}$ of the left- or right-localizing classes is called covering if any set $\{a_\alpha\}_{\alpha \in A}$, $a_\alpha \in M_\alpha$ has a finite number of elements $a_1, a_2, \dots, a_N$ the sum of which is an invertible element.

Systems $\{M^l_\alpha\}_{\alpha \in A}, M^l_\alpha \in \mathfrak{X}_1$ and $\{M^p_\alpha\}_{\alpha \in A}, M^p_\alpha \in \mathfrak{X}_4$ are reciprocally interchangeable with an element $x \in \mathfrak{X}_2$ if:

1. For any $a_\alpha \in M^l_\alpha$ there exist $b_\alpha \in M^p_\alpha$ such that
$$x a_\alpha = b_\alpha x.$$
2. For any $b_\alpha \in M^p_\alpha$ there exist $a_\alpha \in M^l_\alpha$ such that
$$b_\alpha x = x a_\alpha.$$

Lemma 1.9.7. *Let $\{M^l_\alpha\}_{\alpha \in A}$, $M^l_\alpha \in \mathfrak{X}_1$ and $\{M^p_\alpha\}_{\alpha \in A}$, $M^p_\alpha \in \mathfrak{X}_4$ be the systems of localizing classes which are reciprocally interchangeable with respect to an element $x \in \mathfrak{X}_2$, and let $\{M^l_\alpha\}_{\alpha \in A}$ be covering system. Then the element x is left-invertible if and only if it is M^l_α-invertible for all $\alpha \in A$.*

Proof. Necessity is obvious, so let us assume that $x \in \mathfrak{X}_2$ is M^l_α-invertible for any $\alpha \in A$. Then for any $\alpha \in A$ there exist elements $z_\alpha \in \mathfrak{X}_3$ and $a_\alpha \in M^l_\alpha$ such that

$$z_\alpha x a_\alpha = a_\alpha.$$

However $\{M^l_\alpha\}_{\alpha\in A}$ is supposed to be covering. Therefore any set of elements $\{a_\alpha\}_{\alpha\in A}$ contains a finite subsystem $a_{\alpha_1}, a_{\alpha_2}, \dots, a_{\alpha_N}$ such that the element $\sum_{j=1}^n a_{\alpha_j}$ is invertible. Since the systems $\{M^l_\alpha\}_{\alpha\in A}$ and $\{M^p_\alpha\}_{\alpha\in A}$ are reciprocally interchangeable with respect to x, there exist elements $b_{\alpha_j} \in M^p_{\alpha_j}, j = 1, 2, \dots, N$ such that

$$x a_{\alpha_j} = b_{\alpha_j} x, \quad j = 1, 2, \dots, N.$$

Set

$$u := \sum_{j=1}^N z_{\alpha_j} b_{\alpha_j},$$

so

$$ux = \left(\sum_{j=1}^N z_{\alpha_j} b_{\alpha_j} x \right) = \sum_{j=1}^N z_{\alpha_j} x a_{\alpha_j} = \sum_{j=1}^N a_{\alpha_j}.$$

It follows that the element x is left-invertible and $x_l^{-1} = \left(\sum_{j=1}^N a_{\alpha_j}\right)^{-1} u$. The right-invertibility can be proved analogously. □

Lemmas 1.9.6 and 1.9.7 implies the following result.

Theorem 1.9.8. *Let $\{M^l_\alpha\}_{\alpha\in A}$, $M^l_\alpha \in \mathfrak{X}_1$ and $\{M^p_\alpha\}_{\alpha\in A}$, $M^p_\alpha \in \mathfrak{X}_4$ be systems of covering localizing classes which are reciprocally interchangeable with respect to an element $x \in \mathfrak{X}_2$, and let x be $\{M^l_\alpha, M^l_\alpha\}$-equivalent to y_α for all $\alpha \in A$. Then the element x is invertible if and only the elements y_α are $\{M^l_\alpha, M^l_\alpha\}$-invertible for all $\alpha \in A$.*

1.10 Singular Integral and Mellin Operators

We give a brief overview in singular integral and Mellin operators, without proofs and in a form appropriate to the theory of approximation methods for related operator equations. General references to the theory of one-dimensional singular integral operators are [9, 92], although the first book is devoted to very general operators which are not considered in this book. Our exposition mainly follows [102] and [190], where relevant proofs are presented.

1.10.1 Singular Integrals

Throughout this section, let p and α denote fixed real numbers satisfying the inequalities $p > 1$ and $0 < 1/p + \alpha < 1$ and, given a (possibly unbounded) subinterval I of the real axis $\mathbb{R}$, let $L_p(I,\alpha)$ refer to the weighted Lebesgue space endowed with the norm

$$\|f\| = \left(\int_I |f(t)|^p |t|^{\alpha p} dt \right)^{1/p} . \qquad (1.41)$$

Clearly $L_p(I,\alpha)$ can be viewed as a closed subspace of $L_p(\mathbb{R},\alpha)$, and this allows us to identify the identity operator on $L_p(I,\alpha)$ with the operator χ_I of multiplication by the characteristic function χ_I of I acting on $L_p(\mathbb{R},\alpha)$. Analogously, a linear bounded operator A on $L_p(I,\alpha)$ can be identified with the operator $\chi_I A \chi_I$ on $L_p(\mathbb{R},\alpha)$. This identification will be often used without additional comments. Further, we always identify the dual space $L_p(I,\alpha)^*$ with the space $L_q(I,-\alpha)$, where $1/p+1/q=1$, and let $(\cdot,\cdot)$ refer to the sesqui-linear form $(u,v) = \int_I u(t)\overline{v}(t)dt$.

The singular integral operator S_I on I is defined by

$$(S_I f)(t) = \frac{1}{\pi i} \int_I \frac{f(s)}{s-t} ds, \quad t \in I .$$

The kernel singularity as $s \to t$ means that this integral does not exist in the usual sense, but it does exist as a Cauchy principal value integral almost everywhere. If parameters p and α satisfy the above conditions, then the operator S_I is bounded on $L_p(I,\alpha)$. Two important properties of the singular integral operator $S_{\mathbb{R}}$ are as follows:

1. $S_{\mathbb{R}}$ is its own inverse, i.e.,
$$S_{\mathbb{R}}^2 = I .$$

2. $S_{\mathbb{R}}$ is a Fourier convolution operator.

To make the second assertion more accurate, let F denote the Fourier transform acting on the Schwartz space $S(\mathbb{R})$ via

$$(Ff)(z) = \int_{\mathbb{R}} e^{-2\pi i x z} f(x) dx , \; z \in \mathbb{R} ,$$

and write F^{-1} for the inverse Fourier transform

$$(F^{-1}f)(x) = \int_{\mathbb{R}} e^{2\pi i x z} f(z) dz , \; x \in \mathbb{R} .$$

The Fourier transform extends by continuity to a unitary operator on the Hilbert space $L_2(\mathbb{R},0)$, and is denoted by F again. Now one can show that the restriction of the singular integral operator $S_{\mathbb{R}}$ onto $L_p(\mathbb{R},\alpha) \cap L_2(\mathbb{R},0)$ coincides with the restriction of the Fourier convolution operator $F^{-1}a_0F$ with generating function $a_0(z) = \text{sgn} z$ (the sign function) on the same space, viz.,

$$S_{\mathbb{R}} = F^{-1} a_0 F.$$

Moreover, if b is a bounded function, then the operator $F^{-1}bF$ is well defined on $L_p(\mathbb{R},\alpha) \cap L_2(\mathbb{R},0)$. If this operator extends boundedly onto all of $L_p(\mathbb{R},\alpha)$, then this extension is called the Fourier convolution operator and denoted by $W^0(b)$, and the function b is referred to as an $L_p(\mathbb{R},\alpha)$-Fourier multiplier. Each $L_p(\mathbb{R},\alpha)$-multiplier is a bounded function, and

$$\|b\|_\infty \leq \|W^0(b)\|_{\mathcal{L}(L_p(\mathbb{R},\alpha))} \,. \tag{1.42}$$

Conversely, if b is a bounded function with finite total variation $V(b)$, then b is an $L_p(\mathbb{R},\alpha)$-multiplier for each $p > 1$ and $\alpha \in (-1/p, 1-1/p)$, and

$$\|W^0(b)\|_{\mathcal{L}(L_p(\mathbb{R}))} \leq C_{p,\alpha}(\|b\|_\infty + V(b)) \tag{1.43}$$

(Stechkin's inequality, see [11], 9.3(e)). Further, the set of $L_2(\mathbb{R},0)$-multipliers coincides with the algebra $L_\infty(\mathbb{R})$ of all essentially bounded and measurable functions, and

$$\|W^0(a)\|_{\mathcal{L}(L_2(\mathbb{R},\alpha))} = \|a\|_\infty \,.$$

1.10.2 The Algebra $\sum^p(\alpha)$

Let $\sum^p(\alpha)$ denote the smallest closed subalgebra of $\mathcal{L}(L_p(\mathbb{R},\alpha))$ containing the operator $S_{\mathbb{R}^+}$ and the identity operator. There is an alternative description of the algebra $\sum^p(\alpha)$ that dominates the whole theory of this algebra. We define the Mellin transform M (depending on p and α) by

$$(Mf)(z) = \int_0^\infty x^{1/p+\alpha-zi-1} f(x)dx, \quad z \in \mathbb{R}\,,$$

and the inverse Mellin transform M^{-1} by

$$(M^{-1}f)(x) = \frac{1}{2\pi}\int_{-\infty}^\infty x^{zi-1/p-\alpha} f(z)dz, \quad x \in \mathbb{R}^+.$$

Further, we introduce operators $E_{p,\alpha}$ by

$$(E_{p,\alpha}f)(x) = \frac{1}{2\pi} f\left(\frac{1}{2\pi}\ln x\right) x^{-1/p-\alpha}, \quad x \in \mathbb{R}^+.$$

One can easily check that

$$\|E_{p,\alpha}f\|_{L_p(\mathbb{R}^+,\alpha)} = (2\pi)^{1/p-1}\|f\|_{L_p(\mathbb{R},0)},$$

and that $ME_{p,\alpha} = F$. If b is an $L^p_{\mathbb{R}}(0)$-multiplier then the operator

$$M^0(b) := E_{p,\alpha}W^0(b)E^{-1}_{p,\alpha}$$

is bounded on $L^p(\mathbb{R}^+,\alpha)$, and we call $M^0(b)$ the Mellin convolution by b or the Mellin convolution operator, or simply the Mellin operator. The function b is also called the symbol of the corresponding Mellin operator $M^0(b)$. For another representation of Mellin convolutions see Section 1.10.3. As a consequence of (1.42) one has

$$\|b\|_\infty \le \|M^0(b)\|_{\mathcal{L}(L_p(\mathbb{R}^+,\alpha))}$$

for each multiplier, and (1.43) implies the Stechkin inequality for Mellin convolutions:

$$\|M^0(b)\|_{\mathcal{L}(L_p(\mathbb{R}^+,\alpha))} \le C(\|b\|_\infty + V(b))$$

where b is a bounded function with finite total variation. Moreover,

$$\|M^0(b)\|_{L_2(\mathbb{R}^+,0)} = \|b\|_\infty .$$

The algebra $\sum^p(\alpha)$ can be characterized as follows.

Theorem 1.10.1 (I.B. Simonenko, Chin Ngok Min′, [214]). *$\sum^p(\alpha)$ is the smallest closed subalgebra of $\mathcal{L}(L_p(\mathbb{R},\alpha))$ which contains all Mellin convolution operators $M^0(b)$, where b is a continuous function of finite total variation that possesses finite limits at $\pm\infty$.*

Let us mention a few operators belonging to the algebra $\sum^p(\alpha)$ that can be easily verified with the help of Theorem 1.10.1. For each complex number β with $0 < \operatorname{Re}\beta < 2\pi$, the generalized Hankel operator

$$(N_\beta f((t) = \frac{1}{\pi i}\int_{\mathbb{R}^+} \frac{f(s)}{s - e^{i\beta}t}ds, \quad t \in \mathbb{R}^+ ,$$

belongs to $\sum^p(\alpha)$, since N_β is the Mellin convolution operator $M^0(n_\beta)$ with the symbol

$$n_\beta(z) = \frac{e^{(z+i(1/p+\alpha))(\pi-\beta)}}{\sinh \pi(z + i(1/p + \alpha))},$$

cf. [97]. Moreover, the operators $S_{\mathbb{R}^+}$ and N_β are related by the identity

$$S^2_{\mathbb{R}^+} - I = N_\beta N_{2\pi-\beta} .$$

Further, the weighted singular integral operator

$$(S_{\mathbb{R}^+,\gamma} f)(t) = \frac{1}{\pi i}\int_{\mathbb{R}^+} \left(\frac{t}{s}\right)^\gamma \frac{f(s)}{s-t}ds, \quad t \in \mathbb{R}^+,$$

belongs to the algebra $\sum^p(\alpha)$ for parameters γ satisfying the condition $0 < \operatorname{Re}(1/p + \alpha + \gamma) < 1$. In addition, if $0 < \operatorname{Re}\beta < 2\pi$ then the weighted generalized Hankel operator

$$(N_{\beta,\gamma} f)(t) = \frac{1}{\pi i}\int_{\mathbb{R}^+} \left(\frac{t}{s}\right)^\gamma \frac{f(s)}{s - e^{i\beta}t}ds, \quad t \in \mathbb{R}^+,$$

also lies in $\sum^p(\alpha)$, and

$$S_{\mathbb{R}^+,\gamma} = M^0(s_\gamma) \text{ with } s_\gamma(z) = \coth \pi(z + i(1/p + \alpha + \gamma)),$$

$$N_{\beta,\gamma} = M^0(n_{\beta,\gamma}) \text{ with } n_{\beta,\gamma}(z) = \frac{e^{(z+i(1/p+\alpha+\gamma))(\pi-\beta)}}{\sinh \pi(z + i(1/p + \alpha + \gamma))}.$$

The weighted operators $S_{\mathbb{R}^+,\beta}$ and $N_{\pi,\gamma}$ are connected by

$$S_{\mathbb{R}^+,\beta} S_{\mathbb{R}^+,\gamma} - \cos \pi(\beta - \gamma) N_{\pi,\beta} N_{\pi,\gamma} = I.$$

Let us denote the operator N_π simply by N, and let $N^p(\alpha)$ refer to the smallest closed two-sided ideal of $\sum^p(\alpha)$ which contains N. Then:

(a) *An operator $M^0(b) \in \sum^p(\alpha)$ belongs to the ideal $N^p(\alpha)$ if and only if $b(\pm\infty) = 0$.*

(b) (i) *$N^p(\alpha)$ is the smallest closed two-sided ideal of $\sum^p(\alpha)$ which contains any operator N_β with $0 < \operatorname{Re} \beta < 2\pi$.*

(ii) *$N^p(\alpha)$ is the smallest closed two-sided ideal of $\sum^p(\alpha)$ which contains N^2.*

(iii) *$N^p(\alpha)$ is the smallest closed subalgebra of $\sum^p(\alpha)$ which contains the operators N^2 and $S_{\mathbb{R}^+} N^2$.*

(iv) *$N^p(\alpha)$ is the smallest closed subalgebra of $\sum^p(\alpha)$ which contains the operators N and $S_{\mathbb{R}^+} N$.*

(c) *The algebra $\sum^p(\alpha)$ decomposes into the direct sum $\mathbb{C}I + \mathbb{C}S_{\mathbb{R}^+} + N^p(\alpha)$, M. Costabel,* [28].

(d) *The maximal ideal space of commutative Banach algebra $\sum^p(\alpha)$ is homeomorphic to the two-point compactification $\overline{\mathbb{R}}$ of the real axis. In particular, the Mellin convolution operator $M^0(b) \in \sum^p(\alpha)$ is invertible in $\sum^p(\alpha)$ if and only if $b(z) \neq 0$ for all $z \in \mathbb{R}$ and if $b(\pm\infty) \neq 0$. Thus $\sum^p(\alpha)$ is an inverse closed subalgebra of $\mathcal{L}(L_p(\mathbb{R}, \alpha))$.*

(e) *The algebras $\sum^p(\alpha)$ and $\sum^p(0)$ are isometrically isomorphic. The isomorphism sends the singular integral operator $S_{\mathbb{R}^+} \in \sum^p(\alpha)$ into the weighted singular integral operator $S_{\mathbb{R}^+,\alpha} \in \sum^p(0)$.*

1.10.3 Integral Representations

Let $\mathcal{N}$ denote the smallest (not necessarily closed) subalgebra of the algebra of all continuous functions on $\mathbb{R}$ containing all functions b which are, together with their Mellin images $k = M^{-1}b$, continuously differentiable and for which there exist constants M_1 and M_2 and polynomials P_1 and P_2 such that

$$\sum_{x \in \mathbb{R}} |b(z)(1 + |z|) \leq M_1, \; \sup_{z \in \mathbb{R}} |b'(z)(1 + |z|)^2| \leq M_2$$

and

$$|k(x)(1+x)| \le P_1(|\ln x|)\,,\; |k'(x)(1+x)^2| \le P_2(|\ln x|)\,.$$

Obviously, the set $M^{-1}(\mathcal{N})$ is an algebra with respect to the *Mellin convolution*

$$(k_1 \star k_2)(x) = \int_{\mathbb{R}^+} k_1\left(\frac{x}{s}\right) k_2(s) \frac{ds}{s}, \quad x \in \mathbb{R}^+\,,$$

as multiplication, and a function k belongs to $M^{-1}(\mathcal{N})$ if and only if k is the kernel function of a Mellin convolution operator $M^0(b)$ with $b \in \mathcal{N}$, i.e.,

$$(M^0(b)f)(x) = \int_{\mathbb{R}^+} k\left(\frac{x}{s}\right) f(s) \frac{ds}{s}, \quad x \in \mathbb{R}^+\,.$$

1.10.4 The Algebra $\sum^p(\alpha, \beta, \nu)$

Given real numbers $0 \le \beta_1 < \beta_2 < 2\pi$, let Γ be the subset of $\mathbb{C}$ defined by $\Gamma := e^{i\beta_1}\mathbb{R}^+ \cup e^{i\beta_2}\mathbb{R}^+$. Each half-axis $e^{i\beta_j}\mathbb{R}^+$ can be provided with one of the two possible orientations and we put $\nu_j := 1$ if $e^{i\beta_j}\mathbb{R}^+$ is directed away from zero and $\nu_j := -1$ if $e^{i\beta_j}\mathbb{R}^+$ is directed towards 0. Notice that the case where one half-axis is directed away and the other towards 0 is of special interest throughout this book.

We consider the space $L_p(\Gamma, \alpha) := L_p(\Gamma, |t|^\alpha)$. It is well known that if $1 < p < \infty$ and $0 < 1/p + \alpha < 1$, then the singular integral operator

$$(S_\Gamma f)(x) := \frac{1}{\pi i} \int_\Gamma \frac{f(t)}{t-x} dt\,,$$

is well defined and bounded on $L_p(\Gamma, \alpha)$. Notice that S_Γ depends on the chosen orientation (ν_1, ν_2). Denote by $L_p^2(\mathbb{R}^+, \alpha)$ the Banach space of all pairs $(f_1, f_2)^T$ of functions $f_1, f_2 \in L_p(\mathbb{R}^+, \alpha)$ endowed with the norm

$$\|(f_1, f_2)^T\| := (\|f_1\|^p + \|f_2\|^p)^{1/p}\,,$$

and let $\eta : L_p(\Gamma, \alpha) \mapsto L_p(\mathbb{R}^+, \alpha)$ refer to the mapping

$$(\eta f)(t) := (f(e^{i\beta_1 t}), f(e^{i\beta_2 t}))^T\,,\; t \in \mathbb{R}^+\,.$$

It is easily seen that the mapping $A \mapsto \eta^{-1} A \eta$ is an isometrical isomorphism between the algebras $\mathcal{L}(L_p(\Gamma, \alpha))$ and $\mathcal{L}(L_p^2(\mathbb{R}^+, \alpha))$. Write χ_1 for the operator of multiplication by the characteristic function of $e^{i\beta_1}\mathbb{R}^+$, and consider the smallest subalgebra of $\mathcal{L}(L^p(\Gamma, \alpha))$ which contains the operators χ_1, S_Γ, and the identify operator. This subalgebra is denoted by $\sum^p(\alpha, \beta, \nu)$, where $\nu = (\nu_1, \nu_2)$ indicates the chosen orientation, and $\beta = (\beta_1, \beta_2)$ – the angles of the rays $e^{i\beta_j}\mathbb{R}^+, j = 1, 2$. The following proposition describes the image of the algebra $\sum^p(\alpha, \beta, \nu)$ under the isomorphism generated by the mapping η.

Proposition 1.10.2. *$\eta\sum^p(\alpha,\beta,\nu)\eta^{-1}$ is the algebra of all operators $\begin{pmatrix} A & B \\ C & D \end{pmatrix} \in \mathcal{L}(L_p^2(\mathbb{R}^+,\alpha))$ with $A, D \in \sum^p(\alpha)$ and $B, C \in N^p(\alpha)$.*

Example 1.10.3. For the operator $(A_{ij})_{i,j=1}^2 := \eta S_\Gamma \eta^{-1}$, straightforward computation yields

$$A_{ij} = \begin{cases} \nu_j S_{\mathbb{R}^+} & \text{if} \quad i = j, \\ \nu_j M^0(n_{\beta_i - \beta_j}) & \text{if} \quad i > j, \\ \nu_j M^0(n_{2\pi+\beta_i-\beta_j}) & \text{if} \quad i < j. \end{cases}$$

This proposition allows us to conclude that the algebra $\sum^p(\alpha,\beta,\gamma)$ does not depend on the values of the angles β_j or the orientation of the axis $e^{i\beta_j}\mathbb{R}^+$. Moreover, this algebra does not even depend on the weight $|t|^\alpha$, as property (e) of Section 1.10.2 shows.

Since the entries of a matrix $\begin{pmatrix} A & B \\ C & D \end{pmatrix} \in \eta\sum^p(\alpha,\beta,\nu)\eta^{-1}$ commute with each other, one has an effective invertibility criterion for operators from the algebra $\sum^p(\alpha,\beta,\nu)$. Thus an operator $E \in \sum^p(\alpha,\beta,\nu)$ is invertible if and only if $\eta E\eta^{-1} = \begin{pmatrix} A & B \\ C & D \end{pmatrix}$ is invertible or if and only if

$$\det\begin{pmatrix} A & B \\ C & D \end{pmatrix} = AD - BC(\in \Sigma^p(\alpha))$$

is invertible, and it remains to apply property (d) of Section 1.10.2.

1.10.5 Singular Integral Operators with Piecewise Continuous Coefficients

A bounded Lyapunov arc is an oriented bounded curve Γ in the complex plane $\mathbb{C}$ which is homeomorphic to the closed interval $[0,1]$ and which satisfies the Lyapunov condition – i.e., the first derivative of the parameter representation $t : [0,l] \to \Gamma$ by its arc length must be a Hölder continuous function. The points $t(0)$ and $t(l)$ are called the start and endpoint of Γ, respectively. We shall assume that Γ is oriented according to the natural orientation of $[0,l]$. A curve Γ is called a simple closed piecewise Lyapunov curve if:

- Γ is closed, i.e., Γ divides the complex plane into two parts, each having Γ as its boundary and located at one side of Γ.
- Γ is the union of a finite number m of Lyapunov arcs $\Gamma_1,\dots,\Gamma_m$ such that the endpoint of Γ_i coincides with the starting point of Γ_{i+1}, $i = 0,\dots,m-1$ with $\Gamma_{m+1} := \Gamma_0$. Further, we assume that $\Gamma_i \cup \Gamma_{i+1}$ is either Lyapunov or the tangents of Γ_i and Γ_{i+1} at z_i (the endpoint of Γ_i) differ from each other.
- The orientation from the collection $\Gamma_1,\dots,\Gamma_m$ induced for Γ is the "counterclockwise", i.e., if the curve is traced out in accordance to the induced orientation, the bounded part of the plane with boundary Γ lies on the left.

Let Γ be a piecewise Lyapunov curve. The singular integral operator S_Γ is defined by

$$(S_\Gamma f)(t) := \frac{1}{\pi i} \int_\Gamma \frac{f(z)}{z-t}\, dz, \ t \in \Gamma$$

where the integral is understood as the Cauchy principal value. Further, let $t_1, \ldots, t_n \in \Gamma$, $\alpha_1, \ldots, \alpha_n \in \mathbb{R}$, and put

$$w(z) := \prod_{i=1}^{n} |z - t_i|^{\alpha_i} . \tag{1.44}$$

By $L_p(\Gamma, w)$ we denote the weighted Lebesgue space on Γ with the norm

$$\|f\| := \left(\int_\Gamma |f(z)|^p w^p(z) |dz| \right)^{1/p} .$$

A theorem of Khvedelidze [124] claims that the singular integral operator S_Γ is bounded on $L_p(\Gamma, w)$ if and only if

$$1 < p < \infty \text{ and } 0 < \alpha_i + \frac{1}{p} < 1, \ i = 1, \ldots, n .$$

A piecewise continuous function on Γ is a function which at each point z_0 of Γ, has finite one-sided limits. By $\mathbf{PC}(\Gamma)$ we denote the Banach algebra of all piecewise continuous functions on Γ provided with the supremum norm. For each $a \in \mathbf{PC}(\Gamma)$ the operator of multiplication by a is a bounded linear operator on $L_p(\Gamma, w)$, so it makes sense to introduce the smallest closed subalgebra $\mathcal{O}_p(\Gamma, w)$ of $\mathcal{L}(L_p(\Gamma, w))$ containing the operator S_Γ and the algebra $\mathbf{PC}(\Gamma)$.

By $\mathcal{K}(L_p(\Gamma, w))$ we denote the ideal of all compact operators on $L_p(\Gamma, w)$. It is well known that

$$\mathcal{K}(L_p(\Gamma, w)) \subset \mathcal{O}_p(\Gamma, w) .$$

Let π refer to the canonical homomorphism from $\mathcal{O}_p(\Gamma, w)$ onto the quotient algebra

$$\mathcal{O}_p^\pi(\Gamma, w) := \mathcal{O}_p(\Gamma, w)/\mathcal{K}(L_p(\Gamma, w)) .$$

Finally, let us write $\mathbf{C}(\Gamma) \subset \mathbf{PC}(\Gamma)$ for the Banach algebra of all functions which are continuous at each point of Γ.

Recall that if $a \in \mathbf{C}(\Gamma)$ then $aS_\Gamma - S_\Gamma a \in \mathcal{K}(L_p(\Gamma, \alpha))$ cf. [91, 157].

Our next goal is to describe the algebra $\mathcal{O}_p(\Gamma, w)$ and to present a Fredholm criterion for its elements. Let Γ be a closed piecewise Lyapunov curve and let $z \in \Gamma$ be any (fixed) point. Choose a neighbourhood $\mathcal{U} \subset \mathbb{C}$ of z such that $\mathcal{U} \cap \Gamma$ is the union of two Lyapunov arcs L_1 and L_2 which have only one common point z. One arc is directed towards z and the other away from z. By $\beta_i(z)$, $i = 1, 2$ we denote the angle between the tangents L_1 and L_i at the point z. Without loss we can assume that $0 = \beta_1 < \beta_2 < 2\pi$. Let us mention that this definition does not depend on which arc is directed towards z. Given the weight function w by (1.44) we put

$$\alpha(z) = \begin{cases} \alpha_i & \text{if} \quad z = t_i,\ i = 1, 2, \ldots, n, \\ 0 & \text{otherwise.} \end{cases}$$

Now define a new weight function depending on z by $w_z(t) := |t|^{\alpha(z)}$ and set $\nu_i(z) = 1$ if Γ_i is directed away from z and $\nu_i(z) = -1$ if Γ_i is directed to z.

Let $\sum_2^p(\alpha) \subset \mathcal{L}(L_p^2(\mathbb{R}^+, \alpha))$ refer to the Banach algebra of all 2×2 matrices $(A_{ij})_{i,j=1}^2$ with entries $A_{ij} \in \sum^p(\alpha)$, and let $\widetilde{\sum}_2^p(\alpha)$ be the subalgebra of $\sum_2^p(\alpha)$ such that the entries A_{12} and A_{21} belong to the ideal $N_p(\alpha)$.

Theorem 1.10.4. a) *For each $z \in \Gamma$ there exists a homomorphism $A \mapsto smb(A, z)$ from $\mathcal{O}_p(\Gamma, w)$ onto the algebra $\widetilde{\sum}_2^p(\alpha(z))$. In particular, for the generators S_Γ and $a \in \mathbf{PC}(\Gamma)$ of $\mathcal{O}_p(\Gamma, w)$ we have*

$$smb(S_\Gamma, z) = \begin{pmatrix} S_{\mathbb{R}^+} & N_{2\pi+\beta_1-\beta_2} \\ N_{\beta_2-\beta_1} & S_{\mathbb{R}^+} \end{pmatrix} \cdot \operatorname{diag}(1, -1)$$

and, if we abbreviate $a_i(t) := \lim_{\substack{t \to z \\ t \in \Gamma_i}} a(t)$ for each $a \in \mathbf{PC}(\Gamma)$, then

$$smb(aI, z) = \operatorname{diag}(a_1(z)I, a_2(z)I).$$

b) *An operator $A \in \mathcal{O}_p(\Gamma, w)$ is Fredholm if and only if $smb(A, z)$ is invertible for all $z \in \Gamma$.*

c) *If $\pi(A)$ stands for the image of the operator A in the Calkin algebra, then*

$$\|\pi(A)\| = \sup_{z \in \Gamma} \|smb(A, z)\|.$$

d) *The radical of $\mathcal{O}_p^\pi(\Gamma, w)$ is trivial and A is compact if and only if*

$$smb(A, z) = 0 \text{ for all } z \in \Gamma.$$

The proof of this theorem is based on the local principle of Allan. The main point is to show that the local algebra assigned to $z \in \Gamma$ is isometrically isomorphic to $\sum^p(\alpha(z), \beta(z), (+1, -1))$. Here we assume that L_2 is directed towards z and L_1 away from z. Let us also note that we still use the notation β_1 and β_2 despite the fact that $\beta_1 = 0$. This is because we want to keep our notation in accordance with the crucial Example 1.10.3.

Remark 1.10.5. Using results of Sections 1.10.2 and 1.10.4 we obtain effective tools to study invertibility of smb(A, z).

Corollary 1.10.6. *Let Γ_1, Γ_2 be closed piecewise Lyapunov curves. Consider $\mathcal{O}_p(\Gamma_1, w_1)$ and $\mathcal{O}_p(\Gamma_2, w_2)$. Then $\mathcal{O}_p^\pi(\Gamma_1, w_1)$ and $\mathcal{O}_p^\pi(\Gamma_2, w_2)$ are isometrically isomorphic.*

A remarkable consequence of Theorem 1.10.4, b) is whether an operator $aI + bS_\Gamma$ is Fredholm or not does not depend on the values of the angles: indeed, put $\delta := 1/p + \alpha(z)$ and consider the matrix

$$D(t) := \operatorname{diag}\left(1, e^{(t+\delta)\beta_2(z)}\right).$$

Then it is easy to see that $D^{-1}\operatorname{smb}(aI + bS_\Gamma, z)D$ dos not depend on β_2, which gives the assertion; recall that $\beta_1 = 0$.

1.10.6 Singular Integral Operators with Conjugation

The study of the smallest closed algebra $\mathcal{E}_p(\Gamma, w)$ generated by the algebra $\mathcal{O}_p(\Gamma, w)$ and by the operator of conjugation $(Mf)(t) := \overline{f(t)}$ can be reduced to the study of the algebra $\mathcal{O}_p^{2\times 2}(\Gamma, w)$, where $\mathcal{O}_p^{2\times 2}(\Gamma, w)$ is the algebra of all 2×2 matrices with entries from $\mathcal{O}_p(\Gamma, w)$. Notice that $\mathcal{E}_p(\Gamma, w)$ is a real algebra. For simplicity, let us assume that $p = 2$ and note the nontrivial fact that $\mathcal{O}_2(\Gamma, w)$ is a C^*-algebra, so results from Section 1.4 can now be used to examine the algebra $\mathcal{E}_2(\Gamma, w)$. However, first one must check whether the element $M \in \mathcal{L}_{\text{add}}(L_2(\Gamma, w))$ and the complex C^*-algebra $\mathcal{O}_2(\Gamma, w)$ satisfy conditions $(A_1) - (A_5)$ of Section 1.2. The only nontrivial condition is that $MS_\Gamma M$ belongs to the same algebra $\mathcal{O}_2(\Gamma, w)$. This is indeed the case, and its symbol is given by

$$\operatorname{smb}(MS_\Gamma M, z) = -\begin{pmatrix} S_{\mathbb{R}^+} & N_{\beta_2-\beta_1} \\ N_{2\pi+\beta_1-\beta_2} & S_{\mathbb{R}^+} \end{pmatrix} \cdot \operatorname{diag}(+1, -1).$$

Moreover, $S_\Gamma + MS_\Gamma M$ is compact if and only if Γ is a closed Lyapunov curve without corners. The operator $V_\Gamma := \frac{1}{2}(S_\Gamma + MS_\Gamma M)$ is then the double layer potential operator (cf. Section 4.4 for the definition).

Using the terminology of Section 1.2, we have

$$\mathcal{E}_2(\Gamma, w) = \tilde{\mathcal{O}}_2(\Gamma, w).$$

Let us write $\mathcal{A}$ instead $\mathcal{O}_2(\Gamma, w)$ and introduce the set

$$E_{\tilde{\mathcal{A}}}^{2\times 2} := \left\{ \begin{pmatrix} b & d \\ MdM & MbM \end{pmatrix} : b, d \in \mathcal{A} \right\} \subset \mathcal{A}^{2\times 2}.$$

According to Section 1.4, the real algebras $\tilde{\mathcal{A}}$ and $E_{\tilde{\mathcal{A}}}^{2\times 2}$ are isometrically isomorphic, where the isometric isomorphism Ψ is given by

$$\Psi(\tilde{a}) = \Psi(a + bM) := \begin{pmatrix} a & b \\ MbM & MaM \end{pmatrix}.$$

Moreover, $\tilde{a} \in \tilde{\mathcal{A}}$ is invertible if and only if $\Psi(\tilde{a})$ is invertible in $\mathcal{A}^{2\times 2}$ (Corollary 1.4.7), so the following theorem holds.

Theorem 1.10.7. a) *For each* $z \in \Gamma$ *there exists a homomorphism* $A \mapsto smb\,(A,z)$ *from* $\mathcal{E}_2(\Gamma,w)$ *onto the closed subalgebra* $\sum_4^p(\alpha)$ *consisting of all matrices* $(B_{ij})_{i,j=1}^2$ *with* $B_{ij} \in \widetilde{\sum}_2^2$. *In particular,*

$$smb\,(S_\Gamma, z) = (A_{ij})_{i,j=1}^2$$

with $A_{21} = A_{12} = 0$ *and*

$$A_{11} = \begin{pmatrix} S_{\mathbb{R}^+} & N_{2\pi+\beta_1-\beta_2} \\ N_{\beta_2-\beta_1} & S_{\mathbb{R}^+} \end{pmatrix} \operatorname{diag}\{1,-1\},$$

$$A_{22} = -\begin{pmatrix} S_{\mathbb{R}^+} & N_{\beta_2-\beta_1} \\ N_{2\pi+\beta_1-\beta_2} & S_{\mathbb{R}^+} \end{pmatrix} \operatorname{diag}\{1,-1\},$$

$$smb\,(M,z) = \begin{pmatrix} 0 & 0 & I & 0 \\ 0 & 0 & 0 & I \\ I & 0 & 0 & 0 \\ 0 & I & 0 & 0 \end{pmatrix},$$

and if a *is a piecewise continuous function, then*

$$smb\,(a,z) = \operatorname{diag}\{a_1(z), a_2(z), \overline{a_1(z)}, \overline{a_2(z)}\}.$$

b) *The operator* $A \in \mathcal{E}_2(\Gamma,w)$ *is Fredholm if and only if* $smb\,(A,z) \neq 0$ *for all* $z \in \Gamma$.

c) *If* $\pi(A)$ *is the Calkin image of an operator* $A \in \mathcal{E}_2(\Gamma,w)$, *then*

$$\|\pi(A)\| = \sup_{z\in\Gamma} \|smb\,(A,z)\|.$$

Finally, for completeness we mention that unlike the operators without conjugation, the Fredholmness of an operator $A \in \mathcal{E}_2(\Gamma,w)$ essentially depends on the values of the angles of the curve Γ. An example is given in [164], where the operator

$$A = (1+\sqrt{2})\frac{S_\Gamma + I}{2} + (1-\sqrt{2})\frac{S_\Gamma - I}{2} + M$$

is considered on the space $L_2(\Gamma)$. If Γ is a circle, then this operator A is Fredholm, but if Γ is a rectangle, then it is not.

1.11 Comments and References

Real C^*-algebras are treated for instance in [96, 141]. The literature concerning complex C^*-algebras is increasingly extensive, [4, 8, 38, 66, 88, 171, 159, 224]. In the context of these books a special family of real C^* algebras is of interest. These algebras are built as extensions of complex C^*-algebras by an element m which satisfies a few axioms. Such an abstract approach is strongly motivated by numerous applications and originates from [58, 59]. Single blocks of such constructions were used earlier in investigations of one-dimensional singular integral operators with conjugations and Carleman shifts (see, for example [143]).

Section 1.2–Section 1.5: We mainly follow our papers [58, 59]. Theorem 1.3.8, which allows us to represent any space of bounded additive operators $\mathcal{L}_{add}(\mathcal{H})$ as an extension of the complex C^*-algebra $\mathcal{L}(\mathcal{H})$, is new.

Section 1.6–Section 1.7: These sections are devoted to the abstract stability problem and to its algebraization. The fact that the problem of the stability of operator sequences can be translated into an invertibility problem in a given algebra was first recognized by A. Kozak [127]. He successfully applied this idea to study a finite section method for convolution equations with continuous data. The initial approach, however, was too restrictive to tackle the stability of various operator sequences arising in approximation of other operator equations. A further development of this idea is connected with some results similar to Theorem 1.6.6, the first version of which was proposed by one of the authors in [207]. The material of Section 1.7 is based on C^*-algebra techniques and can be found in [192] in the case of complex algebras. For the real extensions of complex C^*-algebras, the corresponding results are presented in [58].

Section 1.8: The use of para-algebras in studying the Fredholm properties of linear operators was initiated by D. Przeworska-Rolewicz and S. Rolewicz in [184]. The fact that this concept is also applicable in numerical analysis is noted in [44].

Section 1.9: Local principles are in a sense a replacement for the Gelfand theory for complex commutative Banach algebras. They are intended to study the invertibility of elements in non-commutative Banach algebras. There are a few versions of such principles – viz., the local principle of Simonenko, of Allan, and the one of Gohberg and Krupnik. A discussion of various aspects of these principles can be found in [10]. The local principle of Gohberg and Krupnik is distinguished by its simplicity and a wide range of applicability. Moreover, if the localization process is carried out with the help of central subalgebras which are C^*-algebras (strictly real C^*-algebras), then all the aforementioned local principles coincide at least. Local principles became a power tool in operator theory and numerical analysis. It is easy to see that the Gohberg-Krupnik local principle can be modified in such a way that it is applicable also in suitable para-algebras. For more information on this topic the reader can consult [189].

Section 1.10: There exists a huge variety of papers dealing with singular integral and Mellin operators. The material of Sections 1.10.1–1.10.4 is based on [102]. Theorems 1.10.4 and 1.10.7 are taken from [191] but these results are the outcome of efforts made by many mathematicians. Let us only mention [27, 28, 75, 138, 143, 214] and of course [91, 92].

Chapter 2

Approximation of Additive Integral Operators on Smooth Curves

We start with polynomial and spline approximation methods for the Cauchy singular integral equation

$$(Au)(t) \equiv a(t)\varphi(t) + \frac{b(t)}{\pi i}\int_\Gamma \frac{\varphi(\tau)\,d\tau}{\tau - t} + \overline{c(t)\varphi(t)} + \overline{\frac{d(t)}{\pi i}\int_\Gamma \frac{\varphi(\tau)\,d\tau}{\tau - t}}$$

$$+ \int_\Gamma k_0(t,\tau)\varphi(\tau)\,d\tau + \overline{\int_\Gamma k_1(t,\tau)\varphi(\tau)\,d\tau} = f(t), \quad t \in \Gamma \tag{2.1}$$

where Γ is a simple closed Lyapunov contour. The coefficients a, b, c, d belong to given functional classes, and conditions imposed on the kernels $k_0(t,\tau), k_1(t,\tau)$ ensure the compactness of the corresponding integral operators. The study of the stability of different approximation methods for equation (2.1) takes into account the smoothness of the coefficients a, b, c, d. Usually this is easier for equations with continuous coefficients. One encounters two main problems for equation (2.1). The first is connected with non-linearity of the operators considered over the field of complex numbers $\mathbb{C}$. However, this non-linearity can be called 'weak', and the corresponding difficulties are not of fundamental importance. Much more important is that the structure of the approximation sequences do not allow us to get a 'good' factorization of these sequences, even in the case where the coefficients a, b, c, d are continuous. Therefore, initially, approximation methods for Cauchy integral equations with conjugation were developed in two directions. One way is not to apply approximation methods to equation (2.1) directly but rather to an associated system of singular integral equations without conjugation [40, 118, 119, 120]. This leads to an unnecessary increase in the size of the systems of algebraic equations obtained. Some efforts were made to construct approximation methods working without passing to associated systems of singular integral equations [225, 226], but those methods are applicable to a restricted number of equations, since analytic functions in factorizations of certain combinations of the coefficients of equation (2.1) must satisfy additional metric relations. Moreover, those analytic functions have to be known to the user. However, the requirement to have the exact factorization of the coefficients of equation (2.1) makes this approach almost non-feasible. On the other hand, systematic use of Banach algebra methods may help to overcome all these difficulties.

2.1 Polynomial Galerkin Method for Equations with Continuous Coefficients

Let Γ_0 be the unit circle with center at the origin. In the space $L_2 := L_2(\Gamma_0)$ consider the operators

$$(M\varphi)(t) = \overline{\varphi(t)},\ (S\varphi)(t) = \frac{1}{2\pi i}\int\limits_{\Gamma_0} \frac{\varphi(\tau)\,d\tau}{\tau - t},\ P = (1/2)(I+S),\ Q = I - P.$$

The operators P and Q are bounded projections in L_2 [91, 124]. For simplicity, let us assume that the functions $k_0(t,\tau) = 0$ and $k_1(t,\tau) = 0$ for all $t,\tau \in \Gamma_0$ and rewrite equation (2.1) in the form

$$(G\varphi)(t) \equiv ((a_0P + b_0Q) + M(a_1P + b_1Q))\varphi(t) = f(t), \quad t \in \Gamma_0, \tag{2.2}$$

where a_m, b_m, $m = 0, 1$ are operators of multiplication by the functions $a_0 = a+b$, $b_0 = a - b$, $a_1 = c + d$, $b_1 = c - d$.

An approximate solution of equation (2.2) is sought in the form

$$\varphi(t) = \sum_{k=-n}^{n} \varphi_k t^k. \tag{2.3}$$

The unknown coefficients $\varphi_k, k = -n, -n+1, \ldots, n$ are defined by the system of algebraic equations

$$\begin{aligned} &\sum_{j=0}^{n} a_{k-j}^{(0)}\varphi_j + \sum_{j=-n}^{-1} b_{k-j}^{(0)}\varphi_j + \sum_{j=0}^{n} \overline{a}_{-k-j}^{(1)}\overline{\varphi_j} + \sum_{j=-n}^{-1} \overline{b}_{-k-j}^{(0)}\overline{\varphi_j} = f_k, \\ &k = -n, -n+1, \ldots, n \end{aligned} \tag{2.4}$$

where f_k are the Fourier coefficients of the function f, viz.,

$$f_k = \frac{1}{2\pi}\int_0^{2\pi} f(\exp(i\theta))\exp(-ik\theta)\,d\theta, \quad k = 0, \pm 1, \ldots,$$

and $a_k^{(m)}, b_k^{(m)}$ are the Fourier coefficients of the functions $a^{(m)}, b^{(m)}$, $m = 0, 1$.

Let P_n denote the projection

$$(P_n\varphi)(t) = P_n\left(\sum_{k=-\infty}^{+\infty} \varphi_k t^k\right) = \sum_{k=-n}^{n} \varphi_k t^k. \tag{2.5}$$

Then system (2.4) can be written as the operator equation

$$G_n\varphi_n \equiv P_nGP_n\varphi_n = P_nf, \quad \varphi_n \in \operatorname{im} P_n. \tag{2.6}$$

Along with the projections P_n, $n \in \mathbb{N}$, consider the operators W_n, $n \in \mathbb{N}$, defined by

$$(W_n\varphi)(t) = W_n\left(\sum_{k=-\infty}^{+\infty} \varphi_k t^k\right) = \sum_{k=-n}^{-1} \varphi_{-n-k-1} t^k + \sum_{k=0}^{n} \varphi_{n-k} t^k. \tag{2.7}$$

The functions $e_k : t \mapsto t^k$, $k \in \mathbb{Z}$ form a complete orthonormal basis in the space $L_2 = L_2(\Gamma_0, \mu)$ where μ is the normalized Lebesgue measure on Γ_0. Thus the operators P_n and W_n satisfy all requirements of Section 1.6, viz., they act as linear operators on both the complex and real Hilbert spaces L_2 and $(L_2)_\mathbb{R}$. Therefore, to study approximation method (2.6) one can use Proposition 1.6.5. However, one still must show that the sequence (G_n) defined by (2.6) belongs to the related algebra $\mathcal{A}_\mathbb{R}$ (see Section 1.6, p. 30). Notice that the adjoint operator to the operator G of (2.2) is

$$G^* = (P\bar{a}_0 + Q\bar{b}_0)I + (P\bar{a}_1 + Q\bar{b}_1)M \tag{2.8}$$

and, in the case at hand, the operator $\widehat{M}$ defined on p. 30, is just MtI, where tI means the operator of multiplication by e_1.

Lemma 2.1.1. *The operators P, Q and $\widehat{M}$ satisfy the following relations:*

a) $\widehat{M}^2 = I, \quad \widehat{M}P = Q\widehat{M}, \quad e_{-1}M = \widehat{M}$.

b) *Let $a \in \mathbf{C}(\Gamma_0)$. Then $MaM = \bar{a}I$ and $\widehat{M}a\widehat{M} = \bar{a}I$.*

c) $P_nM = MP_n$, $P_nP = PP_n$, $W_nP = PW_n$ *for all $n \in \mathbb{N}$.*

d) $s - \lim W_nMW_n = \widehat{M}$*; notice that W_n and M are $\mathbb{R}$-linear self-adjoint operators.*

The assertions a) – c) are easily checked and d) was already mentioned in Section 1.6.

For $a \in \mathbf{C}(\Gamma_0)$, we set $\widetilde{a} := a(\bar{t})$.

Lemma 2.1.2. *Assume that the coefficients a_m, b_m, $m = 0, 1$ are continuous, and let G be the operator defined in (2.2). Then the sequence (G_n), $G_n := P_nGP_n$ belongs to the algebra $\mathcal{A}_\mathbb{R}$. Moreover,*

$$s - \lim W_nG_nW_n = P\widetilde{a}_0P + Q\widetilde{b}_0Q + \widehat{M}(P\widetilde{a}_1P + Q\widetilde{b}_1Q).$$

Proof. Basically, the problem comes down to the question whether the sequences $(P_n a P_n)$ and $(P_n M P_n)$ belong to the algebra $\mathcal{A}_{\mathbb{R}}$. For the sequence $(P_n M P_n)$ this is clear because

$$s - \lim P_n M P_n = \widehat{M}.$$

Further, using the weak convergence of the sequence (W_n) to zero and the compactness of the operators PaQ and QaP, one easily shows that

$$s - \lim W_n a W_n = P\tilde{a}P + Q\tilde{a}Q,$$

and the proof of Lemma 2.1.2 follows. □

Note that the proof of the last relation can also be found in [183].

Lemma 2.1.3. *Let a_m, b_m, $m = 0, 1$ be continuous functions on Γ_0, and $\Delta := a_0\overline{b}_0 - a_1\overline{b}_1$ be invertible in $\mathbf{C}(\Gamma_0)$, i.e., $\Delta(t) \neq 0$ for all $t \in \Gamma_0$. Consider the $\mathbb{R}$-linear operator*

$$F := c_0 P + d_0 Q + M(c_1 P + d_1 Q), \tag{2.9}$$

where $c_0 = \Delta^{-1}\overline{b}_0$, $c_1 = -\Delta^{-1}a_1$, $d_0 = \overline{\Delta^{-1}a_0}$, $d_1 = -\overline{\Delta^{-1}b_1}$. Then there are operators $T_1, T_2 \in \mathcal{K}((L_2)_{\mathbb{R}})$ such that

$$GF = I + T_1, \quad FG = I + T_2.$$

Proof. The proof mimics the well-known proof of the related result for singular integral operators with continuous coefficients and without conjugation. The relations a) and b) should be taken into account.

First of all, let us mention that the condition $\Delta(t) \neq 0$, $t \in \Gamma_0$ is necessary and sufficient for the operator G to be Fredholm. Here we would like to outline basic ideas in the proof of this result because, in a sense, they lie in the foundation of the methods exploited in this book, especially in the use of the homomorphism Ψ defined by (1.26). Let $(L_2^2)_{\mathbb{R}}$ denote the real space $(L_2)_{\mathbb{R}} \times (L_2)_{\mathbb{R}}$. Write $G = A + MB$; thus A and B are singular integral operators with continuous coefficients and without conjugation. Since the operator $-iGi = A - MB$ is Fredholm if and only if the operator G is Fredholm, the operator

$$\begin{pmatrix} G & 0 \\ 0 & -iGi \end{pmatrix} = \begin{pmatrix} A + MB & 0 \\ 0 & A - MB \end{pmatrix}$$

acting on the space $(L_2^2)_{\mathbb{R}}$ is Fredholm if and only if the operator G is Fredholm. Applying the identity

$$\begin{pmatrix} A + MB & 0 \\ 0 & A - MB \end{pmatrix} = \frac{1}{2}\begin{pmatrix} I & M \\ I & -M \end{pmatrix}\begin{pmatrix} A & MBM \\ B & MAM \end{pmatrix}\begin{pmatrix} I & I \\ M & -M \end{pmatrix} \tag{2.10}$$

and using the equations $MAM = \overline{a}_0 Q + \overline{b}_0 P + K_1$, $MBM = \overline{a}_1 Q + \overline{b}_1 P + K_1$, where K_1 and K_1 are $\mathbb{C}$-linear compact operators, one obtains that the operator G is Fredholm if and only if the operator

$$\Psi(G) := \begin{pmatrix} A & MBM \\ B & MAM \end{pmatrix} = \begin{pmatrix} a_0 & \overline{b}_1 \\ a_1 & \overline{b}_0 \end{pmatrix} P + \begin{pmatrix} b_0 & \overline{a}_1 \\ b_1 & \overline{a}_0 \end{pmatrix} Q + K$$

where K is again a compact operator, is Fredholm. Now the general theory of singular integral operators applies. In particular, it gives that the operator $\Psi(G)$ is Fredholm if and only if the function $\Delta = \det(a_0 \overline{b}_0 - a_1 \overline{b}_1)$ does not vanish on Γ_0. Moreover, the operator

$$\begin{pmatrix} a_0 & \overline{b}_1 \\ a_1 & \overline{b}_0 \end{pmatrix}^{-1} P + \begin{pmatrix} b_0 & \overline{a}_1 \\ b_1 & \overline{a}_0 \end{pmatrix}^{-1} Q$$

is a regularizer for the operator $\Psi(G)$. Using this fact, it is now easy to find out that F is a regularizer for G provided $\Delta(t) \neq 0$ for all $t \in \Gamma_0$.

Another important consequence of identity (2.10) is that the index $\operatorname{ind}_{\mathbb{R}} G$ of the operator G considered over the field of real numbers can be expressed via index $\operatorname{ind}_{\mathbb{C}} \Psi(G)$ of the operator $\Psi(G)$ considered over the field of complex numbers, viz., the equation

$$\operatorname{ind}_{\mathbb{R}} G = \frac{1}{2} \operatorname{ind}_{\mathbb{R}} \Psi(G) = \operatorname{ind}_{\mathbb{C}} \psi(G) \tag{2.11}$$

holds. Relation (2.11) is often used in what follows. □

The operator F constructed in the proof of Lemma 2.1.3 can also be used to establish the stability of the polynomial Galerkin method for singular integral equations with conjugation.

Theorem 2.1.4. *Let a_m, b_m, $m = 0, 1$ be continuous functions on Γ_0, and let $G \in \mathcal{L}_{add}(L_2)$ be the operator defined in* (2.2). *Then the sequence $(P_n G P_n)$ is stable if and only if the operators G and*

$$\tilde{G} := P\tilde{a}_0 P + Q\tilde{b}_0 Q + \widehat{M}(P\tilde{a}_1 P + Q\tilde{b}_1 Q)$$

are invertible.

Proof. The sequence $(G_n), G_n = P_n G P_n$ belongs to the algebra $\mathcal{A}_{\mathbb{R}}$, hence the necessity follows from Theorem 1.8.4. In order to show sufficiency, one only needs to establish the invertibility of the coset $G_n^\circ := (G_n) + \mathcal{J}$ in the quotient algebra $\mathcal{A}_{\mathbb{R}}/\mathcal{J}$ (see Proposition 1.6.5). Consider next the sequence $(F_n), F_n := P_n F P_n$ with the operator F defined by (2.9), which also belongs to the algebra $\mathcal{A}_{\mathbb{R}}$. We claim that the coset $(F_n)^\circ$ is the inverse element for $(G_n)^\circ$ in the algebra $\mathcal{A}_{\mathbb{R}}/\mathcal{J}$, i.e., there exist sequences $(J'_n), (J''_n) \in \mathcal{J}$ such that

$$G_n F_n = P_n + J'_n, \quad F_n G_n = P_n + J''_n, \quad n \in \mathbb{N}.$$

Consider for example the product G_nF_n. It can be written as the sum of sixteen summands one of which is

$$E_n = P_n a_0 P P_n M d_1 Q P_n.$$

Using E_n as an example, let us show how to study these terms. This operator can be rewritten as

$$P_n a_0 P P_n M d_1 Q P_n = P_n P a_0 P P_n M d_1 Q P_n + P_n Q a_0 P P_n M d_1 Q P_n \tag{2.12}$$

and $(P_n Q a_0 P P_n M d_1 Q P_n) \in \mathcal{J}$, because the operator $Q a_0 P \in \mathcal{K}(L_2)$.

Consider now the sequence

$$\begin{aligned}(R_n) &= (P_n P a_0 P P_n M d_1 Q P_n)\\ &= (P_n P a_0 M d_1 Q P_n) - (P_n P a_0 P(I - P_n) M d_1 Q P_n).\end{aligned} \tag{2.13}$$

The first term in the right-hand side of (2.13) can be written as

$$(P_n P a_0 M d_1 Q P_n) = (P_n P a_0 \overline{d}_1 P P_n M) + j_n$$

with $j_n \in \mathcal{J}$. The second term in the right-hand side of (2.13) can be written as

$$(P_n P a_0 P(I - P_n) M d_1 Q P_n) = (P_n P a_0 P(I - P_n) P \overline{d}_1 Q P_n) + j_n'$$

with $j_n' \in \mathcal{J}$. Since

$$P_n P a_0 P(I - P_n) P \overline{d}_1 Q P_n = W_n P K P W_n$$

with some compact operator $K \in \mathcal{K}(L_2)$ (see [11]), we can represent the sequence (R_n) of (2.13) as

$$(R_n) = (P_n P a_0 \overline{d}_1 Q P_n M) + j_n'', \quad j_n'' \in \mathcal{J}.$$

Performing analogous computations for the remaining fifteen summands in the product G_nF_n, one can rewrite it in the form

$$\begin{aligned}G_nF_n &= P_n \big(P(a_0c_0 + \overline{b}_1 c_1)P + Q(b_0d_0 + \overline{a}_1 d_1)Q\\ &\qquad + P(a_0\overline{d}_1 + b_1\overline{d}_0)MQ + Q(b_0\overline{c}_1 + \overline{a_1c_0})MP\big) P_n + J_n'\\ &= P_n(P+Q)P_n + J_n' = P_n + J_n',\end{aligned}$$

where $(J_n') \in \mathcal{J}$, i.e., $(G_n)^\circ (F_n)^\circ = (P_n)^\circ$. Analogously, one can show that $(F_n)^\circ (G_n)^\circ = (P_n)^\circ$, which completes the proof. □

Consider now the stability of the polynomial Galerkin method in the space L_2 for the equation

$$(G + T)\varphi = f$$

where the operator G is defined as above, and

$$(T\varphi)(t) = \int_\Gamma k_0(t,\tau)\varphi(\tau)\,d\tau + \overline{\int_\Gamma k_1(t,\tau)\varphi(\tau)\,d\tau}.$$

Theorem 2.1.5. *Let $a, b, c, d \in \mathbf{C}(\Gamma_0)$ and let $k_0, k_1 \in \mathbf{C}(\Gamma_0 \times \Gamma_0)$. The sequence $(P_n(G+T)P_n)$ is stable if and only if the operators $G+T$ and $\widetilde{G}$ are invertible.*

The operator T is compact, hence the sequence (P_nTP_n) does not influence the invertibility of the coset $(P_nGP_n)+\mathcal{J}$ in the quotient algebra $\tilde{\mathcal{A}}/\mathcal{J}$. Moreover, $\widetilde{(G+T)} = \widetilde{G}$.

2.2 Polynomial Collocation Method for Equations with Continuous Coefficients

Consider the set $\mathbf{R} = \mathbf{R}(\Gamma_0)$ of all Riemann integrable functions on Γ_0. When provided with the norm

$$||f||_\infty = \sup_{t\in\Gamma_0} |f(t)|,$$

this set becomes a Banach space. Let π_n be the set of all trigonometrical polynomials

$$\pi_n := \left\{ p \in \mathbf{C}(\Gamma_0) : x_n(t) = \sum_{k=-n}^{n} \alpha_k t^k, \quad \alpha_k \in \mathbb{C},\ k = -n, -n+1, \dots, n \right\},$$

and let L_n denote the operator which to each function $f \in \mathbf{R}$ assigns its Lagrange interpolation polynomial L_nf with nodes

$$t_j = \exp(2\pi i/(2n+1)), \quad j = -n, -n+1, \dots, n. \tag{2.14}$$

It is known [112] that the operator L_n can be represented in the form

$$(L_nf)(t) = \sum_{k=-n}^{n} \beta_k t^k, \quad \beta_k = \frac{1}{2n+1} \sum_{j=-n}^{n} f(t_j)t_j^{-k}. \tag{2.15}$$

The operator L_n is a projection on the space $\mathbf{R}$, and for any function $f \in \mathbf{R}$ and for any polynomial $x_n \in \pi_n$ one has [183]

$$||L_nfx_n||_2 \le ||f||_\infty ||x_n||_2, \tag{2.16}$$

$$||L_nf - f||_2 \to 0 \quad \text{as} \quad n \to \infty. \tag{2.17}$$

On the space $L_2(\Gamma_0)$ we also consider the operators P_n and W_n defined by (2.5) and (2.7) respectively. An approximate solution of equation (2.3) is sought in the form

$$x_n(t) = \sum_{k=-n}^{n} \varphi_k t^k.$$

The unknown coefficients φ_k, $k = -n, -n+1, \ldots, n$ are derived from the system of algebraic equations

$$a_0(t_j)\sum_{k=0}^{n}\varphi_k t_j^k + b_0(t_j)\sum_{k=-n}^{-1}\varphi_k t_j^k + \overline{a_1(t_j)}\sum_{k=0}^{n}\overline{\varphi}_k t_j^{-k} + \overline{b_1(t_j)}\sum_{k=-n}^{-1}\overline{\varphi}_k t_j^{-k} = f(t_j),$$

$j = -n, -n+1, \ldots, n$, which is equivalent to the operator equation

$$G_n' x_n = L_n G P_n x_n = L_n f, \quad x_n \in \pi_n. \tag{2.18}$$

It is more convenient to consider the equation

$$G_n' x_n = P_n f, \quad x_n \in \pi_n \tag{2.19}$$

instead of equation (2.18), which is legitimate because the stability of the sequence (G_n') implies convergence of the collocation method for any right-hand side $f \in \mathbf{R}$. Indeed, let φ, x_n and ψ_n be the respective solutions of equations (2.3), (2.18) and (2.19), and let the sequence (G_n') be stable. Then our claim follows from the inequality

$$||\varphi - x_n||_2 \le ||\varphi - \psi_n||_2 + ||(G_n)^{-1}||\,||L_n f - P_n f||_2$$

and from relation (2.17).

Lemma 2.2.1. *Let functions $a_m, b_m \in \mathbf{R}, m = 0,1$ and let G be the operator defined by (2.2). Then the sequence (G_n') belongs to the algebra $\mathcal{A}_{\mathbb{R}}$.*

Proof. Since $||M|| = 1$ and the projections $P_n, n \in \mathbb{N}$ are uniformly bounded, inequality (2.16) implies that

$$||G_n' P_n \varphi|| \le C(||a_0||_\infty + ||b_0||_\infty + ||a_1||_\infty + ||b_1||_\infty)||\varphi||_2$$

for any $\varphi \in L_2(\Gamma_0)$, hence

$$\sup_n ||G_n' P_n|| < \infty.$$

Moreover, for any $k \le n$, the space π_k is invariant with respect to the operator P_n. Thus

$$G_n' P_n \varphi_k = L_n G \varphi_k$$

and by (2.17) the sequence $(G_n' P_n \varphi_k)$ strongly converges to $G\varphi_k$ for any $k \in \mathbb{N}$. By the Banach-Steinhaus theorem [77], the operator sequence $(G_n' P_n)$ strongly converges to the operator G on the whole space $L_2(\Gamma_0)$. Analogously, one can establish the strong convergence of the sequences $((G_n')^* P_n)$, $(\widetilde{G}_n' P_n)$ and $((\widetilde{G}_n')^* P_n)$ to the operators

$$\begin{aligned} G^* &= P\overline{a}_0 I + Q\overline{b}_0 I + (P\overline{a}_1 + Q\overline{b}_1)M, \\ \widetilde{G}' &= \widetilde{a}_0 P + \widetilde{b}_0 Q + Mt(\widetilde{a}_1 P + \widetilde{b}_1 Q), \\ (\widetilde{G}')^* &= P\overline{\widetilde{a}}_0 I + Q\overline{\widetilde{b}}_0 I + P\overline{\widetilde{a}}_1 + Q\overline{\widetilde{b}}_1)t^{-1}M, \end{aligned}$$

respectively. Indeed, taking into account the relation $(L_nfP_n)^* = L_n\overline{f}P_n$, [117] and equation (2.8) one gets

$$(G'_n)^* = PL_n\overline{a}_0P_n + QL_n\overline{b}_0P_n + (PL_n\overline{a}_1 + QL_n\overline{b}_1)P_nM, \tag{2.20}$$

and the strong convergence of the sequence $(G'_n)^*$ to the operator G^* follows. Considering the operator

$$\widetilde{G}'_n = W_n\left(L_n((a_0P + b_0Q) + M(a_1P + b_1Q))\right)W_n$$

one notes that for any function $f \in \mathbf{R}$,

$$W_nL_nfW_n = L_n\widetilde{f}P_n, \quad W_nL_nMfW_n = L_nMt\widetilde{f}P_n. \tag{2.21}$$

Therefore

$$\widetilde{G}'_n = L_n\left((\widetilde{a}_0P + \widetilde{b}_0Q) + Mt(\widetilde{a}_1P + \widetilde{b}_1Q)\right)P_n$$

and, applying the Banach-Steinhaus Theorem once more, one derives the strong convergence of the sequence $(\widetilde{G}'_n)$ to the operator $\widetilde{G}'$. Combining now relations (2.20) and (2.21) we obtain the corresponding claim for the sequence $((\widetilde{G}'_n)^*)$. This completes the proof of Lemma 2.2.1. □

Lemma 2.2.2. *Let L_n, P_n and W_n be the operators defined by* (2.15), (2.5) *and* (2.7), *respectively, and let $f \in \mathbf{C}(\Gamma_0)$. Then the sequences*

$$(PL_nfQP_n), (QL_nfPP_n), (PL_nMfPP_n), (QL_nMfQP_n)$$

belong to the ideal $\mathcal{J}$ of $\mathcal{A}_{\mathbb{R}}$.

Proof. The inclusions $(PL_nfQP_n) \in \mathcal{J}$, $(QL_nfPP_n) \in \mathcal{J}$ have been established in [117]. To establish the inclusion for the sequence (PL_nMfPP_n), we write

$$\begin{aligned} PL_nMfPP_n &= PL_n\overline{f}MPP_n = PL_n\overline{f}tQ\widehat{M}P_n \\ &= PL_n\overline{f}tQt^{-1}P_nMP_n \\ &= P_nPL_n\overline{f}QP_nMP_n + PL_n\overline{f}Kt^{-1}P_nMP_n, \end{aligned} \tag{2.22}$$

where $K = tQ - Qt$ is a compact operator. It is clear that the first summand in (2.22) belongs to the ideal $\mathcal{J}$. Furthermore, since

$$PL_n\overline{f}Kt^{-1}P_nMP_n = PL_n\overline{f}P_nKt^{-1}P_nMP_n + PL_n\overline{f}(I - P_n)Kt^{-1}P_nMP_n$$

and $||(I - P_n)Kt^{-1}P_n|| \to 0$ as $n \to \infty$, the second summand of (2.22) is also in $\mathcal{J}$.

Analogously one shows that $(QL_nMfQP_n) \in \mathcal{J}$. □

Theorem 2.2.3. *If $a_m, b_m \in \mathbf{C}(\Gamma_0)$, $m = 0, 1$, then the sequence (L_nGP_n) is stable if and only if the operators G and $\widetilde{G}'$ are invertible in $\mathcal{L}_{add}(L_2)$.*

Proof. The sequence $(L_nGP_n) \in \mathcal{A}_{\mathbb{R}}$, hence the necessity follows from Proposition 1.6.5. To establish sufficiency of the above conditions we show the invertibility of the coset $(L_nGP_n) + \mathcal{J}$ in the quotient algebra $\mathcal{A}_{\mathbb{R}}/\mathcal{J}$. Let F be the operator defined by (2.9). Consider the product $(L_nGP_n)(L_nFP_n)$. Rewriting the operator $L_nGL_nFP_n$ in the form

$$\begin{aligned} L_nGL_nFP_n &= L_na_0c_0PL_n + L_nb_0d_0QP_n + L_nMa_1c_0PP_n + L_nMb_1d_0QP_n \\ &\quad + L_na_0Md_1QP_n + L_nb_0Mc_1P_n + L_nMa_1Md_1QP_n \\ &\quad + L_nMb_1Mc_1PP_n \\ &\quad + (L_n(a_0 - b_0)P_n + L_nM(a_1 - b_1)P_n) \\ &\qquad \times (PL_nd_0QP_n - QL_nc_0PP_n + PL_nMc_1PP_n - QL_nMd_1QP_n) \end{aligned} \tag{2.23}$$

we note that the sequence (J'_n) generated by the operators from the last two rows of (2.23) belongs to the ideal $\mathcal{J}$. Indeed, this sequence is the product of two sequences, one of which $((L_n(a_0-b_0)P_n+L_nM(a_1-b_1)P_n))$ belongs to the algebra $\tilde{\mathcal{A}}$, and another is in ideal $\mathcal{J}$ by Lemma 2.2.2. Combining the remaining terms in (2.23) in an appropriate way, we obtain

$$\begin{aligned} L_nGP_nL_nFP_n &= L_n((a_0c_0 + \overline{b}_1c_1)P + (b_0d_0 + \overline{a}_1d_1)Q \\ &\qquad + (\overline{a_1c_0} + b_0\overline{c}_1)MP + (b_1\overline{d}_0 + a_0\overline{d}_1)MQ)P_n + J'_n \\ &= L_n(P+Q)P_n + J'_n = P_n + J'_n. \end{aligned}$$

Thus the coset $(L_nGP_n) + \mathcal{J}$ is right invertible. Analogously one shows that it is also left invertible, and reference to Proposition 1.6.5 finishes the proof. □

Note that for the space $L_2(\Gamma_0)$ there are effective sufficient conditions for stability of the collocation method.

Corollary 2.2.4. *Let $a_m, b_m \in \mathbf{C}(\Gamma_0)$, $m = 0, 1$, $\Delta(t) = a_0(t)\overline{b}_0(t) - a_1(t)\overline{b}_1(t) \neq 0$ for all $t \in \Gamma_0$, $\operatorname{ind}_{\Gamma_0}(\Delta(t)) = 0$ and for all $t \in \Gamma_0$ the inequality*

$$(|a_0(t)| - |a_1(t)|)(|b_0(t)| - |b_1(t)|) > 0$$

holds. Then the sequence (L_nGP_n) is stable.

Indeed, as was mentioned in [143] the above conditions provide the invertibility of the operators G and G' in the space $L_2(\Gamma_0)$, so the result follows from Theorem 2.2.3.

Remark 2.2.5. The proofs in Sections 2.2, 2.3 do not use any specific character of the scalar case considered, so all the results of these sections, with the exception of Corollary 2.2.4, are valid for systems of singular integral equations with conjugation.

2.3 Equations with Differential Operators

In this section we consider the applicability of approximation methods to equations containing differential operators, and particularly polynomial collocation and Galerkin methods for singular integro-differential equations. Moreover, in Section 3.5 we study the stability of approximation methods for the generalized Riemann-Hilbert-Poincaré problem, which also contains the operator of differentiation.

2.3.1 Singular Integro-Differential Equations

Consider first the following singular integro-differential equation

$$(K\varphi)(t) \equiv \sum_{j=0}^{q} \left\{ c_j(t)\varphi^{(j)}(t) + \frac{d_j(t)}{2\pi i} \int_{\Gamma_0} \frac{\varphi^{(j)}(\tau)\,d\tau}{\tau - t} + \int_{\Gamma_0} k_j(t,\tau)\varphi^{(j)}(\tau)\,d\tau \right\} = f(t), \quad (2.24)$$

where φ is an unknown function and $f \in L_2(\Gamma_0)$; $c_j, d_j \in L_\infty(\Gamma_0)$, $k_j \in L_\infty(\Gamma_0 \times \Gamma_0)$ are given functions, and

$$\varphi^{(0)}(t) := \varphi(t), \quad \varphi^{(j)}(t) := \frac{d^j\varphi}{dt^j}(t), \quad j = 1, 2, \ldots, q.$$

Let $H_2^q = H_2^q(\Gamma_0)$ denote the set of functions which are q-times differentiable on Γ_0 and such that all their derivatives up to the order $q-1$ are absolutely continuous, $\varphi^{(q)} \in L_2(\Gamma_0)$ and

$$\int_{\Gamma_0} \varphi^{(-k-1)}(\tau)\,d\tau = 0, \quad k = 0, 1, \ldots, q-1.$$

Moreover, by $L_{2,q} = L_{2,q}(\Gamma_0)$ we denote the image of the space L_2 under the mapping $F := P + t^{-q}Q$.

Lemma 2.3.1 ([195]). *The operator $D^q : H_2^q \mapsto L_{2,q}$ defined by*

$$(D^q\varphi)(t) := \varphi^{(q)}(t)$$

is continuously invertible, and the operator $D^{-q} : L_{2,q} \mapsto H_2^q$ defined by

$$(D^{-q}\varphi)(t) := (N^+\varphi)(t) + (N^-\varphi)(t),$$

where

$$(N^+\varphi)(t) = \frac{(-1)^{q-1}}{2\pi i(q-1)!} \int_{\Gamma_0} (Pf)(\tau)(\tau - t)^{q-1} \ln\left(1 - \frac{t}{\tau}\right) d\tau,$$

$$(N^-\varphi)(t) = \frac{(-1)^{q}}{2\pi i(q-1)!} \int_{\Gamma_0} (Qf)(\tau)(\tau - t)^{q-1} \ln\left(1 - \frac{\tau}{t}\right) d\tau,$$

is the inverse for the operator D^q.

This result implies the invertibility of an auxiliary operator B, which plays an important role in the study of approximation methods under consideration.

Lemma 2.3.2. *The operator* $B : H_2^q \mapsto L_2$ *defined by*

$$B := (P + t^q Q)D^q \tag{2.25}$$

is invertible and

$$B^{-1} = D^{-q}(P + t^{-q}Q).$$

Proof. Considering the products BB^{-1} and $B^{-1}B$ and using the relation $Pt^qQD^q = 0$, one obtains

$$BB^{-1}(P + t^q Q)D^q D^{-q}(P + t^{-q}Q) = P + Q = I$$

and

$$\begin{aligned} B^{-1}B &= D^{-q}(P + Q - t^{-q}Pt^qQ + Pt^qQ)D^q \\ &= D^{-q}(P + Q)D^q - D^{-q}(t^{-q} + 1)Pt^qQD^q = I, \end{aligned}$$

so the operator B is invertible. □

For $\varphi \in H_2^q(\Gamma_0)$, let us consider its Fourier series

$$\varphi(t) = \sum_{k=q}^{+\infty} \varphi_k t^k + \sum_{k=-\infty}^{-1} \varphi_k t^k,$$

and let P_n^q denote the projections

$$(P_n^q)(t) = \sum_{k=q}^{n+q} \varphi_k t^k + \sum_{k=-n}^{-1} \varphi_k t^k.$$

Since $\varphi \in H_2^q(\Gamma_0)$, one can differentiate its Fourier series [6], so that

$$BP_n^q = P_n B P_n^q, \tag{2.26}$$

where $P_n := P_n^0$ is the projection defined on the space $L_2(\Gamma_0)$ by (2.5). Let us also note that the sequence of projections (P_n^q) strongly converges to the identity operator on the space $H_2^q(\Gamma_0)$ [15].

Lemma 2.3.3. *Let* B *be the operator defined by* (2.25). *Then the sequence* $(P_n B P_n^q)$ *is stable.*

Proof. Indeed, since

$$P_n^q B^{-1} P_n = B^{-1} P_n \tag{2.27}$$

for all $n \in \mathbb{N}$, then

$$(P_n B P_n^q)(P_n^q B^{-1} P_n) = (P_n)$$

and

$$(P_n^q B^{-1} P_n)(P_n B P_n^q) = (P_n^q),$$

so the sequence $(P_n B P_n^q)$ is stable and

$$||(P_n B P_n^q)^{-1} P_n|| \le ||B^{-1}||$$

for all $n \in \mathbb{N}$. □

Let T be the operator defined by

$$(T\varphi)(t) := \sum_{j=0}^{q-1} \left\{ c_j(t)\varphi^{(j)}(t) + \frac{d_j(t)}{2\pi i} \int_{\Gamma_0} \frac{\varphi^{(j)}(\tau)\, d\tau}{\tau - t} + \int_{\Gamma_0} k_j(t,\tau)\varphi^{(j)}(\tau)\, d\tau \right\} + \int_{\Gamma_0} k_q(t,\tau)\varphi^{(q)}(\tau)\, d\tau.$$

The embedding theorems and properties of integral operators show that $T \in \mathcal{K}(H_2^q, L_2)$.

An approximate solution of equation (2.24) is sought in the form

$$x_n(t) = \sum_{k=q}^{n+q} x_k t^k + \sum_{k=-n}^{-1} x_k t^k. \tag{2.28}$$

The coefficients x_k are defined from the system of linear algebraic equations

$$a(t_j) \sum_{k=0}^{n} \frac{(k+q)!}{k!} x_{k+q} t_j^k + (-1)^k b(t_j) \sum_{k=-n}^{-1} \frac{(-k-1+q)!}{(-k-1)!} x_k t_j^{k-q} + (Tx_n)(t_j) = f(t_j), \tag{2.29}$$

$j = -n, -n+1, \dots, n$, where $a = c_q + d_q$, $b = c_q - d_q$ and the mesh points t_j are defined by (2.14).

The system of linear algebraic equations (2.29) is equivalent to the operator equation

$$K_n x_n = L_n(aP + t^{-q} bQ + T_1) B P_n^q x_n = L_n f,$$

with an operator $T_1 \in \mathcal{K}(L_2(\Gamma_0))$. Recall that the operators L_n, $n \in \mathbb{N}$, are defined by (2.15).

Theorem 2.3.4. *Let $a, b \in \mathbf{C}(\Gamma_0)$, and the coefficients c_j, d_j, $j = 0, \dots, q-1$ and the kernels k_j, $j = 0, \dots, q$ be such that the operator $T_1 : L_2(\Gamma_0) \mapsto \mathbf{R}(\Gamma_0)$ is compact.*

Then:

i) *The sequence* (K_n) *is stable if and only if the operator* $K \in \mathcal{L}(H_2^q, L_2)$ *is invertible;*

ii) *If the operator* $K \in \mathcal{L}(H_2^q, L_2)$ *is invertible, then the collocation method of* (2.28) – (2.29) *converges for any right-hand side* $f \in \mathbf{R}(\Gamma_0)$.

Proof. Using Lemma 2.3.3 and relation (2.27) one reduces the stability problem for the collocation method (2.29) to the stability of the sequence $L_n(aP+t^{-q}bQ+T_1)P_n$. Now Theorem 2.3.4 follows from known results on stability of the polynomial collocation method for Cauchy singular integral equations [183]. □

Lemma 2.3.3 can also be applied to study the polynomial Galerkin method for equation (2.24). An approximate solution of this equation is again sought in the form (2.28), but the unknown coefficients x_k are defined from the system of linear algebraic coefficients

$$\sum_{k=0}^{n} \frac{(k+q)!}{k!} a_{s-k} x_{k+q} + (-1)^q \sum_{k=q+1}^{n+q} \frac{k!}{(k-q-1)!} b_{s+k} x_{-k+q} \\ + \sum_{k=q}^{n+q} T_{ks}^1 x_k + \sum_{k=-n}^{-1} T_{ks}^1 x_k = f_s, \quad s=-n,\dots,n, \tag{2.30}$$

where a_s, b_s, f_s and T_{ks}^1 are the Fourier coefficients of the functions a, b, f and $T^1(\tau^k)$, respectively.

The system of linear algebraic equations (2.30) is equivalent to the operator equation

$$P_n K P_n^q x_n \equiv P_n((aP + t^{-q} bQ + T_1)B) P_n^q x_n = P_n f,$$

where $T_1 \in \mathcal{K}(L_2(\Gamma_0))$ and $x_n \in \operatorname{im} P_n^q$.

Theorem 2.3.5. *Let* $c_j, d_j \in \mathbf{C}(\Gamma_0)$, *and* $k_j \in \mathbf{C}(\Gamma_0)$, $j = 0, \dots, q$. *The sequence* $(P_n K P_n^q)$ *is stable if and only if the operators* $K : H_2^q \mapsto L_2$ *and* $\widetilde{K} = (P\widetilde{a}P + Qt^q\widetilde{b}Q) : L_2 \mapsto L_2$ *are invertible.*

Remark 2.3.6. Polynomial approximation methods for systems of singular integro-differential equations can be studied analogously.

2.3.2 Approximate Solution of Singular Integro-Differential Equations with Conjugation

Let us now consider singular integro-differential equations with conjugation, viz.,

$$K\varphi \equiv ((a_0P + b_0Q) + M(a_1P + b_1Q))D^q\varphi + T\varphi = g \tag{2.31}$$

where $(M\varphi)(t) = \overline{\varphi(t)}$ and $T \in \mathcal{K}_{add}(H_2^q, L_2)$. Such equations often arise in elasticity theory [161, 168, 170, 221, 222], but methods to obtain their approximate solutions remain little studied. Below we present stability results for two approximation methods for equation (2.31), based on projections P_n and L_n.

Theorem 2.3.7. *Let $a_j, b_j \in \mathbf{C}(\Gamma_0)$, $i = 0, 1$ and $T \in \mathcal{K}_{add}(H_2^q, L_2)$. The sequence $(L_n K P_n^q)$ is stable if and only if the operators $K : H_2^q(\Gamma_0) \mapsto L_2(\Gamma_0)$ and*

$$\widetilde{K} = ((\widetilde{a}_0 P + t^q \widetilde{b}_0) + M(t\widetilde{a}_1 P + t^{q+1}\widetilde{b}_1 Q) : L_2(\Gamma_0) \mapsto L_2(\Gamma_0)$$

are invertible.

Analogously, the stability of the polynomial Galerkin method can be described as follows.

Theorem 2.3.8. *Let $a_j, b_j \in \mathbf{C}(\Gamma_0)$, $j = 0, 1$, and $T \in \mathcal{K}_{add}(H_2^q, L_2)$. The sequence $(P_n K P_n^q)$ is stable if and only if the operators $K : H_2^q(\Gamma_0) \mapsto L_2(\Gamma_0)$ and*

$$\widetilde{K} = ((P\widetilde{a}_0 P + Q\widetilde{b}_0 t^q Q) + (QMt\widetilde{a}_1 P + PM\widetilde{b}_1 t^{q+1} Q) : L_2(\Gamma_0) \mapsto L_2(\Gamma_0)$$

are invertible.

To prove both the above results one has to represent the operator K of (2.31) in the form

$$K = ((a_0 P + t^{-q} b_0 Q) + M(a_1 P + t^{-q} b_1 Q) + T_1)B,$$

where $T_1 \in \mathcal{K}_{add}(L_2, L_2)$, and use relation (2.27), Lemma 2.3.3, and the results of Sections 2.1 and 2.2.

2.4 A C^*-Algebra of Operator Sequences

Here we study an algebra of operator sequences generated by paired circulants and by the operator of complex conjugation. In particular, we obtain a criteria for stability of elements which belong to such an algebra. Applying these results in different situations we obtain necessary and sufficient conditions for the stability of variety of spline approximation methods for singular integral equations with conjugation in the case of piecewise continuous coefficients a, b, c and d. Let $l_2(n), n \in \mathbb{N}$ denote the n-dimensional complex Hilbert space with the scalar product

$$(x, y) = \sum_{j=0}^{n-1} x_j \overline{y}_j, \quad x, y \in l_2(n),$$
$$x = (x_0, \ldots, x_{n-1}), \quad y = (y_0, \ldots, y_{n-1}).$$

By $\mathbb{C}^{n\times n}$ we denote the set of all $n \times n$ matrices with entries from $\mathbb{C}$. A matrix $A_n \in \mathbb{C}^{n\times n}$ of the form $A_n = (a_{j-k})_{j,k=0}^{n-1}$ is called circulant if

$$a_{n-k} = a_{-k}, \quad k = 0, 1, 2, \ldots, n-1.$$

It is well known [151] that any circulant can be represented in the form

$$A_n = U_n D_n U_n^{-1} \tag{2.32}$$

where U_n and D_n are, correspondingly, unitary and diagonal matrices

$$U_n = h^{1/2}(\exp(i2\pi jkh))_{j,k=0}^{n-1}, \quad D_n = (\lambda_j \delta_{jk})_{j,k=0}^{n-1}$$

with

$$\lambda_j = \sum_{k=0}^{n-1} a_k \exp(-i2\pi jkh), \quad j = 0, 1, \dots, n-1; \quad h = 1/n.$$

On the other hand, any matrix $A_n \in \mathbb{C}^{n\times n}$ of the form (2.32) is a circulant. In the following each linear operator acting on the space $l_2(n)$ is identified with a matrix $A_n \in \mathbb{C}^{n\times n}$, and $||A_n||$ means the operator norm on the space $\mathcal{L}(l_2(n))$.

Given a bounded function α on Γ_0, let α_n and $\widetilde{\alpha}_n$ denote the diagonal matrices

$$\alpha_n := (\alpha(t_k)\delta_{jk})_{j,k=0}^{n-1}, \quad \widetilde{\alpha}_n := (\alpha(\tau_k)\delta_{jk})_{j,k=0}^{n-1}, \tag{2.33}$$

where $t_k := \exp(i2\pi k/n)$, $\tau_k := \exp(i2\pi(k+\epsilon)/n)$, $0 < \epsilon < 1$, $k = 0, 1, \dots, n-1$. Consider now a circulant

$$\widehat{\alpha}_n := U_n \alpha_n U_n^{-1}. \tag{2.34}$$

It is clear that eigenvalues of circulant (2.34) coincide with values of the function α at the points t_k, $k = 0, 1, \dots, n-1$, hence

$$||\widehat{\alpha}|| = ||\alpha|| \leq \sup_{t\in\Gamma_0} ||\alpha(t)||, \quad ||\widetilde{\alpha}|| \leq \sup_{t\in\Gamma_0} ||\alpha(t)||.$$

Let us also note that multiplying the matrices in (2.34) yields the representation

$$\widehat{\alpha} = \left(\frac{1}{n} \sum_{l=0}^{n-1} \alpha(t_l) \exp(i2\pi l(k-j)/n \right)_{k,j=0}^{n-1} \tag{2.35}$$

for the circulant $\widehat{\alpha}$. We shall consider also an antilinear operator $\overline{M}_n$ on the space $l_2(n)$ defined as

$$\overline{M}_n(\xi^{(n)}) := (\overline{\xi}_0, \overline{\xi}_1, \dots, \overline{\xi}_{n-1}), \quad \xi^{(n)} = (\xi_0, \xi_1, \dots, \xi_{n-1}),$$

where the over-bar again denotes the operation of complex conjugation. Let C_n be additive operators on the spaces $l_2(n)$ such that the sequence $(||C_n||)$ converges to zero. Below we study the stability of operator sequences (B_n) with the operator $B_n : l_2(n) \to l_2(n)$ having the form

$$B_n := \widetilde{a}_n \widehat{\alpha}_n + \widetilde{b}_n \widehat{\beta}_n + (\widetilde{c}_n \widehat{\gamma}_n + \widetilde{d}_n \widehat{\theta}_n)\overline{M}_n + C_n, \quad n \in \mathbb{N}, \tag{2.36}$$

where $a, b, c, d, \alpha, \beta, \gamma, \theta \in \mathbf{PC}(\Gamma_0)$.

Let us now study how circulants interact with the operator $\overline{M}_n$.

Lemma 2.4.1. *Assume that a circulant* $A_n = (a_{j-k})_{k,j=0}^{n-1}$ *is represented in the form* (2.34) *and that its eigenvalues are equal to* λ_j, $j = 0, 1, \ldots, n-1$. *Then the corresponding eigenvalues* $\lambda_j^\star$, $j = 0, 1, \ldots, n-1$ *of the circulant* $\overline{A}_n = (\overline{a}_{j-k})_{k,j=0}^{n-1}$ *are equal to*

$$\lambda_0^\star = \overline{\lambda}_0, \quad \lambda_j^\star = \overline{\lambda}_{n-j}, \quad j = 1, 2, \ldots, n-1. \tag{2.37}$$

Proof. From (2.34), (2.35) the eigenvalues of the matrix A_n have the form

$$\lambda_j = \sum_{k=0}^{n-1} a_k e^{-i2\pi jkh}, \quad j = 0, 1, \ldots, n-1 \tag{2.38}$$

where $h = 1/n$. For the eigenvalues $\lambda_j^\star$, $j = 1, 2, \ldots, n-1$ of the matrix $\overline{A}_n$ one obtains

$$\begin{aligned} \lambda_j^\star &= \sum_{k=0}^{n-1} \overline{a}_k e^{-i2\pi jkh} = \overline{\sum_{k=0}^{n-1} a_k e^{i2\pi jkh}} \\ &= \overline{\sum_{k=0}^{n-1} a_k e^{-i2\pi knh + i2\pi jkh}} = \overline{\sum_{k=0}^{n-1} a_k e^{-i2\pi(n-j)kh}}, \end{aligned} \tag{2.39}$$

so (2.38) and (2.39) yields (2.37). □

Lemma 2.4.2. *Let* $a, \alpha \in \mathbf{PC}(\Gamma_0)$ *and let* $\widetilde{a}_n$ *and* $\widehat{\alpha}_n$ *be the matrices defined in* (2.33) *and* (2.34), *respectively. Then*

$$\overline{M}_n \widetilde{a}_n = \widetilde{\overline{a}}_n \overline{M}_n, \quad \overline{M}_n \widehat{\alpha}_n = \widehat{\alpha}_n^\circ \overline{M}_n, \tag{2.40}$$

where $\widetilde{\overline{a}}_n$ *is the matrix of the form* (2.33) *generated by the function* $\overline{a(t)}$, $t \in \Gamma_0$ *and* $\widehat{\alpha}_n^\circ$ *is the circulant of the form* (2.34) *generated by the function* $\alpha^\circ(t) := \overline{\alpha(\overline{t})}$, $t \in \Gamma_0$.

Proof. The first of relations (2.40) is obvious. Let us compute the eigenvalues $\alpha_k^\star$, $k = 0, 1, \ldots, n-1$ of the circulant $\widehat{\overline{\alpha}}_n$. Taking into account Lemma 2.4.1 and the definition of $\widehat{\alpha}_n$, for any $k = 1, 2, \ldots, n-1$ one obtains

$$\begin{aligned} \alpha_k^\star &= \overline{\alpha(t_{n-k})} = \overline{\alpha(\exp(i2\pi(n-k)/n))} \\ &= \overline{\alpha(\exp(-i2\pi k/n))} = \overline{\alpha(\overline{t}_k)} = \alpha^\circ(t_k), \end{aligned}$$

and the proof is completed. □

Consider now the set $\mathcal{R}$ of all bounded sequences (A_n) of bounded additive operators $A_n : l_2(n) \to l_2(n)$. Provided with natural operations of addition, multiplication and multiplication by real numbers, the set $\mathcal{R}$ becomes a real algebra. Let us also consider the smallest *complex* C^*-algebra $\mathcal{C}$ of $\mathcal{R}$ which contains all sequences of the form $(\widetilde{a}_n)$ and $(\widehat{\alpha}_n)$, where $\widetilde{a}_n, \widehat{\alpha}_n$ are circulants of the form (2.33) and (2.34) generated respectively by the functions $a, \alpha \in \mathbf{PC}(\Gamma_0)$ and all sequences (C_n) of $\mathbb{C}$-linear operators with $||C_n|| \to 0$ as $n \to \infty$. Notice that the collection of sequences in $\mathcal{C}$ tending to zero in the norm, actually forms a two-sided closed ideal in $\mathcal{C}$ which we denote by $G_{\mathcal{C}}$. Let us also recall that the above-mentioned circulants are identified with the corresponding linear operators on $l_2(n)$. By m we denote the sequence $(\overline{M}_n) \in \mathcal{R}$. Then Lemmas 2.4.1 and 2.4.2 imply the following result.

Proposition 2.4.3. *The real algebra $\mathcal{R}$, complex C^*-algebra $\mathcal{C}$ and the element m satisfy axioms $(A_1) - (A_5)$ of Section* 1.2.

Proof. We check axiom (A_1) and axiom (A_5), because three others are obvious. It suffices to verify these axioms only for the generators of the algebra $\mathcal{C}$. The corresponding relations can be extended on the whole algebra $\mathcal{C}$ by continuity. Thus let $\widetilde{a}$ and $\widehat{\alpha}$ denote the sequences $(\widetilde{a}_n)$ and $(\widehat{\alpha}_n)$ with $a, \alpha \in \mathbf{PC}(\Gamma_0)$. By (2.40) one obtains

$$m\widetilde{a}m = \widetilde{\widetilde{a}}, \quad m\widehat{\alpha}m = \widehat{\alpha}^\circ,$$

and since $\widetilde{\widetilde{a}}, \widehat{\alpha}^\circ \in \mathbf{PC}(\Gamma_0)$, the axiom (A_1) is satisfied. Now we check axiom (A_5) for the element $\widehat{\alpha}$. Note that (2.40) implies the equalities

$$\widehat{\alpha}_n = \overline{M}_n \widehat{\alpha}_n^\circ \overline{M}_n, \quad \overline{M}_n \widehat{\alpha}_n \overline{M}_n = \widehat{\alpha}_n^\circ. \tag{2.41}$$

Using now (2.34) and the first equality in (2.41) one obtains

$$(\widehat{\alpha}_n)^* = (U_n \alpha_n U_n^{-1})^* = \widehat{\overline{\alpha}}_n, \tag{2.42}$$

$$(\overline{M}_n \widehat{\alpha}_n \overline{M}_n)^* = (\widehat{\alpha}_n^\circ)^* = \widehat{\overline{\alpha}}_n^\circ \tag{2.43}$$

where A_n^* means the adjoint operator to the operator A_n. Thus taking into account (2.42) and applying the second of the equalities (2.41) one arrives at the relation

$$\overline{M}_n(\widehat{\alpha}_n)^* \overline{M}_n = \overline{M}_n \widehat{\overline{\alpha}}_n \overline{M}_n = \widehat{\overline{\alpha}}_n^\circ.$$

Comparing this with (2.43) we get

$$(m\alpha m)^* = \widehat{\overline{\alpha}}^\circ = m\alpha^* m.$$

For the element $\widetilde{a}$, the axiom (A_5) can be verified analogously. □

From this result one can consider the extension $\widetilde{\mathcal{C}}$ of the real algebra $\mathcal{C}$ by the element m and use results of Chapter 1 to study the stability problem for sequences generated by paired circulants and complex conjugation. For example, combining Proposition 2.4.3 with Corollary 1.4.7 one obtains a stability result for elements from the algebra $\widetilde{\mathcal{C}}$.

Theorem 2.4.4. *The sequence (B_n),*

$$B_n = \widetilde{a}_n \widehat{\alpha}_n + \widetilde{b}_n \widehat{\beta}_n + (\widetilde{c}_n \widehat{\gamma}_n + \widetilde{d}_n \widehat{\theta}_n)\overline{M}_n, \quad n \in \mathbb{N}$$

is stable if and only if the sequence $(B_n^{(1)}) \in \mathcal{L}^{2\times 2}(l_2(n))$,

$$B_n^{(1)} = \begin{pmatrix} \widetilde{a}_n\widehat{\alpha}_n + \widetilde{b}_n\widehat{\beta}_n & \widetilde{c}_n\widehat{\gamma}_n + \widetilde{d}_n\widehat{\theta}_n \\ \widetilde{\overline{c}}_n\widehat{\gamma}_n^\circ + \widetilde{\overline{d}}_n\widehat{\theta}_n^\circ & \widetilde{\overline{a}}_n\widehat{\alpha}_n^\circ + \widetilde{\overline{b}}_n\widehat{\beta}_n^\circ \end{pmatrix} \tag{2.44}$$

is stable.

Thus the problem of stability of sequences (2.36) is reduced to the stability of matrix sequences (2.44). In this section, such sequences are studied straightforwardly. On the other hand, the reader familiar with the tensor products of C^*-algebras can consult Section 2.11 where another approach to the problem is presented.

Let us now describe operators which play a useful role in studying the algebra $\widetilde{\mathcal{C}}$. Let $\chi_j^{(n)} = \chi_j^{(n)}(t)$, $t \in \Gamma_0$ be the characteristic function of the circular arc $[\exp(i2\pi j/n), \exp(i2\pi(j+1)/n))$, and let $g_j^{(n)} := \sqrt{n}\chi_j^{(n)}$. By L_n we denote the orthogonal projection which projects the Hilbert space $L_2(\Gamma_0)$ onto the linear span of the functions $g_j^{(n)}$, $j = 0, 1, \ldots, n-1$, viz.,

$$(L_n f)(t) = \sum_{k=0}^{n-1} f_k g_k^{(n)}(t), \quad f_k = (f, g_k^{(n)}),$$

where $(\cdot,\cdot)$ denotes the scalar product on $L_2(\Gamma_0)$. In addition, let us define another sequence (P_n) of projections on the space $L_2(\Gamma_0)$ by

$$(P_n f)(t) = \sum_{k=-[n/2]}^{[(n-1)/2]} f_k t^k, \quad f_k = \frac{1}{2\pi}\int_0^{2\pi} f(\exp(i\theta))\exp(-ik\theta)\, d\theta.$$

Consider also the operators $K_n^\varepsilon : L_2(\Gamma_0) \mapsto \operatorname{im} L_n$ and $U_n : \operatorname{im} L_n \mapsto \operatorname{im} L_n$ defined by

$$(K_n^\varepsilon f)(t) := \sum_{j=0}^{n-1} f(\exp(i2\pi(j+\varepsilon)/n))\chi_j^{(n)}(t), \quad 0 \le \varepsilon < 1,$$

$$U_n\left(\sum_{j=0}^{n-1} \xi_j g_j^{(n)}\right) := \sum_{k=0}^{n-1} \frac{1}{\sqrt{n}}\left(\sum_{j=0}^{n-1} \exp(-i2\pi jk/n)\xi_j\right) g_k^{(n)}.$$

We also identify the operators that act on the space $\operatorname{im} L_n$ with their matrices in the basis $\chi_j^{(n)}, j = 0, 1, \ldots, n-1$. There is for example the following correspondence:

$$\begin{aligned} \widetilde{a}_n &\sim K_n^\varepsilon a L_n : \operatorname{im} L_n \mapsto \operatorname{im} L_n, \\ \widehat{\alpha}_n &\sim U_n K_n^0 \alpha L_n U_n^{-1} L_n : \operatorname{im} L_n \mapsto \operatorname{im} L_n, \end{aligned} \tag{2.45}$$

which implies that the matrix sequence (2.44) can be identified with the operator sequence

$$B_n^{(1)} = \begin{pmatrix} K_n^\varepsilon a L_n U_n K_n^0 \alpha L_n U_n^{-1} L_n & K_n^\varepsilon c L_n U_n K_n^0 \gamma L_n U_n^{-1} L_n \\ K_n^\varepsilon \overline{c} L_n U_n K_n^0 \gamma^\circ L_n U_n^{-1} L_n & K_n^\varepsilon \overline{a} L_n U_n K_n^0 \alpha^\circ L_n U_n^{-1} L_n \end{pmatrix}$$
$$+ \begin{pmatrix} K_n^\varepsilon b L_n U_n K_n^0 \beta L_n U_n^{-1} L_n & K_n^\varepsilon d L_n U_n K_n^0 \theta L_n U_n^{-1} L_n \\ K_n^\varepsilon \overline{d} L_n U_n K_n^0 \theta^\circ L_n U_n^{-1} L_n & K_n^\varepsilon \overline{a} L_n U_n K_n^0 \beta^\circ L_n U_n^{-1} L_n \end{pmatrix}. \quad (2.46)$$

Let $\omega \in \Gamma_0$. By p_ω we denote an integer from the set $\{0, 1, \dots, n-1\}$ such that $\omega \in [\exp(i2\pi(p_\omega - 1)/n), \exp(i2\pi p_\omega/n)]$, and let $t_\omega := \exp(i2\pi p_\omega/n)$. For $\omega, v \in \Gamma_0$, consider two families of isomorphisms $\widetilde{E}_n^{(\omega,1)}$ and $\widetilde{E}_n^{(v,2)}$, where $\widetilde{E}_n^{(\omega,1)} : \operatorname{im} L_n \mapsto \operatorname{im} L_n$ and $\widetilde{E}_n^{(v,2)} : \operatorname{im} L_n \mapsto \operatorname{im} P_n$ are defined by

$$\widetilde{E}_n^{(\omega,1)} : \sum_{j=0}^{n-1} \xi_j g_j^{(n)} \mapsto \sum_{j=0}^{n-1} (t_\omega)^{-j} \xi_j g_j^{(n)},$$
$$\widetilde{E}_n^{(v,2)} = E_n U_n^{-1} \widetilde{E}_n^{(v_\varepsilon,1)} U_n L_n,$$

where $v_\varepsilon := v \exp(-i2\pi\varepsilon/n)$, $0 \le \varepsilon < 1$ and $E_n : \operatorname{im} L_n \mapsto \operatorname{im} P_n$,

$$E_n : \sum_{j=0}^{n-1} \xi_j g_j^{(n)} \mapsto \sum_{j=0}^{[(n-1)/2]} \xi_j t^j + \sum_{j=[(n-1)/2]+1}^{n-1} \xi_j t^{j-n}.$$

Then the operators $\widetilde{E}_n^{(\omega,1)}, \widetilde{E}_n^{(v,2)}$ satisfy conditions imposed on the operators E_n^t from Section 1.6. On the space $(\operatorname{im} L_n) \times (\operatorname{im} L_n)$ we also define operators $E_n^{(\omega,1)}, E_n^{(v,2)}$ by

$$E_n^{(\omega,1)} := \begin{pmatrix} \widetilde{E}_n^{(\omega,1)} & 0 \\ 0 & \widetilde{E}_n^{(\omega,1)} \end{pmatrix}, \quad E_n^{(v,2)} := \begin{pmatrix} \widetilde{E}_n^{(v,2)} & 0 \\ 0 & \widetilde{E}_n^{(v,2)} \end{pmatrix}.$$

Let S be the Cauchy singular integral operator and let P and Q denote the corresponding Riesz projections. Consider the matrix-functions

$$\begin{aligned} \mathcal{A}(t) &:= \begin{pmatrix} a(t) & 0 \\ 0 & \overline{a(t)} \end{pmatrix}, & \mathcal{B}(t) &:= \begin{pmatrix} b(t) & 0 \\ 0 & \overline{b(t)} \end{pmatrix}, \\ \mathcal{C}(t) &:= \begin{pmatrix} c(t) & 0 \\ 0 & \overline{c(t)} \end{pmatrix}, & \mathcal{D}(t) &:= \begin{pmatrix} d(t) & 0 \\ 0 & \overline{d(t)} \end{pmatrix}, \\ \boldsymbol{\alpha}(t) &:= \begin{pmatrix} \alpha(t) & 0 \\ 0 & \alpha^\circ(t) \end{pmatrix}, & \boldsymbol{\beta}(t) &:= \begin{pmatrix} \beta(t) & 0 \\ 0 & \beta^\circ(t) \end{pmatrix}, \\ \boldsymbol{\gamma}(t) &:= \begin{pmatrix} \gamma(t) & 0 \\ 0 & \gamma^\circ(t) \end{pmatrix}, & \boldsymbol{\theta}(t) &:= \begin{pmatrix} \theta(t) & 0 \\ 0 & \theta^\circ(t) \end{pmatrix} \end{aligned} \quad (2.47)$$

and the constant matrix

$$\Lambda := \begin{pmatrix} 0 & 1 \\ 1 & 0 \end{pmatrix}. \quad (2.48)$$

Lemma 2.4.5. *If* $(B_n^{(1)})$ *is the sequence defined by* (2.44), *then for all* $\omega, v \in \Gamma_0$ *the strong limits*

$$W_{(\omega,1)}(B_n^{(1)}) = s - \lim_{n\to\infty} E_n^{(\omega,1)} B_n^{(1)} (E_n^{(\omega,1)})^{-1},$$
$$W_{(v,2)}(B_n^{(1)}) = s - \lim_{n\to\infty} E_n^{(v,2)} B_n^{(1)} (E_n^{(v,2)})^{-1}$$

exist and

$$\begin{aligned} W_{(\omega,1)}(B_n^{(1)}) &= [\mathcal{A}\boldsymbol{\alpha}(\omega+0) + \mathcal{B}\boldsymbol{\beta}(\omega+0) + (\mathcal{C}\boldsymbol{\gamma}(\omega+0) + \mathcal{D}\boldsymbol{\theta}(\omega+0))\Lambda]\, P \\ &\quad + [\mathcal{A}\boldsymbol{\alpha}(\omega-0) + \mathcal{B}\boldsymbol{\beta}(\omega-0) + (\mathcal{C}\boldsymbol{\gamma}(\omega-0) + \mathcal{D}\boldsymbol{\theta}(\omega-0))\Lambda]\, Q, \end{aligned} \tag{2.49}$$

$$\begin{aligned} W_{(v,2)}(B_n^{(1)}) &= P\big[\mathcal{A}(v+0)\widetilde{\boldsymbol{\alpha}} + \mathcal{B}(v+0)\widetilde{\boldsymbol{\beta}} + (\mathcal{C}(v+0)\widetilde{\boldsymbol{\gamma}} + \mathcal{D}(v+0)\widetilde{\boldsymbol{\theta}})\Lambda\big] \\ &\quad + Q\big[\mathcal{A}(v-0)\widetilde{\boldsymbol{\alpha}} + \mathcal{B}(v-0)\widetilde{\boldsymbol{\beta}} + (\mathcal{C}(v-0)\widetilde{\boldsymbol{\gamma}} + \mathcal{D}(v-0)\widetilde{\boldsymbol{\theta}})\Lambda\big], \end{aligned} \tag{2.50}$$

where $\widetilde{\boldsymbol{\alpha}}(t) := \boldsymbol{\alpha}(\overline{t})$, $\widetilde{\boldsymbol{\beta}}(t) := \boldsymbol{\beta}(\overline{t})$, $\widetilde{\boldsymbol{\gamma}}(t) := \boldsymbol{\gamma}(\overline{t})$, $\widetilde{\boldsymbol{\theta}}(t) := \boldsymbol{\theta}(\overline{t})$, $t \in \Gamma_0$.

Proof. We only consider the operators $E_n^{(v,2)} B_n^{(1)} (E_n^{(v,2)})^{-1}$. Using results of [105] (see also [183, Section 10.34]) we obtain that the strong limit of the sequence $\left(E_n^{(v,2)} B_n^{(1)} (E_n^{(v,2)})^{-1}\right)$ is:

$$\begin{aligned} W_{(v,2)}(B_n^{(1)}) &= \begin{pmatrix} (a(v+0)P + a(v-0)Q)\widetilde{\alpha} & (c(v+0)P + c(v-0)Q)\widetilde{\gamma} \\ (\widetilde{c}(v+0)P + \widetilde{c}(v-0)Q)\widetilde{\gamma}^{\circ} & (\widetilde{a}(v+0)P + \widetilde{a}(v-0)Q)\widetilde{\alpha}^{\circ} \end{pmatrix} \\ &\quad + \begin{pmatrix} (b(v+0)P + b(v-0)Q)\widetilde{\beta} & (d(v+0)P + d(v-0)Q)\widetilde{\theta} \\ (\widetilde{d}(v+0)P + \widetilde{d}(v-0)Q)\widetilde{\theta}^{\circ} & (\widetilde{b}(v+0)P + \widetilde{b}(v-0)Q)\widetilde{\beta}^{\circ} \end{pmatrix}. \end{aligned} \tag{2.51}$$

Simple computations show that (2.50) and (2.51) coincide. The sequence $(E_n^{(\omega,1)} B_n^{(1)} (E_n^{(\omega,1)})^{-1})$ is considered analogously. □

Note that due to previously mentioned results of [105, 183], the sequences of the adjoint operators $\left((E_n^{(\omega,1)} B_n^{(1)} (E_n^{(\omega,1)})^{-1} L_n)^*\right)$ and $\left((E_n^{(v,2)} B_n^{(1)} (E_n^{(v,2)})^{-1} L_n)^*\right)$ satisfy the conditions of Theorem 1.6.6. Moreover, if K is a compact operator on the space $L_2(\Gamma_0) \times L_2(\Gamma_0)$, then condition (1.33) is also satisfied, hence the sequence $(B_n^{(1)})$ belongs to the C^*-algebra $\mathcal{B}^{2\times 2}$ generated by the sequences $(E_n^{(\omega,1)})$ and $(E_n^{(v,2)})$.

Let (D_n) be a sequence of the form

$$D_n = \begin{pmatrix} K_n^{\varepsilon} a L_n U_n K_n^0 \alpha L_n U_n^{-1} L_n & K_n^{\varepsilon} b L_n U_n K_n^0 \beta L_n U_n^{-1} L_n \\ K_n^{\varepsilon} c L_n U_n K_n^0 \gamma L_n U_n^{-1} L_n & K_n^{\varepsilon} d L_n U_n K_n^0 \theta L_n U_n^{-1} L_n \end{pmatrix} \tag{2.52}$$

where the functions $a, b, c, d, \alpha, \beta, \gamma, \theta \in \mathbf{PC}(\Gamma_0)$. Define $\mathcal{U}_1^{2\times 2}$ as the smallest closed C^*-algebra which contains all sequences

$$\sum_{i=1}^{l} \prod_{j=1}^{m} \left(D_n^{(ij)}\right)_{n=1}^{\infty},$$

where $l, m \in \mathbb{N}$ and $(D_n^{(ij)})_{n=1}^{\infty}$ are sequences of the form (2.52). Since the closure of the set

$$\mathcal{J}^{2\times 2} := \Bigg\{ \sum_{k=1}^{p} \left((E_n^{(\omega_k,1)})^{-1} L_n T_k L_n E_n^{(\omega_k,1)} \right)_{n=1}^{\infty} + \sum_{j=1}^{s} \left((E_n^{(v_j,2)})^{-1} L_n \widetilde{T}_j L_n E_n^{(v_j,2)}) \right)_{n=1}^{\infty} + (C_n)_{n=1}^{\infty} : \quad p, s \in \mathbb{N},$$
$$\omega_k, v_j \in \Gamma_0,\, T_k, \widetilde{T}_j \in \mathcal{K}^2(L_2(\Gamma_0)), \text{ and } ||C_n|| \to 0 \quad \text{as} \quad n \to \infty \Bigg\}$$

is a closed two-sided ideal in $\mathcal{B}^{2\times 2}$, one can consider the set $\mathcal{U}_1^{2\times 2} + \mathcal{J}^{2\times 2}$. This set is a C^*-subalgebra of the C^*-algebra $\mathcal{B}^{2\times 2}$ [67, 163]. Put $\mathcal{U}^{2\times 2} := \mathcal{U}_1^{2\times 2} + \mathcal{J}^{2\times 2}$ and $\overset{\circ}{\mathcal{U}}{}^{2\times 2} := \mathcal{U}^{2\times 2} / \mathcal{J}^{2\times 2}$.

Corollary 2.4.6. *A sequence $(B_n) \in \mathcal{U}^{2\times 2}$ is stable if and only if for all $\omega, v \in \Gamma_0$ the operators $W_{(\omega,1)}(B_n), W_{(v,2)}(B_n)$ are invertible in $\mathcal{L}(L_2^2(\Gamma_0))$ and the coset $(B_n)^\circ := (B_n) + \mathcal{J}^{2\times 2}$ is invertible in $\overset{\circ}{\mathcal{U}}{}^{2\times 2}$.*

Proof. Notice that $\overset{\circ}{\mathcal{U}}{}^{2\times 2}$ is a C^*-subalgebra of the C^*-algebra $\overset{\circ}{\mathcal{B}}{}^{2\times 2} := \mathcal{B}^{2\times 2} / \mathcal{J}^{2\times 2}$. Therefore the invertibility of an element $a \in \overset{\circ}{\mathcal{U}}{}^{2\times 2}$ in $\overset{\circ}{\mathcal{B}}{}^{2\times 2}$ implies its invertibility in $\overset{\circ}{\mathcal{U}}{}^{2\times 2}$, [67, 163]. □

Thus the stability problem is again reduced to the study of invertibility of some elements in C^*-algebras. Note that the operators $W_{(\omega,1)}(B_n), W_{(v,2)}(B_n)$ belong to algebras of singular integral operators with piecewise continuous coefficients. To check their invertibility one might try to use results available in the literature [91, 92, 144], but this problem is far from being solved. On the other hand, the invertibility of the coset $(B_n)^\circ$ can be studied in more detail, because in this case one can use localization techniques.

Lemma 2.4.7. *Let $a, \alpha \in \mathbf{C}(\Gamma_0)$, and let $(I_n^{(1)})^\circ, (I_n^{(2)})^\circ$ be the cosets of $\overset{\circ}{\mathcal{U}}{}^{2\times 2}$ which respectively contain the diagonal sequences*

$$(I_n^{(1)}) = \begin{pmatrix} (K_n^\varepsilon a L_n) & 0 \\ 0 & (K_n^\varepsilon a L_n) \end{pmatrix},$$
$$(I_n^{(2)}) = \begin{pmatrix} (U_n K_n \alpha L_n U_n^{-1} L_n) & 0 \\ 0 & (U_n K_n \alpha L_n U_n^{-1} L_n) \end{pmatrix}.$$

Then $(I_n^{(1)})^\circ, (I_n^{(2)})^\circ$ belong to the center of the algebra $\overset{\circ}{\mathcal{U}}{}^{2\times 2}$.

The proof of this result follows immediately from results in [175, 183].

Consider now the subalgebra $\mathrm{alg}_{\overset{\circ}{\mathcal{U}}{}^{2\times 2}}\left((I_n^{(1)})^\circ, (I_n^{(2)})^\circ\right)$ of the algebra $\overset{\circ}{\mathcal{U}}{}^{2\times 2}$ generated by the cosets $(I_n^{(1)})^\circ, (I_n^{(2)})^\circ$ described in Lemma 2.4.7 with $a, \alpha \in \mathbf{C}(\Gamma_0)$. This algebra is in the center of $\overset{\circ}{\mathcal{U}}{}^{2\times 2}$. The space of its maximal ideals can be identified with the set $\Gamma_0 \times \Gamma_0$, and for each $(v, \omega) \in \Gamma_0 \times \Gamma_0$, let $\overset{\circ}{\mathcal{J}}{}^{2\times 2}_{v,\omega}$ denote the closed two-sided ideal of $\overset{\circ}{\mathcal{U}}{}^{2\times 2}$ which is generated by the ideal (v, ω). Moreover, we set $\overset{\circ}{\mathcal{U}}{}^{2\times 2}_{v,\omega} := \overset{\circ}{\mathcal{U}}{}^{2\times 2}/\overset{\circ}{\mathcal{J}}{}^{2\times 2}_{v,\omega}$ and $(A_n)^\circ_{v,\omega} := (A_n)^\circ + \overset{\circ}{\mathcal{J}}{}^{2\times 2}_{v,\omega}$.

Let p_v be the characteristic function of the arc $(-v, v) \in \Gamma_0$ and let q_ω be the characteristic function of the arc $(-\omega, \omega) \in \Gamma_0$. Then by p and q we denote the sequences $p = (K_n^\varepsilon p_v L_n)$ and $q = (U_n K_n q_\omega L_n U_n^{-1} L_n)$.

Lemma 2.4.8. *If the coset $(B_n^{(1)})^\circ_{v_0,\omega_0} \in \overset{\circ}{\mathcal{U}}{}^{2\times 2}_{v_0,\omega_0}$ contains the sequence (2.46), then it admits the representation*

$$\begin{aligned}
(B_n^{(1)})^\circ_{v_0,\omega_0} = {} & \left[\mathcal{A}(v_0+0) \otimes p^\circ_{v_0,\omega_0} + \mathcal{A}(v_0-0) \otimes ((L_n) - p)^\circ_{v_0,\omega_0}\right] \\
& \times \left[\boldsymbol{\alpha}(\omega_0+0) \otimes q^\circ_{v_0,\omega_0} + \boldsymbol{\alpha}(\omega_0-0) \otimes ((L_n) - q)^\circ_{v_0,\omega_0}\right] \\
& + \left[\mathcal{B}(v_0+0) \otimes p^\circ_{v_0,\omega_0} + \mathcal{B}(v_0-0) \otimes ((L_n) - p)^\circ_{v_0,\omega_0}\right] \\
& \times \left[\boldsymbol{\beta}(\omega_0+0) \otimes q^\circ_{v_0,\omega_0} + \boldsymbol{\beta}(\omega_0-0) \otimes ((L_n) - q)^\circ_{v_0,\omega_0}\right] \\
& + \left[(\mathcal{C}(v_0+0)\Lambda) \otimes p^\circ_{v_0,\omega_0} + (\mathcal{C}(v_0-0)\Lambda) \otimes ((L_n) - p)^\circ_{v_0,\omega_0}\right] \\
& \times \left[(\Lambda\boldsymbol{\gamma}(\omega_0+0)\Lambda) \otimes q^\circ_{v_0,\omega_0} + (\Lambda\boldsymbol{\gamma}(\omega_0-0)\Lambda) \otimes ((L_n) - q)^\circ_{v_0,\omega_0}\right] \\
& + \left[(\mathcal{D}(v_0+0)\Lambda) \otimes p^\circ_{v_0,\omega_0} + (\mathcal{D}(v_0-0)\Lambda) \otimes ((L_n) - p)^\circ_{v_0,\omega_0}\right] \\
& \times \left[(\Lambda\boldsymbol{\theta}(\omega_0+0)\Lambda) \otimes q^\circ_{v_0,\omega_0} + (\Lambda\boldsymbol{\theta}(\omega_0-0)\Lambda) \otimes ((L_n) - q)^\circ_{v_0,\omega_0}\right],
\end{aligned} \tag{2.53}$$

where $A \otimes B$ denotes the tensor product of matrices A and B.

Proof. Consider first the sequence $(D_n) = (K_n^\varepsilon a(t) L_n U_n K_n \alpha(t) L_n U_n^{-1} L_n)$. From the representation

$$\begin{aligned}
D_n = {} & K_n^\varepsilon[a(v_0+0)p_{v_0}(t) + a(v_0-0)(1 - p_{v_0}(t))]L_n \\
& \times L_n U_n K_n[\alpha(\omega_0+0)q_{v_0}(t) + \alpha(\omega_0-0)(1 - q_{v_0}(t))]L_n U_n^{-1} L_n \\
& + K_n^\varepsilon[a(t) - (a(v_0+0)p_{v_0}(t) + a(v_0-0)(1 - p_{v_0}(t)))]L_n U_n K_n \alpha(t) L_n U_n^{-1} L_n \\
& + K_n^\varepsilon[a(v_0+0)p_{v_0}(t) + a(v_0-0)(1 - p_{v_0}(t))]L_n \\
& \times L_n U_n K_n[\alpha(t) - (\alpha(\omega_0+0)q_{v_0}(t) + \alpha(\omega_0-0)(1 - q_{v_0}(t)))]L_n U_n^{-1} L_n \\
= {} & D_n^{(1)} + D_n^{(2)} + D_n^{(3)},
\end{aligned}$$

one notices that the sequences $(D_n^{(2)})$ and $(D_n^{(3)})$ belong to the ideal $\mathcal{J}_{v_0,\omega_0}$. Therefore

$$(D_n)^\circ_{v_0,\omega_0} = \left[a(v_0+0)\otimes p^\circ_{v_0,\omega_0} + a(v_0-0)\otimes((L_n)-p)^\circ_{v_0,\omega_0}\right] \\ \times\left[\alpha(\omega_0+0)\otimes q^\circ_{v_0,\omega_0} + \alpha(\omega_0-0)\otimes((L_n)-q)^\circ_{v_0,\omega_0}\right].$$

Consider now the sequence (2.46) which consists of two matrix sequences of operators (A_n) and (C_n), i.e., $B_n^{(1)} = A_n + C_n$. Therefore $(B_n^{(1)})^\circ_{v_0,\omega_0} = (A_n)^\circ_{v_0,\omega_0} + (C_n)^\circ_{v_0,\omega_0}$. Here we will only compute $(C_n)^\circ_{v_0,\omega_0}$. For this coset one has

$$\begin{aligned}
(C_n)^\circ_{v_0,\omega_0} &= \begin{pmatrix} (C_n^{11})^\circ_{v_0,\omega_0} & (C_n^{12})^\circ_{v_0,\omega_0} \\ (C_n^{21})^\circ_{v_0,\omega_0} & (C_n^{22})^\circ_{v_0,\omega_0} \end{pmatrix} \\
&= \begin{pmatrix} (C_n^{11})^\circ_{v_0,\omega_0} & 0 \\ 0 & (C_n^{22})^\circ_{v_0,\omega_0} \end{pmatrix} + \begin{pmatrix} 0 & (C_n^{12})^\circ_{v_0,\omega_0} \\ (C_n^{21})^\circ_{v_0,\omega_0} & \end{pmatrix} \\
&= \left[\begin{pmatrix} b(v_0+0) & 0 \\ 0 & \overline{b(v_0+0)} \end{pmatrix}\begin{pmatrix} p^\circ_{v_0,\omega_0} & \\ & p^\circ_{v_0,\omega_0} \end{pmatrix} + \right. \\
&\quad \left. + \begin{pmatrix} b(v_0-0) & 0 \\ 0 & \overline{b(v_0-0)} \end{pmatrix}\begin{pmatrix} ((L_n)-p)^\circ_{v_0,\omega_0} & 0 \\ 0 & ((L_n)-p)^\circ_{v_0,\omega_0} \end{pmatrix}\right] \\
&\quad \times\left[\begin{pmatrix} \beta(\omega_0+0) & 0 \\ 0 & \overline{\beta(\omega_0+0)} \end{pmatrix}\begin{pmatrix} q^\circ_{v_0,\omega_0} & \\ & q^\circ_{v_0,\omega_0} \end{pmatrix} + \right. \\
&\quad \left. + \begin{pmatrix} \beta(\omega_0-0) & 0 \\ 0 & \overline{\beta(\omega_0-0)} \end{pmatrix}\begin{pmatrix} ((L_n)-q)^\circ_{v_0,\omega_0} & 0 \\ 0 & ((L_n)-q)^\circ_{v_0,\omega_0} \end{pmatrix}\right] \\
&\quad + \left[\begin{pmatrix} 0 & d(v_0+0) \\ \overline{d(v_0+0)} & 0 \end{pmatrix}\begin{pmatrix} p^\circ_{v_0,\omega_0} & \\ & p^\circ_{v_0,\omega_0} \end{pmatrix} + \right. \\
&\quad \left. + \begin{pmatrix} 0 & d(v_0-0) \\ \overline{b(v_0-0)} & 0 \end{pmatrix}\begin{pmatrix} ((L_n)-p)^\circ_{v_0,\omega_0} & 0 \\ 0 & ((L_n)-p)^\circ_{v_0,\omega_0} \end{pmatrix}\right] \\
&\quad \times\left[\begin{pmatrix} \theta^\circ(\omega_0+0) & 0 \\ 0 & \theta(\omega_0+0) \end{pmatrix}\begin{pmatrix} q^\circ_{v_0,\omega_0} & \\ & q^\circ_{v_0,\omega_0} \end{pmatrix} + \right. \\
&\quad \left. + \begin{pmatrix} \theta^\circ(\omega_0-0) & 0 \\ 0 & \beta(\omega_0-0) \end{pmatrix}\begin{pmatrix} ((L_n)-q)^\circ_{v_0,\omega_0} & 0 \\ 0 & ((L_n)-q)^\circ_{v_0,\omega_0} \end{pmatrix}\right] \\
&= \left[\mathcal{B}(v_0+0)\otimes p^\circ_{v_0,\omega_0} + \mathcal{B}(v_0-0)\otimes((L_n)-p)^\circ_{v_0,\omega_0}\right] \\
&\quad \times\left[\boldsymbol{\beta}(\omega_0+0)\otimes q^\circ_{v_0,\omega_0} + \boldsymbol{\beta}(\omega_0-0)\otimes((L_n)-q)^\circ_{v_0,\omega_0}\right] \\
&\quad + \left[(\mathcal{D}(v_0+0)\Lambda)\otimes p^\circ_{v_0,\omega_0} + (\mathcal{D}(v_0-0)\Lambda)\otimes((L_n)-p)^\circ_{v_0,\omega_0}\right] \\
&\quad \times\left[(\Lambda\boldsymbol{\theta}\Lambda)(\omega_0+0)\otimes q^\circ_{v_0,\omega_0} + (\Lambda\boldsymbol{\theta}\Lambda)(\omega_0-0)\otimes((L_n)-q)^\circ_{v_0,\omega_0}\right].
\end{aligned} \tag{2.54}$$

Representation (2.54) and the corresponding representation for the coset $(A_n)^\circ_{v_0,\omega_o}$ implies formula (2.53), so the algebra $\overset{\circ}{\mathcal{U}}{}^{2\times 2}_{v_0,\omega_0}$ is generated by the cosets $p^\circ_{v_0,\omega_0}$, $q^\circ_{v_0,\omega_0}$ and $(L_n)^\circ_{v_0,\omega_0}$. □

To proceed, an auxiliary result is needed.

Lemma 2.4.9 ([105]; [183], Lemma 10.38). *With the notation of Lemma* 2.4.8, *the following properties hold:*

1. $p^\circ_{v_0,\omega_0} \cdot p^\circ_{v_0,\omega_0} = p^\circ_{v_0,\omega_0}$;
2. $q^\circ_{v_0,\omega_0} \cdot q^\circ_{v_0,\omega_0} = q^\circ_{v_0,\omega_0}$;
3. $\operatorname{sp}_{\overset{\circ}{\mathcal{U}}_{v_0,\omega_0}}((qpq)^\circ_{v_0,\omega_0}) = [0,1]$.

Let $\mathbf{C}([0,1],\mathcal{L}(\mathbb{C}^2))$ be the algebra of continuous 2×2-matrix-functions on $[0,1]$ and define $\overline{e},\overline{p}$ and $\overline{q}$ by

$$\overline{e} = \begin{pmatrix} 1 & 0 \\ 0 & 1 \end{pmatrix}, \quad \overline{p} = \begin{pmatrix} x & \sqrt{x(1-x)} \\ \sqrt{x(1-x)} & 1-x \end{pmatrix}, \quad \overline{q} = \begin{pmatrix} 1 & 0 \\ 0 & 0 \end{pmatrix}. \tag{2.55}$$

Then from the Halmos two projection theorem [106] and Lemmas 2.4.8 and 2.4.9 we get:

Theorem 2.4.10. *Let $\overline{e},\overline{p},\overline{q}$ be defined by* (2.55). *The C^*-algebras $\overset{\circ}{\mathcal{U}}{}^{2\times 2}_{v_0,\omega_0}$ and $\operatorname{alg}^{2\times 2}(\overline{e},\overline{p},\overline{q})$ are isometrically *-isomorphic, and there is an isomorphism which maps the cosets $p^\circ_{v_0,\omega_0}$ and $q^\circ_{v_0,\omega_0}$ respectively into the elements $\overline{p}$ and $\overline{q}$. Denoting this isomorphism by Ψ°, one obtains*

$$\begin{aligned}
\Psi^\circ((B_n^{(1)})^\circ_{v_0,\omega_0}) = {} & [(\mathcal{A}(v_0+0)-\mathcal{A}(v_0-0))\otimes\overline{p} + \mathcal{A}(v_0-0)\otimes\overline{e}] \\
& \times [(\boldsymbol{\alpha}(\omega_0+0)-\boldsymbol{\alpha}(\omega_0-0)]\otimes\overline{q} + \boldsymbol{\alpha}(\omega_0-0)\otimes\overline{e}] \\
& + [(\mathcal{B}(v_0+0)-\mathcal{B}(v_0-0))\otimes\overline{p} + \mathcal{B}(v_0-0)\otimes\overline{e}] \\
& \times [(\boldsymbol{\beta}(\omega_0+0)-\boldsymbol{\beta}(\omega_0-0))\otimes\overline{q} + \boldsymbol{\beta}(\omega_0-0)\otimes\overline{e}] \\
& + [((\mathcal{C}(v_0+0)-\mathcal{C}(v_0-0))\Lambda)\otimes\overline{p} + (\mathcal{C}(v_0-0)\Lambda)\otimes\overline{e}] \\
& \times [(\Lambda(\boldsymbol{\gamma}(\omega_0+0)-\boldsymbol{\gamma}(\omega_0-0))\Lambda)\otimes\overline{q} + (\Lambda\boldsymbol{\gamma}(\omega_0-0)\Lambda)\otimes\overline{e}] \\
& + [((\mathcal{D}(v_0+0)-\mathcal{D}(v_0-0))\Lambda)\otimes\overline{p} + (\mathcal{D}(v_0-0)\Lambda)\otimes\overline{e}] \\
& \times [(\Lambda(\boldsymbol{\theta}(\omega_0+0)-\boldsymbol{\theta}(\omega_0-0))\Lambda)\otimes\overline{q} + (\Lambda\boldsymbol{\theta}(\omega_0-0)\Lambda)\otimes\overline{e}] .
\end{aligned} \tag{2.56}$$

Definition 2.4.11. *For a given sequence (B_n) of the form* (2.36), *the matrix $\Psi^\circ((B_n^{(1)})^\circ_{v_0,\omega_0})$ is called the local symbol of (B_n) at the point $(v_0,\omega_0)\in\Gamma_0\times\Gamma_0$.*

Combining now Theorem 2.4.10 with Allan's local principle (cf. Theorem 1.9.5), we can formulate the following result.

Theorem 2.4.12. *The coset $(B_n^{(1)})^\circ \in \overset{\circ}{\mathcal{U}}{}^{2\times 2}$ is invertible in $\overset{\circ}{\mathcal{U}}{}^{2\times 2}$ if and only if the local symbol $\Psi^\circ((B_n)^\circ_{v,\omega}) \in \operatorname{alg}^{2\times 2}(\overline{e},\overline{p},\overline{q})$ of the sequence* (2.36) *is invertible for each $(v,\omega)\in\Gamma_0\times\Gamma_0$.*

Lemma 2.4.13. *Let (B_n) be a matrix sequence of the form (2.36). If $W_{(\omega,1)}(B_n^{(1)})$ is a Fredholm operator for all $\omega \in \Gamma_0$, then for each $(v,w) \in \Gamma_0 \times \Gamma_0$, the local symbol $\Psi^\circ((B_n^{(1)})^\circ_{v,\omega})$ of (B_n) is invertible.*

Proof. The operator $W_{(\omega,1)}(B_n^{(1)})$ defined in (2.49) can be written in the form

$$W_{(\omega,1)}(B_n^{(1)}) = A_\omega P + B_\omega Q = \sum_{i=1}^{4}(A_\omega^i P + B_\omega^i Q),$$

where

$$\begin{aligned} A_\omega^1 P + B_\omega^1 Q &= \mathcal{A} \cdot (\boldsymbol{\alpha}(\omega+0)P + \boldsymbol{\alpha}(\omega-0)Q), \\ A_\omega^2 P + B_\omega^2 Q &= \mathcal{B} \cdot (\boldsymbol{\beta}(\omega+0)P + \boldsymbol{\beta}(\omega-0)Q), \\ A_\omega^3 P + B_\omega^3 Q &= \mathcal{C} \cdot (\boldsymbol{\gamma}(\omega+0)P + \boldsymbol{\gamma}(\omega-0)Q), \\ A_\omega^4 P + B_\omega^4 Q &= \mathcal{D} \cdot (\boldsymbol{\theta}(\omega+0)P + \boldsymbol{\theta}(\omega-0)Q). \end{aligned} \tag{2.57}$$

Let $a \in \mathbf{PC}(\Gamma_0)$. Consider the singular integral operator

$$R = a\alpha(\omega+0)P + a\alpha(\omega-0)Q,$$

and compute its Gohberg-Krupnik symbol $\sigma(R)$. According to [90, 93], it is

$$\sigma(R) = \begin{pmatrix} a_{11}(v,\omega,x) & a_{12}(v,\omega,x) \\ a_{21}(v,\omega,x) & a_{22}(v,\omega,x) \end{pmatrix}$$

where

$$\begin{aligned} a_{11}(v,\omega,x) &= xa(v+0)\alpha(\omega+0) + (1-x)a(v-0)\alpha(\omega+0), \\ a_{12}(v,\omega,x) &= \sqrt{x(1-x)}[a(v+0) - a(v-0)\alpha(\omega-0)], \\ a_{21}(v,\omega,x) &= \sqrt{x(1-x)}[a(v+0) - a(v-0)\alpha(\omega+0)], \\ a_{22}(v,\omega,x) &= xa(v-0)\alpha(\omega-0) + (1-x)a(v+0)\alpha(\omega-0). \end{aligned}$$

Computing analogously the Gohberg-Krupnik symbol $\sigma(\widehat{R})$ for the operator

$$\widehat{R} = \bar{a}\alpha^\circ(\omega+0)P + \bar{a}\alpha^\circ(\omega-0)Q,$$

we obtain the symbol of the first operator in (2.57), viz.,

$$\sigma(A_\omega^1 P + B_\omega^1 Q) = \begin{pmatrix} \sigma(R) & 0_{2\times2} \\ 0_{2\times2} & \sigma(\widehat{R}) \end{pmatrix}, \tag{2.58}$$

where $0_{2\times2}$ is the 2×2 zero matrix. It remains to note that matrix (2.58) coincides with the matrix from the first two rows of (2.56). Thus the local symbol $\Psi^\circ_{v,\omega}((B_n^{(1)}))$ of the sequence (B_n) coincides with the Gohberg-Krupnik symbol of the operator $W_{\omega,1}(B_n^{(1)})$ everywhere on $\Gamma_0 \times \Gamma_0$. However by [90, 93] this operator is Fredholm if and only if its symbol is invertible. This completes the proof. □

Now we are able to formulate the main result of this section. Let $\mathcal{A} = \mathcal{A}(t), \mathcal{B} = \mathcal{B}(t), \mathcal{C} = \mathcal{C}(t), \mathcal{D} = \mathcal{D}(t), \boldsymbol{\alpha} = \boldsymbol{\alpha}(t), \boldsymbol{\beta} = \boldsymbol{\beta}(t), \boldsymbol{\gamma} = \boldsymbol{\gamma}(t), \boldsymbol{\theta} = \boldsymbol{\theta}(t), t \in \Gamma_0$ be the matrix-functions introduced in (2.47). With each sequence of the form (2.36) and with each point $\tau \in \Gamma_0$ we associate two singular integral operators B_τ and B^τ defined on the space $L_2^2(\Gamma_0)$ by

$$\begin{aligned} B_\tau &:= [\mathcal{A}\boldsymbol{\alpha}(\tau+0) + \mathcal{B}\boldsymbol{\beta}(\tau+0) + (\mathcal{C}\boldsymbol{\gamma}(\tau+0) + \mathcal{D}\boldsymbol{\theta}(\tau+0))\Lambda]\, P \\ &\quad + [\mathcal{A}\boldsymbol{\alpha}(\tau-0) + \mathcal{B}\boldsymbol{\beta}(\tau-0) + (\mathcal{C}\boldsymbol{\gamma}(\tau-0) + \mathcal{D}\boldsymbol{\theta}(\tau-0))\Lambda]\, Q, \qquad (2.59) \\ B^\tau &:= P\big[\mathcal{A}(\tau+0)\widetilde{\boldsymbol{\alpha}} + \mathcal{B}(\tau+0)\widetilde{\boldsymbol{\beta}} + (\mathcal{C}(\tau+0)\widetilde{\boldsymbol{\gamma}} + \mathcal{D}(\tau+0)\widetilde{\boldsymbol{\theta}})\Lambda\big] \\ &\quad + Q\big[\mathcal{A}(\tau-0)\widetilde{\boldsymbol{\alpha}} + \mathcal{B}(\tau-0)\widetilde{\boldsymbol{\beta}} + (\mathcal{C}(\tau-0)\widetilde{\boldsymbol{\gamma}} + \mathcal{D}(\tau-0)\widetilde{\boldsymbol{\theta}})\Lambda\big]. \qquad (2.60) \end{aligned}$$

Theorem 2.4.14. *Let $a, b, c, d, \alpha, \beta, \gamma, \theta \in \mathbf{PC}(\Gamma_0)$. Then the following assertions hold:*

1. *The sequence* (2.36) *is stable if and only if for each $\tau \in \Gamma_0$ the operators B_τ and B^τ are invertible on the space $L_2^2(\Gamma_0)$.*

2. *Let $a, b, c, d \in \mathbf{PC}(\Gamma_0)$ and let $\alpha, \beta, \gamma, \theta \in \mathbf{C}(\Gamma_0 \setminus \{1\})$ (or $\alpha, \beta, \gamma, \theta \in \mathbf{C}(\Gamma_0 \setminus \{-1, 1\})$). Then the sequence* (2.36) *is stable if and only if the operator B_1 (respectively, the operators B_{-1}, B_1) and all operators B^τ, $\tau \in \Gamma_0$ are invertible on the space $L_2^2(\Gamma_0)$.*

3. *Let $a, b, c, d \in \mathbf{C}(\Gamma_0)$ and let $\alpha, \beta, \gamma, \theta \in \mathbf{C}(\Gamma_0 \setminus \{1\})$ (or $\alpha, \beta, \gamma, \theta \in \mathbf{C}(\Gamma_0 \setminus \{-1, 1\})$). Then the sequence* (2.36) *is stable if and only if the operator B_1 (respectively, the operators B_{-1}, B_1) is (are) invertible on the space $L_2^2(\Gamma_0)$ and*

$$\det\left[\mathcal{A}(t)\boldsymbol{\alpha}(\tau) + \mathcal{B}(t)\boldsymbol{\beta}(\tau) + \mathcal{C}(t)\boldsymbol{\gamma}(\tau)\Lambda + \mathcal{D}(t)\boldsymbol{\theta}(\tau)\Lambda\right] \neq 0$$

   for all $t, \tau \in \Gamma_0$.

Proof. It follows from Theorem 2.4.12, Corollary 2.4.6 and Lemma 2.4.13. □

This result allows us to introduce the following definition.

Definition 2.4.15. *Let (B_n) be a sequence of the form* (2.36), *and let B_τ, B^τ be the operators defined in* (2.59), (2.60). *An operator function $smb_{B_n} : \Gamma_0 \mapsto \mathcal{L}^2(L_2^2(\Gamma_0))$ defined by*

$$\operatorname{smb}_{B_n}(\tau) = \operatorname{diag}\,(B_\tau, B^\tau)$$

is called the symbol of the sequence (2.36).

Thus Theorem 2.4.14 shows that the sequence (2.36) is stable if and only if its symbol is invertible.

In the subsequent sections we will study a variety of sequences of the form (2.36) originating from various approximation methods for singular integral equations with conjugation; and we also will find the functions $\alpha, \beta, \gamma, \theta$ responsible for the stability of the methods under consideration.

2.4.1 Moore-Penrose Invertibility

In conclusion let us present a result concerning Moore-Penrose invertibility in the real algebra $\tilde{\mathcal{C}}$ generated by sequences of paired circulants and by the sequence $(\overline{M}_n)$.

Theorem 2.4.16. *Let $a, b, c, d, \alpha, \beta, \gamma, \theta \in \mathbf{PC}(\Gamma_0)$. Then the following assertions are equivalent:*

1. *The sequence*

$$\left(\widetilde{a}_n\widehat{\alpha}_n + \widetilde{b}_n\widehat{\beta}_n + (\widetilde{c}_n\widehat{\gamma}_n + \widetilde{d}_n\widehat{\theta}_n)\overline{M}_n + C_n\right) \tag{2.61}$$

is weakly asymptotically More-Penrose invertible;

2. *For any $\tau \in \Gamma_0$, the operators $B^\tau, B_\tau \in \mathcal{L}(L_2^2(\Gamma_0))$ defined by (2.59), (2.60) are normally solvable and the norms of their Moore-Penrose inverses are uniformly bounded.*

3. *The sequence (2.61) can be represented in the form (1.39).*

The proof of this theorem follows from Theorems 1.7.2, 2.4.14, and 1.7.10. Note that if the operators B_τ and B^τ are Fredholm, then the norms of their Moore-Penrose inverses are uniformly bounded [193].

Remark 2.4.17. In order to regularize the sequence $(\tilde{A}_n)$ of (2.61) practically, we have to consider the matrix sequence $(\Psi(\tilde{A}_n))$,

$$\Psi(\tilde{A}_n) = \begin{pmatrix} \widetilde{a}_n\widehat{\alpha}_n + \widetilde{b}_n\widehat{\beta}_n & \widetilde{c}_n\widehat{\gamma}_n + \widetilde{d}_n\widehat{\theta}_n \\ \overline{M}_n(\widetilde{c}_n\widehat{\gamma}_n + \widetilde{d}_n\widehat{\theta}_n)\overline{M}_n & \overline{M}_n(\widetilde{a}_n\widehat{\alpha}_n + \widetilde{b}_n\widehat{\beta}_n)\overline{M}_n \end{pmatrix}.$$

These matrices can be regularized (in the sense of (26)) using their singular value decompositions (see [194]). We can then apply the mapping Ψ^{-1} to regularized matrices in order to get a corresponding regularization for the initial sequence (2.61).

2.5 The ε-Collocation Method

In this and the next sections of this chapter, Theorem 2.4.14 is applied to investigate the stability of spline approximation methods for singular integral equations with conjugation. Let us start with the ε-collocation method.

For a positive integer n, consider the uniform partition of $\mathbb{R}$ by the mesh points $x_j := j/n$, $j \in \mathbb{Z}$. By $\underline{S}_n^\delta$ we denote the space of all 1-periodic smoothest splines of degree δ subordinate to the above partition, i.e., each function f of $\underline{S}_n^\delta$ and all its derivatives of order $\delta - 1$ are continuous and 1-periodic on $\mathbb{R}$ and the restriction of f to any interval $(j/n, (j+1)/n)$, $j \in \mathbb{Z}$ is a polynomial of degree at most δ. By $\underline{S}_n^0$ we denote the set of all 1-periodic step functions corresponding to the above partition of $\mathbb{R}$ into subintervals. We consider spline space $\widetilde{\underline{S}}_n^\delta$ which depends on ε and is defined by [183]

$$\widetilde{\underline{S}}_n^\delta := \begin{cases} \underline{S}_n^{\delta+1} & \text{if} \quad \varepsilon = 0 \quad \text{and} \quad \delta \quad \text{is even,} \\ \underline{S}_n^{\delta+1} & \text{if} \quad \varepsilon = 0.5 \quad \text{and} \quad \delta \quad \text{is odd,} \\ \underline{S}_n^{\delta} & \text{otherwise.} \end{cases}$$

Let the circle Γ_0 be given by the parametric representation

$$\Gamma_0 := \{e^{i2\pi s} : 0 \le s \le 1\}.$$

Via this parametrization we have one-to-one correspondence between functions φ on Γ_0 and the 1-periodic functions on $\mathbb{R}$, namely $\varphi : \Gamma \mapsto \mathbb{C}$ corresponds to $\varphi_0 : \mathbb{R} \mapsto \mathbb{C}$, $\varphi_0(s) = \varphi(e^{i2\pi s})$. Throughout this chapter, we shall identify the functions φ and φ_0. In particular, the spline spaces $\widetilde{\underline{S}}_n^\delta$ are assumed to be defined on Γ_0.

Consider now the following integral equation with conjugation

$$A\varphi \equiv a\varphi + bS\varphi + cM\varphi + dSM\varphi + k^0\varphi + k^1M\varphi = f \tag{2.62}$$

where

$$(k^0\varphi)(t) = \int_{\Gamma_0} k_0(t,\tau)\varphi(\tau)\,d\tau, \quad (k^1\varphi)(t) = \int_{\Gamma_0} k_1(t,\tau)\varphi(\tau)\,d\tau,$$

and the operators S and M are defined in Section 2.1.

Let $\mathbf{R} = \mathbf{R}(\Gamma_0)$ denote the set of all Riemann integrable functions on Γ_0. For a fixed $\varepsilon \in [0,1)$, the interpolation projection $K_n^\varepsilon : \mathbf{R} \mapsto \underline{S}_n^\delta$ is defined by (see [2, 3, 183])

$$(K_n^\varepsilon x - x)((j+\varepsilon)/n) = 0, \quad j = 0, 1, \ldots, n-1.$$

To simplify notation we will often write K_n instead of K_n^ε. The collocation equations

$$(Au_n)((j+\varepsilon)/n) = f((j+\varepsilon)/n), \quad u_n \in \underline{S}_n^\delta, \quad j = 0, 1, \ldots, n-1 \tag{2.63}$$

are equivalent to the operator equation

$$A_n u_n = K_n f, \quad A_n = K_n A\Big|_{\underline{S}_n^\delta} \in \mathcal{L}(\underline{S}_n^\delta, \widetilde{\underline{S}}_n^\delta).$$

We shall analyze the structure of the collocation operator A_n,

$$\begin{aligned} A_n = {} & (K_n aI\Big|_{\widetilde{\underline{S}}_n^\delta})(K_n\Big|_{\underline{S}_n^\delta}) + (K_n b\Big|_{\widetilde{\underline{S}}_n^\delta})(K_n S\Big|_{\underline{S}_n^\delta} \\ & + (K_n cI\Big|_{\widetilde{\underline{S}}_n^\delta})(K_n M\Big|_{\underline{S}_n^\delta}) + (K_n d\Big|_{\widetilde{\underline{S}}_n^\delta})(K_n SM\Big|_{\underline{S}_n^\delta} + K_n k^0\Big|_{\underline{S}_n^\delta} + K_n k^1\Big|_{\underline{S}_n^\delta}. \end{aligned} \tag{2.64}$$

Here we used the well-known property of the interpolation operator

$$K_n ab = K_n a K_n b, \quad a, b \in \mathbf{R}(\Gamma_0).$$

Let $\varepsilon \in [0,1)$. For $j = 0, 1, \ldots n-1$ consider the points

$$t_j = \exp(i2\pi j/n), \quad v_j = \exp(i2\pi(j+1/2)/n), \quad \tau_j = \exp(i2\pi(j+\varepsilon)/n),$$

and let $\{\chi_k^{(n)}\}_{k=0}^{n-1}$ and $\{\Psi_k^{(n)}\}_{k=0}^{n-1}$ be the respective interpolation bases from $\underline{S}_n^\delta$ and $\underline{\widetilde{S}}_n^\delta$, satisfying the conditions

$$\begin{aligned} \chi_k^{(n)}(t_j) &= \delta_{jk} \quad \text{if} \quad \delta \quad \text{is even;} \\ \chi_k^{(n)}(v_j) &= \delta_{jk} \quad \text{if} \quad \delta \quad \text{is odd;} \\ \Psi_k^{(n)}(\tau_j) &= \delta_{jk} \quad \text{for} \quad j, k = 0, 1, \ldots, n-1. \end{aligned}$$

Let us first assume that δ is odd. In accordance with [178, 183], the matrix of the first two operators in (2.64) in the bases $\{\chi_k^{(n)}\}_{k=0}^{n-1}$, $\{\Psi_k^{(n)}\}_{k=0}^{n-1}$ can be represented in the form $\widetilde{a}_n \widehat{\alpha}_n + \widetilde{b}_n \widehat{\beta}_n$, where

$$\begin{aligned} \widetilde{a}_n &= (a(\tau_k)\delta_{jk})_{k,j=0}^{n-1}, & \widetilde{b}_n &= (b(\tau_k)\delta_{jk})_{k,j=0}^{n-1}, \\ \widehat{\alpha}_n &= U_n \alpha_n U_n^{-1}, & \widehat{\beta}_n &= U_n \beta_n U_n^{-1}, \\ \alpha_n &= (\alpha_{\varepsilon,\delta}(t_k)\delta_{jk})_{j,k=0}^{n-1}, & \beta_n &= (\beta_{\varepsilon,\delta}(t_k)\delta_{jk})_{k,j=0}^{n-1}, \end{aligned} \tag{2.65}$$

and the functions $\alpha_{\varepsilon,\delta}, \beta_{\varepsilon,\delta} \in \mathbf{PC}(\Gamma_0)$ are defined by

$$\alpha_{\varepsilon,\delta}(e^{i2\pi s}) = \begin{cases} 1 & \text{if} \quad s = 0, \\ \dfrac{\Phi_{\varepsilon,\delta}(s)\sigma_{\varepsilon,\delta}^+(s)}{\Phi_{0,\delta}(s)\sigma_{0,\delta}^+(s)} & \text{if} \quad 0 < s < 1, \end{cases} \tag{2.66}$$

$$\beta_{\varepsilon,\delta}(e^{i2\pi s}) = \begin{cases} 1 & \text{if} \quad s = 0, \\ \dfrac{\Phi_{\varepsilon,\delta}(s)\sigma_{\varepsilon,\delta}^-(s)}{\Phi_{0,\delta}(s)\sigma_{0,\delta}^+(s)} & \text{if} \quad 0 < s < 1, \end{cases} \tag{2.67}$$

$$\Phi_{\varepsilon,\delta}(s) := 2^{-\delta}\pi^{-\delta-1}\sin(\pi s)e^{i\pi s(2\varepsilon-1)}, \tag{2.68}$$

$$\sigma_{\varepsilon,\delta}^{\pm}(s) := \sum_{k=0}^{+\infty}(k+s)^{-\delta-1}e^{i2\pi k\varepsilon} \pm \sum_{k=1}^{+\infty}(k-s)^{-\delta-1}e^{i2\pi k\varepsilon}. \tag{2.69}$$

On the other hand, a description for the second pair of terms in (2.64) can be obtained from Lemma 2.4.2. More precisely, one can show that the matrix sequence $(\widetilde{c}_n\widehat{\gamma}_n + \widetilde{d}_n\widehat{\theta}_n)\overline{M}_n$ corresponds to the third and fourth terms in (2.64). The matrices

$\widetilde{c}_n, \widetilde{d}_n$ are defined analogously to $\widetilde{a}_n, \widetilde{b}_n$, and $\gamma_{\varepsilon,\delta}, \theta_{\varepsilon,\delta}$ are the following functions:

$$\gamma_{\varepsilon,\delta}(e^{i2\pi s}) = \begin{cases} 1 & \text{if} \quad s=0, \\ \dfrac{\overline{\Phi_{\varepsilon,\delta}(1-s)\sigma^{+}_{\varepsilon,\delta}(1-s)}}{\overline{\Phi_{0,\delta}(1-s)\sigma^{+}_{0,\delta}(1-s)}} & \text{if} \quad 0<s<1, \end{cases}$$

$$\theta_{\varepsilon,\delta}(e^{i2\pi s}) = \begin{cases} 1 & \text{if} \quad s=0, \\ \dfrac{\overline{\Phi_{\varepsilon,\delta}(1-s)\sigma^{-}_{\varepsilon,\delta}(1-s)}}{\Phi_{0,\delta}(1-s)\sigma^{+}_{0,\delta}(1-s)} & \text{if} \quad 0<s<1. \end{cases}$$

Hence the sequence (A_n), where

$$A_n := \widetilde{a}_n\widehat{\alpha}_n + \widetilde{b}_n\widehat{\beta}_n + (\widetilde{c}_n\widehat{\gamma}_n + \widetilde{d}_n\widehat{\theta}_n)\overline{M}_n + K_n k^0\Big|_{\underline{S}^\delta_n} + K_n k^1 M\Big|_{\underline{S}^\delta_n}, \quad n\in\mathbb{N}. \tag{2.70}$$

corresponds to the sequence (2.64).

If δ is even, then similar to the previous considerations one arrives at a sequence of the form (2.70) with the generating functions $\alpha_{\varepsilon,\delta}, \beta_{\varepsilon,\delta}, \gamma_{\varepsilon,\delta}, \theta_{\varepsilon,\delta}$ defined by

$$\alpha_{\varepsilon,\delta}(e^{i2\pi s}) = \begin{cases} 1 & \text{if} \quad s=0, \\ \dfrac{\Phi_{\varepsilon,\delta}(s)\sigma^{-}_{\varepsilon,\delta}(s)}{\Phi_{1/2,\delta}(s)\sigma^{-}_{1/2,\delta}(s)} & \text{if} \quad 0<s<1, \end{cases} \tag{2.71}$$

$$\beta_{\varepsilon,\delta}(e^{i2\pi s}) = \begin{cases} 1 & \text{if} \quad s=0, \\ \dfrac{\Phi_{\varepsilon,\delta}(s)\sigma^{+}_{\varepsilon,\delta}(s)}{\Phi_{1/2,\delta}(s)\sigma^{-}_{1/2,\delta}(s)} & \text{if} \quad 0<s<1, \end{cases} \tag{2.72}$$

$$\gamma_{\varepsilon,\delta}(e^{i2\pi s}) = \begin{cases} 1 & \text{if} \quad s=0, \\ \dfrac{\overline{\Phi_{\varepsilon,\delta}(1-s)\sigma^{-}_{\varepsilon,\delta}(1-s)}}{\overline{\Phi_{1/2,\delta}(1-s)\sigma^{-}_{1/2,\delta}(1-s)}} & \text{if} \quad 0<s<1, \end{cases}$$

$$\theta_{\varepsilon,\delta}(e^{i2\pi s}) = \begin{cases} 1 & \text{if} \quad s=0, \\ \dfrac{\overline{\Phi_{\varepsilon,\delta}(1-s)\sigma^{+}_{\varepsilon,\delta}(1-s)}}{\overline{\Phi_{0,\delta}(1-s)\sigma^{-}_{0,\delta}(1-s)}} & \text{if} \quad 0<s<1, \end{cases}$$

and with $\Phi_{\varepsilon,\delta}(s), \sigma^{\pm}_{\varepsilon,\delta}(s)$ defined in (2.68), (2.69).

Notice that in all cases the functions $\gamma_{\varepsilon,\delta} = \alpha^{\circ}_{\varepsilon,\delta} \in \mathbf{C}^\infty(\Gamma_0)$ and $\theta_{\varepsilon,\delta} = \beta^{\circ}_{\varepsilon,\delta} \in \mathbf{PC}^\infty(\Gamma_0 \setminus \{1\})$. Moreover, it is easily seen that the matrices $\boldsymbol{\alpha}_{\varepsilon,\delta}, \boldsymbol{\beta}_{\varepsilon,\delta}, \boldsymbol{\gamma}_{\varepsilon,\delta}, \boldsymbol{\theta}_{\varepsilon,\delta}$ are connected by the relations

$$\boldsymbol{\alpha}_{\varepsilon,\delta} = \Lambda\boldsymbol{\gamma}_{\varepsilon,\delta}\Lambda, \quad \boldsymbol{\beta}_{\varepsilon,\delta} = \Lambda\boldsymbol{\theta}_{\varepsilon,\delta}\Lambda.$$

Let us find the limit values $\beta_{\varepsilon,\delta}(1\pm 0)$, $\theta_{\varepsilon,\delta}(1\pm 0)$, $\beta^{\circ}_{\varepsilon,\delta}(1\pm 0)$ and $\theta^{\circ}_{\varepsilon,\delta}(1\pm 0)$. Assume for definiteness that δ is odd and consider the one-sided limits for the function $\beta_{\varepsilon,\delta}$ as $t\in\Gamma_0$ tend to 1. Considering the right limit $\beta_{\varepsilon,\delta}(1+0)$ one obtains

$$\beta_{\varepsilon,\delta}(1+0)=\lim_{s\to+0}\frac{\Phi_{\varepsilon,\delta}(s)\sigma^{-}_{\varepsilon,\delta}(s)}{\Phi_{0,\delta}(s)\sigma^{+}_{0,\delta}(s)}.$$

Since $\dfrac{\Phi_{\varepsilon,\delta}(s)}{\Phi_{0,\delta}(s)}=\exp(i2\pi\varepsilon s)$, it suffices to find the limit $\lim\limits_{s\to+0}\dfrac{\sigma^{-}_{\varepsilon,\delta}(s)}{\sigma^{+}_{0,\delta}(s)}$. The last fraction can be transformed as follows:

$$\begin{aligned}\frac{\sigma^{-}_{\varepsilon,\delta}(s)}{\sigma^{+}_{0,\delta}(s)}&=\frac{\sum\limits_{k=0}^{+\infty}(k+s)^{-\delta-1}e^{i2\pi k\varepsilon}-\sum\limits_{k=1}^{+\infty}(k-s)^{-\delta-1}e^{i2\pi k\varepsilon}}{\sum\limits_{k=0}^{+\infty}(k+s)^{-\delta-1}+\sum\limits_{k=1}^{+\infty}(k-s)^{-\delta-1}}\\
&=\frac{s^{-\delta-1}+\sum\limits_{k=1}^{+\infty}\left[(k+s)^{-\delta-1}e^{i2\pi k\varepsilon}-(k-s)^{-\delta-1}e^{i2\pi k\varepsilon}\right]}{s^{-\delta-1}+\sum\limits_{k=1}^{+\infty}\left[(k+s)^{-\delta-1}+(k-s)^{-\delta-1}\right]}\\
&=\frac{1+s^{\delta+1}\sum\limits_{k=1}^{+\infty}\left[(k+s)^{-\delta-1}e^{i2\pi k\varepsilon}-(k-s)^{-\delta-1}e^{i2\pi k\varepsilon}\right]}{1+s^{\delta+1}\sum\limits_{k=1}^{+\infty}\left[(k+s)^{-\delta-1}+(k-s)^{-\delta-1}\right]}.\end{aligned}$$

Therefore

$$\lim_{s\to+0}\frac{\sigma^{-}_{\varepsilon,\delta}(s)}{\sigma^{+}_{0,\delta}(s)}=1,$$

consequently, $\beta_{\varepsilon,\delta}(1+0)=1$. To find $\beta_{\varepsilon,\delta}(1-0)$, it suffices to consider the limit of the function $\sigma^{-}_{\varepsilon,\delta}(1-s)/\sigma^{+}_{0,\delta}(1-s)$ as s tends to $+0$. Analogous considerations show that

$$\lim_{s\to+0}\frac{\sigma^{-}_{\varepsilon,\delta}(1-s)}{\sigma^{+}_{0,\delta}(1-s)}=-1.$$

Thus

$$\begin{aligned}\beta_{\varepsilon,\delta}(1+0)&=1, &\beta_{\varepsilon,\delta}(1-0)&=-1,\\ \theta_{\varepsilon,\delta}(1-0)&=1, &\theta_{\varepsilon,\delta}(1+0)&=-1.\end{aligned}\tag{2.73}$$

Combining this with Theorem 2.4.14 one gets the following result.

Theorem 2.5.1. *If $a, b, c, d \in \mathbf{PC}$, $k^0, k^1 \in \mathcal{K}(L_2(\Gamma_0), \mathbf{R}(\Gamma_0))$, $\varepsilon \in [0, 1)$, and the operator $B\varphi \equiv a\varphi + bS\varphi + cM\varphi + dSM\varphi$ is invertible in $\mathcal{L}(L_2(\Gamma_0))$, then the ε-collocation method (2.63) is stable if and only if the operator A is invertible in $\mathcal{L}(L_2(\Gamma_0))$, the operator*

$$\widetilde{B} := \begin{pmatrix} a+b & c-d \\ \overline{c}+\overline{d} & \overline{a}-\overline{b} \end{pmatrix} P + \begin{pmatrix} a-b & c+d \\ \overline{c}-\overline{d} & \overline{a}+\overline{b} \end{pmatrix} Q, \tag{2.74}$$

and the operators

$$\begin{aligned} B^\tau &:= P\Big[(\mathcal{A}(\tau+0) + \mathcal{C}(\tau+0)\Lambda)\widetilde{\boldsymbol{\alpha}}_{\boldsymbol{\varepsilon},\boldsymbol{\delta}} + (\mathcal{B}(\tau+0) + \mathcal{D}(\tau+0)\Lambda)\widetilde{\boldsymbol{\beta}}_{\boldsymbol{\varepsilon},\boldsymbol{\delta}})\Big] \\ &\quad + Q\big[(\mathcal{A}(\tau-0) + \mathcal{C}(\tau-0)\Lambda)\widetilde{\boldsymbol{\alpha}}_{\boldsymbol{\varepsilon},\boldsymbol{\delta}} + (\mathcal{B}(\tau-0) + \mathcal{D}(\tau-0)\Lambda)\widetilde{\boldsymbol{\beta}}_{\boldsymbol{\varepsilon},\boldsymbol{\delta}})\big] \end{aligned}$$

are invertible in $\mathcal{L}(L_2(\Gamma_0))$ for all $\tau \in \Gamma_0$.

Proof. The proof follows immediately from Theorem 2.4.14, perturbation arguments for stable sequences and from properties of the functions $\alpha_{\varepsilon,\delta}$ and $\beta_{\varepsilon,\delta}$. □

Remark 2.5.2. If $c \equiv d \equiv 0$, $k^1 = 0$, then the conditions of stability of the ε-collocation method are equivalent to the conditions obtained earlier for singular integral equations without conjugation [183].

2.6 Spline Galerkin Method

Let $\underline{S}_n^\delta$ be the spline space defined in the previous section. By $L_n, n \in \mathbb{N}$ we denote the orthoprojections from $L_2(\Gamma_0)$ onto $\underline{S}_n^\delta$ and such that

$$(x - L_n x, \varphi) = 0$$

for each $x \in L_2(\Gamma_0)$ and for each $\varphi \in \underline{S}_n^\delta$. Furthermore, let us consider the operators $A_n = L_n A\Big|_{\underline{S}_n^\delta}$, where A is defined by (2.62). Suppose first that the coefficients a, b, c, d are continuous on Γ_0. Since $||L_n f(I - L_n)|| \to 0$ as $n \to \infty$ for each $f \in \mathbf{C}(\Gamma_0)$, one gets the following representation for the operators A_n:

$$\begin{aligned} A_n &= (L_n aI\Big|_{\underline{S}_n^\delta})(L_n\Big|_{\underline{S}_n^\delta}) + (L_n b\Big|_{\underline{S}_n^\delta})(L_n S\Big|_{\underline{S}_n^\delta} + (L_n cI\Big|_{\underline{S}_n^\delta})(L_n M\Big|_{\underline{S}_n^\delta}) \\ &\quad + (L_n d\Big|_{\underline{S}_n^\delta})(L_n SM\Big|_{\underline{S}_n^\delta} + L_n k^0\Big|_{\underline{S}_n^\delta} L_n k^1\Big|_{\underline{S}_n^\delta} + R_n^{(1)}, \end{aligned} \tag{2.75}$$

where $R_n^{(1)} : \underline{S}_n^\delta \to \underline{S}_n^\delta$ are additive operators with $||R_n^{(1)}|| \to 0$ as $n \to \infty$. Define the basis $\{\varphi_k\}_{k=0}^{n-1}$ for $\underline{S}_n^\delta$ according to [39, 178, 183] and examine the matrices corresponding to the operators A_n in this basis. As shown in [178], the matrices $((\varphi_j, \varphi_k))_{j,k=0}^{n-1}$ and $((S\varphi_j, \varphi_k))_{j,k=0}^{n-1}$ are circulants and

$$((\varphi_j, \varphi_k))_{j,k=0}^{n-1} = n^{-1}\widehat{\zeta}_n, \quad ((S\varphi_j, \varphi_k))_{j,k=0}^{n-1} = n^{-1}\widehat{\rho}_n$$

with

$$\zeta_n(e^{i2\pi s}) = 2^{-\delta}\pi^{-3\delta-2}\sin^{2(\delta+1)}(\pi s)\sigma^+_{0,2\delta+1}(s),$$
$$\rho_n(e^{i2\pi s}) = 2^{-\delta}\pi^{-3\delta-2}\sin^{2(\delta+1)}(\pi s)\sigma^-_{0,2\delta+1}(s).$$

Consequently, the matrix of the operator $L_nS|_{\underline{S}_n^\delta}$ has the form

$$(L_nS\Big|_{\underline{S}_n^\delta}) = \widehat{\zeta}_n^{-1}\widehat{\rho}_n = \widehat{\beta}_n,$$

where

$$\widehat{\beta}_n(e^{i2\pi s}) = \begin{cases} 1 & \text{if } s = 0, \\ (\sigma^-_{0,2\delta+1}(s))/(\sigma^+_{0,2\delta+1}(s)) & \text{if } s \neq 0. \end{cases}$$

Furthermore,

$$(L_nc\Big|\underline{S}_n^\delta) = \widehat{\zeta}_n^{-1}\widetilde{c}_n\widehat{\zeta}_n + R_n^{(2)},$$

with $||R_n^{(2)}||$ tends to 0 as $n \to \infty$.

Let us also note the commutation property

$$\overline{M}_n\widehat{\zeta}_n = \widehat{\zeta}_n\overline{M}_n.$$

Therefore, from Lemmas 2.4.1 and 2.4.2 one obtains

$$A_n = \widehat{\zeta}_n^{-1}[(\widetilde{a}_n + \widetilde{b}_n\widehat{\beta}_n) + (\widetilde{c}_n + \widetilde{d}_n\widehat{\theta}_n)\overline{M}_n]\widehat{\zeta}_n + K_nk^0\Big|_{\underline{S}_n^\delta} + K_nk^1M\Big|_{\underline{S}_n^\delta} + R_n,\ n \in \mathbb{N}$$

where $||R_n||$ tends to 0 as $n \to \infty$ and

$$\widehat{\theta}_n(e^{i2\pi s}) = \begin{cases} 1 & \text{if } s = 0, \\ (\sigma^-_{0,2\delta+1}(1-s))/(\sigma^+_{0,2\delta+1}(1-s)) & \text{if } s \neq 0. \end{cases}$$

The functions β and $\theta_{\varepsilon,\delta}$ belong to $\mathbf{C}(\Gamma_0 \setminus \{1\})$ and $\beta(1+0) = 1$, $\beta(1-0) = -1$, $\theta_{\varepsilon,\delta}(1+0) = -1$, $\theta_{\varepsilon,\delta}(1-0) = 1$.

The following theorem is an immediate consequence of Theorem 2.4.14.

Theorem 2.6.1. *Assume that $a, b, c, d \in \mathbf{C}(\Gamma_0)$ and $k^0, k^1 \in \mathcal{K}(L_2(\Gamma_0))$. The Galerkin method (2.75) is applicable to the operator A if and only if the operator A is invertible in $\mathcal{L}(L_2(\Gamma_0))$, the operator $\widetilde{B}$ defined by (2.74) is invertible in $\mathcal{L}(L_2^2(\Gamma_0))$, and*

$$\det[(\mathcal{A}(t) + \mathcal{C}(t)\Lambda) + (\mathcal{B}(t) + \mathcal{D}(t)\Lambda)\beta(\tau)] \neq 0$$

on $\Gamma_0 \times \Gamma_0$.

Finally, let us state a result related to the Galerkin method for equations with piecewise continuous coefficients. We assume the coefficients of the integral equation under consideration are discontinuous at the points $t_k = \exp(i2\pi j/m)$, $j = 0, 1, \ldots, m-1$, and consider the spline Galerkin method based on spline space $\underline{S}_n^0$.

Theorem 2.6.2. *Let* $a, b, c, d \in \mathbf{PC}(\Gamma_0) \cap \mathbf{C}(\Gamma_0 \setminus \{t_0, t_1, \dots, t_m - 1\})$, $k^0, k^1 \in \mathcal{K}(L_2(\Gamma_0))$, *and let* $n = ml$, $l \in \mathbb{N}$. *The Galerkin method* (2.75) *based on the splines of* $\underline{S}_n^{\delta}$ *is stable if and only if the operator* A *is invertible in* $\mathcal{L}(L_2(\Gamma_0))$, *the operator* $\widetilde{B}$ *defined by* (2.74) *and the operators*

$$\begin{aligned}B^{\tau} &= P[(\mathcal{A}(\tau+0)) + \mathcal{C}(\tau+0)\Lambda) + (\mathcal{B}(\tau+0) + \mathcal{D}(\tau+0)\Lambda)\boldsymbol{\beta}] \\ &\quad + Q[(\mathcal{A}(\tau-0)) + \mathcal{C}(\tau-0)\Lambda) + (\mathcal{B}(\tau-0) + \mathcal{D}(\tau-0)\Lambda)\widetilde{\boldsymbol{\beta}}]\end{aligned}$$

are invertible in $\mathcal{L}(L_2^2(\Gamma_0))$ *for all* $\tau \in \Gamma_0$.

2.7 Quadrature Methods for Equations on the Unit Circle

Quadrature methods play an important role in approximate solution of singular integral equations, because they can be easily implemented. Presenting the stability analysis for equations with conjugation, we only consider the equation

$$Bu \equiv (aI + bS + cM + dSM)u = f, \tag{2.76}$$

where $a, b, c, d \in \mathbf{PC}(\Gamma_0)$. In order to discretize a regular integral, one may use the following quadrature rules:

$$\int_{\Gamma_0} f(\tau)\,d\tau \approx \sum_{j=0}^{n-1} f(t_j)\frac{2\pi i}{n}\,t_j, \tag{2.77}$$

$$\int_{\Gamma_0} f(\tau)\,d\tau \approx \sum_{\substack{j=0\\ j\equiv 0 \pmod 2}}^{n-1} f(t_j)\frac{4\pi i}{n}\,t_j, \tag{2.78}$$

$$\int_{\Gamma_0} f(\tau)\,d\tau \approx \sum_{\substack{j=0\\ j\equiv 1 \pmod 2}}^{n-1} f(t_j)\frac{4\pi i}{n}\,t_j, \tag{2.79}$$

where $t_j = \exp(i2\pi j/n)$, $j = 0, 1, \dots, n-1$.

Writing the singular integral S in the form

$$(Su)(t) = u(t) + \frac{1}{\pi i}\int_{\Gamma_0} \frac{u(\tau) - u(t)}{\tau - t}\,d\tau$$

and using one of the formulas (2.77) – (2.79) we compute the values of Su at the points $\tau_k = \exp(i2\pi(k+\varepsilon)/n)$, $k = 0, 1, \dots, n-1$, $0 \le \varepsilon < 1$. Thus formula (2.77) implies

$$(Su)(\tau_k) \approx u(\tau_k) + \frac{2}{n}\sum_{j=0}^{n-1} \frac{u(t_j) - u(\tau_k)}{t_j - \tau_k}\,t_j \quad \text{if} \quad \varepsilon \neq 0, \tag{2.80}$$

$$(Su)(t_k) \approx u(t_k) + \frac{2}{n}\left(u'(t_k)t_k + \sum_{\substack{j=0\\ j\neq k}}^{n-1} \frac{u(t_j)-u(t_k)}{t_j - t_k} t_j\right) \quad \text{if} \quad \varepsilon = 0, \tag{2.81}$$

while (2.78), (2.79) give us

$$(Su)(\tau_k) \approx u(\tau_k) + \frac{4}{n} \sum_{\substack{j=0\\ j\equiv k+1 (\mathrm{mod}\, 2)}}^{n-1} \frac{u(t_j)-u(\tau_k)}{t_j - \tau_k} t_j. \tag{2.82}$$

Let us consider the expression (2.80) and (2.81). Since [183, 185]

$$\frac{2}{n}\sum_{j=0}^{n-1} \frac{t_j}{t_j - \tau_k} = 1 + i\cot(\pi\varepsilon),$$

the expressions (2.80) and (2.81) can be, respectively, rewritten as

$$(Su)(\tau_k) \approx -i\cot(\pi\varepsilon)u(\tau_k) + \frac{2}{n}\sum_{j=0}^{n-1} \frac{u(t_j)}{t_j - \tau_k} t_j, \quad \text{if} \quad \varepsilon \neq 0, \tag{2.83}$$

and

$$(Su)(t_k) \approx \frac{1}{n}u(t_k) + \frac{2}{n}u'(t_k)t_k + \sum_{\substack{j=0\\ j\neq k}}^{n-1} \frac{u(t_j)t_j}{t_j - t_k} \quad \text{if} \quad \varepsilon = 0. \tag{2.84}$$

Similarly

$$(Su)(t_k) \approx u(\tau_k) + \frac{4}{n} \sum_{\substack{j=0\\ j\equiv k+1 (\mathrm{mod}\, 2)}}^{n-1} \frac{u(t_j)}{t_j - \tau_k} t_j. \tag{2.85}$$

The right-hand side of expressions (2.83) – (2.85) contains the terms $u(\tau_k)$ and $u'(t_k)$, which will be replaced by other terms as follows:

1. If $\varepsilon \neq 0$, we set

$$u(\tau_k) \approx \begin{cases} u(t_k), & \text{if} \quad \varepsilon \neq \frac{1}{2}, \\ \frac{1}{2}(u(t_k) + u(t_{k+1})), & \text{if} \quad \varepsilon = \frac{1}{2}. \end{cases} \tag{2.86}$$

2. If $\varepsilon = 0$, we set

$$\frac{1}{n}u(t_k) + \frac{2}{n}u'(t_k)t_k = 0, \tag{2.87}$$

or

$$u'(t_k) = \frac{u(t_{k+1}) - u(t_k)}{t_{k+1} - t_k}. \tag{2.88}$$

Let us now evaluate the function Bu at the points τ_k, $k = 0, 1, \ldots, n-1$. Thus

$$a(\tau_k)u(\tau_k) + b(\tau_k)(Su)(\tau_k) + c(\tau_k)\overline{u(\tau_k)} + d(\tau_k)(S\overline{u})(\tau_k).$$

To compute $(Su)(\tau_k)$ and $(S\overline{u})(\tau_k)$ we use formulas (2.83) – (2.85) and replace $u(\tau_k), u'(t_k)$ or $(1/n)u(t_k)+(2/n)u'(t_k)t_k$ by one of the expressions (2.86) – (2.88). Equating $Bu(\tau_k)$ with $f(\tau_k)$, yields the systems of algebraic equations with respect to unknown values of the function u at the points t_k, $k = 0, 1, \ldots, n-1$. Let us now denote $u(t_j)$ by ξ_j, $j = 0, 1, \ldots, n-1$ and write down the resulting equations:

$$\begin{aligned}(a(\tau_k) - i\cot(\pi\varepsilon)b(\tau_k))\xi_k + b(\tau_k)\frac{2}{n}\sum_{j=0}^{n-1}\frac{t_j}{t_j - \tau_k}\xi_j \\ + (c(\tau_k) - i\cot(\pi\varepsilon)d(\tau_k))\overline{\xi}_k + d(\tau_k)\frac{2}{n}\sum_{j=0}^{n-1}\frac{t_j}{t_j - \tau_k}\overline{\xi}_j = f(\tau_k),\end{aligned} \tag{2.89}$$

$$\begin{aligned}a(v_k)\frac{1}{2}(\xi_k + \xi_{k+1}) + b(v_k)\frac{2}{n}\sum_{j=0}^{n-1}\frac{t_j}{t_j - v_k}\xi_j \\ + c(v_k)\frac{1}{2}(\overline{\xi}_k + \overline{\xi}_{k+1}) + d(v_k)\frac{2}{n}\sum_{j=0}^{n-1}\frac{t_j}{t_j - v_k}\overline{\xi}_j = f(v_k),\end{aligned} \tag{2.90}$$

$$\begin{aligned}a(t_k)\xi_k + b(t_k)\frac{2}{n}\sum_{\substack{j=0\\ j\neq k}}^{n-1}\frac{t_j}{t_j - t_k}\xi_j \\ + c(t_k)\overline{\xi}_k + d(t_k)\frac{2}{n}\sum_{\substack{j=0\\ j\neq k}}^{n-1}\frac{t_j}{t_j - t_k}\overline{\xi}_j = f(t_k),\end{aligned} \tag{2.91}$$

$$\begin{aligned}(a(t_k) + \frac{1}{n}b(t_k))\xi_k + b(t_k)\frac{2}{n}t_k\frac{\xi_{k+1} - \xi_{k-1}}{t_{k+1} - t_{k-1}} + b(t_k)\frac{2}{n}\sum_{\substack{j=0\\ j\neq k}}^{n-1}\frac{t_j}{t_j - t_k}\xi_j + (c(t_k)\overline{\xi}_k \\ + \frac{1}{n}d(t_k))\overline{\xi}_k + d(t_k)\frac{2}{n}t_k\frac{\overline{\xi}_{k+1} - \overline{\xi}_{k-1}}{t_{k+1} - t_{k-1}} + d(t_k)\frac{2}{n}\sum_{\substack{j=0\\ j\neq k}}^{n-1}\frac{t_j}{t_j - t_k}\overline{\xi}_j = f(t_k),\end{aligned} \tag{2.92}$$

$$(a(t_k) + \frac{1}{n}b(t_k))\xi_k + b(t_k)\frac{2}{n}t_k\frac{\xi_{k+1} - \xi_{k-1}}{t_{k+1} - t_{k-1}} + b(t_k)\frac{2}{n}\sum_{\substack{j=0\\ j\neq k}}^{n-1}\frac{t_j}{t_j - t_k}\,\xi_j$$
$$+ c(t_k)\overline{\xi}_k + d(t_k)\frac{2}{n}\sum_{\substack{j=0\\ j\neq k}}^{n-1}\frac{t_j}{t_j - t_k}\,\overline{\xi}_j = f(t_k), \tag{2.93}$$

$$a(t_k)\xi_k + b(t_k)\frac{2}{n}\sum_{\substack{j=0\\ j\neq k}}^{n-1}\frac{t_j}{t_j - t_k}\,\xi_j + (c(t_k)\overline{\xi}_k + \frac{1}{n}d(t_k))\,\overline{\xi}_k$$
$$+ d(t_k)\frac{2}{n}t_k\frac{\overline{\xi}_{k+1} - \overline{\xi}_{k-1}}{t_{k+1} - t_{k-1}} + d(t_k)\frac{2}{n}\sum_{\substack{j=0\\ j\neq k}}^{n-1}\frac{t_j}{t_j - t_k}\,\overline{\xi}_j = f(t_k), \tag{2.94}$$

$$a(t_k)\xi_k + b(t_k)\frac{4}{n}\sum_{\substack{j=0\\ j\equiv k+1 (\mathrm{mod}\, 2)}}^{n-1}\frac{t_j}{t_j - t_k}\,\xi_j$$
$$+ c(t_k)\overline{\xi}_k + d(t_k)\frac{4}{n}\sum_{\substack{j=0\\ j\equiv k+1 (\mathrm{mod}\, 2)}}^{n-1}\frac{t_j}{t_j - t_k}\,\overline{\xi}_j = f(t_k), \tag{2.95}$$

$$a(t_k)\xi_k + b(t_k)\frac{2}{n}\sum_{\substack{j=0\\ j\neq k}}^{n-1}\frac{t_j}{t_j - t_k}\,\xi_j$$
$$+ c(t_k)\overline{\xi}_k + d(t_k)\frac{4}{n}\sum_{\substack{j=0\\ j\equiv k+1 (\mathrm{mod}\, 2)}}^{n-1}\frac{t_j}{t_j - t_k}\,\overline{\xi}_j = f(t_k), \tag{2.96}$$

$$(a(t_k) + \frac{1}{n}b(t_k))\xi_k + b(t_k)\frac{2}{n}t_k\frac{\xi_{k+1} - \xi_{k-1}}{t_{k+1} - t_{k-1}} + b(t_k)\frac{2}{n}\sum_{\substack{j=0\\ j\neq k}}^{n-1}\frac{t_j}{t_j - t_k}\,\xi_j$$
$$+ c(t_k)\overline{\xi}_k + d(t_k)\frac{4}{n}\sum_{\substack{j=0\\ j\equiv k+1 (\mathrm{mod}\, 2)}}^{n-1}\frac{t_j}{t_j - t_k}\,\overline{\xi}_j = f(t_k), \tag{2.97}$$

$$a(t_k)\xi_k + b(t_k)\frac{4}{n}\sum_{\substack{j=0\\ j\equiv k+1 (\mathrm{mod}\, 2)}}^{n-1} \frac{t_j}{t_j - t_k}\xi_j$$

$$+\, c(t_k)\overline{\xi}_k + d(t_k)\frac{2}{n}\sum_{\substack{j=0\\ j\neq k}}^{n-1}\frac{t_j}{t_j-t_k}\overline{\xi}_j = f(t_k), \tag{2.98}$$

where $\xi_n := \xi_0$, $\xi_{-1} := \xi_{n-1}$ and $k = 0, 1, \ldots, n-1$ for all systems (2.89) – (2.97). Moreover, in the system (2.90), the symbol v_k denotes the point $\exp(i2\pi(k+\frac{1}{2})/n)$. If any of the above systems of algebraic equations has a solution $\xi_k^{(n)} := \xi_k$, $k = 0, 1, \ldots, n-1$, then an approximate solution of the integral equation (2.76) can be obtained via trigonometric approximation

$$u_n = \sum_{k=0}^{n-1}\xi_k^{(n)}\Psi_k^{(n)}, \quad \Psi_k^{(n)} := \frac{1}{n}\sum_{j=-[n/2]}^{[(n-1)/2]}(t_k^{(n)})^{-j}t^j, \quad t\in\Gamma_0.$$

We define an interpolation projection $K_n = K_n^{\varepsilon}$ by [183]

$$K_n u_n := \sum_{k=0}^{n-1} u(\tau_k^{(n)})\widetilde{\Psi}_k^{(n)}, \quad \widetilde{\Psi}_k^{(n)}(t) := \frac{1}{n}\sum_{j=-[n/2]}^{[(n-1)/2]}(\tau_k^{(n)})^{-j}t^j, \quad t\in\Gamma_0.$$

The systems of algebraic equations (2.89) – (2.97) are equivalent to the operator equations $B_n = K_n f$, where B_n is generated by the operator B and some well-known operator sequences connected with quadrature methods (2.89) – (2.97). Thus, using corresponding circulant representations for approximations of the Cauchy singular integral operator S, one can write the operator B_n in the form [183]

$$B_n = \widetilde{a}_n\widehat{\alpha}_n + \widetilde{b}_n\widehat{\beta}_n + (\widetilde{c}_n\widehat{\gamma}_n + \widetilde{d}_n\widehat{\theta}_n)\overline{M}_n + C_n$$

where $||C_n|| \to 0$ as $n\to\infty$.

The functions $\alpha, \beta, \gamma, \theta$ that generate the corresponding circulants depend on the quadrature method used. These functions are given in Table 2.1 below. Computation of these functions is straightforward, and we refer the reader to [183] for details. However, let us mention that

$$\alpha(1) = \beta(1) = \gamma(1) = \theta(1) = 1$$

for all quadrature methods (2.89) – (2.97). We also use the notation

$$\alpha = \alpha(t) = \alpha(e^{i2\pi s}), \quad \beta = \beta(t) = \beta(e^{i2\pi s}),$$
$$\gamma = \gamma(t) = \gamma(e^{i2\pi s}), \quad \theta = \theta(t) = \theta(e^{i2\pi s})$$

for $s \in [0, 1)$.

Table 2.1:

Method	α	β	γ	θ
(2.89)	1	$-i\cot(\pi\varepsilon)+\frac{2\exp(i2\pi\varepsilon s)}{1-\exp(i2\pi\varepsilon)}$	1	$-i\cot(\pi\varepsilon)+\frac{2\exp(i2\pi\varepsilon s)}{1-\exp(i2\pi\varepsilon)}$
(2.90)	$\frac{1+t}{2}$	$\exp(i\pi s)$	$\frac{1+t}{2}$	$\exp(i\pi s)$
(2.91)	1	$1-2s$	1	$1-2s$
(2.92)	1	$1-2s+\pi^{-1}\sin(2\pi s)$	1	$1-2s+\pi^{-1}\sin(2\pi s)$
(2.93)	1	$1-2s+\pi^{-1}\sin(2\pi s)$	1	$1-2s$
(2.94)	1	$1-2s$	1	$1-2s+\pi^{-1}\sin(2\pi s)$
(2.95)	1	$1,\ \text{if } 0\le s<1/2$ $-1,\ \text{if } 1/2\le s<1$	1	$1,\ \text{if } 0\le s<1/2$ $-1,\ \text{if } 1/2\le s<1$
(2.96)	1	$1-2s$	1	$1,\ \text{if } 0\le s<1/2$ $-1,\ \text{if } 1/2\le s<1$
(2.97)	1	$1-2s+\pi^{-1}\sin(2\pi s)$	1	$1,\ \text{if } 0\le s<1/2$ $-1,\ \text{if } 1/2\le s<1$
(2.98)	1	$1,\ \text{if} 0\le s<1/2$ $-1,\ \text{if} 1/2\le s<1$	1	$1-2s$

It should be noted that for all methods the functions generating the corresponding circulants have the limit values

$$\begin{aligned}\alpha(1\pm0)=\alpha^*(1\pm0)=\gamma(1\pm0)=\gamma^*(1\pm0)&=1,\\ \beta(1+0)=\theta(1+0)=\beta^*(1-0)=\theta^*(1-0)&=1,\\ \beta(1-0)=\beta^*(1+0)=\theta(1-0)=\theta^*(1+0)&=-1,\end{aligned}$$

and consequently the operator B_1 has the form

$$B_1=\begin{pmatrix} a+b & c+d\\ \overline{c}-\overline{d} & \overline{a}-\overline{b}\end{pmatrix}P+\begin{pmatrix} a-b & c-d\\ \overline{c}+\overline{d} & \overline{a}+\overline{b}\end{pmatrix}Q. \tag{2.99}$$

The functions α and γ are continuous everywhere on Γ_0, and for the methods (2.89) – (2.94) the functions β,θ are continuous on $\Gamma_0\setminus\{1\}$. Thus we are left with computing the limits of the functions $\beta,\theta,\beta^*,\theta^*$ at the point $t=-1$ for the methods (2.95) – (2.98). The corresponding left and right limits are given in Table 2.2.

It follows that for methods (2.95) and (2.98) the respective operators are

$$B_{-1}=\begin{pmatrix} a-b & c-d\\ \overline{c}+\overline{d} & \overline{a}+\overline{b}\end{pmatrix}P+\begin{pmatrix} a+b & c+d\\ \overline{c}-\overline{d} & \overline{a}-\overline{b}\end{pmatrix}Q \tag{2.100}$$

Table 2.2:

Method	$\beta(-1+0)$ $\beta^*(-1-0)$	$\beta(-1-0)$ $\beta^*(-1+0)$	$\theta(-1+0)$ $\theta^*(-1-0)$	$\theta(-1-0)$ $\theta^*(-1+0)$
(2.95)	-1	$+1$	-1	$+1$
(2.96)	0	0	-1	$+1$
(2.97)	0	0	-1	$+1$
(2.98)	-1	$+1$	0	0

and

$$B_{-1} = \begin{pmatrix} a-b & c \\ \overline{c} & \overline{a}+\overline{b} \end{pmatrix} P + \begin{pmatrix} a+b & c \\ \overline{c} & \overline{a}-\overline{b} \end{pmatrix} Q, \tag{2.101}$$

whereas for methods (2.96), (2.97) the operator is

$$B_{-1} = \begin{pmatrix} a & c-d \\ \overline{c}+\overline{d} & \overline{a} \end{pmatrix} P + \begin{pmatrix} a & c+d \\ \overline{c}-\overline{d} & \overline{a} \end{pmatrix} Q. \tag{2.102}$$

Now we can formulate the main result concerning the stability of quadrature methods (2.89) – (2.98).

Theorem 2.7.1. *Assume that the coefficients a, b, c, d of equation* (2.76) *belong to* $\mathbf{PC}(\Gamma_0)$.

1. *The sequence (B_n) corresponding to one of the methods* (2.89) – (2.94) *is stable if and only if the following two conditions are satisfied:*

   (a) *the operator B_1 is invertible on $\mathcal{L}(L_2^2(\Gamma_0))$;*

   (b) *for each $\tau \in \Gamma_0$ the operator*

   $$\begin{aligned} B^\tau &= P[\mathcal{A}(\tau+0)\widetilde{\boldsymbol{\alpha}} + \mathcal{B}(\tau+0)\widetilde{\boldsymbol{\beta}} + \mathcal{C}(\tau+0)\widetilde{\boldsymbol{\gamma}}\Lambda + \mathcal{D}(\tau+0)\widetilde{\boldsymbol{\theta}}\Lambda] \\ &\quad + Q[\mathcal{A}(\tau-0)\widetilde{\boldsymbol{\alpha}} + \mathcal{B}(\tau-0)\widetilde{\boldsymbol{\beta}} + \mathcal{C}(\tau-0)\widetilde{\boldsymbol{\gamma}}\Lambda + \mathcal{D}(\tau-0)\widetilde{\boldsymbol{\theta}}\Lambda] \end{aligned}$$

   is invertible on $\mathcal{L}(L_2^2(\Gamma_0))$.

2. *The sequence (B_n) corresponding to one of the methods* (2.95) – (2.98) *is stable if and only if the conditions* 1(a), 1(b) *are satisfied and the operator B_{-1} is invertible on $\mathcal{L}(L_2^2(\Gamma_0))$.*

Recall that the operators B_1 and B_{-1} are defined by (2.99) – (2.102).

Of course, for continuous coefficients a, b, c, d the condition 1(b) can be simplified.

Corollary 2.7.2. *Let* $a, b, c, d \in \mathbf{C}(\Gamma_0)$. *The quadrature methods* (2.89) – (2.94) *are stable if and only if the operator* (2.99) *is invertible on* $\mathcal{L}(L_2^2(\Gamma_0))$ *and the determinant*

$$\det[\mathcal{A}(\tau)\widetilde{\boldsymbol{\alpha}}(t) + \mathcal{B}(\tau)\widetilde{\boldsymbol{\beta}}(t) + \mathcal{C}(\tau)\widetilde{\boldsymbol{\gamma}}(t) + \mathcal{D}(\tau)\widetilde{\boldsymbol{\theta}}(t)] \neq 0$$

for all $t, \tau \in \Gamma_0$.

The corresponding result can be also formulated for the quadrature methods (2.95) – (2.98).

Remark 2.7.3. If the coefficients of the above integral equation satisfy one of the conditions $a = \pm b$ or $c = \pm d$, then the coefficients of the operators B_{-1} and B_1 are triangle matrices. Henceforth one may obtain effective conditions for invertibility of the operators B_{-1} and B_1.

2.8 Polynomial Qualocation Method

Let $X(\Gamma_0)$ be a functional space on Γ_0. Consider the operator equation

$$Bu = f, \tag{2.103}$$

where $f \in X(\Gamma_0)$ and $B \in \mathcal{L}_{add}(X(\Gamma_0))$. The qualocation method for solving (2.103) is characterized by the choice of a pair (S_h, T_h), $\dim S_h = \dim T_h = n_h$ of finite-dimensional subspaces of $X(\Gamma_0)$ and also by the choice of a quadrature rule Q_h. The method is to find an element $u_h \in S_h$ such that

$$Q_h(v_h \cdot Bu_h) = Q_h(v_h \cdot f) \tag{2.104}$$

for all $v_h \in T_h$. This method can be viewed as a discrete version of the Galerkin method or as a generalization of the collocation method based on quadrature formulas. However, in comparison with Galerkin methods this approach has simpler numerical implementation. On the other hand, the convergence of this method is usually better than the convergence of collocation approximations.

In this section we will use the quadrature formula

$$Q_h g = \frac{1}{\sqrt{n}} \sum_{j=0}^{n-1} [wg((j+\varepsilon_1)/n) + (1-w)g((j+\varepsilon_2)/n)], \quad h = \frac{1}{n} \tag{2.105}$$

where $0 \leq \varepsilon_1 < \varepsilon_2 < 1$, $w \in (0,1)$, and the function $g(\exp(i2\pi s))$ is identified with $g(s)$.

Let $\underline{S}_n^\delta$ be the spline space introduced in Section 2.5. By T_h we denote the space

$$T_h := \operatorname{span}\{v_p = \exp(i2\pi ps), \quad 0 \leq p \leq n-1, \quad p \in \mathbb{N}, \quad s \in [0,1)\}.$$

Now we are looking for a function $u_h \in \underline{S}_n^\delta$ which satisfies relation (2.104) with the operator B defined in (2.76). Let us assume that the coefficients a, b, c, d of the operator B are piecewise continuous on Γ_0 and continuous on $\Gamma_0 \backslash \{t_0, t_1, \ldots, t_{m-1}\}$ where $t_j = \exp(i2\pi j/m)$. We also assume that $n = ml$, $l \in \mathbb{N}$. We can show that the stability of approximation method (2.104) is equivalent to the stability of an operator sequence (B_n) from the real algebra $\widetilde{\mathcal{C}}$ generated by the sequences of circulants $(\widehat{a}_n)$, diagonal matrices $(\widetilde{a}_n)$, and by the sequence $(\overline{M}_n)$. Recall that algebra $\widetilde{\mathcal{C}}$ was studied in Section 2.3.

Let $\{\chi_k^{(n)}\}_{k=0}^{n-1}$ be the interpolation basis in the space $\underline{S}_n^\delta$ introduced in Section 2.5 and let

$$u^{(n)}(t) = \sum_{k=0}^{n-1} u_k \chi_k^{(n)}(t). \tag{2.106}$$

The qualocation method (2.104) for the operator (2.76) determines a system of algebraic equations of the form

$$B_n u^{(n)} = R_n^{(1)} u^{(n)} + R_n^{(2)} \overline{M}_n u^{(n)} = f^{(n)} \tag{2.107}$$

with the matrices $R_n^{(1)} = (r_{p,k}^{(1)})_{p,k=0}^{n-1}$, $R_n^{(2)} = (r_{p,k}^{(2)})_{p,k=0}^{n-1}$ and with the right-hand side $f^{(n)} = (f_p^{(n)})_{p=0}^{n-1}$, where

$$\begin{aligned}
r_{p,k}^{(1)} &= Q_h(\exp(-i2\pi ps)(aI + bS)\chi_k^{(n)}(t)),\\
r_{p,k}^{(2)} &= Q_h(\exp(-i2\pi ps)(cI + dS)\chi_k^{(n)}(t)),\\
f_p^{(n)} &= Q_h(\exp(-i2\pi ps)f(t)).
\end{aligned}$$

In the following lemma the notation of Section 2.5 is used.

Lemma 2.8.1. *The operator B_n from* (2.107) *admits the representation*

$$\begin{aligned}
B_n = {} & wK_n^0 t^{-\varepsilon_1} L_n U_n^{-1} [K_n^{\varepsilon_1} a L_n U_n K_n^0 \alpha_{\varepsilon_1,\delta} L_n U_n^{-1} L_n \\
& \qquad + K_n^{\varepsilon_1} b L_n U_n K_n^0 \beta_{\varepsilon_1,\delta} L_n U_n^{-1} L_n] \\
& + (1-w) K_n^0 t^{-\varepsilon_2} L_n U_n^{-1} [K_n^{\varepsilon_2} a L_n U_n K_n^0 \alpha_{\varepsilon_2,\delta} L_n U_n^{-1} L_n \\
& \qquad + K_n^{\varepsilon_2} b L_n U_n K_n^0 \beta_{\varepsilon_2,\delta} L_n U_n^{-1} L_n] \\
& + \{ wK_n^0 t^{-\varepsilon_1} L_n U_n^{-1} [K_n^{\varepsilon_1} c L_n U_n K_n^0 \alpha_{\varepsilon_1,\delta}^* L_n U_n^{-1} L_n \\
& \qquad + K_n^{\varepsilon_1} d L_n U_n K_n^0 \beta_{\varepsilon_1,\delta}^* L_n U_n^{-1} L_n] \\
& + (1-w) K_n^0 t^{-\varepsilon_2} L_n U_n^{-1} [K_n^{\varepsilon_2} c L_n U_n K_n^0 \alpha_{\varepsilon_2,\delta}^* L_n U_n^{-1} L_n \\
& \qquad + K_n^{\varepsilon_2} d L_n U_n K_n^0 \beta_{\varepsilon_2,\delta}^* L_n U_n^{-1} L_n] \} M.
\end{aligned} \tag{2.108}$$

Note that $t^{-\varepsilon}$ means that branch of the function $t^{-\varepsilon}$ which tends to 1 as t tends to $1+0$.

Proof. Let $u^{(n)}$ be as in (2.106). Then

$$Bu^{(n)} = \sum_{k=0}^{n-1} \left((a\chi_k^{(n)} + b(S\chi_k^{(n)}))u_k + (c\chi_k^{(n)} + b(S\chi_k^{(n)}))\overline{u}_k \right).$$

Moreover,

$$\overline{v}_p \cdot Bu^{(n)} = \exp(-i2\pi s)Bu^{(n)}.$$

Using the linearity of the quadrature rule (2.105) one can rewrite the expression $Q_h(\overline{v}_p \cdot Bu^{(n)})$ as a sum of four terms,

$$\begin{aligned} Q_h(\overline{v}_p \cdot Bu^{(n)}) &= Q_h\left(\overline{v}_p \cdot \sum_{k=0}^{n-1} a\chi_k^{(n)} u_k\right) + Q_h\left(\overline{v}_p \cdot \sum_{k=0}^{n-1} b(S\chi_k^{(n)}) u_k\right) \\ &\quad + Q_h\left(\overline{v}_p \cdot \sum_{k=0}^{n-1} c\chi_k^{(n)} \overline{u}_k\right) + Q_h\left(\overline{v}_p \cdot \sum_{k=0}^{n-1} d(S\chi_k^{(n)}) \overline{u}_k\right). \end{aligned}$$

We check relation (2.108) for the third term only. Thus

$$\begin{aligned} &Q_h\left(\overline{v}_p \cdot \sum_{k=0}^{n-1} c\chi_k^{(n)} \overline{u}_k\right) \\ &= \frac{1}{\sqrt{n}} \sum_{j=0}^{n-1} \sum_{l=1}^{2} w_l \exp(-i2\pi p(j+\varepsilon_l)/n) \sum_{k=0}^{n-1} c((j+\varepsilon_l)/n)\chi_k^{(n)}((j+\varepsilon_l)/n)\overline{u}_k \end{aligned}$$

where $w_1 = w$, $w_2 = 1 - w$. The matrix of the above operator is the sum of two matrices $M_n^{(1)} = (m_{pk}^{(1)})_{p,k=0}^{n-1}$ and $M_n^{(2)} = (m_{pk}^{(2)})_{p,k=0}^{n-1}$ where

$$\begin{aligned} m_{pk}^{(1)} &= \frac{1}{\sqrt{n}} w_1 \exp(-i2\pi p(j+\varepsilon_1)/n) \sum_{j=0}^{n-1} c((j+\varepsilon_1)/n)\chi_k^{(n)}((j+\varepsilon_l)/n), \\ m_{pk}^{(2)} &= \frac{1}{\sqrt{n}} w_2 \exp(-i2\pi p(j+\varepsilon_2)/n) \sum_{j=0}^{n-1} c((j+\varepsilon_2)/n)\chi_k^{(n)}((j+\varepsilon_2)/n). \end{aligned}$$

However, similar expressions arose earlier in Section 2.5 while studying the collocation equations. Using appropriate representations from this section, we find that the operators $w_1 K_n^0 t^{-\varepsilon_1} L_n U_n^{-1} K_n^{\varepsilon_1} c L_n U_n \alpha^{\circ}_{\varepsilon_1,n} L_n U_n^{-1} L_n$ and $w_2 K_n^0 t^{-\varepsilon_2} L_n U_n^{-1} K_n^{\varepsilon_2} c L_n U_n \alpha^{\circ}_{\varepsilon_2,n} L_n U_n^{-1} L_n$ respectively correspond to the matrices $M_n^{(1)}$ and $M_n^{(2)}$, and the proof is complete. □

Lemma 2.8.2. *The operator sequence* (B_n) *with the operators* B_n *defined in* (2.108) *is stable if and only if the sequence* (B'_n) *with*

$$\begin{aligned} B'_n = w_1 U_n K_n^0 t^{-\varepsilon_1} L_n U_n^{-1} [&K_n^0 a L_n U_n K_n^0 \alpha_{\varepsilon_1,\delta} L_n U_n^{-1} L_n \\ &+ K_n^0 b L_n U_n K_n^0 \beta_{\varepsilon_1,\delta} L_n U_n^{-1} L_n] \end{aligned}$$

$$
\begin{aligned}
&+ w_2 K_n^0 t^{-\varepsilon_2} L_n U_n^{-1} [K_n^0 a L_n U_n K_n^0 \alpha_{\varepsilon_2,\delta} L_n U_n^{-1} L_n \\
&\qquad\qquad + K_n^0 b L_n U_n K_n^0 \beta_{\varepsilon_2,\delta} L_n U_n^{-1} L_n] \\
&+ \{ w_1 K_n^0 t^{-\varepsilon_1} L_n U_n^{-1} [K_n^0 c L_n U_n K_n^0 \alpha^*_{\varepsilon_1,\delta} L_n U_n^{-1} L_n \\
&\qquad\qquad + K_n^0 d L_n U_n K_n^0 \beta^*_{\varepsilon_1,\delta} L_n U_n^{-1} L_n] \\
&+ w_2 K_n^0 t^{-\varepsilon_2} L_n U_n^{-1} [K_n^0 c L_n U_n K_n^0 \alpha^*_{\varepsilon_2,\delta} L_n U_n^{-1} L_n \\
&\qquad\qquad + K_n^0 d L_n U_n K_n^0 \beta^*_{\varepsilon_2,\delta} L_n U_n^{-1} L_n] \} M \qquad (2.109)
\end{aligned}
$$

is stable.

Proof. Multiplication by the isometric operator U_n does not influence the stability of any operator sequence, so the result follows from the fact that under our assumptions on the coefficients a, b, c, d the norms of the operators $(K_n^{\varepsilon_l} - K_n^0) a L_n$, $(K_n^{\varepsilon_l} - K_n^0) b L_n$, $(K_n^{\varepsilon_l} - K_n^0) c L_n$, $(K_n^{\varepsilon_l} - K_n^0) d L_n$ tend to 0 as $n \to \infty$. □

Let us rewrite the sequence (2.109) in matrix form. It is

$$
\begin{aligned}
B'_n = {} & [w_1 \widehat{t}_{-\varepsilon_1,n} (\widetilde{a}_n \widehat{\alpha}_{\varepsilon_1,n} + \widetilde{b}_n \widehat{\beta}_{\varepsilon_1,n}) + w_2 \widehat{t}_{-\varepsilon_2,n} (\widetilde{a}_n \widehat{\alpha}_{\varepsilon_2,n} + \widetilde{b}_n \widehat{\beta}_{\varepsilon_2,n})] \\
& + [w_1 \widehat{t}_{-\varepsilon_1,n} (\widetilde{c}_n \widehat{\alpha}_{\varepsilon_1,n} + \widetilde{d}_n \widehat{\beta}_{\varepsilon_1,n}) + w_2 \widehat{t}_{-\varepsilon_2,n} (\widetilde{c}_n \widehat{\alpha}_{\varepsilon_2,n} + \widetilde{d}_n \widehat{\beta}_{\varepsilon_2,n})] \overline{M}_n,
\end{aligned}
$$

where the functions $\alpha_\varepsilon = \alpha_{\varepsilon,\delta}$, $\beta_\varepsilon = \beta_{\varepsilon,\delta}$, $\gamma_\varepsilon = \gamma_{\varepsilon,\delta}$, $\theta_\varepsilon = \theta_{\varepsilon,\delta}$ are defined in Section 2.5, and $\widehat{t}_{-\varepsilon_l,n}$, $l = 1, 2$ are the circulants corresponding to the functions $t^{-\varepsilon_l}$, $l = 1, 2$. By Theorem 2.4.4 the sequence (B'_n) is stable if and only if the sequence

$$
\begin{aligned}
B_n^{(1)} = {} & w_1 \begin{pmatrix} \widehat{t}_{-\varepsilon_1,n} & 0 \\ 0 & \widehat{t}_{-\varepsilon_1,n} \end{pmatrix} \begin{pmatrix} \widetilde{a}_n \widehat{\alpha}_{\varepsilon_1,n} + \widetilde{b}_n \widehat{\beta}_{\varepsilon_1,n} & \widetilde{c}_n \widehat{\alpha}^\circ_{\varepsilon_1,n} + \widetilde{d}_n \widehat{\beta}^\circ_{\varepsilon_1,n} \\ \widetilde{\overline{c}}_n \widehat{\alpha}_{\varepsilon_1,n} + \widetilde{\overline{d}}_n \widehat{\beta}_n & \widetilde{\overline{a}}_n \widehat{\alpha}^\circ_{\varepsilon_1,n} + \widetilde{\overline{b}}_n \widehat{\beta}^\circ_{\varepsilon_1,n} \end{pmatrix} \\
& + w_2 \begin{pmatrix} \widehat{t}_{-\varepsilon_2,n} & 0 \\ 0 & \widehat{t}_{-\varepsilon_2,n} \end{pmatrix} \begin{pmatrix} \widetilde{a}_n \widehat{\alpha}_{\varepsilon_2,n} + \widetilde{b}_n \widehat{\beta}_{\varepsilon_2,n} & \widetilde{c}_n \widehat{\alpha}^\circ_{\varepsilon_2,n} + \widetilde{d}_n \widehat{\beta}^\circ_{\varepsilon_2,n} \\ \widetilde{\overline{c}}_n \widehat{\alpha}_{\varepsilon_2,n} + \widetilde{\overline{d}}_n \widehat{\beta}_n & \widetilde{\overline{a}}_n \widehat{\alpha}^\circ_{\varepsilon_2,n} + \widetilde{\overline{b}}_n \widehat{\beta}^\circ_{\varepsilon_2,n} \end{pmatrix}
\end{aligned}
\qquad (2.110)
$$

is stable. Note that the sequence (2.110) belongs to the C^*-algebra $\mathcal{U}_1^{2\times 2}$ considered in Section 2.4, so Theorem 2.4.14 is applicable to the sequence $(B_n^{(1)})$. Of course, adjustments concerning the invertibility of local symbols $\Psi^\circ_{v,\omega}(B_n^{(1)})$, $v, \omega \in \Gamma_0$ must be made, but one can follow the proof of Lemma 2.4.13. Thus the stability result for the polynomial qualocation method can be formulated as follows.

Theorem 2.8.3. *Let* $a, b, c, d \in \mathbf{PC}(\Gamma_0)$ *be continuous on the set* $\Gamma_0 \setminus \{t_0, t_1, \ldots, t_{m-1}\}$ *where* $t_j = \exp(i2\pi j/m)$. *The qualocation method* (2.104) *for the operator* (2.76) *considered on the space* $L_2(\Gamma_0)$ *is stable if and only if the* *operators*

$$\begin{aligned}B_\tau := w_1 &\left((\tau+0)^{-\varepsilon_1} P + (\tau-0)^{-\varepsilon_1} Q\right)\\ &\times \left\{\left[(\mathcal{A}+\mathcal{C}\Lambda)\boldsymbol{\alpha}_{\varepsilon_1,\delta}(\tau+0) + (\mathcal{B}+\mathcal{D}\Lambda)\boldsymbol{\beta}_{\varepsilon_1,\delta}(\tau+0)\right] P\right.\\ &\quad \left. + \left[(\mathcal{A}+\mathcal{C}\Lambda)\boldsymbol{\alpha}_{\varepsilon_1,\delta}(\tau-0) + (\mathcal{B}+\mathcal{D}\Lambda)\boldsymbol{\beta}_{\varepsilon_1,\delta}(\tau-0)\right] Q\right\}\\ + w_2 &\left((\tau+0)^{-\varepsilon_2} P + \tau-0)^{-\varepsilon_2} Q\right)\\ &\times \left\{\left[(\mathcal{A}+\mathcal{C}\Lambda)\boldsymbol{\alpha}_{\varepsilon_2,\delta}(\tau+0) + (\mathcal{B}+\mathcal{D}\Lambda)\boldsymbol{\beta}_{\varepsilon_2,\delta}(\tau+0)\right] P\right.\\ &\quad \left. + \left[(\mathcal{A}+\mathcal{C}\Lambda)\boldsymbol{\alpha}_{\varepsilon_2,\delta}(\tau-0) + (\mathcal{B}+\mathcal{D}\Lambda)\boldsymbol{\beta}_{\varepsilon_2,\delta}(\tau-0)\right] Q\right\},\end{aligned}$$

$$\begin{aligned}B^\tau := w_1 &t^{\varepsilon_1}\\ &\times \left\{P\left[(\mathcal{A}(\tau+0)+\mathcal{C}(\tau+0)\Lambda)\widetilde{\boldsymbol{\alpha}}_{\varepsilon_1,\delta} + (\mathcal{B}(\tau+0)+\mathcal{D}(\tau+0)\Lambda)\widetilde{\boldsymbol{\beta}}_{\varepsilon_1,\delta}\right]\right.\\ &\quad \left. + Q\left[(\mathcal{A}(\tau-0)+\mathcal{C}(\tau-0)\Lambda)\widetilde{\boldsymbol{\alpha}}_{\varepsilon_1,\delta} + (\mathcal{B}(\tau-0)+\mathcal{D}(\tau-0)\Lambda)\widetilde{\boldsymbol{\beta}}_{\varepsilon_1,\delta}\right]\right\}\\ + w_1 &t^{\varepsilon_2}\\ &\times \left\{P\left[(\mathcal{A}(\tau+0)+\mathcal{C}(\tau+0)\Lambda)\widetilde{\boldsymbol{\alpha}}_{\varepsilon_2,\delta} + (\mathcal{B}(\tau+0)+\mathcal{D}(\tau+0)\Lambda)\widetilde{\boldsymbol{\beta}}_{\varepsilon_2,\delta}\right]\right.\\ &\quad \left. + Q\left[(\mathcal{A}(\tau-0)+\mathcal{C}(\tau-0)\Lambda)\widetilde{\boldsymbol{\alpha}}_{\varepsilon_2,\delta} + (\mathcal{B}(\tau-0)+\mathcal{D}(\tau-0)\Lambda)\widetilde{\boldsymbol{\beta}}_{\varepsilon_2,\delta}\right]\right\}.\end{aligned}$$

are invertible in $\mathcal{L}(L_2^2(\Gamma_0))$ *for all* $\tau \in \Gamma_0$.

Recall that $\mathcal{A}, \mathcal{B}, \mathcal{C}, \mathcal{D}, \widetilde{\boldsymbol{\alpha}}_{\varepsilon_l,\delta}\widetilde{\boldsymbol{\beta}}_{\varepsilon_l,\delta}$, $l = 1, 2$ are the operators of multiplication by corresponding matrix functions.

2.9 Spline Qualocation Method

Let us again consider equation (2.104), but replace the polynomial space T_h by a spline space $\underline{S}_n^\mu$, $\mu \in \mathbb{N}$, while the space $S_h = \underline{S}_n^\delta$ remains unchanged. The formula (2.105) is still used as a quadrature rule Q_h. The stability conditions for the corresponding spline-qualocation method are simpler than the stability conditions for the polynomial qualocation method. For example, for the singular integral operator without conjugation,

$$A = aI + bS, \quad a, b \in \mathbf{C}(\Gamma_0),$$

the stability of the spline qualocation method follows from the invertibility of a certain characteristic singular integral operator and can be verified effectively. On the other hand, considerations of Section 2.8 show that for the polynomial qualocation, one deals with operators from the algebra generated by the identity operator, the operator S and operators of multiplication by piecewise continuous functions. However, for such operators there are no effective invertibility conditions.

Let us describe a special basis in the spline space $\underline{S}_n^\mu$. First we note that there is a one-to-one correspondence between functions on Γ_0 and on $[0, n)$, viz., each function φ on Γ_0 can be identified with the function

$$\widetilde{\varphi} : s \mapsto \varphi\left(\frac{i2\pi s}{n}\right)$$

on the interval $[0, n)$. Thus we assume that $n \geq \mu + 1$, and instead of splines from $\underline{S}_n^\mu$ we consider n-periodic polynomial splines $\widehat{\varphi}$. Moreover, we assume that on $[0, n)$ such splines satisfy the following conditions:

1) Each spline is a polynomial of order at most μ on each of the intervals $[0, 1], [1, 2], \ldots, [n-1, n]$.
2) $\widehat{\varphi} \in \mathbf{C}^{\mu-1}((0, n))$.
3) $\widehat{\varphi}^{(k)}(0) = \widehat{\varphi}^{(k)}(n)$ for all $k = 0, 1, \ldots, \mu - 1$.

The corresponding spline space is denoted by $\widehat{S}_n^\mu$. Suppose that $n \geq \mu + 1$ and choose an appropriate basis in $\widehat{S}_n^\mu$. To this end, we construct a spline $\widehat{f}_\mu$ with support in $[0, \mu + 1]$ such that

$$\widehat{f}_\mu(s) = P_0(s) = a_0 s^\mu, \quad a_0 \in \mathbb{C}$$

for all $s \in [0.1]$. Let P_k be the restriction of $\widehat{f}_\mu$ onto the interval $[k, k+1]$, $k = 0, 1, \ldots, n-1$. Then P_k is a polynomial and condition 3) implies that

$$P_{k+1}^{(r)}(k+1) = P_k^{(r)}(k+1), \quad r = 0, 1, \ldots, \mu - 1; \quad k = 0, 1, \ldots, n-2,$$

i.e.,

$$P_{k+1}(s) - P_k(s) = a_{k+1}(s - (k+1))^\mu, \quad a_{k+1} \in \mathbb{C},$$

hence

$$\widehat{f}_\mu(s) = a_0 s^\mu + a_1(s-1)^\mu + \ldots + a_k(s-k)^\mu, \quad 0 \leq k \leq s \leq k+1 \leq \mu + 1.$$

If we take into account condition 3) and the requirement $\operatorname{supp} \widehat{f}_\mu \subset [0, \mu + 1]$, we obtain $P_\mu(s) = c(\mu + 1 - s)^\mu$. It remains to find the coefficients $a_0, a_1, \ldots, a_k$ so that the equation

$$a_0 s^\mu + a_1(s-1)^\mu + \ldots + a_\mu(s-\mu)^\mu = c(\mu + 1 - s)^\mu$$

is an identity for some $c \in \mathbb{C}$. From the binomial expansion, the last equation can be represented in the form

$$\sum_{k=0}^{\mu}\sum_{j=0}^{\mu}(-1)^{\mu-k}\binom{\mu}{k} j^{\mu-k} s^k = c(-1)^\mu \sum_{k=0}^{\mu}(-1)^{\mu-k}\binom{\mu}{k}(\mu+1)^{\mu-k} s^k,$$

so equating the coefficients at the powers of s on both sides one obtains the system of linear algebraic equations

$$\sum_{j=0}^{\mu} a_j j^{\mu-k} = c(-1)^\mu(\mu+1)^{\mu-k}, \quad k = 0, 1, \ldots, \mu,$$

i.e.,

$$
\begin{array}{lllll}
a_0 & +\ldots+ & a_\mu & = c(-1)^\mu & \\
0a_0 & +\ldots+ & \mu a_\mu & = c(-1)^\mu(\mu+1) & \\
\cdot & \cdot & \cdot & \cdot & \cdot \\
0^\mu a_0 & +\ldots+ & \mu^\mu a_\mu & = c(-1)^\mu(\mu+1)^\mu &
\end{array}
\tag{2.111}
$$

Thus the system of linear equations (2.111) has the determinant $\Delta = \Delta_V(0,1,\ldots,\mu)$, where $\Delta_V(s_1,s_2\ldots,s_r)$ means the Vandermonde determinant

$$
\begin{vmatrix}
1 & 1 & \ldots & 1 \\
s_1 & s_2 & \ldots & s_r \\
\cdot & \cdot & \cdot & \cdot \\
s_1^{r-1} & s_2^{r-1} & \ldots & s_r^{r-1}
\end{vmatrix}
= \prod_{l<j}(s_j - s_l).
$$

If we replace the j-th column by the column of the right-hand sides of (2.111), then the resulting determinant is

$$
\begin{aligned}
\Delta_j &= c(-1)^\mu \Delta(0,1,\ldots,j-1,\mu+1,j+1,\ldots,\mu) \\
&= c(-1)^\mu \Delta \frac{(\mu+1-0)\ldots(\mu+1-(j-1))}{(j-0)\ldots(j-(j-1))} \\
&\quad \times \frac{((j+1)-(\mu+1))\ldots(\mu-(\mu+1))}{(j+1-j)\ldots(\mu-j)} \\
&= c(-1)^\mu \Delta \binom{\mu+1}{j}(-1)^{j-\mu} = c(-1)^j \Delta \binom{\mu+1}{j}.
\end{aligned}
$$

Setting $c = 1$ we obtain

$$
a_j = (-1)^j \binom{\mu+1}{j}, \quad j = 0,1,\ldots,\mu,
$$

which yields

$$
\widehat{f}_\mu(s) = \sum_{j=0}^{k}(-1)^j \binom{\mu+1}{j}(s-j)^\mu, \quad 0 \le k \le s \le k+1 \le \mu+1,
$$

so that

$$
\widehat{f}_\mu(s) = \mu \int_{s-1}^{s} \widehat{f}_{\mu-1}(t)dt, \quad \mu \in \mathbb{N}. \tag{2.112}
$$

From (2.112) it is easy to derive some known properties of splines. Thus

- $\widehat{f}_\mu(s) = \widehat{f}_\mu(\mu+1-s)$, $s \ge 0$;
- The function $\widehat{f}_\mu$ increases on the interval $[0,(\mu+1)/2]$ and decreases on $[(\mu+1)/2,\mu+1]$.

If $s_0, s_1, \dots, \mu$ are different real numbers, then it follows from the linear independence of the polynomial system

$$\psi_j(s) = (s + s_j)^\mu, \quad j = 0, 1, \dots, \mu$$

that the system

$$\{\widehat{f}_\mu(s), \widehat{f}_\mu(s-1), \dots, \widehat{f}_\mu(s-\mu)\}$$

is a basis in the n-dimensional space $\widehat{S}_n^\mu$. Furthermore, it can be extended n-periodically to the entire axis $\mathbb{R}$, so the following lemma holds.

Lemma 2.9.1. *Let $n \geq \mu + 1$. Then the system $\{f_k^{(n,\mu)}\}_{k=0}^{n-1}$, such that*

$$f_k^{(n,\mu)}(\exp(i2\pi s)) = \widehat{f}_\mu(ns - k), \quad 0 \leq s < 1, \tag{2.113}$$

is a basis in the space $\underline{S}_n^\mu$.

This lemma completes the preparatory work to study the spline qualocation method. Consider first the stability of this method for singular integral equations without conjugation:

$$A\varphi(t) := (aI + bS)u(t) = f(t), \quad a, b \in \mathbf{PC}(\Gamma_0). \tag{2.114}$$

Let Q_h be the quadrature rule (2.105), and let $\{\chi_k^{(n)}\}_{k=0}^{n-1}$ be the interpolation basis for the spline-space $\underline{S}_n^\delta$ defined in 2.5. We are looking for a function $u^{(n)} \in \underline{S}_n^\delta$, which satisfies the conditions

$$\begin{aligned} &Q_h(f_p^{(n,\mu)} \cdot Au^{(n)}) = Q_h(f_p^{(n,\mu)} \cdot f), \\ &p = 0, 1, \dots, n-1. \end{aligned} \tag{2.115}$$

Using the interpolation basis $\{\chi_k^{(n)}\}_{k=0}^{n-1}$ and the subsequent representation (2.106)

$$u^{(n)}(t) = \sum_{k=0}^{n-1} u_k \chi_k^{(n)}(t),$$

one rewrites equations (2.115) as

$$\sum_{k=0}^{n-1} Q_h(f_p^{(n,\mu)} \cdot A\chi_k^{(n)})u_k = Q_h(f_p^{(n,\mu)} \cdot f), \quad p = 0, 1, \dots, n-1.$$

Thus an approximate solution by the spline qualocation method is found by solving a system of linear algebraic equations with the matrix

$$M_n^{(1)} := (m_{pk})_{k,p=0}^{n-1},$$

where $m_{pk} = Q_h(f_p^{(n,\mu)} \cdot A\chi_k^{(n)})$.

If $t^{(n)}_{j,\varepsilon_l}$, $j = 0, 1, \ldots, n-1$ denote the points

$$t^{(n)}_{j,\varepsilon_l} = \exp\left(2\pi i \frac{j+\varepsilon_l}{n}\right), \quad l = 1,2; \quad t^{(n)}_j := t^{(n)}_{j,0},$$

then the entries of the matrices $M^{(1)}_n$, $n \in \mathbb{N}$ are

$$m_{pk} = \sum_{j=0}^{n-1}\sum_{l=1}^{2} w_l f_p^{(n,\mu)}(t^{(n)}_{j,\varepsilon_l}) \left[a(t^{(n)}_{j,\varepsilon_l})\chi_k^{(n)}(t^{(n)}_{j,\varepsilon_l}) + b(t^{(n)}_{j,\varepsilon_l})(S\chi_k^{(n)})(t^{(n)}_{j,\varepsilon_l})\right].$$

Writing

$$c^{(\mu,\varepsilon)}_{k+rn} = \widetilde{f}_\mu(k+\varepsilon), \quad k = 0,1,\ldots,n-1; \quad r \in \mathbb{Z}, \quad 0 \leq \varepsilon < 1, n \geq \mu+1, \tag{2.116}$$

it follows from (2.113) that

$$f_p^{(n,\mu)}(t^{(n)}_{j,\varepsilon_l}) = f_p^{(n,\mu)}\left(\exp\left(2\pi i \frac{j+\varepsilon_l}{n}\right)\right) = \widetilde{f}_\mu(j-p+\varepsilon) = c^{(\mu,\varepsilon_l)}_{j-p},$$

and

$$m_{pk} = \sum_{j=0}^{n-1}\sum_{l=1}^{2} w_l c^{(\mu,\varepsilon_l)}_{j-p} a(t^{(n)}_{j,\varepsilon_l})\chi_k^{(n)}(t^{(n)}_{j,\varepsilon_l}) + \sum_{j=0}^{n-1}\sum_{l=1}^{2} w_l c^{(\mu,\varepsilon_l)}_{j-p} b(t^{(n)}_{j,\varepsilon_l})(S\chi_k^{(n)})(t^{(n)}_{j,\varepsilon_l}). \tag{2.117}$$

It follows from (2.65) and (2.45) that

$$(\chi_k^{(n)}(t^{(n)}_{j,\varepsilon_l}))_{j,k=0}^{n-1} \backsim U_n K_n \alpha_{\varepsilon,\delta} L_n U_N^{-1} L_n \tag{2.118}$$

$$((S\chi_k^{(n)})(t^{(n)}_{j,\varepsilon_l}))_{j,k=0}^{n-1} \backsim U_n K_n \beta_{\varepsilon,\delta} L_n U_N^{-1} L_n, \tag{2.119}$$

where $\alpha_{\varepsilon,\delta}, \beta_{\varepsilon,\delta} \in \mathbf{PC}(\Gamma_0)$ are defined by (2.66), (2.67) if δ is odd and by (2.71), (2.72) if δ is even. Moreover,

$$(a(t^{(n)}_{j,\varepsilon_l})\delta_{jk})_{j,k=0}^{n-1} \sim K_n^\varepsilon a L_n, \tag{2.120}$$

so for any n-periodic sequence $\{c_k\}_{k=-\infty}^{+\infty}$ one has

$$(c_{j-p})_{j,p=0}^{n-1} \sim U_n K_n \xi L_n U_n^{-1}$$

where ξ is the operator of multiplication by the polynomial

$$\xi(t) = \sum_{k=0}^{n} c_k t^k, \quad t \in \Gamma_0.$$

Thus if we set $c_k := c_k^{(\mu,\varepsilon)}$, $k \in \mathbb{Z}$ where $c_k^{(\mu,\varepsilon)}$ are defined in (2.116), we have

$$(c_{j-p}^{(\mu,\varepsilon)})_{j,p=0}^{n-1} \sim U_n K_n F_{\varepsilon,\mu} L_n U_n^{-1} \tag{2.121}$$

with the polynomial

$$F_{\varepsilon,\mu}(t) = \sum_{k=0}^{n-1} \widetilde{f}_\mu(k+\varepsilon)t^k.$$

Combining (2.117)–(2.121) yields the representations

$$\begin{aligned} M_n^{(1)} = \sum_{l=1}^{2} & w_l U_n K_n F_{\varepsilon_l,\mu} L_n U_n^{-1} \\ & \times \left(K_n^{\varepsilon_l} a L_n U_n K_n \alpha_{\varepsilon_l,\delta} L_n U_n^{-1} L_n + K_n^{\varepsilon_l} b L_n U_n K_n \beta_{\varepsilon_l,\delta} L_n U_n^{-1} L_n\right) \end{aligned}$$

for the operators $M_n^{(1)}$.

Assume that the coefficients $a, b \in \mathbf{PC}(\Gamma_0)$ are continuous on the set $\Gamma_0 \setminus \{t_0^{(m)}, t_1^{(m)}, \dots, t_{m-1}^{(m)}\}$ and choose $n = mn_1$, $n_1 \in \mathbb{N}$. Then

$$\lim_{n\to\infty} ||(K_n^\varepsilon - K_n)L_n|| = 0,$$

so the sequence $(M_n^{(1)})$ is stable if and only if the sequence

$$\begin{aligned} \widehat{M}_n^{(1)} = \sum_{l=1}^{2} & w_l U_n K_n F_{\varepsilon_l,\mu} L_n U_n^{-1} \\ & \times \left(K_n a L_n U_n K_n \alpha_{\varepsilon_l,\delta} L_n U_n^{-1} L_n + K_n b L_n U_n K_n \beta_{\varepsilon_l,\delta} L_n U_n^{-1} L_n\right) \end{aligned} \tag{2.122}$$

is stable. The stability of the sequence (2.122) can be treated similarly to the stability of the sequence (2.109), under the following result.

Theorem 2.9.2. *Let $a, b \in \mathbf{PC}(\Gamma_0)$ be continuous functions on the set $\Gamma_0 \setminus \{t_0^{(m)}, t_1^{(m)}, \dots, t_{m-1}^{(m)}\}$. The spline qualocation method (2.115) is stable if and only if for any $\tau \in \Gamma_0$ the operators*

$$\begin{aligned} A^\tau = \sum_{l=1}^{2} w_l \widetilde{F}_{\varepsilon_l,\mu}[(a(\tau+0)P + (a(\tau+0)Q)]\widetilde{\alpha}_{\varepsilon_l,\delta} \\ + (b(\tau+0)P + (a(\tau+0)Q)\widetilde{\beta}_{\varepsilon_l,\delta}], \end{aligned} \tag{2.123}$$

and

$$A_\tau = A_\tau^+ P + A_\tau^- Q, \tag{2.124}$$

where

$$A_\tau^\pm(t) = \sum_{l=1}^{2} w_l F_{\varepsilon_l,\mu}(\tau)[a(t)\alpha_{\varepsilon_l,\delta}(\tau \pm 0) + b(t)\beta_{\varepsilon_l,\delta}(\tau \pm 0)],$$

are invertible in $\mathcal{L}(L_2(\Gamma_0))$ *for any* $\tau \in \Gamma_0$.

There is a difference between stability conditions for the polynomial [104] and spline qualocation methods for singular integral equations without conjugation. In polynomial qualocation, even in the case where the coefficients of the initial equation are continuous, one still has to check the invertibility of the operators that belong to the algebra of singular integral operators generated by the Cauchy singular integral operator and piecewise continuous functions. However, there are no effective invertibility criteria for such operators. In contrast, the invertibility of at least one family of operators responsible for stability of the spline qualocation method (viz., (2.124)) that consists of singular integral operators is well studied, so corresponding criteria can be easily verified. Moreover, if the coefficients of the initial equation are continuous, then the invertibility of all the auxiliary operators (2.123) and (2.124) can be studied effectively. Below we formulate the corresponding results for cases of piecewise continuous and continuous coefficients a and b.

Let $\Delta_\tau : \Gamma_0 \times [0,1] \mapsto \mathbb{C}$ denote the matrix function

$$\Delta_\tau(t,x) :=$$
$$\begin{pmatrix} A_\tau^+(t+0)x + A_\tau^+(t-0)(1-x) & (A_\tau^-(t+0) - A_\tau^-(t-0))\sqrt{x(1-x)} \\ (A_\tau^+(t+0) - A_\tau^+(t-0))\sqrt{x(1-x)} & A_\tau^-(t+0)(1-x) + A_\tau^-(t-0)x \end{pmatrix},$$

and set $c_\tau := (A_\tau^-)^{-1}A_\tau^+$. Combining Theorem 2.9.2 with results of [90] and [94], one rewrites the stability conditions in a more convenient form.

Corollary 2.9.3. *The spline-qualocation method* (2.115) *for the operator* (2.114) *is stable if and only if the following four conditions are satisfied:*

1. $\det \Delta_\tau(t,x) \neq 0$ *for all* $\tau \in \Gamma_0$ *and for all* $x \in [0,1]$*;*
2. *For every* $\tau \in \Gamma_0$*, the winding number of the curve*

$$\Gamma_{c_\tau} := \{c_\tau(t+0)x + c_\tau(t-0)(1-x),\ (t,x) \in (\Gamma_0, [0,1])\}$$

   is equal to zero;
3. *For every* $\tau \in \Gamma_0$*, the operators* A^τ *defined by* (2.123) *are invertible in* $\mathcal{L}(L_2)$.

If we assume that the coefficients a and b are continuous on Γ, then for every $\tau \in \Gamma_0$, the operator A^τ becomes just the operator of multiplication by the function

$$A^\tau = \sum_{l=1}^{2} w_l \widetilde{F}_{\varepsilon_l,\mu}(a\widetilde{\alpha}_{\varepsilon_l,\delta} + b\widetilde{\beta}_{\varepsilon_l,\delta}).$$

Moreover, since $\alpha_{\varepsilon,\delta} \in \mathbf{C}(\Gamma_0)$ and $\beta_{\varepsilon,\delta} \in \mathbf{C}(\Gamma_0 \setminus \{1\})$ the functions A_τ^+ and A_τ^- coincide for every $\tau \in \Gamma_0 \setminus \{1\}$, i.e.,

$$A_\tau^+(t) = A_\tau^-(t) = \sum_{l=1}^{2} w_l F_{\varepsilon_l,\mu}(\tau)(a(t)\widetilde{\alpha}_{\varepsilon_l,\delta}(\tau) + b(t)\widetilde{\beta}_{\varepsilon_l,\delta}(\tau)).$$

Thus for $\tau \neq 1$ the operator A_τ is again a multiplication operator. More precisely,

$$A_\tau = A_\tau^+ I, \quad \tau \neq 1.$$

Recall that for (2.73) we also have $\beta_{\varepsilon_l,\delta}(1 \pm 0) = \pm 1$ and that $\alpha_{\varepsilon_l,\delta}(1) = 1$.

Corollary 2.9.4. *Let $a, b \in \mathbf{C}(\Gamma_0)$. The spline qualocation method for the operator A is stable if and only if:*

1. *The operator*

$$A_1 = \sum_{l=1}^{2} w_l F_{\varepsilon_l,\delta}(1)(a+b)P + \sum_{l=1}^{2} w_l F_{\varepsilon_l,\delta}(1)(a-b)Q$$

is invertible in $\mathcal{L}(L_2)$;

2. *The operators of multiplication A^τ, $\tau \in \Gamma_0$ and A_τ, $\tau \in \Gamma_0 \setminus \{1\}$ are invertible in $\mathcal{L}(L_2)$.*

All conditions of Corollary 2.9.4 are easily verified.

Using the previous considerations and the results of Section 2.4 one can study the stability of the spline qualocation method for singular integral equations with conjugation. Let us treat the approximation method (2.115). For simplicity, we now consider the operator B defined by (2.76). Following the arguments of Section 2.5 and using representations (2.122), one notes that the stability of spline qualocation method for the operator B can be studied via the stability of the operator sequence

$$\begin{aligned} B_n' = &\sum_{l=1}^{2} w_l U_n K_n F_{\varepsilon_l,\mu} L_n U_n^{-1} \\ &\times \left(K_n a L_n U_n K_n \alpha_{\varepsilon_l,\delta} L_n U_n^{-1} L_n + K_n b L_n U_n K_n \beta_{\varepsilon_l,\delta} L_n U_n^{-1} L_n\right) \\ &+ \sum_{l=1}^{2} w_l U_n K_n F_{\varepsilon_l,\mu} L_n U_n^{-1} \\ &\times \left(K_n c L_n U_n K_n \alpha^\circ_{\varepsilon_l,\delta} L_n U_n^{-1} L_n + K_n d L_n U_n K_n \beta^\circ_{\varepsilon_l,\delta} L_n U_n^{-1} L_n\right) M. \end{aligned}$$

Rewriting this sequence in matrix form, one obtains

$$\begin{aligned} B_n' = &[w_1 \widehat{F}_{-\varepsilon_1,n}(\widetilde{a}_n \widehat{\alpha}_{\varepsilon_1,n} + \widetilde{b}_n \widehat{\beta}_{\varepsilon_1,n}) + w_2 \widehat{F}_{-\varepsilon_2,n}(\widetilde{a}_n \widehat{\alpha}_{\varepsilon_2,n} + \widetilde{b}_n \widehat{\beta}_{\varepsilon_2,n})] \\ &+ [w_1 \widehat{F}_{-\varepsilon_1,n}(\widetilde{c}_n \widehat{\alpha}_{\varepsilon_1,n} + \widetilde{d}_n \widehat{\beta}_{\varepsilon_1,n}) + w_2 \widehat{F}_{-\varepsilon_2,n}(\widetilde{c}_n \widehat{\alpha}_{\varepsilon_2,n} + \widetilde{d}_n \widehat{\beta}_{\varepsilon_2,n})] \overline{M}_n, \end{aligned} \tag{2.125}$$

where the functions $\alpha_\varepsilon = \alpha_{\varepsilon,\delta}$, $\beta_\varepsilon = \beta_{\varepsilon,\delta}$, $\gamma_\varepsilon = \gamma_{\varepsilon,\delta}$, $\theta_\varepsilon = \theta_{\varepsilon,\delta}$ are defined in Section 2.5, and $\widehat{F}_{\varepsilon_l,n}$, $l = 1,2$ are the circulants corresponding to the functions $F_{\varepsilon_l,\mu}$, $l = 1,2$. Note that the sequence (2.125) belongs to the real algebra $\widetilde{\mathcal{C}}$ studied in Section 2.4. This again provides the possibility to apply results of Section 2.4 to study the stability of (2.125).

For any $\tau \in \Gamma_0$, let us introduce the matrices $B^\tau_{\pm,l} = B^\pm_\tau(t)$, $l = 1,2$ and $B^\pm_\tau = B^\tau_{\pm,l}(t)$, $t \in \Gamma_0$ by

$$B^\tau_{\pm,l} = (\mathcal{A}(\tau \pm 0) + \mathcal{C}(\tau \pm 0)\Lambda)\boldsymbol{\alpha}_{\varepsilon_l,\delta}(t) + (\mathcal{B}(\tau \pm 0) + \mathcal{D}(\tau \pm 0)\Lambda)\boldsymbol{\beta}_{\varepsilon_l,\delta}(t),$$

$$B^\pm_\tau = \sum_{l=1}^{2} w_l F_{\varepsilon_l,\mu}(\tau)[(\mathcal{A}(t) + \mathcal{C}(t)\Lambda)\boldsymbol{\alpha}_{\varepsilon_l,\delta}(\tau \pm 0) + (\mathcal{B}(t) + \mathcal{D}(t)\Lambda)\boldsymbol{\beta}_{\varepsilon_l,\delta}(\tau \pm 0)],$$

where $\mathcal{A}, \mathcal{B}, \mathcal{C}, \mathcal{D}, \boldsymbol{\alpha}_{\varepsilon_l,\delta}, \boldsymbol{\beta}_{\varepsilon_l,\delta}$ and Λ are defined in accordance with (2.47) and (2.48). Then we have at once

Theorem 2.9.5. *Let $a,b,c,d \in \mathbf{PC}(\Gamma_0)$ and let them be continuous on the set $\Gamma_0 \setminus \{t_0^{(m)}, t_1^{(m)}, \ldots, t_{m-1}^{(m)}\}$. The spline qualocation method (2.115) for the operator $B \in \mathcal{L}_{add}(L_2(\Gamma_0))$ is stable if and only if for any $\tau \in \Gamma_0$, the operators*

$$B^\tau = \sum_{l=1}^{2} w_l \widetilde{F}_{\varepsilon_l,\mu}(PB^\tau_{+,l} + QB^\tau_{-,l}), \tag{2.126}$$

and

$$B_\tau = B^+_\tau P + B^-_\tau Q, \tag{2.127}$$

are invertible in $\mathcal{L}(L^2_2(\Gamma_0))$.

In particular, if $a,b,c,d \in \mathbf{C}(\Gamma_0)$ then

$$B^\tau_{+,l} = B^\tau_{-,l} \quad \text{for every} \quad \tau \in \Gamma_0,$$

and

$$B^+_\tau = B^-_\tau \quad \text{for every} \quad \tau \in \Gamma_0 \setminus \{1\},$$

so for these $\tau \in \Gamma_0$, the operators (2.126) and (2.127) become the operators of multiplication by the corresponding matrix functions, so the stability conditions for the spline qualocation method are much simpler.

Corollary 2.9.6. *Let $a,b,c,d \in \mathbf{C}(\Gamma_0)$. The spline qualocation method for the operator B is stable if and only if:*

1. *The operator*

$$B_1 = \sum_{l=1}^{2} w_l F_{\varepsilon_l,\delta}(1)\left[\begin{pmatrix} a+b & c-d \\ \overline{c}+\overline{d} & \overline{a}-\overline{b} \end{pmatrix} P + \begin{pmatrix} a-b & c+d \\ \overline{c}-\overline{d} & \overline{a}+\overline{b} \end{pmatrix} Q\right]$$

is invertible in $\mathcal{L}(L^2_2)$;

2. *The operators of multiplication* B^{τ}, $\tau \in \Gamma_0$ *and* B_{τ}, $\tau \in \Gamma_0 \backslash \{1\}$ *are invertible in* $\mathcal{L}(L_2^2)$.

Note that Condition 2 can be easily verified. However, in contrast to Corollary 2.9.4, there are no effective criteria to check the invertibility of the operator B_1, and this can be achieved only in special cases.

Remark 2.9.7. The results of Sections 2.5, 2.6, 2.8, 2.9 are also valid for singular integral equations with conjugation on an arbitrary simple closed Lyapunov curve Γ. Indeed, let $\gamma = \gamma(s)$, $s \in \mathbb{R}$ be a 1-periodic parametrization of Γ. If we identify points of Γ and Γ_0 in such a way that $\Gamma \ni \gamma(s) \sim \exp(i2\pi s) \in \Gamma_0$ and proceed analogously to the previous considerations, we can establish stability conditions for the corresponding approximation methods on Γ. However, such an approach fails for quadrature methods, because the quadratures employed here are based on specific relations between the points of the unit circle.

2.10 Quadrature Methods for Equations on Closed Smooth Curves

Let $\gamma : \mathbb{R} \to \Gamma$ be a 1-periodic parametrization of simple closed smooth curve Γ such that $\gamma'(s) \neq 0$, $s \in [0, 1)$ and such that the second derivative γ'' satisfies the Hölder condition. Fix an $\varepsilon \in (-1/2, 1/2]$, and for any $n \in \mathbb{N}$ introduce the points

$$t_j^{(n)} := \gamma\left(\frac{j+1/2}{n}\right), \quad \tau_j^{(n)} := \gamma\left(\frac{j+\varepsilon+1/2}{n}\right),$$

where $j = 0, 1, \ldots, n-1$. The quadrature methods for the operator B considered below are based on the quadrature rules

$$\int_{\Gamma} \varphi(t)dt \approx \sum_{j=0}^{n-1} \varphi(t_j^{(n)})(t_{j+1}^{(n)} - t_j^{(n)}),$$

$$\int_{\Gamma} \varphi(t)dt \approx \sum_{j=0}^{n-1} \varphi(t_j^{(n)})(t_{j+1}^{(n)} - t_{j-1}^{(n)}).$$

These quadratures approximate integrals of smooth functions with the corresponding rates of error $1/n$ and $1/n^2$.

An approximate solution of the equation

$$(Bx)(t) \equiv a(t)x(t) + \frac{b(t)}{\pi i}\int_{\Gamma} \frac{x(\tau)\,d\tau}{\tau - t} + c(t)\overline{x(t)} + \frac{d(t)}{\pi i}\int_{\Gamma} \frac{\overline{x(\tau)\,d\tau}}{\tau - t} = f(t), \quad (2.128)$$

is sought in the form

$$x_n(t) = \sum_{k=0}^{n-1} \xi_k \chi_{k,0}^{(n)}(t),$$

where

$$\chi_{k,0}^{(n)}(\gamma(s)) = \begin{cases} ns - k + \frac{1}{2} & \text{if} \quad s \in \left[\dfrac{k-1/2}{n}, \dfrac{k+1/2}{n}\right], \\ k - ns + \frac{3}{2} & \text{if} \quad s \in \left[\dfrac{k+1/2}{n}, \dfrac{k+3/2}{n}\right], \\ 0 & \text{if} \quad s \notin \left[\dfrac{k-1/2}{n}, \dfrac{k+3/2}{n}\right]. \end{cases}$$

The coefficients ξ_k, $k = 0, 1, \ldots, n-1$ are the solutions of the algebraic equations

$$\begin{aligned} &(a(\tau_k^{(n)}) - i\cot(\pi\varepsilon)b(\tau_k^{(n)}))\,\xi_k + \frac{b(\tau_k^{(n)})}{\pi i}\sum_{j=0}^{n-1}\frac{t_{j+1}^{(n)} - t_j^{(n)}}{t_j^{(n)} - \tau_k^{(n)}}\xi_j \\ &+ (c(\tau_k^{(n)}) - i\cot(\pi\varepsilon)d(\tau_k^{(n)}))\overline{\xi}_k + \frac{d(\tau_k^{(n)})}{\pi i}\sum_{j=0}^{n-1}\frac{t_{j+1}^{(n)} - t_j^{(n)}}{t_j^{(n)} - \tau_k^{(n)}}\,\overline{\xi}_j = f(\tau_k^{(n)}), \end{aligned} \tag{2.129}$$

$$\begin{aligned} &\frac{a(\tau_k^{(n)})}{2}(\xi_k + \xi_{k+1}) + \frac{b(\tau_k^{(n)})}{2\pi i}\sum_{j=0}^{n-1}\frac{t_{j+1}^{(n)} - t_{j-1}^{(n)}}{t_j^{(n)} - \tau_k^{(n)}}\,\xi_j \\ &\quad + \frac{c(\tau_k^{(n)})}{2}(\overline{\xi}_k + \overline{\xi}_{k+1}) + \frac{d(\tau_k^{(n)})}{2\pi i}\sum_{j=0}^{n-1}\frac{t_{j+1}^{(n)} - t_{j-1}^{(n)}}{t_j^{(n)} - \tau_k^{(n)}}\,\overline{\xi}_j = f(\tau_k^{(n)}), \end{aligned} \tag{2.130}$$

$$\begin{aligned} &a(t_k)\xi_k + \frac{b(t_k)}{\pi i}\sum_{\substack{j=0\\ j\neq k}}^{n-1}\frac{t_{j+1}^{(n)} - t_j^{(n)}}{t_j^{(n)} - t_k^{(n)}}\,\xi_j \\ &\qquad + c(\tau_k^{(n)})\overline{\xi}_k + \frac{d(\tau_k^{(n)})}{2\pi i}\sum_{\substack{j=0\\ j\neq k}}^{n-1}\frac{t_{j+1}^{(n)} - t_j^{(n)}}{t_j^{(n)} - t_k^{(n)}}\,\overline{\xi}_j = f(t_k), \end{aligned} \tag{2.131}$$

$$\begin{aligned} &a(t_k)\xi_k + \frac{b(t_k)}{\pi i}\sum_{\substack{j=0\\ j\equiv k+1 (\mathrm{mod}\, 2)}}^{n-1}\frac{t_{j+1}^{(n)} - t_{j-1}^{(n)}}{t_j^{(n)} - t_k^{(n)}}\,\xi_j \\ &\qquad + c(\tau_k^{(n)})\overline{\xi}_k + \frac{d(\tau_k^{(n)})}{2\pi i}\sum_{\substack{j=0\\ j\equiv k+1 (\mathrm{mod}\, 2)}}^{n-1}\frac{t_{j+1}^{(n)} - t_{j-1}^{(n)}}{t_j^{(n)} - t_k^{(n)}}\,\overline{\xi}_j = f(t_k), \end{aligned} \tag{2.132}$$

where $\xi_n := \xi_0$, $\xi_{-1} := \xi_{n-1}$ and $k = 0, 1, \ldots, n-1$ for all systems (2.129) – (2.132).

Consider the function

$$\chi_{k,\varepsilon}^{(n)}(\gamma(s)) = \begin{cases} ns - k + \frac{1}{2} + \varepsilon & \text{if} \quad s \in \left[\dfrac{k-1/2+\varepsilon}{n}, \dfrac{k+1/2+\varepsilon}{n}\right], \\ k - ns + \frac{3}{2} + \varepsilon & \text{if} \quad s \in \left[\dfrac{k+1/2+\varepsilon}{n}, \dfrac{k+3/2+\varepsilon}{n}\right], \\ 0 & \text{if} \quad s \notin \left[\dfrac{k-1/2+\varepsilon}{n}, \dfrac{k+3/2+\varepsilon}{n}\right]. \end{cases}$$

We denote the orthogonal projection from $L_2(\Gamma_0)$ onto subspace $X_n :=$ span $\{\chi_{0,\varepsilon}^{(n)}, \chi_{1,\varepsilon}^{(n)}, \dots, \chi_{n-1,\varepsilon}^{(n)}\}$ by L_n^ε, and the interpolation projection by

$$N_n^\varepsilon u = \sum_{k=0}^{n-1} u(\tau_k^{(n)})\chi_{k,\varepsilon}^{(n)}.$$

Let $B_n \in \mathcal{L}_{add}(\operatorname{im} L_n^0, \operatorname{im} L_n^\varepsilon)$ be the operator the matrix representation of which in the bases $(\chi_{k,0}^{(n)}, \chi_{k,\varepsilon}^{(n)})$ is $B_n = R_n^{(1)} + R_n^{(2)}\overline{M}_n$, where the matrices $R_n^{(1)}$ and $R_n^{(2)}$ are defined by the corresponding parts of systems (2.129) – (2.132). Thus each of the systems (2.129) – (2.132) is equivalent to the operator equation

$$B_n x_n = N_n^\varepsilon f.$$

We identify any function φ defined on Γ with a function $\widetilde{\varphi}$ on the unit circle Γ_0 by

$$\widetilde{\varphi} = \varphi \circ \gamma \circ \gamma_0^{-1}, \quad \gamma_0(s) = \exp(i2\pi s), \quad s \in \mathbb{R},$$

and let $t = \gamma_0(s)$. The stability conditions for these quadrature methods are then given by the following two theorems.

Theorem 2.10.1. *Let $a, b, c, d \in \mathbf{PC}(\Gamma)$ and operator B defined by (2.128) be invertible in $\mathcal{L}_{add}(L_2(\Gamma))$. The quadrature methods (2.129) – (2.131) are stable if and only if the operator*

$$B_1 = \begin{pmatrix} \widetilde{a}+\widetilde{b} & \widetilde{c}+\widetilde{d} \\ \overline{\widetilde{c}}-\overline{\widetilde{d}} & \overline{\widetilde{a}}-\overline{\widetilde{b}} \end{pmatrix} P_{\Gamma_0} + \begin{pmatrix} \widetilde{a}-\widetilde{b} & \widetilde{c}-\widetilde{d} \\ \overline{\widetilde{c}}+\overline{\widetilde{d}} & \overline{\widetilde{a}}+\overline{\widetilde{b}} \end{pmatrix} Q_{\Gamma_0}$$

and the operators

$$B^\tau = P_{\Gamma_0} A_+^\tau + Q_{\Gamma_0} A_-^\tau \tag{2.133}$$

are invertible in $\mathcal{L}(L_2^2(\Gamma))$ for all $\tau \in \Gamma_0$, where $A_\pm^\tau$ are the operators of multiplication by the matrix functions

$$A_+^\tau(t) = [\mathcal{A}(\tau+0)\boldsymbol{\alpha}(\overline{t}) + \mathcal{B}(\tau+0)\boldsymbol{\beta}(\overline{t}) + (\mathcal{C}(\tau+0)\boldsymbol{\gamma}(\overline{t}) + \mathcal{D}(\tau+0)\boldsymbol{\theta}(\overline{t}))],$$
$$A_-^\tau(t) = [\mathcal{A}(\tau-0)\boldsymbol{\alpha}(\overline{t}) + \mathcal{B}(\tau-0)\boldsymbol{\beta}(\overline{t}) + (\mathcal{C}(\tau-0)\boldsymbol{\gamma}(\overline{t}) + \mathcal{D}(\tau-0)\boldsymbol{\theta}(\overline{t}))],$$

and the functions $\alpha, \beta, \gamma, \theta$ are defined by:

1. *For the method* (2.129):

$$\alpha(t) = \gamma(t) = 1, \quad \beta(t) = \theta(t) = -i\cot(\pi\varepsilon) + \frac{2\exp(i2\pi\varepsilon s)}{1-\exp(i\pi\varepsilon)};$$

2. *For the method* (2.130):

$$\alpha(t) = \gamma(t) = \frac{1+t}{2}, \quad \beta(t) = \theta(t) = \exp(i\pi s);$$

3. *For the method* (2.131):

$$\alpha(t) = \gamma(t) = 1, \quad \beta(t) = \theta(t) = 1 - 2s.$$

Theorem 2.10.2. *Let $a, b, c, d \in \mathbf{PC}(\Gamma)$ and operator B defined by* (2.128) *be invertible in $\mathcal{L}_{add}(L_2(\Gamma))$. The quadrature method* (2.132) *is stable if and only if the operator*

$$B_{-1} = \begin{pmatrix} \widetilde{a}-\widetilde{b} & \widetilde{c}-\widetilde{d} \\ \overline{\widetilde{c}}+\overline{\widetilde{d}} & \overline{\widetilde{a}}+\overline{\widetilde{b}} \end{pmatrix} P_{\Gamma_0} + \begin{pmatrix} \widetilde{a}+\widetilde{b} & \widetilde{c}+\widetilde{d} \\ \overline{\widetilde{c}}-\overline{\widetilde{d}} & \overline{\widetilde{a}}-\overline{\widetilde{b}} \end{pmatrix} Q_{\Gamma_0}$$

and the operators (2.133) *are invertible in $\mathcal{L}(L_2^2(\Gamma))$ for all $\tau \in \Gamma_0$, where*

$$\alpha(t) = \gamma(t) = 1, \quad \beta(t) = \theta(t) = \begin{cases} 1 & \text{if } 0 \le s < 1/2, \\ -1 & \text{if } 1/2 \le s < 1. \end{cases}$$

The proof of stability mainly follows the corresponding proofs of [183, pp. 366–370] for singular integral operators without conjugation, but leads to operator sequences studied in Section 2.4. Thus let us consider the operator $J : L_2(\Gamma) \mapsto L_2(\Gamma_0)$ defined by

$$J\varphi = \widetilde{\varphi}$$

where $\widetilde{\varphi} = \varphi \circ \omega$ and $\omega = \gamma \circ \gamma_0^{-1}$. Following [143, 157] one can represent the operator B as a product $B = J^{-1}B^{(0)}J$ where

$$B^{(0)} = \widetilde{a}I + \widetilde{b}S_{\Gamma_0} + \widetilde{c}M + \widetilde{d}S_{\Gamma_0}M + k^{(0)} + k^{(1)}M,$$

and where the operators $k^{(0)}$ and $k^{(1)}$ are correspondingly defined by

$$(k^{(0)}\widetilde{x})(t) = \frac{\widetilde{b}(t)}{\pi i}\int_{\Gamma_0}\left[\frac{\omega'(\tau)}{\omega(\tau)-\omega(t)} - \frac{1}{\tau - t}\right]\widetilde{x}(\tau)\,d\tau, \tag{2.134}$$

$$(k^{(1)}\widetilde{x})(t) = \frac{\widetilde{d}(t)}{\pi i}\int_{\Gamma_0}\left[\frac{\omega'(\tau)}{\omega(\tau)-\omega(t)} - \frac{1}{\tau - t}\right]\widetilde{x}(\tau)\,d\tau. \tag{2.135}$$

Thus the study of quadrature methods (2.129) – (2.132) can be reduced to the study of quadrature methods for the operator $B^{(0)}$. However, the corresponding approximation sequences for the operator $B^{(0)}$ satisfy all conditions of Theorem 2.7.1. For more details the reader can consult [56].

Remark 2.10.3. Let $H^r(\Gamma)$ be the Sobolev space of order r [122], and let x and x_n be exact and approximate solutions of the equation (2.128). Assuming the invertibility of the operator B, the stability of quadrature methods (2.129) – (2.132), and from the approximation properties of splines [39, 183] and (1.30), one can show that:

1. If $a, b, c, d \in H^1(\Gamma)$, then the approximate solutions of (2.128) obtained by method (2.129) or (2.130) satisfy the inequality

$$||x - x_n||_{L_2(\Gamma)} \leq \frac{d_5}{n} ||x||_{H^1(\Gamma)}.$$

2. If $a, b, c, d \in H^2(\Gamma)$ and the derivative γ''' exists and satisfies the Hölder condition, the approximate solutions of (2.128) obtained by method (2.131) or (2.132) satisfy the inequality

$$||x - x_n||_{L_2(\Gamma)} \leq \frac{d_6}{n^2} ||x||_{H^2(\Gamma)}.$$

2.11 A Remark on Tensor Product Techniques

In this section we present another approach to problems from Sections 2.1–2.2 and 2.4. This approach is based on the theory of tensor products of C^*-algebras. Let $\mathfrak{C}_k$ denote the C^*-algebra $\mathbb{C}^{k\times k}$. Let us first consider the complex C^*-algebra $\mathcal{C}$ and the element m introduced after the proof of Lemma 2.4.2. By Proposition 2.4.3, these objects satisfy all the axioms $(A_1) - (A_5)$ of Section 1.2, so Corollary 1.4.7 yields that stability in the algebra $\widetilde{\mathcal{C}}$ can be reduced to stability in the algebra $\mathcal{C}^{2\times 2}$. However, the latter problem can be studied with the help of stability results for sequences from the algebra $\mathcal{C}$ which are known [183, Chapter 10]. Note that the algebra $\mathcal{C}$ can also be generated by the sequences (B_n), where

$$B_n := \widetilde{a}_n \widehat{\alpha}_n + \widetilde{b}_n \widehat{\beta}_n + C_n,$$

where $a, b \in \mathbf{PC}(\Gamma_0)$ and (C_n) belongs to the ideal $G_{\mathcal{C}}$ of all sequences that uniformly converge to zero. To each such sequence (B_n) we assign two families $W_{(w,1)}(B_n)$ and $W_{(w,2)}(B_n)$, $w \in \Gamma_0$ of singular integral operators acting on the space $L_2(\Gamma_0)$ and defined by

$$W_{(w,1)}(B_n) := [\alpha(w+0)a + \beta(w+0)b]P + [\alpha(w-0)a + \beta(w-0)b]Q,$$
$$W_{(w,2)}(B_n) := P[a(w+0)\widetilde{\alpha} + b(w+0)\widetilde{\beta}] + Q[a(w-0)\widetilde{\alpha} + b(w-0)\widetilde{\beta}],$$

where $\widetilde{\alpha}(t) = \alpha(\overline{t})$ and $\widetilde{\beta}(t) = \beta(\overline{t})$.

The mappings $W_{(\omega,1)}$, $W_{(\omega,2)}$, $\omega \in \Gamma_0$ can be extended to *-homomorphisms acting on $\mathcal{C}$ with images in $\mathcal{O}_2(\Gamma_0, 1)$ (cf. Section 1.10.5). If these extended homomorphisms are denoted by $W_{(\omega,1)}$ and $W_{(\omega,2)}$ again, the following result holds [183, Theorem 10.41].

Theorem 2.11.1. *A sequence* $(B_n) \in \mathcal{C}$ *is stable if and only if the operators* $W_{(\omega,1)}(B_n)$ *and* $W_{(\omega,2)}(B_n)$ *are invertible for each* $w \in \Gamma_0$.

In passing we note one interesting result. Consider the C^*-algebra of all bounded functions on Γ_0 that take values in $\mathcal{O}_2(\Gamma_0, 1) \times \mathcal{O}_2(\Gamma_0, 1)$ and its subalgebra $\operatorname{smb}\mathcal{C}$ which is constituted by all functions of the form $(W_{(\omega,1)}(B_n), W_{(\omega,2)}(B_n))_{\omega\in\Gamma_0}$, $(B_n) \in \mathcal{C}$. It is easily seen that $\operatorname{smb}\mathcal{C}$ actually forms a C^*-algebra and the mapping $\operatorname{smb} : \mathcal{C} \mapsto \operatorname{smb}\mathcal{C}$ defined by

$$(B_n) \mapsto (W_{(\omega,1)}(B_n), W_{(\omega,2)}(B_n))_{\omega\in\Gamma_0}$$

is again a *-homomorphism with kernel $G_{\mathcal{C}}$. Since any injective *-isomorphism preserves the norm [66], one obtains that for any sequence (B_n) from the algebra $\mathcal{C}$ the relation

$$\lim_{n\to\infty} ||B_n|| = ||\operatorname{smb}(B_n)|| = \sup_{w\in\Gamma_0} \max\{||W_{(\omega,1)}(B_n)||, ||W_{(\omega,2)}(B_n)||\}.$$

Thus if the sequence (B_n) belongs to the algebra $\mathcal{C}$ and is stable, then for the condition numbers $\operatorname{cond} B_n := ||B_n||||B_n^{-1}||$ one has

$$\lim_{n\to\infty} (\operatorname{cond} B_n) = ||\operatorname{smb}(B_n)||\, ||(\operatorname{smb}(B_n))^{-1}||.$$

Now we shall demonstrate how the stability problem for the sequences $(B_n) \in \mathcal{C}^{2\times 2}$ can be studied. Let us denote by $G_{\mathcal{C}^{2\times 2}}$ the ideal of all sequences in $\mathcal{C}^{2\times 2}$ tending in the norm to zero. It is easily seen that $G_{\mathcal{C}^{2\times 2}} = G_{\mathcal{C}}^{2\times 2}$. Let us also recall the following result [68, 69].

Lemma 2.11.2. *If*

$$(0) \mapsto \mathcal{A} \mapsto \mathcal{D} \mapsto \mathcal{E} \mapsto (0)$$

is an exact sequence of C^**-algebras, then the sequence*

$$(0) \mapsto \mathcal{A} \otimes \mathfrak{C}_k \mapsto \mathcal{D} \otimes \mathfrak{C}_k \mapsto \mathcal{E} \otimes \mathfrak{C}_k \mapsto (0)$$

is also exact, where $\mathfrak{C}_k := \mathbb{C}^{k\times k}$.

The proof of Lemma 2.11.2 is based upon the fact that the tensor product $\mathcal{A} \otimes \mathfrak{C}_k$ of a C^*-algebra $\mathcal{A}$ and $\mathfrak{C}_k$ is naturally isomorphic to the C^*-algebra of $(k \times k)$-matrices with entries from $\mathcal{A}$. If we now set $\mathcal{A} := G_{\mathcal{C}}$, $\mathcal{D} := \mathcal{C}$, then $\mathcal{E}$ is isometrically isomorphic to $\mathcal{C}/G_{\mathcal{C}}$. This algebra is nothing else than the algebra $\operatorname{smb}\mathcal{C}$, and the sequence of C^*-algebras

$$(0) \mapsto G_{\mathcal{C}} \mapsto \mathcal{C} \mapsto \operatorname{smb}\mathcal{C} \mapsto (0)$$

is exact. Thus by Lemma 2.11.2 the sequence

$$(0) \mapsto G_{\mathcal{C}}^{2\times 2} \mapsto \mathcal{C}^{2\times 2} \mapsto (\operatorname{smb}\mathcal{C})^{2\times 2} \mapsto (0)$$

is also exact, so

$$\mathcal{C}^{2\times 2}/G_{\mathcal{C}}^{2\times 2} \cong (\operatorname{smb}\mathcal{C})^{2\times 2},$$

which implies Theorem 2.4.14.

Consider now another example of application of Lemma 2.11.2. Let $\mathcal{A}$ be the C^*-algebra defined in Section 2.1. Consider the collocation method studied in Section 2.2, and introduce the smallest C^*-subalgebra $\mathcal{A}(\mathcal{C})$ of $\mathcal{A}$ that contains all sequences of the form $(L_n(aI + bS)P_n)$, where a and b are continuous function on Γ_0, and the ideal $G_{\mathcal{C}}$ of all sequences (C_n), $C_n : \operatorname{im} P_n \mapsto \operatorname{im} P_n$ that converge uniformly to zero. Recall that for any sequence $(A_n) = (L_n(aI+bS)P_n)$ the strong limits $W_1(A_n) := s-\lim A_n P_n$ and $W_2(A_n) := s-\lim(W_n A_n W_n)P_n$ exist and can be extended to *-homomorphisms on the whole $\mathcal{A}(\mathcal{C})$, since for any generator of this algebra the above strong limits exist. By $\operatorname{smb}\mathcal{A}(\mathcal{C})$ we denote the C^*-algebra constituted by all pairs $(W_1(A_n), W_2(A_n))$, $(A_n) \in \mathcal{A}(\mathcal{C})$, so

$$\mathcal{A}(\mathcal{C})/G_{\mathcal{C}} \cong \operatorname{smb}\mathcal{A}(\mathcal{C}),$$

and Lemma 2.11.2 again yields

$$\mathcal{A}(\mathcal{C})^{2\times 2}/G_{\mathcal{C}}^{2\times 2} \cong (\operatorname{smb}\mathcal{A}(\mathcal{C}))^{2\times 2}.$$

Clearly, this argumentation is as before. The details are left to the reader. Note that the polynomial Galerkin method from Section 2.1 can be studied similarly.

2.12 Comments and References

Sections 2.1 – 2.2: It is worth mentioning that the initial achievements in studying approximation methods for singular integral operators without conjugation

$$A = a_0 P + b_0 Q \tag{2.136}$$

are connected with special representations of the operator A. Thus for certain classes of coefficients a_0, b_0 this operator can be represented in the form

$$A = (a_+ P + b_- Q)(Pa_- + Qb_+) + T,$$

where $a_\pm$ and $b_\pm$ are analytic functions in special factorizations

$$a_0 = a_+ a_-, \quad b = b_- b_+$$

of the coefficients a and b, and T is a compact operator. The operators $a_+P + b_-Q$ and $Pa_- + Qb_+$ interact well with projections onto polynomial subspaces, which allows us to establish the stability of approximation methods for the operator (2.136), [85, 86, 89, 112, 174]. Such an approach does not work for the operator

$$G = a_0 P + b_0 Q + M(a_1 P + b_1 Q)$$

that contains the operator of the complex conjugation M, therefore for operators similar to G, other approximation methods were developed in two directions. One of the first results was obtained for the method of simple iteration for the Markushevich boundary problem

$$\varphi^+(t) = a(t)\varphi^-(t) + \overline{\varphi^-(t)} + g(t), \quad t \in \Gamma, \tag{2.137}$$

where $\varphi^\pm$ are the boundary values of an unknown function analytic in two complementary domains of the complex plane $\mathbb{C}$ divided by a closed contour Γ [156]. Note that convergence of the corresponding approximation method was proved under strict metrical conditions imposed on the coefficients a and b. Some direct approximation methods for the operator G and the boundary problem (2.137) are considered in [225, 226], assuming that certain coefficients are analytic in the designated domains.

On the other hand, approximation methods were not applied to the operator G or equation (2.137), but rather to associated systems of equations without conjugation, with successive use of factorization to establish the stability of the method. For singular integral operators with conjugation, this approach was used in [118, 119, 120]; for similar operators but also containing Carleman shift, the corresponding approximation methods are studied in [40, 42, 227], and for boundary problems similar to problem (2.137) in [41, 43]. Such an approach leads to an unnecessary size increase in the systems of algebraic equations obtained, and also requires very strong assumptions about partial indices of the matrix coefficients of associated systems of integral equations. However, the direct application of polynomial projection methods to singular integral equations with conjugation used here does not face such difficulties [50, 51]. Notice also that Theorems 2.1.4 and 2.2.3 can be generalized on the case of piecewise continuous coefficients. The related proofs are much more involved and require new tools. In particular, if one uses ideas similar to Section 2.4, then the stability of the polynomial collocation and Galerkin methods for singular integral operators with piecewise continuous matrix-valued coefficients is crucial. However these results are known (see, e.g., [183, Chapter 7]).

Note that a direct approach to approximate solution of singular integral equations with simple shifts was first used in [155].

Section 2.3: These results are taken from [47, 48, 223], although the formulations and proofs here are slightly different from [47, 48, 223].

Sections 2.4 – 2.8: There are numerous investigations devoted to spline approximation methods for equation (2.62) in the special case $c = d = k^1 = 0$ (see, for example, the papers [2, 3, 82, 142, 176, 177, 179, 182, 185, 197, 199, 200, 215, 216] and books [7, 102, 183, 196]). In particular, it was discovered that in many cases, investigation of the approximation method under consideration leads to the study of matrix sequences having a special structure, viz., sequences of paired circulants [178, 182]. The stability problem for such matrix sequences was completely solved in [178] (see also notes and comments to Chapter 10 in [183]). The Banach algebra $\mathcal{C}$ generated by such sequences was studied later [105], and those results were used here to study real extension $\widetilde{\mathcal{C}}$ of $\mathcal{C}$ and various spline approximation methods for singular integral equations with conjugation on smooth curves [57].

The qualocation method (2.104) was first studied in [215] and [216]. It was observed in [104] that the corresponding approximation sequences belong to the above-mentioned algebra $\mathcal{C}$, which allows one to obtain the necessary and sufficient conditions of its stability for singular integral equations with piecewise continuous coefficients.

Section 2.9: The spline-qualocation method for boundary integral equations was studied in [20]. The results presented here are taken from [52]. It turns out that the stability conditions for the spline qualocation method are essentially simpler than for the polynomial qualocation method.

Section 2.10: The results on stability of the corresponding quadrature methods for singular integral equations with conjugation on simple closed contour are taken from [56].

Chapter 3

Approximation Methods for the Riemann-Hilbert Problem

Approximation methods for the Riemann-Hilbert boundary problem are considered in this chapter. A particular feature of these problems is that the operators studied act in a pair of spaces, so the corresponding operator spaces do not have any multiplication operation which makes the use of algebraic techniques more difficult. However, by introducing appropriate para-algebras, one can obtain necessary and sufficient stability conditions for the approximation methods considered. Note that there are two formulations for the Riemann-Hilbert boundary problem: one is concerned with the solutions from a chosen space of functions analytic inside a given domain D of the complex plane $\mathbb{C}$, whereas another contains an additional requirement that the corresponding solutions take real values at a fixed point of the domain D. Here, we deal with the second formulation because it arises more often in applications (cf. [162, Problem P]).

3.1 Galerkin Method

Let Γ_0 be the unit circle and let D stand for the open unit disc bounded by Γ_0. By $\overset{\circ}{L}{}_p^+ = \overset{\circ}{L}{}_p^+(\Gamma_0)$, $1 < p < \infty$, we denote the subset of the space $L_p(\Gamma_0)$ consisting of the boundary values of the functions Φ that are analytic in D and such that $\operatorname{Im}\Phi(0) = 0$. Note that $\overset{\circ}{L}{}_p^+$, $1 < p < \infty$ is a linear space over the field of real numbers. By $\widetilde{L}_p = \widetilde{L}_p(\Gamma_0)$, $1 < p < \infty$ we denote the Banach space which consists of all real-valued elements of $L_p(\Gamma_0)$ equipped with the L_p norm. In spaces $\overset{\circ}{L}{}_2^+$

and $\widetilde{L}_2$, inner products can respectively be introduced as

$$< x, y > = \operatorname{Re}\left(\frac{1}{2\pi}\int_0^{2\pi} x(e^{is})\overline{y(e^{is})}\, ds\right), \quad x, y \in \overset{\circ}{L_2^+} \tag{3.1}$$

and

$$[x, y] = \frac{1}{2\pi}\int_0^{2\pi} x(e^{is})y(e^{is})\, ds, \quad x, y \in \widetilde{L}_2. \tag{3.2}$$

It is clear that the sets $\overset{\circ}{L_2^+}$ and $\widetilde{L}_2$ provided with the inner products (3.1) and (3.2), become Hilbert spaces.

The matrix Riemann-Hilbert problem can now be formulated as follows:

Let $G \in L_\infty^{m\times m}(\Gamma_0)$ be an $(m \times m)$-matrix function, $f \in \widetilde{L}_2^m$, and let M be the operator of complex conjugation. Find a vector function $\Phi \in \overset{\circ}{L_2^{+,m}}$ which satisfies the boundary condition

$$\frac{1}{2}(I + M)G(t)\Phi(t) = f(t), \quad t \in \Gamma_0. \tag{3.3}$$

As in Section 2.1, consider the projection operator $P_n : L_2(\Gamma_0) \to L_2(\Gamma_0)$ defined by

$$(P_n\varphi)(t) = P(\sum_{k=-\infty}^{+\infty} \varphi_k t^k) := \sum_{k=-n}^{+n} \varphi_k t^k,$$

where φ_k, $k \in \mathbb{Z}$ are the Fourier coefficients of φ. Let us introduce the finite-dimensional operator

$$K_0 := \frac{1}{2}(I + M)tQt^{-1}P$$

and note that the operator $P^\circ := P - K_\circ$ projects the space L_2 onto the subspace $\overset{\circ}{L_2^+}$. Let Q° denote the complementary projection, so $Q^\circ = I - P^\circ$. The operator $(1/2)(I+M)$ projects the space L_2 onto the subspace $\widetilde{L}_2$. Moreover, if $(1/2)(I+M)$ is considered as an operator acting from the space $\overset{\circ}{L_2^+}$ to $\widetilde{L}_2$, then it is invertible and the inverse operator $V : \widetilde{L}_2 \to \overset{\circ}{L_2^+}$ is defined by

$$(V\varphi)(t) = V(\sum_{k=-\infty}^{-1} \varphi_k t^k + \varphi_0 + \sum_{k=1}^{+\infty} \varphi_k t^k) = \varphi_0 + 2\sum_{k=1}^{+\infty} \varphi_k t^k. \tag{3.4}$$

To prove invertibility, one has to multiply the operators V and $(1/2)(I + M)$ and recall that a function $\varphi \in \widetilde{L}_2(\Gamma_0)$ if and only if its Fourier coefficients satisfy the relation

$$\varphi_k = \overline{\varphi}_{-k}, \quad k = 0, 1, \ldots .$$

In spaces $\overset{\circ}{L}{}_2^+$ and $\widetilde{L}_2$ we consider the approximation operators P_n° and $\widetilde{P}_n$ defined by

$$(P_n^\circ\varphi)(t) = P_n^\circ(\sum_{k=0}^{+\infty}\varphi_k t^k) = \varphi_0 + \varphi_1 t + \ldots + \varphi_{n-1}t^{n-1} + (\operatorname{Re}\varphi_n)t^n,$$

$$(\widetilde{P}_n\varphi)(t) = \widetilde{P}_n(\sum_{k=-\infty}^{+\infty}\varphi_k t^k) = (\operatorname{Re}\varphi_{-n})t^{-n} + \varphi_{-n+1} + \ldots + \varphi_{-1}t^{-1}$$
$$+ \varphi_0 + \varphi_1 t + \ldots + \varphi_{n-1}t^{n-1} + (\operatorname{Re}\varphi_n)t^n.$$

It follows immediately from the definitions that

$$(P_n^\circ)^2 = P_n^\circ, \quad (\widetilde{P}_n)^2 = \widetilde{P}_n,$$
$$(P_n^\circ)^* = P_n^\circ, \quad (\widetilde{P}_n)^* = \widetilde{P}_n,$$

for all $n = 1, 2, \ldots$. Moreover, the strong convergence of the projections P_n to the identity operator I in $\mathcal{L}(L_2)$ implies that the operator sequences (P_n°) and $(\widetilde{P}_n)$ converge strongly to the identity operators $I_{\overset{\circ}{L}{}_2^+}$ and $I_{\widetilde{L}_2}$, respectively, so the sequences (P_n°) and $(\widetilde{P}_n)$ satisfy all requirements of Section 1.6.

Along with equation (3.2) we consider the equation

$$\widetilde{P}_n\frac{1}{2}(I+M)GP^\circ P_n^\circ\Phi_n^\circ = \widetilde{P}_n f, \tag{3.5}$$

where $\Phi_n^\circ \in \operatorname{im} P_n^\circ$. Writing $\Phi_n^\circ(t) = \sum_{k=0}^n c_k t^k$, equation (3.5) is equivalent to the system of algebraic equations

$$\begin{aligned}
\operatorname{Re}\left[\sum_{k=0}^{n-1}(g_{-n-k}c_k + \overline{g_{n-k}c_k})\right] &= \operatorname{Re} f_{-n},\\
\sum_{k=0}^{n-1}(g_{p-k}c_k + \overline{g_{-p-k}c_k}) &= f_p,\\
\operatorname{Im} c_0 = \operatorname{Im} c_n &= 0,\\
p = -n+1, \ldots, n-1.&
\end{aligned}$$

This system can be reduced to a system of linear algebraic equations.

In the spaces $\overset{\circ}{L}{}_2^+$, $\widetilde{L}_2$ and L_2 we consider the operator sequences $(W_n^\circ)_{n=1}^\infty$,

$(\widetilde{W}_n)_{n=1}^{\infty}$ and $(W_n)_{n=1}^{\infty}$, where

$$(W_n^\circ\varphi)(t) = W_n^\circ(\sum_{k=0}^{+\infty}\varphi_k t^k) = (\operatorname{Re}\varphi_n) + \varphi_{n-1}t + \ldots + \varphi_1 t^{n-1} + \varphi_0 t^n,$$

$$(\widetilde{W}_n\varphi)(t) = \widetilde{W}_n(\sum_{k=-\infty}^{+\infty}\varphi_k t^k) = \frac{1}{2}\varphi_0 t^{-n} + \varphi_{-1}t^{-n+1} + \ldots + \varphi_{-n+1}t^{-1}$$
$$+ 2\operatorname{Re}\varphi_n + \varphi_1 t + \ldots + vp_1 t^{n-1} + \frac{1}{2}\varphi_0 t^n,$$

$$(W_n\varphi)(t) = W_n(\sum_{k=-\infty}^{+\infty}\varphi_k t^k) = \varphi_{-1}t^{-n} + \varphi_{-2}t^{-n+1} + \ldots + \varphi_{-n}t^{-1}$$
$$+ \varphi_n + \varphi_{n-1}t + \ldots + \varphi_0 t^n.$$

It is easy to check that the operators W_n° and $\widetilde{W}_n$ are connected with the operators P_n° and $\widetilde{P}_n$ by the following relations:

$$(W_n^\circ)^2 = P_n^\circ, \quad (\widetilde{W}_n)^2 = \widetilde{P}_n, \quad n = 1, 2, \ldots,$$
$$W_n^\circ P_n^\circ = W_n^\circ, \quad \widetilde{W}_n\widetilde{P}_n = \widetilde{W}_n, \quad n = 1, 2, \ldots.$$

Moreover, one also has the following result for the adjoint operators.

Lemma 3.1.1. *For any $n = 1, 2, \ldots$ the operators W_n° and $\widetilde{W}_n$ are self-adjoint, i.e.,*

$$(W_n^\circ)^* = W_n^\circ, \quad (\widetilde{W}_n)^* = \widetilde{W}_n, \tag{3.6}$$

and the sequences (W_n°) and $(\widetilde{W}_n)$ weakly converge to zero.

Proof. Equations (3.6) follow immediately from the definition of inner products in the spaces $\overset{\circ}{L}{}_2^+$ and $\widetilde{L}_2$, cf. (3.1), (3.2). To show the weak convergence of the sequences (W_n°) and $(\widetilde{W}_n)$ one can use the representations

$$W_n^\circ = Wt^{-n}P_n^\circ, \quad \widetilde{W}_n = Wt^n\overline{Q}\widetilde{P}_n + Wt^{-n}\overline{P}\widetilde{P}_n,$$

where $\overline{P}, \overline{Q}$ and W are continuous operators defined by

$$(\overline{P}\varphi)(t) = \overline{P}(\sum_{k=-\infty}^{+\infty}\varphi_k t^k) = \frac{\varphi_0}{2} + \sum_{k=0}^{\infty}\varphi_k t^k,$$
$$\overline{Q} = I - \overline{P}, \quad (W\varphi)(t) = \varphi(\overline{t}),$$

and the weak convergence of the sequences of the multiplication operators (t^n) and (t^{-n}) to zero. □

Thus the sequences $(P_n^\circ), (\widetilde{P}_n), (W_n^\circ), (\widetilde{W}_n)$ define the para-algebra $\mathcal{A} = (\mathcal{A}^{\overset{\circ}{L}{}_2^+}, \mathcal{A}^{\overset{\circ}{L}{}_2^{+,m}\widetilde{L}_2^m}, \mathcal{A}^{\widetilde{L}_2^m\overset{\circ}{L}{}_2^{+,m}}, \mathcal{A}^{\widetilde{L}_2^m})$. We mention a number of useful relations connecting the operators $(1/2)(I+M), V, P_n^\circ$, $\widetilde{P}_n, W_n^\circ, \widetilde{W}_n, W_n$, viz.,

$$P_n^\circ V = V\widetilde{P}_n, \quad \widetilde{P}_n\frac{1}{2}(I+M)P^\circ = \frac{1}{2}(I+M)P_n^\circ P^\circ, \tag{3.7}$$

$$W_n^\circ V = V\widetilde{W}_n, \quad \widetilde{W}_n\frac{1}{2}(I+M)P^\circ = \frac{1}{2}(I+M)W_n^\circ P^\circ, \tag{3.8}$$

$$W_n^\circ = PW_n - K_n = W_nP - K_n, \tag{3.9}$$

where

$$K_n\Big(\sum_{k=-\infty}^{+\infty}\varphi_k t^k\Big) = i\mathrm{Im}\,\varphi_n \quad \text{and} \quad K_0 = K^\circ.$$

Now we can show that operator sequences associated with approximation methods for the Riemann-Hilbert problem (3.1) belong to the para-algebra $\mathcal{A}$.

Lemma 3.1.2. *If $G \in L_\infty^{m\times m}(\Gamma_0)$, then the sequence $(\widetilde{P}_n\frac{1}{2}(I+M)GP^\circ P_n^\circ)$ is in $\mathcal{A}^{\overset{\circ}{L}{}_2^{+,m}\widetilde{L}_2^m}$.*

Proof. Since the sequences $(\widetilde{P}_n)$ and (P_n°) converge strongly to the identity operators,

$$s-\lim_{n\to\infty}\widetilde{P}_n\frac{1}{2}(I+M)GP^\circ P_n^\circ = \frac{1}{2}(I+M)GP^\circ.$$

Then, taking into account the equality

$$(\widetilde{P}_n\frac{1}{2}(I+M)GP^\circ P_n^\circ)^* = P_n^\circ P^\circ\frac{1}{2}(I+M)G^*\widetilde{P}_n$$

one also obtains

$$s-\lim_{n\to\infty}(\widetilde{P}_n\frac{1}{2}(I+M)GP^\circ P_n^\circ)^*\widetilde{P}_n = P^\circ G^*\frac{1}{2}(I+M) = (\frac{1}{2}(I+M)GP^\circ)^*.$$

Consider now the sequence $(\widetilde{W}_n\frac{1}{2}(I+M)GP^\circ W_n^\circ)$. We can represent the operator $\widetilde{W}_n\frac{1}{2}(I+M)GP^\circ W_n^\circ$ in the form

$$\begin{aligned}\widetilde{W}_n\frac{1}{2}&(I+M)GP^\circ W_n^\circ \\ &= \widetilde{W}_n\frac{1}{2}(I+M)P^\circ GP^\circ W_n^\circ + \widetilde{W}_n\frac{1}{2}(I+M)Q^\circ GP^\circ W_n^\circ \\ &\qquad = R_n^{(1)} + R_n^{(2)}.\end{aligned}$$

According to (3.8) and (3.9) the operator $R_n^{(1)}$ can be rewritten as

$$R_n^{(1)} = \frac{1}{2}(I+M)W_nPGPW_nP^\circ - \frac{1}{2}(I+M)W_nK^\circ GPW_nP^\circ.$$

Since the operator K° is compact, the sequence $(\frac{1}{2}(I+M)W_nK^\circ GPW_nP^\circ)$ converges to zero uniformly. Recalling from [207] that for any $G \in L_\infty^{m\times m}$ one has the strong limit

$$s-\lim_{n\to\infty} W_nPGPW_n = P\widetilde{G}P$$

where $\widetilde{G}(t) = G(\bar{t})$, it follows that

$$s-\lim_{n\to\infty} R_n^{(1)} = \frac{1}{2}(I+M)P^\circ\widetilde{G}P^\circ.$$

Before studying the sequence $(R_n^{(2)})$ we note the easily verified equation

$$\widetilde{W_n}\frac{1}{2}(I+M)Q = \frac{1}{2}(I+M)tW_nQ, \tag{3.10}$$

so using (3.10) we obtain

$$\begin{aligned} R_n^{(2)} = {} & \frac{1}{2}(I+M)tW_nQGPW_n \\ & - \frac{1}{2}(I+M)tW_nQGK^\circ W_n - \frac{1}{2}(I+M)tW_nQGP^\circ K_n. \end{aligned} \tag{3.11}$$

Since W_nQGPW_n tends to zero as $n\to\infty$ and the operator K° is compact, all operators in the right-hand side of (3.11) strongly converge to zero. Therefore

$$s-\lim_{n\to\infty} \widetilde{W_n}\frac{1}{2}(I+M)Q^\circ GP^\circ W_n^\circ = \frac{1}{2}(I+M)P\widetilde{G}P^\circ = \frac{1}{2}(I+M)P^\circ\widetilde{G}P^\circ.$$

The sequence $(\widetilde{W_n}\frac{1}{2}(I+M)Q^\circ GP^\circ W_n^\circ)^*$ can be studied analogously. That completes the proof. □

Consider next the sequence $(P_n^\circ P^\circ GP^\circ V\widetilde{P_n})$ with $G \in L_\infty^{m\times m}$.

Lemma 3.1.3. *Let $G \in L_\infty^{m\times m}$. Then $(P_n^\circ P^\circ GP^\circ V\widetilde{P_n}) \in \mathcal{A}^{\widetilde{L}_2^m \overset{\circ}{L}{}_2^{+,m}}$.*

Proof. As previously, the most demanding part of the proof is the study of the sequence $(W_n^\circ P^\circ GP^\circ V\widetilde{W_n})$. Let us represent the operator $W_n^\circ P^\circ GP^\circ V\widetilde{W_n}$ as the sum of four operators, viz.,

$$\begin{aligned} W_n^\circ P^\circ GP^\circ V\widetilde{W_n} &= W_nPGPW_nP^\circ V - K_nPGPW_nP_n^\circ V \\ &\quad - W_n^\circ P^\circ GK_nP^\circ V - W_nK^\circ GPW_nP^\circ V. \end{aligned}$$

It is clear that the last two terms strongly converge to zero as $n \to \infty$. Now $K_n PGPW_n V$ tends to $K^\circ PGP^\circ V$,

$$s - \lim_{n\to\infty} W_n^\circ P^\circ GP^\circ V\widetilde{W}_n = P\widetilde{G}P^\circ V - K^\circ P\widetilde{G}P^\circ V = P^\circ \widetilde{G}P^\circ V.$$

The same approach also works for the sequence $(W_n^\circ P^\circ GP^\circ V\widetilde{W}_n)^*$, and hence $(P_n^\circ P^\circ GP^\circ V\widetilde{P}_n) \in \mathcal{A}^{\widetilde{L}_2^m \overset{\circ}{L}{}_2^{+,m}}$. □

We need an auxiliary result.

Lemma 3.1.4. *Let $a \in L_\infty$. The operator $P^\circ aQ^\circ$ ($Q^\circ aP^\circ$) is compact on the space L_p, $1 < p < \infty$, if and only if the operator PaQ (QaP) is compact on the same space L_p.*

Proof. Since the projections P and Q are bounded and K° is a finite-dimensional operator, all the three operators $K^\circ aQ, PaK^\circ$ and $K^\circ aK^\circ$ are compact. Thus for the operator $P^\circ aQ^\circ$ the result follows from the equation

$$P^\circ aQ^\circ = PaQ + K^\circ aQ - PaK^\circ - K^\circ aK^\circ. \qquad \square$$

Now we can prove the main result of this section.

Theorem 3.1.5. *Let matrices $G, G^{-1} \in \mathbf{C}^{m\times m}$. The sequence $(\widetilde{P}_n \frac{1}{2}(I+M)GP^\circ P_n^\circ)$ is stable if and only if the operator $\frac{1}{2}(I+M)GP^\circ \in \mathcal{L}(\overset{\circ}{L}{}_2^{+,m}, \widetilde{L}_2^m)$ and the operator $P^\circ \widetilde{G}P^\circ \in \mathcal{L}(\overset{\circ}{L}{}_2^{+,m})$ are invertible.*

Proof. Necessity immediately follows from Lemma 3.1.2 and Theorem 1.8.4. For sufficiency we consider the operators $K = \frac{1}{2}(I + M)GP^\circ$ and $R := P^\circ G^{-1}P^\circ V$, and show that

$$(\widetilde{P}_n KP^\circ)(P^\circ R\widetilde{P}_n) - (\widetilde{P}_n) \in \mathcal{J}^{\widetilde{L}_2^m}, \tag{3.12}$$

$$(P^\circ R\widetilde{P}_n)(\widetilde{P}_n KP^\circ) - (P_n^\circ) \in \mathcal{J}^{\overset{\circ}{L}{}_2^{+,m}}. \tag{3.13}$$

Let us consider the sequence

$$\begin{aligned}(\widetilde{P}_n KP^\circ)(P^\circ R\widetilde{P}_n) = (\widetilde{P}_n) &- (\widetilde{P}_n\frac{1}{2}(I + M)GQ^\circ G^{-1}P^\circ V\widetilde{P}_n)\\ &- (\widetilde{P}_n\frac{1}{2}(I + M)GP^\circ Q_n^\circ P^\circ G^{-1}P^\circ V\widetilde{P}_n),\end{aligned}$$

where $Q_n^\circ = I - P_n^\circ$. For the continuous matrix G^{-1} the operator $QG^{-1}P$ is compact, so it follows from Lemma 3.1.4 that the operator $Q^\circ G^{-1}P^\circ$ is compact and therefore the sequence $(\widetilde{P}_n \frac{1}{2}(I + M)GQ^\circ G^{-1}P^\circ V\widetilde{P}_n)$ belongs to the ideal

$\mathcal{J}^{\widetilde{L}_2^m}$. It remains to consider the sequence $(\widetilde{P}_n\frac{1}{2}(I+M)GP^\circ Q_n^\circ P^\circ G^{-1}P^\circ V\widetilde{P}_n)$. Let $a, b \in L_\infty$. We now show that

$$\widetilde{P}_n\frac{1}{2}(I+M)aP^\circ Q_n^\circ P^\circ bP^\circ V\widetilde{P}_n = \widetilde{P}_n\frac{1}{2}(I+M)Q^\circ aP^\circ Q_n^\circ P^\circ bP^\circ V\widetilde{P}_n + \widetilde{W}_n\frac{1}{2}(I+M)P^\circ\widetilde{a}Q^\circ\widetilde{b}P^\circ V\widetilde{P}_n. \tag{3.14}$$

Due to the "commutation" relations (3.7), (3.8), one only needs to establish relation

$$P_n^\circ P^\circ aP^\circ Q_n^\circ P^\circ bP^\circ P_n^\circ = W_n^\circ P^\circ\widetilde{a}Q^\circ P^\circ\widetilde{b}P^\circ W_n^\circ. \tag{3.15}$$

This follows by using the Fourier expansions of the corresponding elements. Thus

$$b(t) = \sum_{j=-\infty}^{+\infty} b_j t^j, \quad a(t) = \sum_{l=-\infty}^{+\infty} a_l t^l, \quad x_n(t) = \sum_{k=0}^{n} f_k^* t^k$$

where

$$f_k^* := \begin{cases} f_k & \text{if } \ 0 \le k < n, \\ \operatorname{Re} f_n & \text{if } \ k = n, \end{cases}$$

so for every $x_n \in \operatorname{im} P_n^\circ$ one has

$$\begin{aligned} &P_n^\circ P^\circ aP^\circ Q_n^\circ P^\circ bP^\circ P_n^\circ x_n(t) \\ &= \operatorname{Re}\left(\sum_{r=n+1}^{\infty}\sum_{k=0}^{n} a_{-r}b_{r-k}f_k^* + ia_{-n}\operatorname{Im}\left(\sum_{k=0}^{n} b_{r-k}f_k^*\right)\right) \\ &+ \sum_{p=1}^{n-1}\left(\sum_{r=n+1}^{\infty}\sum_{k=0}^{n} a_{p-r}b_{r-k}f_k^* + ia_{p-n}\operatorname{Im}\left(\sum_{k=0}^{n} b_{r-k}f_k^*\right)\right)t^p \\ &+ \operatorname{Re}\left(\sum_{r=n+1}^{\infty}\sum_{k=0}^{n} a_{n-r}b_{r-k}f_k^* + ia_0\operatorname{Im}\left(\sum_{k=0}^{n} b_{r-k}f_k^*\right)\right)t^n \\ &= W_n^\circ P^\circ\widetilde{a}Q^\circ\widetilde{b}P^\circ W_n^\circ x_n(t). \end{aligned}$$

This and relation (3.14) implies

$$\widetilde{P}_n\frac{1}{2}(I+M)GP^\circ Q_n^\circ P^\circ G^{-1}P^\circ V\widetilde{P}_n = \widetilde{P}_n\frac{1}{2}(I+M)Q^\circ GP^\circ Q_n^\circ P^\circ G^{-1}P^\circ V\widetilde{P}_n + \widetilde{W}_n\frac{1}{2}(I+M)P^\circ\widetilde{G}Q^\circ\widetilde{G}^{-1}P^\circ V\widetilde{W}_n,$$

and since $G^{\pm 1} \in \mathbf{C}^{m\times m}$, the sequence $(\widetilde{W}_n\frac{1}{2}(I+M)P^\circ\widetilde{G}Q^\circ\widetilde{G}^{-1}P^\circ V\widetilde{W}_n) \in \mathcal{J}^{\widetilde{L}_2^m}$. Let us show that the sequence $(\widetilde{P}_n\frac{1}{2}(I+M)Q^\circ GP^\circ Q_n^\circ P^\circ G^{-1}P^\circ V\widetilde{P}_n)$ belongs to the same ideal $\mathcal{J}^{\widetilde{L}_2^m}$. We estimate the norms of the corresponding operators; thus

$$\begin{aligned}&||\widetilde{P}_n\frac{1}{2}(I+M)Q^\circ GP^\circ Q_n^\circ P^\circ G^{-1}P^\circ V\widetilde{P}_n||\\&\leq ||\widetilde{P}_n\frac{1}{2}(I+M)QGPQ_nPG^{-1}P^\circ V\widetilde{P}_n||\\&\quad+||\widetilde{P}_n\frac{1}{2}(I+M)QGPK_nPG^{-1}P^\circ V\widetilde{P}_n||,\end{aligned}$$

where $Q_n = I - P_n$ and $(K_n x)(t) = i(\operatorname{Im} x_n)t^n$. Let $\widetilde{K}_n$ denote the operator defined on the space L_2 by

$$(\widetilde{K}_n x)(t) = x_n t^n.$$

Since $|a\operatorname{Im} b| \leq |ab|$ for any $a, b \in \mathbb{C}$, one has

$$||QGPK_n|| \leq ||QGP\widetilde{K}_n||.$$

The strong convergence of the operators Q_n and K_n to zero and the compactness of the operator QGP implies that $(\widetilde{P}_n\frac{1}{2}(I+M)Q^\circ GP^\circ Q_n^\circ P^\circ G^{-1}P^\circ V\widetilde{P}_n) \in \mathcal{J}^{\widetilde{L}_2^m}$. Inclusion (3.13) can be verified analogously. Relations (3.12) and (3.13) show that the element $(\widetilde{P}_n K P_n^\circ)$ is $\mathcal{J}$-invertible, which completes the proof. □

3.2 Interpolation method for the Riemann-Hilbert problem with continuous coefficients

In this section an approximation method based on modified Lagrange operators $\widetilde{L}_n, n \in \mathbb{N}$ is studied. Note that these operators do not preserve the values of functions on the grid under consideration, so the usual methods to study the stability of projection methods [86, 112] connected with Lagrange operators are not effective. However, the algebraic approach used in Section 3.1 works well.

Let $\mathbf{R}$ again be the set of the Riemann integrable functions on the unit circle Γ_0, and $\widetilde{\mathbf{R}}$ be the subset of the real-valued functions from $\mathbf{R}$. Let us recall that the interpolation Lagrange operator L_n over the uniform grid is defined by

$$(L_n f)(t) = \sum_{k=-n}^{n} \beta_k t^k, \quad \beta_k = \frac{1}{2n+1}\sum_{j=-n}^{n} f(t_j^{(n)})t_j^{-k}. \tag{3.16}$$

The operator L_n acts from the space $\widetilde{\mathbf{R}}$ to $\widetilde{\mathbf{R}}$, because for any real-valued function f the coefficients β_k in (3.16) satisfy the relation

$$\overline{\beta}_k = \beta_{-k}, \quad k = 0, 1, \ldots, n, \tag{3.17}$$

and this means that $L_n f \in \widetilde{\mathbf{R}}$. Consider further an operator $\widetilde{L}_n$ defined on the linear space $\mathbf{R}$ by

$$\widetilde{\beta}_k = \begin{cases} \beta_k & \text{if} \quad k = -n+1, -n+2, \dots, n-1, \\ \operatorname{Re}\left(\dfrac{1}{2n+1} \displaystyle\sum_{j=-n}^{n} f(t_j^{(n)}) t_j^{-n} \right) & \text{if} \quad k = n, \\ \operatorname{Re}\left(\dfrac{1}{2n+1} \displaystyle\sum_{j=-n}^{n} f(t_j^{(n)}) t_j^{n} \right) & \text{if} \quad k = -n, \end{cases}$$

where the coefficients β_k are given by (3.16). If $f \in \widetilde{\mathbf{R}}$, then from (3.17) the polynomial $\widetilde{L}_n f$ belongs to the subspace $\operatorname{im} \widetilde{P}_n$. Let us list some properties of the operators $\widetilde{L}_n$ inherited from the Lagrange operator [86, 113, 228].

Lemma 3.2.1. *For any $f \in \mathbf{R}$ the limit*

$$\lim_{n\to\infty} ||f - \widetilde{L}_n f||_2 = 0.$$

Lemma 3.2.2. *For any $g \in \mathbf{R}$ and for any polynomial $x_n(t) = \sum_{j=-n}^{n} a_j t^j$, the inequality*

$$||\widetilde{L}_n(g x_n)||_2 \le ||g||_\infty ||x_n||_2 \tag{3.18}$$

holds.

Proof. To verify inequality (3.18), one has to apply the Parseval formula twice. Thus, if β_k, $k = -n, \dots, n$ are the coefficients of the Lagrange polynomial $L_n(g x_n)(t)$, then

$$\begin{aligned} ||\widetilde{L}_n(g x_n)||_2^2 &= \frac{1}{2\pi} \int_0^{2\pi} |(\widetilde{L}_n(g x_n))(e^{i\theta})|^2 \, d\theta = \sum_{k=-n}^{n} |\widetilde{\beta}_k|^2 \\ &\le \sum_{k=-n}^{n} |\beta_k|^2 = \frac{1}{2\pi} \int_0^{2\pi} |(L_n(g x_n))(e^{i\theta})|^2 = ||L_n(g x_n)||_2^2, \end{aligned}$$

and it remains to use the known inequality [86, 113]

$$||L_n(g x_n)||_2 \le ||g||_\infty ||x_n||_2,$$

to complete the proof. □

An approximate solution for the Riemann-Hilbert problem (3.3) is sought as a polynomial $\Phi_n^\circ \in \operatorname{im} P_n^\circ$, viz.,

$$\Phi_n^\circ(t) = \sum_{k=0}^{n} a_k t^k.$$

The operator equation

$$\widetilde{L}_n \frac{1}{2}(I+M)GP^\circ \overset{\circ}{P}_n \overset{\circ}{\Phi}_n = \widetilde{L}_n f \tag{3.19}$$

is then equivalent to a system of algebraic equations with respect to the unknown coefficients a_k, $k = 0, 1, \ldots, n$, on equating the coefficients of the polynomials from the left- and right-hand sides of (3.19).

As mentioned in Section 2.2, it is more convenient to consider the equation

$$\widetilde{L}_n \frac{1}{2}(I+M)GP^\circ \overset{\circ}{P}_n \overset{\circ}{\Phi}_n = \widetilde{P}_n f$$

instead of (3.19). Of course, this does not make any difference for the stability of the method. To study the stability of the sequence $(\widetilde{L}_n \frac{1}{2}(I+M)GP^\circ \overset{\circ}{P}_n \overset{\circ}{\Phi}_n)$, we again invoke the para-algebra $\mathcal{A} = (\mathcal{A}^{\overset{\circ}{L}{}_2^+}, \mathcal{A}^{\overset{\circ}{L}{}_2^{+,m} \widetilde{L}_2^m}, \mathcal{A}^{\widetilde{L}_2^m \overset{\circ}{L}{}_2^{+,m}}, \mathcal{A}^{\widetilde{L}_2^m})$.

Lemma 3.2.3. *If $G \in \mathbf{R}^{m\times m}$, the sequence $(A_n) = (\widetilde{L}_n \frac{1}{2}(I+M)GP^\circ \overset{\circ}{P}_n \overset{\circ}{\Phi}_n)$ belongs to the para-algebra $\mathcal{A}$.*

Proof. It suffices to show that

$$s - \lim_{n\to\infty} A_n \overset{\circ}{P}_n = A, \quad s - \lim_{n\to\infty} \widetilde{W_n A_n \overset{\circ}{W}_n} = \widetilde{A} \tag{3.20}$$

and that

$$s - \lim_{n\to\infty} (A_n)^* \widetilde{P}_n = A^*, \quad s - \lim_{n\to\infty} (\widetilde{W_n A_n \overset{\circ}{W}_n})^* \widetilde{P}_n = \widetilde{A}^*. \tag{3.21}$$

The first of relations (3.20) can be verified similarly to the corresponding relation from Section 2.2 and is a consequence of the strong convergence of the sequence $(\overset{\circ}{P}_n)$ to the identity operator on the space $\overset{\circ}{L}{}_2^+$ and Lemmas 3.2.1 and 3.2.2. Indeed, if $G \in \mathbf{R}^{m\times m}$, then

$$\begin{aligned}||\widetilde{L}_n \frac{1}{2}(I+M)GP^\circ \overset{\circ}{P}_n x||_2 &\le \frac{1}{2}||(I+M)\widetilde{L}_n GP^\circ \overset{\circ}{P}_n x||_2 \\ &\le ||\widetilde{L}_n GP^\circ \overset{\circ}{P}_n x||_2 \le ||G||_\infty ||\overset{\circ}{P}_n x||_2 \le ||G||_\infty ||x||_2,\end{aligned}$$

i.e., the sequence $(\widetilde{L}_n \frac{1}{2}(I+M)GP^\circ \overset{\circ}{P}_n x)$ is uniformly bounded. Let us now take $n \ge l$ and let $x_l \in \operatorname{im} \overset{\circ}{P}_n$. Then $P^\circ \overset{\circ}{P}_n x_l = x_l$ and $\frac{1}{2}(I+M)GP^\circ \overset{\circ}{P}_n x_l$ is Riemann integrable, so by Lemma 3.2.1, $\widetilde{L}_n \frac{1}{2}(I+M)GP^\circ \overset{\circ}{P}_n x_l$ tends to $\frac{1}{2}(I+M)GP^\circ x_l$ as n tends to 0. Therefore, by the Banach Steinhaus theorem [77]

$$s - \lim_{n\to\infty} \widetilde{L}_n \frac{1}{2}(I+M)GP^\circ \overset{\circ}{P}_n = \frac{1}{2}(I+M)GP^\circ.$$

Let us now consider the sequence $(\widetilde{W}_n \widetilde{P}_n \frac{1}{2}(I+M)GP^\circ P_n^\circ W_n^\circ)$. It can be represented as the sum of two sequences $(R_n^{(1)}) = (\widetilde{W}_n \frac{1}{2}(I+M)P^\circ \widetilde{L}_n GP^\circ W_n^\circ)$ and $(R_n^{(2)}) = (\widetilde{W}_n \frac{1}{2}(I+M)Q^\circ \widetilde{L}_n GP^\circ W_n^\circ)$. To study these sequences, we use equation (3.8) and the equation

$$\widetilde{W}_n \frac{1}{2}(I+M)Q^\circ = \frac{1}{2}(I+M)tW_nQ.$$

Thus for $(R_n^{(1)})$ one obtains

$$R_n^{(1)} = \frac{1}{2}(I+M)W_nP\widetilde{L}_nGPW_nP^\circ P_n^\circ - \frac{1}{2}(I+M)W_nK^\circ GP^\circ W_n^\circ.$$

Since the operator K° is finite-dimensional and the sequence (W_n) weakly converges to zero, the limit

$$s-\lim_{n\to\infty} \frac{1}{2}(I+M)W_nK^\circ GP^\circ W_n^\circ = 0,$$

so by [117] one has

$$s-\lim_{n\to\infty} R_n^{(1)} = \frac{1}{2}(I+M)P^\circ \widetilde{G}P^\circ.$$

Analogously,

$$s-\lim_{n\to\infty} R_n^{(2)} = \frac{1}{2}(I+M)tQ\widetilde{G}P^\circ.$$

The last two equations lead to the relation

$$s-\lim_{n\to\infty} \widetilde{W}_nA_nW_n^\circ = \frac{1}{2}(I+M)(P^\circ + tQ)\widetilde{G}P^\circ.$$

It remains to show the limit relations (3.21). Consider for example the sequence $(A_n^*\widetilde{P}_n)$. The identity

$$[\frac{1}{2}(I+M)GP^\circ\varphi, \psi] = <\varphi, P^\circ G^* \frac{1}{2}(I+M)\psi>,$$

is valid for all $\varphi \in \overset{\circ}{L}{}_2^{+,m}$ and for all $\psi \in \widetilde{L}_2^m$, therefore

$$(\frac{1}{2}(I+M)GP^\circ)^* = P^\circ G^* \frac{1}{2}(I+M). \tag{3.22}$$

Straightforward computation of the adjoint operator to the operator $\widetilde{L}_n \frac{1}{2}(I+M)GP^\circ P_n^\circ$ gives

$$(\widetilde{L}_n \frac{1}{2}(I+M)GP^\circ P_n^\circ)^* \widetilde{P}_n = P^\circ \widetilde{L}_n G^* \frac{1}{2}(I+M)\widetilde{P}_n, \tag{3.23}$$

so from (3.22) and (3.23) we notice that

$$s-\lim_{n\to\infty}(\widetilde{L}_n\frac{1}{2}(I+M)GP^\circ P_n^\circ)^*\widetilde{P}_n=(\frac{1}{2}(I+M)GP^\circ)^*.$$

Another sequence in (3.21) is considered analogously. Thus for any $G\in\mathbf{R}^{m\times m}$, the sequence $(\widetilde{L}_n\frac{1}{2}(I+M)GP^\circ P_n^\circ)\in\mathcal{A}$. □

Theorem 3.2.4. *If $G\in\mathbf{C}^{m\times m}$, then the sequence $(\widetilde{L}_n\frac{1}{2}(I+M)GP^\circ P_n^\circ)$ is stable if and only if the operators $A=\frac{1}{2}(I+M)GP^\circ$ and $\widetilde{A}=\frac{1}{2}(I+M)(P^\circ+tQ)\widetilde{G}P^\circ$ are invertible in $\mathcal{L}(\overset{\circ}{L}{}_2^{+,m},\widetilde{L}_2^m)$.*

The proof of this result is based on two lemmas.

Lemma 3.2.5. *If $G\in\mathbf{C}^{m\times m}$, then the sequences $(A_n)=(\widetilde{L}_n\frac{1}{2}(I+M)GP^\circ P_n^\circ)$ and $(B_n)=(\widetilde{P}_n\frac{1}{2}(I+M)GP^\circ P_n^\circ)$ belong to the same coset of the quotient para-algebra $\mathcal{A}/\mathcal{J}$.*

Proof. Consider the difference

$$A_n-B_n=(\widetilde{L}_n-\widetilde{P}_n)\frac{1}{2}(I+M)GP^\circ P_n^\circ.$$

From now on the symbol P_n° is used to denote an operator defined on the space L_2 by

$$\begin{aligned}(P_n^\circ\varphi)(t)=P_n^\circ\left(\sum_{k=-\infty}^{+\infty}\varphi_k t^k\right)&=(\operatorname{Re}\varphi_{-n})t^{-n}+\varphi_{-n+1}t^{-n+1}+\ldots+\varphi_{-1}\\&+\operatorname{Re}\varphi_0+\varphi_1 t+\ldots+\varphi_{n-1}t^{n-1}+(\operatorname{Re}\varphi_n)t^n.\end{aligned}\quad(3.24)$$

It is obvious that this operator is an extension on the whole space L_2 of the operator P_n° considered earlier on the space $\overset{\circ}{L}{}_2^+$ only. Moreover, the restriction of this operator on the space $\widetilde{L}_2$ coincides with the operator $\widetilde{P}_n$, i.e., $P_n^\circ\big|_{\widetilde{L}_2}=\widetilde{P}_n$. Therefore

$$A_n-B_n=\widetilde{L}_nQ_n^\circ\frac{1}{2}(I+M)P^\circ GP^\circ P_n^\circ+\widetilde{L}_nQ_n^\circ\frac{1}{2}(I+M)Q^\circ GP^\circ P_n^\circ.\quad(3.25)$$

Let us first assume that the entries of the matrix G are the polynomials. Then the operator $Q^\circ GP^\circ:\overset{\circ}{L}{}_2^{+,m}\to\mathbf{C}^m$ is compact and, since Q_n° strongly converges to zero, the sequence $(\widetilde{L}_nQ_n^\circ\frac{1}{2}(I+M)Q^\circ GP^\circ P_n^\circ)\in\mathcal{J}^{\overset{\circ}{L}{}_2^{+,m}\widetilde{L}_2^m}$. For the first term in (3.25), we use the identity

$$Q_n^\circ P^\circ GP^\circ W_n^\circ=t^nWQ\widetilde{G}P^\circ P_n^\circ$$

to obtain

$$\widetilde{L}_n Q^\circ_n \frac{1}{2}(I+M)P^\circ G P^\circ P^\circ_n = \widetilde{W}_n \widetilde{K}_n W^\circ_n,$$

where $\widetilde{K}_n = \widetilde{W}_n \frac{1}{2}(I+M)Q^\circ \widetilde{L}_n t^n W Q^\circ G P^\circ P^\circ_n$. Further, since G is a polynomial matrix, the operator $WQ^\circ \widetilde{G} P^\circ : \overset{\circ}{L}{}_2^{+,m} \to \mathbf{C}^m$ is compact. Moreover, for each $l_0 \in \mathbb{Z}$, $0 \le l_0 \le n$ one has

$$\widetilde{W}_n \frac{1}{2}(I+M)Q^\circ \widetilde{L}_n t^n t^{l_0} = \frac{1}{2}(I+M)tt^{-l_0},$$

i.e., on the set $1, t, t^2, \ldots$ the sequence $(\widetilde{W}_n \frac{1}{2}(I+M)Q^\circ \widetilde{L}_n t^n)$ strongly converges to the operator $\frac{1}{2}(I+M)tW$. Therefore

$$\widetilde{L}_n Q^\circ_n \frac{1}{2}(I+M)P^\circ G P^\circ P^\circ_n = \widetilde{W}_n K W^\circ_n + \widetilde{W}_n(\widetilde{K}_n - K)W^\circ_n, \tag{3.26}$$

where $K = \frac{1}{2}(I+M)tQ^\circ \widetilde{G} P^\circ$ is a compact operator and the sequence K_n uniformly converges to the operator K. Thus for any polynomial matrix G, both terms in the right-hand side of (3.26) belong to the ideal $\mathcal{J}$. The case where G is a continuous matrix function is reduced to this in an obvious way. □

Lemma 3.2.6. *If $G \in \mathbf{C}^{m \times m}$ and $\det G(t) \neq 0$ everywhere on Γ_0, then the coset of $\mathcal{A}/\mathcal{J}$ that contains the sequence $\widetilde{P}_n \frac{1}{2}(I+M)GP^\circ P^\circ_n$ is invertible.*

This result was established in the proof of Theorem 3.1.5, cf. (3.12), (3.13).

Combination of Lemmas 3.2.5, 3.2.6 with Theorem 1.8.4 gives the proof of Theorem 3.2.4.

3.3 Local Principle for the Para-Algebra $\mathcal{A}/\mathcal{J}$

Consider again the quotient para-algebra

$$\mathcal{A}/\mathcal{J} := (\mathcal{A}^{\overset{\circ}{L}{}_2^+}/\mathcal{J}^{\overset{\circ}{L}{}_2^+}, \mathcal{A}^{\overset{\circ}{L}{}_2^{+,m}\widetilde{L}_2^m}/\mathcal{J}^{\overset{\circ}{L}{}_2^{+,m}\widetilde{L}_2^m}, \mathcal{A}^{\widetilde{L}_2^m \overset{\circ}{L}{}_2^{+,m}}/\mathcal{J}^{\widetilde{L}_2^m \overset{\circ}{L}{}_2^{+,m}}, \mathcal{A}^{\widetilde{L}_2^m}/\mathcal{J}^{\widetilde{L}_2^m}),$$

and recall that the coset of this para-algebra that contains the sequence (A_n) is denoted by $(A_n)^\circ$. Now we are going to study the stability of the Galerkin and approximation methods for the Riemann-Hilbert problem with discontinuous coefficient G. This can be done by means of the local principle described in Section 1.9.3. For that we construct two systems of elements, one of which belongs to the quotient set $\mathcal{A}^{\widetilde{L}_2^m}/\mathcal{J}^{\widetilde{L}_2^m}$ and the other to the quotient set $\mathcal{A}^{\overset{\circ}{L}{}_2^+}/\mathcal{J}^{\overset{\circ}{L}{}_2^+}$ such that they are reciprocally interchangeable with the cosets $(\widetilde{P}_n \frac{1}{2}(I+M)GP^\circ P^\circ_n)^\circ$ for $G \in L_\infty^{m \times m}$. Further we will show that these systems contain sub-systems of localizing covering classes, which are reciprocally interchangeable with the cosets $(\widetilde{P}_n \frac{1}{2}(I+M)GP^\circ P^\circ_n)^\circ$. To proceed, we need auxiliary identities. Let P°_n be the operator defined by (3.24), and let $\overline{P}_n := P_n^\circ + K^\circ$. The operators P_n° and $\overline{P}_n$ are connected with the operator $\widetilde{P}_n$ by simple relations, viz.,

$$\widetilde{P}_n \frac{1}{2}(I+M) = \frac{1}{2}(I+M)P_n^\circ = \frac{1}{2}(I+M)\overline{P}_n.$$

We also use the identities

$$\frac{1}{2}(I+M)Q^\circ P_n^\circ a P_n^\circ P^\circ = \frac{1}{2}(I+M)Q^\circ t^{-n} P_{2n}^\circ P^\circ a P^\circ P_{2n}^\circ t^n P^\circ, \tag{3.27}$$

and

$$\frac{1}{2}(I+M)P_n^\circ Q^\circ t^{-n} P_{2n}^\circ P^\circ d P^\circ e P_{2n}^\circ t^n P^\circ P_n^\circ = \frac{1}{2}(I+M)P_n^\circ Q^\circ d\overline{P}_n e P^\circ P_n^\circ, \tag{3.28}$$

which are valid for any $a, d, e \in L_\infty$. The proof of (3.27) and (3.28) is similar to the proof of (3.15).

Thus the operator $\frac{1}{2}(I+M)Q^\circ P_n^\circ de P_n^\circ P^\circ$ can be rewritten as

$$\begin{aligned}
\frac{1}{2}(I+M)Q^\circ P_n^\circ de P_n^\circ P^\circ &= \frac{1}{2}(I+M)Q^\circ t^{-n} P_{2n}^\circ P^\circ de P^\circ P_{2n}^\circ t^n P^\circ \\
&= \frac{1}{2}(I+M)Q^\circ t^{-n} P_{2n}^\circ P^\circ d Q^\circ e P^\circ P_{2n}^\circ t^n P^\circ \\
&\quad + \frac{1}{2}(I+M)Q^\circ t^{-n} P_{2n}^\circ P^\circ d P^\circ P_{2n}^\circ P^\circ e P^\circ P_{2n}^\circ t^n P^\circ \\
&\quad + \frac{1}{2}(I+M)Q^\circ t^{-n} P_{2n}^\circ P^\circ d P^\circ Q_{2n}^\circ P^\circ e P^\circ P_{2n}^\circ t^n P^\circ .
\end{aligned}$$

Now applying (3.28) one obtains

$$\begin{aligned}
&\left(\widetilde{P}_n \frac{1}{2}(I+M)Q^\circ de P^\circ P_n^\circ\right) \\
&= \left(\widetilde{P}_n \frac{1}{2}(I+M)Q^\circ d\overline{P}_n e P^\circ P_n^\circ\right) - \left(\widetilde{P}_n \frac{1}{2}(I+M)Q^\circ d K^\circ e P^\circ P_n^\circ\right) \\
&\quad + \left(\widetilde{P}_n \frac{1}{2}(I+M)Q^\circ t^{-n} W_{2n}^\circ P^\circ \widetilde{d} Q^\circ \widetilde{e} P^\circ W_{2n}^\circ t^n P^\circ P_n^\circ\right) \\
&\quad + \left(\widetilde{P}_n \frac{1}{2}(I+M)Q^\circ t^{-n} P_{2n}^\circ P^\circ d Q^\circ e P^\circ P_{2n}^\circ t^n P^\circ P_n^\circ\right).
\end{aligned} \tag{3.29}$$

If at least one of the functions e or d is continuous, then the third sequence in the right-hand side of (3.29) belongs to the ideal $\mathcal{J}$. Indeed, the operator $P^\circ \widetilde{d} Q^\circ \widetilde{e} P^\circ$ is compact and $Q^\circ t^{-n} W_{2n}^\circ$ strongly converges to zero, so $\widetilde{P}_n \frac{1}{2}(I+M)Q^\circ t^{-n} W_{2n}^\circ P^\circ \widetilde{d} Q^\circ \widetilde{e} P^\circ W_{2n}^\circ t^n P^\circ P_n^\circ$ converges to zero uniformly. The sequence $\left(\widetilde{P}_n \frac{1}{2}(I+M)Q^\circ d K^\circ e P^\circ P_n^\circ\right)$ also belongs to the ideal $\mathcal{J}$ because the operator K° is compact. Let us now show that the continuity of d or e implies the inclusion

$$\left(\widetilde{P}_n \frac{1}{2}(I+M)Q^\circ t^{-n} P^\circ_{2n} P^\circ dQ^\circ eP^\circ P^\circ_{2n} t^n P^\circ P^\circ_n\right) \in \overset{\circ}{\mathcal{J}}{}^{L_2^+ \widetilde{L}_2}.$$

Recall that the operator $W \in \mathcal{L}(L_2)$ is defined by $(W\varphi)(t) = \varphi(\bar{t})$, so

$$\widetilde{W}_n \frac{1}{2}(I+M)Q^\circ t^{-n} P^\circ_{2n} P^\circ = \frac{1}{2}(I+M)P_{n-1} W P^\circ,$$

hence one can rewrite the operator under consideration as

$$\begin{aligned}
&\widetilde{P}_n \frac{1}{2}(I+M)Q^\circ t^{-n} P^\circ_{2n} P^\circ dQ^\circ eP^\circ P^\circ_{2n} t^n P^\circ P^\circ_n \\
&= (\widetilde{W}_n)^2 \frac{1}{2}(I+M)Q^\circ t^{-n} P^\circ_{2n} P^\circ dQ^\circ eP^\circ P^\circ_{2n} t^n P^\circ P^\circ_n \\
&= \widetilde{W}_n \frac{1}{2}(I+M)P_{n-1} W P^\circ dQ^\circ eP^\circ P^\circ_{2n} t^n P^\circ P^\circ_n \\
&= \widetilde{W}_n \frac{1}{2}(I+M) W P^\circ dQ^\circ eP^\circ P^\circ_{2n} t^n P^\circ P^\circ_n + R_n, \qquad (3.30)
\end{aligned}$$

where the operators $R_n = -\widetilde{W}_n \frac{1}{2}(I+M)Q_{n-1} W P^\circ dQ^\circ eP^\circ P^\circ_{2n} t^n P^\circ P^\circ_n$ uniformly converge to zero. It remains to consider the first term in (3.30). Since $(I+M)WK^\circ = 0$ and $K^\circ P^\circ_{2n} t^n P^\circ = 0$, we can write

$$\begin{aligned}
&\left(\widetilde{W}_n \frac{1}{2}(I+M) W P^\circ dQ^\circ eP^\circ P^\circ_{2n} t^n P^\circ P^\circ_n\right) \\
&= \left(\widetilde{W}_n \frac{1}{2}(I+M) W PdQePP^\circ_{2n} t^n P^\circ P^\circ_n\right) \\
&\quad + \left(\widetilde{W}_n \frac{1}{2}(I+M) W PdK^\circ ePP^\circ_{2n} t^n P^\circ P^\circ_n\right). \qquad (3.31)
\end{aligned}$$

We next show that both sequences in the right-hand side of (3.31) are in $\overset{\circ}{\mathcal{J}}{}^{L_2^+ \widetilde{L}_2}$. Consider the sequence $\left(\widetilde{W}_n \frac{1}{2}(I+M) W PdK^\circ ePP^\circ_{2n} t^n P^\circ P^\circ_n\right)$. If $\widetilde{K}^\circ$ denotes the operator

$$(\widetilde{K}^\circ \varphi)(t) = \widetilde{K}^\circ \left(\sum_{k=-\infty}^{+\infty} \varphi_k t^k\right) \varphi_0,$$

then

$$\begin{aligned}
&\|\widetilde{W}_n \frac{1}{2}(I+M) W PdK^\circ ePP^\circ_{2n} t^n P^\circ P^\circ_n\| \\
&= \|\widetilde{W}_n \frac{1}{2}(I+M) W PdK^\circ ePt^n P^\circ P^\circ_n\| \\
&\le \|K^\circ ePt^n P\| \\
&\le M_1 \|\widetilde{K}^\circ ePt^n P\| \le M_1 \|Pt^{-n} P\bar{e} \widetilde{K}^\circ\|,
\end{aligned}$$

where $M_1 = \sup_n(||\widetilde{W}_n||_2||W||_2||d||_\infty)$. The operator $\widetilde{K}^\circ$ is compact and $Pt^{-n}P$ strongly converges to zero. Thus the last inequality implies

$$\lim_{n\to\infty} ||\widetilde{W}_n \frac{1}{2}(I+M)WPdK^\circ ePP_{2n}^\circ t^n P^\circ P_n^\circ||_2 = 0,$$

so

$$\left(\widetilde{W}_n \frac{1}{2}(I+M)WPdK^\circ ePP_{2n}^\circ t^n P^\circ P_n^\circ\right) \in \mathcal{J}^{\overset{\circ}{L_2^+}\widetilde{L}_2}.$$

Note that the remaining sequence in the right-hand side of (3.31) also converges to zero uniformly. Estimating the norm of the corresponding operator, one obtains

$$\begin{aligned}
&||\widetilde{W}_n \frac{1}{2}(I+M)WPdQePP_{2n}^\circ t^n P^\circ P_n^\circ||\\
&\le ||\widetilde{W}_n||\,||W||\,||PdQePP_{2n}^\circ t^n P^\circ P_n^\circ||\\
&\le ||\widetilde{W}_n||\,||W||\,||(PdQePP_{2n}^\circ t^n P^\circ P_n^\circ)^*||,
\end{aligned}$$

but the norm $||(PdQePP_{2n}^\circ t^n P^\circ P_n^\circ)^*||$ tends to zero as $n \to \infty$. Thus if at least one of the functions d or e is continuous, the first term in the right-hand side of (3.31) belongs to the ideal $\mathcal{J}^{\overset{\circ}{L_2^+}\widetilde{L}_2}$.

In the quotient sets $\mathcal{A}^{\widetilde{L}_2}/\mathcal{J}^{\widetilde{L}_2}$ and $\mathcal{A}^{\overset{\circ}{L_2^+}}/\mathcal{J}^{\overset{\circ}{L_2^+}}$ we consider the systems $(\widetilde{P}_n f \widetilde{P}_n)^\circ$ and $(P_n^\circ P^\circ f P^\circ P_n^\circ)^\circ$ respectively, where f runs throughout the set $\widetilde{\mathbf{C}}$ of real-valued continuous functions.

Theorem 3.3.1. *If $G \in L_\infty$, then the systems $\{(\widetilde{P}_n f \widetilde{P}_n)^\circ : f \in \widetilde{\mathbf{C}}\}$ and $\{(P_n^\circ P^\circ f P^\circ P_n^\circ)^\circ : f \in \widetilde{\mathbf{C}}\}$ are reciprocally interchangeable with the coset $(\widetilde{P}_n \frac{1}{2}(I + M)P^\circ G P^\circ P_n^\circ)^\circ \in \mathcal{A}^{\overset{\circ}{L_2^+}\widetilde{L}_2}/\mathcal{J}^{\overset{\circ}{L_2^+}\widetilde{L}_2}$.*

Proof. Let $G \in L_\infty$ and $f \in \widetilde{\mathbf{C}}$. Since

$$\begin{aligned}
&\widetilde{P}_n \frac{1}{2}(I+M)P^\circ G f P^\circ P_n^\circ\\
&= \widetilde{P}_n \frac{1}{2}(I+M)P^\circ G P^\circ P_n^\circ P^\circ f P^\circ P_n^\circ + \widetilde{P}_n \frac{1}{2}(I+M)P^\circ G Q^\circ f P^\circ P_n^\circ\\
&\quad + \widetilde{P}_n \frac{1}{2}(I+M)P^\circ G Q_n^\circ f P^\circ P_n^\circ\\
&= \widetilde{P}_n \frac{1}{2}(I+M)P^\circ G P^\circ P_n^\circ P^\circ f P^\circ P_n^\circ + \widetilde{P}_n \frac{1}{2}(I+M)P^\circ G Q^\circ f P^\circ P_n^\circ\\
&\quad + \frac{1}{2}(I+M)P_n^\circ P^\circ G Q_n^\circ f P^\circ P_n^\circ,
\end{aligned}$$

from (3.15) we get

$$\begin{aligned}
&\widetilde{P}_n\frac{1}{2}(I+M)P^\circ GfP^\circ P_n^\circ\\
&=\widetilde{P}_n\frac{1}{2}(I+M)P^\circ GP^\circ P_n^\circ P^\circ fP^\circ P_n^\circ+\widetilde{P}_n\frac{1}{2}(I+M)P^\circ GQ^\circ fP^\circ P_n^\circ\\
&\quad+\widetilde{W}_n\frac{1}{2}(I+M)P^\circ\widetilde{G}Q^\circ\widetilde{f}P^\circ W_n^\circ. \qquad (3.32)
\end{aligned}$$

Taking into account (3.32) and (3.29), one can write

$$\begin{aligned}
&\widetilde{P}_n\frac{1}{2}(I+M)GP^\circ P_n^\circ\cdot P_n^\circ P^\circ fP^\circ P_n^\circ=\widetilde{P}_n\frac{1}{2}(I+M)GP^\circ P_n^\circ P^\circ fP^\circ P_n^\circ\\
&=\widetilde{P}_n\frac{1}{2}(I+M)Q^\circ GP_n^\circ P^\circ fP^\circ P_n^\circ+\widetilde{P}_n\frac{1}{2}(I+M)P^\circ GP^\circ P_n^\circ P^\circ fP^\circ P_n^\circ\\
&\quad-\widetilde{P}_n\frac{1}{2}(I+M)Q^\circ GP_n^\circ Q^\circ fP^\circ P_n^\circ\\
&=\widetilde{P}_n\frac{1}{2}(I+M)Q^\circ GfP^\circ P_n^\circ+\widetilde{P}_n\frac{1}{2}(I+M)P^\circ GfP^\circ P_n^\circ+R_n^{(1)}\\
&=\widetilde{P}_n\frac{1}{2}(I+M)GfP^\circ P_n^\circ+R_n^{(1)}, \qquad (3.33)
\end{aligned}$$

where the sequence $(R_n^{(1)})\in\mathcal{J}^{\overset{\circ}{L_2^+}\widetilde{L}_2}$ because f is continuous on Γ_0.

Analogously, for real-valued continuous function f it is possible to write

$$\begin{aligned}
&\widetilde{P}_nf\widetilde{P}_n\cdot\widetilde{P}_n\frac{1}{2}(I+M)GP^\circ P_n^\circ\\
&=\widetilde{P}_nf\widetilde{P}_n\frac{1}{2}(I+M)GP^\circ P_n^\circ=\widetilde{P}_n\frac{1}{2}(I+M)fP_n^\circ GP^\circ P_n^\circ\\
&=\widetilde{P}_n\frac{1}{2}(I+M)Q^\circ fP_n^\circ GP^\circ P_n^\circ+\widetilde{P}_n\frac{1}{2}(I+M)P^\circ fP^\circ P_n^\circ P^\circ GP^\circ P_n^\circ\\
&\quad+\widetilde{P}_n\frac{1}{2}(I+M)P^\circ fQ^\circ P_n^\circ GP^\circ P_n^\circ=\widetilde{P}_n\frac{1}{2}(I+M)fGP^\circ P_n^\circ+R_n^{(2)}, \qquad (3.34)
\end{aligned}$$

where the sequence $(R_n^{(2)})$ belongs to $\mathcal{J}^{\overset{\circ}{L_2^+}\widetilde{L}_2}$. It follows from relations (3.33) and (3.34) that the above systems are reciprocally interchangeable with respect to any element $(\widetilde{P}_n\frac{1}{2}(I+M)GP^\circ P_n^\circ)^\circ\in\mathcal{A}^{\overset{\circ}{L_2^+}\widetilde{L}_2}/\mathcal{J}^{\overset{\circ}{L_2^+}\widetilde{L}_2}$, $G\in L_\infty$. Moreover, for any $f^*\in\widetilde{\mathbf{C}}$, the coset $(\widetilde{P}_nf^*\widetilde{P}_n)^\circ\in\mathcal{A}^{\widetilde{L}_2}/\mathcal{J}^{\widetilde{L}_2}$ corresponds to the coset $(P_n^\circ P^\circ f^*P^\circ P_n^\circ)^\circ\in\mathcal{A}^{\overset{\circ}{L_2^+}}/\mathcal{J}^{\overset{\circ}{L_2^+}}$ and vice versa. □

Let us now construct systems of localizing classes for the para-algebra $\mathcal{A}/\mathcal{J}$. For any point $\tau\in\Gamma_0$, by N_τ we denote the set of all continuous functions on Γ_0 with the following properties:

1. For any function $f \in N_\tau$ there is a neighbourhood V_τ of τ such that $f(t) = 1$ for all $t \in V_\tau$.

2. $0 \le f(t) \le 1$ for all $t \in \Gamma_0$.

Defining

$$\widetilde{M}_\tau := \{(\widetilde{P}_n f \widetilde{P}_n)^\circ \in \mathcal{A}^{\widetilde{L}_2}/\mathcal{J}^{\widetilde{L}_2} : f \in N_\tau\},$$

$$M_\tau^\circ := \{(P_n^\circ f P_n^\circ)^\circ \in \mathcal{A}^{\overset{\circ}{L}{}_2^+}/\mathcal{J}^{\overset{\circ}{L}{}_2^+} : f \in N_\tau\},$$

we can establish the following result.

Lemma 3.3.2. *The systems $\widetilde{M}_\tau$ and M_τ° are respectively left- and right-localizing classes for the para-algebra $\mathcal{A}/\mathcal{J}$.*

Proof. For the system M_τ° the proof in not difficult. If f_1, f_2 are any two functions from N_τ, then invoking (3.15) one gets

$$\begin{aligned} P_n^\circ P^\circ f_1 P_n^\circ P^\circ f_2 P^\circ P_n^\circ &= P_n^\circ P^\circ f_1 f_2 P^\circ P_n^\circ \\ &\quad - P_n^\circ P^\circ f_1 Q^\circ f_2 P^\circ P_n^\circ - W_n^\circ P^\circ \widetilde{f}_1 Q^\circ \widetilde{f}_2 P^\circ W_n^\circ, \end{aligned}$$

which implies that M_τ° is a left-localizing class for $\mathcal{A}/\mathcal{J}$.

The proof is more involved for the set $\widetilde{M}_\tau$. Recalling that $P_n^\circ \big|_{\widetilde{L}_2} = \widetilde{P}_n$, we rewrite the product $\widetilde{P}_n f_1 \widetilde{P}_n f_2 \widetilde{P}_n$ as

$$\begin{aligned} \widetilde{P}_n f_1 \widetilde{P}_n f_2 \widetilde{P}_n &= \widetilde{P}_n \frac{1}{2}(I+M) f_1 P_n^\circ f_2 P_n^\circ \\ &= \widetilde{P}_n \frac{1}{2}(I+M) P^\circ f_1 P^\circ P_n^\circ P^\circ f_2 P^\circ P_n^\circ + \widetilde{P}_n \frac{1}{2}(I+M) Q^\circ f_1 P_n^\circ f_2 P^\circ P_n^\circ \\ &\quad + \widetilde{P}_n \frac{1}{2}(I+M) P^\circ f_1 P_n^\circ f_2 Q^\circ P_n^\circ + \widetilde{P}_n \frac{1}{2}(I+M) Q^\circ f_1 Q^\circ P_n^\circ Q^\circ f_2 Q^\circ P_n^\circ \\ &\quad + \widetilde{P}_n \frac{1}{2}(I+M) P^\circ f_1 Q^\circ P_n^\circ Q^\circ f_2 P^\circ P_n^\circ + \widetilde{P}_n \frac{1}{2}(I+M) P^\circ f_1 P^\circ P_n^\circ P^\circ f_2 Q^\circ P_n^\circ \\ &= \sum_{j=1}^{6} B_n^{(j)}. \end{aligned} \tag{3.35}$$

The sequences $(B_n^{(5)})$ and $(B_n^{(6)})$ belong to the ideal $\mathcal{J}^{\widetilde{L}_2}$ because the function f_2 is continuous. Consider each of the first four terms in the right-hand side of (3.35) and denote by $\overline{P}$ the operator defined on $\widetilde{L}_2(\Gamma_0)$ with values in $L_2(\Gamma_0)$ such that

$$\overline{P}\left(\sum_{k=-\infty}^{+\infty} \varphi_k t^k\right) = \frac{\varphi_0}{2} + \sum_{k=1}^{\infty} \varphi_k t^k.$$

Then

$$P^\circ W_n^\circ \Big|_{\widetilde{L}_2} = \overline{P}\widetilde{W}_n + \widetilde{K}_n,$$

where

$$\widetilde{K}_n\left(\sum_{k=-\infty}^{+\infty} \varphi_k t^k\right) = \frac{1}{2}\varphi_0 t^n.$$

Combining this with (3.32), we obtain

$$B_n^{(1)} = \widetilde{P}_n\frac{1}{2}(I+M)P^\circ f_1 f_2 P^\circ P_n^\circ - \widetilde{P}_n\frac{1}{2}(I+M)P^\circ f_1 Q^\circ f_2 P^\circ P_n^\circ$$
$$- \widetilde{W}_n\frac{1}{2}(I+M)P^\circ \widetilde{f}_1 Q^\circ \widetilde{f}_2 \overline{P}\widetilde{W}_n - \widetilde{W}_n\frac{1}{2}(I+M)P^\circ \widetilde{f}_1 Q^\circ \widetilde{f}_2 \widetilde{K}_n.$$

Since f_1 and f_2 are continuous, the sequences $(\widetilde{P}_n\frac{1}{2}(I+M)P^\circ f_1 Q^\circ f_2 P^\circ P_n^\circ)$ and $(\widetilde{W}_n\frac{1}{2}(I+M)P^\circ \widetilde{f}_1 Q^\circ \widetilde{f}_2 \overline{P}\widetilde{W}_n)$ are in the ideal $\mathcal{J}^{\widetilde{L}_2}$. Let us rewrite

$$\widetilde{W}_n\frac{1}{2}(I+M)P^\circ \widetilde{f}_1 Q^\circ \widetilde{f}_2 \widetilde{K}_n = \widetilde{W}_n\frac{1}{2}(I+M)P^\circ \widetilde{f}_1 Q^\circ \widetilde{f}_2 \widetilde{K}_n \widetilde{P}_n$$
$$= \widetilde{W}_n\frac{1}{2}(I+M)P\widetilde{f}_1 Q\widetilde{f}_2 \widetilde{R}_n \widetilde{W}_n + \widetilde{W}_n\frac{1}{2}(I+M)P\widetilde{f}_1 K^\circ \widetilde{f}_2 P\widetilde{R}_n \widetilde{W}_n,$$

where $(\widetilde{R}_n\varphi)(t) = (\widetilde{K}_n\widetilde{W}_n\varphi)(t) = (\operatorname{Re}\varphi_n)t^n$, and let R_n be the operator defined on the space L_2 by

$$(R_n\varphi)(t) = \varphi_n t^n.$$

Since $|a\operatorname{Re} b| \leq |ab|$ for any complex numbers a and b, the norm of the operator $\widetilde{W}_n\frac{1}{2}(I+M)P\widetilde{f}_1 Q\widetilde{f}_2 \widetilde{R}_n \widetilde{W}_n$ can be estimated as follows

$$||\widetilde{W}_n\frac{1}{2}(I+M)P\widetilde{f}_1 Q\widetilde{f}_2 \widetilde{R}_n \widetilde{W}_n|| \leq ||\widetilde{W}_n||^2\, ||P\widetilde{f}_1 Q\widetilde{f}_2 \widetilde{R}_n|| = ||R_n \overline{\widetilde{f}}_2 Q \overline{\widetilde{f}}_1 P||.$$

However, the operator $Q\overline{\widetilde{f}}_1 P$ is compact and R_n strongly converges to zero, so

$$\lim_{n\to\infty} ||R_n \overline{\widetilde{f}}_2 Q \overline{\widetilde{f}}_1 P|| = 0,$$

and hence the sequence $(\widetilde{W}_n\frac{1}{2}(I+M)P\widetilde{f}_1 Q\widetilde{f}_2 \widetilde{R}_n \widetilde{W}_n)$ is in the ideal $\mathcal{J}^{\widetilde{L}_2}$. Let us now consider the sequence $(\widetilde{W}_n\frac{1}{2}(I+M)P\widetilde{f}_1 K^\circ \widetilde{f}_2 P\widetilde{R}_n \widetilde{W}_n)$. As $\widetilde{f}_2$ is continuous on Γ_0, for any $\varepsilon > 0$ there is a polynomial $\widetilde{f}_N^{(2)}(t) = \sum_{k=-N}^{+N} a_k^{(2)} t^k$ such that

$$||\widetilde{f}_2 - \widetilde{f}_N^{(2)}||_{\mathbf{C}(\Gamma_0)} < \varepsilon,$$

so

$$K^\circ \widetilde{f}_2 \widetilde{R}_n = K^\circ(\widetilde{f}_2 - \widetilde{f}_N^{(2)})\widetilde{R}_n + K^\circ \widetilde{f}_N^{(2)} \widetilde{R}_n.$$

If we now choose $n > N$, then $K^\circ \widetilde{f}_N^{(2)} \widetilde{R}_n = 0$ and consequently

$$||K^\circ \widetilde{f}_2 \widetilde{R}_n|| < \varepsilon$$

so $(\widetilde{W}_n \frac{1}{2}(I+M)P\widetilde{f}_1 K^\circ \widetilde{f}_2 P \widetilde{R}_n \widetilde{W}_n) \in \mathcal{J}^{\widetilde{L}_2}$ and we finally get

$$(B_n^{(1)})^\circ = (\widetilde{P}_n \frac{1}{2}(I+M)P^\circ f_1 f_2 P^\circ P_n^\circ)^\circ.$$

For the sequence $(B_n^{(2)})$, we use (3.29) to represent the corresponding coset as

$$(B_n^{(2)})^\circ = (\widetilde{P}_n \frac{1}{2}(I+M)Q^\circ f_1 f_2 P^\circ P_n^\circ)^\circ.$$

To study the sequence $(B_n^{(3)})$ we invoke two identities that are similar to (3.27), (3.29) and are valid for any $d, e \in L_\infty$, viz.,

$$\frac{1}{2}(I+M)P^\circ P_n^\circ d e P_n^\circ Q^\circ = \frac{1}{2}(I+M)P^\circ t^{-n} P_{2n}^\circ P^\circ d e P^\circ P_{2n}^\circ t^n Q^\circ$$

and

$$\widetilde{P}_n \frac{1}{2}(I+M)P^\circ d \overline{P}_n e Q^\circ P_n^\circ = \widetilde{P}_n \frac{1}{2}(I+M)P^\circ t^{-n} P_{2n}^\circ P^\circ d P^\circ P_{2n}^\circ P^\circ e P^\circ P_{2n}^\circ t^n Q^\circ P_n^\circ.$$

Thus we first obtain

$$\begin{aligned}
&\widetilde{P}_n \frac{1}{2}(I+M)P^\circ f_1 f_2 Q^\circ P_n^\circ \\
&= \frac{1}{2}(I+M)P^\circ t^{-n} P_{2n}^\circ P^\circ f_1 P^\circ P_{2n}^\circ P^\circ f_2 P^\circ t^n Q^\circ \\
&\quad + \frac{1}{2}(I+M)P^\circ t^{-n} P_{2n}^\circ P^\circ f_1 Q^\circ f_2 P^\circ P_{2n}^\circ t^n Q^\circ \\
&\quad + \frac{1}{2}(I+M)P^\circ t^{-n} P_{2n}^\circ P^\circ f_1 P^\circ Q_{2n}^\circ P^\circ f_2 P^\circ P_{2n}^\circ t^n Q^\circ,
\end{aligned}$$

and then

$$\begin{aligned}
&\widetilde{P}_n \frac{1}{2}(I+M)P^\circ f_1 f_2 Q^\circ P_n^\circ \\
&= \widetilde{P}_n \frac{1}{2}(I+M)P^\circ f_1 P_n^\circ f_2 Q^\circ P_n^\circ + \widetilde{P}_n \frac{1}{2}(I+M)P^\circ f_1 K^\circ f_2 Q^\circ P_n^\circ \\
&\quad + \widetilde{P}_n \frac{1}{2}(I+M)P^\circ t^{-n} P_{2n}^\circ P^\circ f_1 Q^\circ f_2 P^\circ P_{2n}^\circ t^n Q^\circ P_n^\circ \\
&\quad + \widetilde{P}_n \frac{1}{2}(I+M)P^\circ t^{-n} W_{2n}^\circ P^\circ \widetilde{f}_1 Q^\circ \widetilde{f}_2 P^\circ W_{2n}^\circ t^n Q^\circ P_n^\circ. \qquad (3.36)
\end{aligned}$$

From the compactness of the operator K° and the continuity of f_1 and f_2 one notices that $(\widetilde{P}_n \frac{1}{2}(I+M)P^\circ f_1 K^\circ f_2 Q^\circ P_n^\circ)$, $(\widetilde{P}_n \frac{1}{2}(I+M)P^\circ t^{-n} P_{2n}^\circ P^\circ f_1 Q^\circ f_2 P^\circ P_{2n}^\circ t^n Q^\circ P_n^\circ) \in \mathcal{J}^{\widetilde{L}_2}$.

Consider the fourth term in the right-hand side of (3.36). Using the connection between the operators Q and Q_n° we find

$$\begin{aligned}
&\widetilde{P}_n \frac{1}{2}(I+M)P^\circ t^{-n} W_{2n}^\circ P^\circ \widetilde{f}_1 Q^\circ \widetilde{f}_2 P^\circ W_{2n}^\circ t^n Q^\circ P_n^\circ \\
&= \widetilde{W}_n \frac{1}{2}(I+M)P^\circ \widetilde{f}_1 Q^\circ \widetilde{f}_2 P^\circ t^n PWQP_n^\circ \\
&= \widetilde{W}_n \frac{1}{2}(I+M)P\widetilde{f}_1 Q\, \widetilde{f}_2 P t^n PWQP_n^\circ + \widetilde{W}_n \frac{1}{2}(I+M)P\widetilde{f}_1 K^\circ \widetilde{f}_2 P t^n PWQP_n^\circ .
\end{aligned}$$

It is clear now that

$$\lim_{n\to\infty} ||\widetilde{P}_n \frac{1}{2}(I+M)P^\circ t^{-n} W_{2n}^\circ P^\circ \widetilde{f}_1 Q^\circ \widetilde{f}_2 P^\circ W_{2n}^\circ t^n Q^\circ P_n^\circ|| = 0,$$

so the sequence $(\widetilde{P}_n \frac{1}{2}(I+M)P^\circ t^{-n} W_{2n}^\circ P^\circ \widetilde{f}_1 Q^\circ \widetilde{f}_2 P^\circ W_{2n}^\circ t^n Q^\circ P_n^\circ) \in \mathcal{J}^{\widetilde{L}_2}$, and (3.36) implies

$$(B_n^{(3)})^\circ = (\widetilde{P}_n \frac{1}{2}(I+M)P^\circ f_1 f_2 Q^\circ P_n^\circ)^\circ .$$

It remains to study the structure of the coset $(B_n^{(4)})^\circ$. We have

$$\begin{aligned}
&\widetilde{P}_n \frac{1}{2}(I+M)Q^\circ f_1 Q^\circ P_n^\circ Q^\circ f_2 Q^\circ P_n^\circ \\
&= \widetilde{P}_n \frac{1}{2}(I+M)Q^\circ f_1 f_2 Q^\circ P_n^\circ - \widetilde{P}_n \frac{1}{2}(I+M)Q^\circ f_1 P^\circ f_2 Q^\circ P_n^\circ \\
&- \widetilde{P}_n \frac{1}{2}(I+M)Q^\circ f_1 Q^\circ Q_n^\circ Q^\circ f_2 Q^\circ P_n^\circ ,
\end{aligned}$$

and continuity of the functions f_1, f_2 again leads to the inclusion $(\widetilde{P}_n \frac{1}{2}(I+M) Q^\circ f_1 P^\circ f_2 Q^\circ P_n^\circ) \in \mathcal{J}^{\widetilde{L}_2}$. Consider now the sequence $(\widetilde{P}_n \frac{1}{2}(I+M)Q^\circ f_1 Q^\circ Q_n^\circ Q^\circ f_2 Q^\circ P_n^\circ)$. Introducing the operators

$$\begin{aligned}
&\overline{Q} : \widetilde{L}_2 \mapsto L_2, \quad \overline{Q} = I - \overline{P}, \\
&\widetilde{P} : L_2 \mapsto L_2, \quad \widetilde{P}\left(\sum_{k=-\infty}^{+\infty} \varphi_k t^k\right) = i\mathrm{Im}\,\varphi_0 + \sum_{k=1}^{+\infty} \varphi_k t^k, \\
&\widetilde{Q} : L_2 \mapsto L_2, \quad \widetilde{Q} = I - \widetilde{P} = Q + T^\circ, \quad T^\circ\left(\sum_{k=-\infty}^{+\infty} \varphi_k t^k\right) = \mathrm{Re}\,\varphi_0,
\end{aligned}$$

we can rewrite the operator under consideration as

$$\begin{aligned}
&\widetilde{P}_n \frac{1}{2}(I+M)Q^\circ f_1 Q^\circ Q_n^\circ Q^\circ f_2 Q^\circ P_n^\circ = \widetilde{P}_n \frac{1}{2}(I+M)Qf_1 Q^\circ Q_n^\circ Q^\circ f_2 Q^\circ P_n^\circ \\
&= \widetilde{P}_n \frac{1}{2}(I+M)\widetilde{Q} f_1 Q^\circ Q_n^\circ Q^\circ f_2 Q^\circ P_n^\circ - \widetilde{P}_n \frac{1}{2}(I+M)T^\circ f_1 Q^\circ Q_n^\circ Q^\circ f_2 Q^\circ P_n^\circ ,
\end{aligned}$$

and notice that

$$\lim_{n\to\infty} ||T^\circ f_1 Q^\circ Q_n^\circ|| = 0.$$

Indeed, from the uniform approximation of the continuous function f_1 by polynomials $\widetilde{f}_N^{(1)}(t) = \sum_{k=-N}^{+N} a_k^{(1)} t^k$ we get

$$T^\circ f_1 Q^\circ Q_n^\circ = T^\circ (f_1 - f_N^{(1)}) Q^\circ Q_n^\circ + T^\circ f_N^{(1)} Q^\circ Q_n^\circ,$$

so

$$||T^\circ f_1 Q^\circ Q_n^\circ|| \le \sup_{t\in\Gamma_0} |f_1(t) - f_N^{(1)}(t)| + ||T^\circ f_N^{(1)} Q^\circ Q_n^\circ||.$$

However, if $n > N+1$ then $T^\circ f_N^{(1)} Q^\circ Q_n^\circ = 0$, leading to the inclusion

$$(\widetilde{P}_n \frac{1}{2}(I+M) T^\circ f_1 Q^\circ Q_n^\circ Q^\circ f_2 Q^\circ P_n^\circ) \in \mathcal{J}^{\widetilde{L}_2}.$$

Further let us consider the sequence $(\widetilde{P}_n \frac{1}{2}(I+M)\widetilde{Q} f_1 Q^\circ Q_n^\circ Q^\circ f_2 Q^\circ P_n^\circ)$. Simple algebraic transformations yield

$$\widetilde{P}_n \frac{1}{2}(I+M)\widetilde{Q} f_1 Q^\circ Q_n^\circ Q^\circ f_2 Q^\circ P_n^\circ = \widetilde{W}_n \frac{1}{2}(I+M)\widetilde{Q}\widetilde{f}_1 \widetilde{P}\widetilde{f}_2 t P_n t^{-1} \overline{Q}\widetilde{W}_n,$$

so

$$\begin{aligned}
&(\widetilde{P}_n \frac{1}{2}(I+M)\widetilde{Q} f_1 Q^\circ Q_n^\circ Q^\circ f_2 Q^\circ P_n^\circ) \\
&= (\widetilde{W}_n \frac{1}{2}(I+M)\widetilde{Q}\widetilde{f}_1 \widetilde{P}\widetilde{f}_2 \overline{Q}\widetilde{W}_n) - (\widetilde{W}_n \frac{1}{2}(I+M)\widetilde{Q}\widetilde{f}_1 P\widetilde{f}_2 t Q Q_n t^{-1}\overline{Q}\widetilde{W}_n) \\
&\quad + (\widetilde{W}_n \frac{1}{2}(I+M)\widetilde{Q}\widetilde{f}_1 T^\circ \widetilde{f}_2 t Q Q_n t^{-1}\overline{Q}\widetilde{W}_n).
\end{aligned} \tag{3.37}$$

It is easily seen that each sequence on the right-hand side of (3.37) is indeed in the ideal $\mathcal{J}^{\widetilde{L}_2}$. For the first sequence this follows from the compactness of the operator $\widetilde{Q}\widetilde{f}_1\widetilde{P}$, whereas for the second and third the inclusion to $\mathcal{J}^{\widetilde{L}_2}$ is the consequence of the limit relations

$$\begin{aligned}
&\lim_{n\to\infty} ||P\widetilde{f}_2 t Q Q_n|| = \lim_{n\to\infty} ||Q_n Q t^{-1}\overline{\widetilde{f}}_2 P|| = 0, \\
&\lim_{n\to\infty} ||T^\circ \widetilde{f}_2 t Q Q_n|| = 0
\end{aligned}$$

respectively, whence

$$(B_n^{(4)})^\circ = (\widetilde{P}_n \frac{1}{2}(I+M) Q^\circ f_1 f_2 Q^\circ P_n^\circ)^\circ.$$

Finally, combining all representations for $(B_n^{(j)})^\circ, j = 1,2,3,4$, we obtain

$$\begin{aligned}
&(\widetilde{P}_n f_1 \widetilde{P}_n f_2 \widetilde{P}_n)^\circ \\
&= (\widetilde{P}_n \frac{1}{2}(I+M)P^\circ f_1 f_2 P^\circ P_n^\circ)^\circ + (\widetilde{P}_n \frac{1}{2}(I+M)Q^\circ f_1 f_2 P^\circ P_n^\circ)^\circ \\
&+ (\widetilde{P}_n \frac{1}{2}(I+M)P^\circ f_1 f_2 Q^\circ P_n^\circ)^\circ (\widetilde{P}_n \frac{1}{2}(I+M)Q^\circ f_1 f_2 Q^\circ P_n^\circ)^\circ \\
&= (\widetilde{P}_n \frac{1}{2}(I+M) f_1 f_2 \widetilde{P}_n)^\circ = (\widetilde{P}_n f_1 f_2 \widetilde{P}_n)^\circ. \qquad (3.38)
\end{aligned}$$

This means that the system $(\widetilde{P}_n f \widetilde{P}_n)^\circ, f \in N_\tau$ is a right-localizing class in the para-algebra $\mathcal{A}/\mathcal{J}$, and the proof is completed. □

To end with the local principle for the para-algebra $\mathcal{A}/\mathcal{J}$ we need one more result.

Lemma 3.3.3. *$\{M_\tau^\circ\}_{\tau\in\Gamma_0}$ and $\{\widetilde{M}_\tau\}_{\tau\in\Gamma_0}$ are covering systems of localizing classes.*

Proof. Consider first the system $\{\widetilde{M}_\tau\}_{\tau\in\Gamma_0}$. Since Γ_0 is compact, from any subset $\{f_\tau\}_{\tau\in\Gamma_0}$ of $f_\tau \in N_\tau$ one can choose a finite number of elements $f_{\tau_1}, f_{\tau_2}, \dots, f_{\tau_l}$ such that

$$\widehat{f}(t) = \sum_{j=1}^{l} f_{\tau_j}(t) \geq 1 \quad \text{for all} \quad t \in \Gamma_0.$$

Let us first study the stability of the Galerkin method $(\widetilde{P}_n \widehat{f} \widetilde{P}_n)$ for the operator of multiplication by the function $\widehat{f}$. Choose a positive number γ such that

$$\sup_{t\in\Gamma_0} |\gamma \widehat{f}(t) - 1| < 1, \qquad (3.39)$$

which is always possible because $\widehat{f}(t) \geq 1$ everywhere on Γ_0. Then

$$\widetilde{P}_n \widehat{f} \widetilde{P}_n = \frac{1}{\gamma}(\widetilde{P}_n - \widetilde{P}_n(1-\gamma\widehat{f})\widetilde{P}_n).$$

However, inequality (3.39) leads to the estimate

$$||\widetilde{P}_n(1-\gamma\widehat{f})\widetilde{P}_n|| < 1,$$

therefore the Galerkin method applies to the operator of multiplication by $\widehat{f}$, and by Theorem 1.8.4 the coset $(\widetilde{P}_n \widehat{f} \widetilde{P}_n)^\circ$ is invertible in the para-algebra $\mathcal{A}/\mathcal{J}$, so $\{\widetilde{M}_\tau\}_{\tau\in\Gamma_0}$ is a covering system.

The proof that the system $\{M_\tau^\circ\}_{\tau\in\Gamma_0}$ is also covering follows from the relation (3.15). Thus considering the product

$$\begin{aligned}
&P_n^\circ P^\circ \widehat{f} P^\circ P_n^\circ \cdot P_n^\circ P^\circ \widehat{f}^{-1} P^\circ P_n^\circ \\
&\qquad = P_n^\circ P^\circ \widehat{f} \widehat{f}^{-1} P^\circ P_n^\circ + P_n^\circ P^\circ \widehat{f} Q^\circ \widehat{f}^{-1} P^\circ P_n^\circ - P_n^\circ P^\circ \widehat{f} Q_n^\circ \widehat{f}^{-1} P^\circ P_n^\circ
\end{aligned}$$

and using (3.15), we obtain

$$P_n^\circ P^\circ \widehat{f} P^\circ P_n^\circ \cdot P_n^\circ P^\circ \widehat{f}^{-1} P^\circ P_n^\circ = P_n^\circ + P_n^\circ P^\circ \widehat{f} Q^\circ \widehat{f}^{-1} P^\circ P_n^\circ - W_n^\circ P^\circ \widetilde{\widehat{f}} Q^\circ \widetilde{\widehat{f}}^{-1} P^\circ P_n^\circ.$$

However, $\widehat{f} \in \widetilde{\mathbf{C}}(\Gamma_0)$, so we conclude that the sequences $(P_n^\circ P^\circ \widehat{f} Q^\circ \widehat{f}^{-1} P^\circ P_n^\circ)$ and $(W_n^\circ P^\circ \widetilde{\widehat{f}} Q^\circ \widetilde{\widehat{f}}^{-1} P^\circ P_n^\circ)$ belong to the ideal $\mathcal{J}^{\mathring{L}_2^+}$, hence

$$(P_n^\circ P^\circ \widehat{f} P^\circ P_n^\circ)^\circ \cdot (P_n^\circ P^\circ \widehat{f}^{-1} P^\circ P_n^\circ)^\circ = (P_n^\circ)^\circ,$$

i.e., the coset $(P_n^\circ P^\circ \widehat{f} P^\circ P_n^\circ)^\circ$ is invertible and the construction of the system of covering localizing classes for $\mathcal{A}/\mathcal{J}$ is completed. □

Remark 3.3.4. It is easily seen that all results of this section also remain valid for the para-algebra

$$\mathcal{A}_p/\mathcal{J}_p := (\mathcal{A}^{\mathring{L}_p^{+,m}}/\mathcal{J}^{\mathring{L}_p^{+,m}}, \mathcal{A}^{\mathring{L}_p^{+,m}\widetilde{L}_p^m}/\mathcal{J}^{\mathring{L}_p^{+,m}\widetilde{L}_p^m}, \mathcal{A}^{\widetilde{L}_p^m\mathring{L}_p^{+,m}}/\mathcal{J}^{\widetilde{L}_p^m\mathring{L}_p^{+,m}}, \mathcal{A}^{\widetilde{L}_p^m}/\mathcal{J}^{\widetilde{L}_p^m}),$$

for $1 < p < \infty$. However, another proof of the invertibility of the coset $(\widetilde{P}_n \widehat{f}\, \widetilde{P}_n)^\circ$ is needed. It can be obtained from relation (3.38) which is also valid for the para-algebra $\mathcal{A}_p/\mathcal{J}_p$.

3.4 Interpolation and Galerkin methods for the Riemann-Hilbert problem with discontinuous coefficients

The applicability of Galerkin and interpolation methods to the operators

$$K := \frac{1}{2}(I+M)GP^\circ : \mathring{L}_2^{+,m} \mapsto \widetilde{L}_2^m, \quad G \in L_\infty^{m\times m},$$
$$K^\circ := \frac{1}{2}(I+M)P^\circ GP^\circ : \mathring{L}_2^{+,m} \mapsto \widetilde{L}_2^m, \quad G \in L_\infty^{m\times m}$$

is considered in this section. We need the following readily verified statements.

Lemma 3.4.1. *If $\lambda = \lambda_1 + i\lambda_2 \in \mathbb{C}^{m\times m}$, where λ_1, λ_2 are real matrices such that*

$$\det \lambda_1 \neq 0, \tag{3.40}$$

then the operator $B_\lambda = P^\circ \lambda P^\circ \in \Pi\{P_n^\circ, \widetilde{P}_n\}$.

Let $A_0^{m\times m} = A_0^{m\times m}(\Gamma_0)$ denote the set of matrix functions $G \in L_\infty^{m\times m}$ such that

$$\operatorname{Re} G(t) = \frac{1}{2}(G(t) + G^*(t)) \geq \nu_0 > 0, \quad \text{for all} \quad t \in \Gamma_0. \tag{3.41}$$

Lemma 3.4.2. *If $G \in A_0^{m\times m}$, the operator $K^\circ : \overset{\circ}{L}{}_2^{+,m} \mapsto \widetilde{L}_2^m$ is invertible.*

Proof. If $G \in A_0^{m\times m}$, then by [212] there is a positive number γ such that

$$||E - \gamma G|| < 1, \tag{3.42}$$

where E is the identity matrix. Writing

$$K^\circ = \frac{1}{2\gamma}(I + M)(P^\circ - P^\circ(E - \gamma G)P^\circ),$$

we note that the operator $P^\circ - P^\circ(E - \gamma G)P^\circ$ is invertible in $\overset{\circ}{L}{}_2^{+,m}$. However, the operator $(1/2\gamma)(I + M) : \overset{\circ}{L}{}_2^{+,m} \mapsto \widetilde{L}_2^m$ is also invertible, and that completes the proof. □

The inequality (3.42) is also used in the proof of a stability result for the Galerkin method.

Lemma 3.4.3. *If $G \in A_0^{m\times m}$, then $K^\circ \in \Pi\{P_n^\circ, \widetilde{P}_n\}$.*

Proof. Since

$$P_n^\circ P^\circ \gamma G P^\circ P_n^\circ = P_n^\circ - P_n^\circ P^\circ(E - \gamma G)P^\circ P_n^\circ$$

the projection method $(P_n^\circ P^\circ G P^\circ P_n^\circ)$ is applicable to the operator $P^\circ G P^\circ$. Since the operator $(1/2\gamma)(I + M) : \overset{\circ}{L}{}_2^{+,m} \mapsto \widetilde{L}_2^m$ is also invertible and

$$\widetilde{P}_n \frac{1}{2\gamma}(I + M)P^\circ = \frac{1}{2\gamma}(I + M)P_n^\circ P^\circ,$$

one obtains the result. □

Let $A_2^{m\times m}$ denote the set of matrix functions G that can be represented as a product of two matrices

$$G = G_0 h, \tag{3.43}$$

where $G_0 \in A_0^{m\times m}$ and h is a continuous $(m \times m)$-matrix function.

Theorem 3.4.4. *Let $G \in A_2^{m\times m}$. The operator $K^\circ \in \Pi\{P_n^\circ, \widetilde{P}_n\}$ if and only if the operators K° and*

$$\widetilde{K}^\circ := \frac{1}{2}(I + M)P^\circ \widetilde{G} P^\circ \tag{3.44}$$

are invertible.

Proof. The necessity is an immediate consequence of Theorem 1.8.4. Let us show that these conditions are sufficient. For any point $\tau \in \Gamma_0$, we set

$$G_\tau(t) = G_0(t)h(\tau).$$

Relations (3.33), (3.34) and straightforward estimates show that the elements $(\widetilde{P}_n K^\circ P_n^\circ)^\circ$ and $(\widetilde{P}_n K_\tau^\circ P_n^\circ)^\circ := (\widetilde{P}_n \frac{1}{2}(I+M)P^\circ G_\tau P^\circ P_n^\circ)^\circ$ are $\{M_\tau^\circ, \widetilde{M_\tau}\}$-equivalent. A little geometrical thought shows that there exist a complex number μ such that the matrices $h_\mu = \mu h(\tau)$ and $G_\mu(t) = G_0(t)\mu^{-1}$ satisfy condition (3.40) and (3.41), respectively. Moreover, it follows from (3.33) that the cosets $(\widetilde{P}_n K_\tau^\circ P_n^\circ)^\circ$ and $(\widetilde{P}_n \frac{1}{2}(I+M)P^\circ G_\mu P^\circ P_n^\circ)^\circ \times (P_n^\circ P^\circ h_\mu P^\circ P_n^\circ)^\circ$ coincide. However, by Lemmas 3.4.1 - 3.4.2 and Theorem 1.8.4 the cosets $(\widetilde{P}_n \frac{1}{2}(I+M)P^\circ G_\mu P^\circ P_n^\circ)^\circ$ and $(P_n^\circ P^\circ h_\mu P^\circ P_n^\circ)^\circ$ are invertible. Thus the coset $(\widetilde{P}_n K_\tau^\circ P_n^\circ)^\circ$ is also invertible, and to complete the proof one can successively apply Theorem 1.9.8 and Theorem 1.8.4. □

Remark 3.4.5. For completeness, one must also verify that the systems $\{M_\tau^\circ\}$ and $\{\widetilde{M_\tau}\}$ are reciprocally interchangeable with respect to the coset $(\widetilde{P}_n K^\circ P_n^\circ)^\circ$. However, this follows immediately from Theorem 3.3.1.

Let us next mention a result on the applicability of the Galerkin method for the scalar Riemann-Hilbert problem with piecewise continuous coefficients.

Theorem 3.4.6. *Let $G \in \mathbf{PC}$ and suppose that for any point $\tau \in \Gamma_0$ the inequality*

$$\left| \arg \frac{G(\tau+0)}{G(\tau-0)} \right| < \pi \tag{3.45}$$

holds. The operator $K^\circ \in \Pi\{P_n^\circ, \widetilde{P}_n\}$ if and only if the operator K° and the operator $\widetilde{K}^\circ$ defined by (3.44) are invertible.

Proof. One only has to observe that if a scalar coefficient G satisfies (3.45), then it can be represented in form (3.43) [89]. □

Consider now the applicability of the Galerkin method to the operator $\frac{1}{2}(I+M)GP^\circ : \overset{\circ}{L}{}_2^{+,m} \mapsto \widetilde{L}_2^m$. Let us start with an estimate of the norm of the operator $V\frac{1}{2}(I+M) : L_2 \mapsto \overset{\circ}{L}{}_2^{+}$, where V is defined by (3.4).

Let $x \in L_2$, and let $x(t) = \sum_{k=-\infty}^{+\infty} x_k t^k$ be the Fourier series for x. Using the Parseval equation twice we obtain

$$\begin{aligned} ||V\frac{1}{2}(I+M)x||^2 &= ||x_0 + \sum_{k=1}^{+\infty}(x_k + \overline{x}_{-k})t^k||^2 = |x_0|^2 \sum_{1}^{+\infty} |x_k + \overline{x}_{-k}|^2 \\ &\le |x_0|^2 \sum_{k=1}^{+\infty}(|x_k|^2 + |\overline{x}_{-k}|^2) \le 2\left(\sum_{k=-\infty}^{+\infty} |x_k|^2 \right) = 2||x||^2, \end{aligned}$$

which implies the inequality

$$||V\frac{1}{2}(I+M)|| \le \sqrt{2}.$$

Let $\widetilde{A}_0^{m\times m} = \widetilde{A}_0^{m\times m}(\Gamma_0)$ denote the set of matrix functions $G \in L_\infty^{m\times m}$ with the property that there is a positive number ν_0 such that

$$\operatorname{Re} G(t) \geq \frac{||G||_\infty}{\sqrt{2}} + \nu_0 \quad \text{for all} \quad t \in \Gamma_0.$$

Lemma 3.4.7. *If $G \in \widetilde{A}_0^{m\times m}$, then there is a positive number γ^* such that*

$$||E - \gamma^* G|| < \frac{1}{\sqrt{2}}. \tag{3.46}$$

Proof. For any $x \in L_2^m$ and for any $\gamma > 0$, the inequality

$$||(E-\gamma G)x||^2 \leq \left(1 - 2\gamma\left(\frac{||G||_\infty}{\sqrt{2}} + \nu_0\right) + \gamma^2 ||G||_\infty^2\right) ||x||^2.$$

holds. Now one can find a $\gamma^* > 0$ such that

$$1 - 2\gamma^*\left(\frac{||G||_\infty}{\sqrt{2}} + \nu_0\right) + (\gamma^*)^2 2||G||_\infty^2 < \frac{1}{2},$$

which yields inequality (3.46). □

Denote by $\widetilde{A}^{m\times m} = \widetilde{A}^{m\times m}(\Gamma_0)$ the set of matrix functions $G \in L_\infty^{m\times m}$ which can be represented as the product

$$G = G_0 h$$

where $G_0 \in \widetilde{A}_0^{m\times m}$ and $h \in \mathbf{C}^{m\times m}$.

Theorem 3.4.8. *Let $G \in \widetilde{A}^{m\times m}$. The operator $K \in \Pi\{P_n^\circ, \widetilde{P}_n\}$ if and only if the operator K and the operator $\widetilde{K}^\circ$ defined by (3.44) are invertible.*

The proof of this result is based on the Gohberg-Krupnik local principle for the para-algebra $\mathcal{A}/\mathcal{J}$ and is similar to the proof of Theorem 3.4.6.

Theorem 3.4.9. *Assume $G \in \mathbf{PC}$ and for any point $\tau \in \Gamma_0$ the inequality*

$$\left|\arg \frac{G(\tau+0)}{G(\tau-0)}\right| < \frac{\pi}{2} \tag{3.47}$$

holds. The operator $K \in \Pi\{P_n^\circ, \widetilde{P}_n\}$ if and only if the operator K and the operator $\widetilde{K}^\circ$ defined by (3.44) are invertible.

Proof. If the inequality (3.47) holds, then analogous to [89] one can show that the coefficient G admits the representation $G = \widehat{G}_0 \widehat{h}$, where $\widehat{h} \in \mathbf{C}$ and $\widehat{G}_0 \in \mathbf{PC}$ is such that $|\widehat{G}_0| = 1$ and $\operatorname{Re} \widehat{G}_0(t) > 1/2 + \nu_0$ for all $t \in \Gamma_0$. As before, ν_0 is a positive number, so Lemma 3.4.7 and the local principle of Section 3.4 implies the result. □

Our next goal is to study the applicability of the interpolation method to the Hilbert-Riemann problem with discontinuous coefficient. Recall that the symbol X_R is reserved for the set $X \cap \mathbf{R}$, where $\mathbf{R}$ denotes the set of all Riemann integrable functions.

Lemma 3.4.10. *If $G \in \widetilde{A}_0^{m\times m}$, then the operators*

$$A_n = \widetilde{L}_n \frac{1}{2}(I+M)GP^\circ P_n^\circ, \quad A_n : \operatorname{im} P_n^\circ \mapsto \operatorname{im} \widetilde{P}_n$$

are invertible for all $n \in \mathbb{N}$.

Proof. We represent the operator A_n in the form

$$A_n = \frac{1}{2\gamma_0}\widetilde{L}_n \frac{1}{2}(I+M)GP^\circ P_n^\circ - \frac{1}{2\gamma_0}\widetilde{L}_n(I+M)(E-\gamma_0 G)P^\circ P_n^\circ,$$

where γ_0 is chosen in order to satisfy inequality (3.46). If we multiply the operators $V_n = \gamma_0 P_n^\circ V \widetilde{P}_n$ and A_n, we obtain

$$\begin{aligned} V_n A_n &= P_n^\circ V \widetilde{P}_n \widetilde{L}_n \frac{1}{2}(I+M)P^\circ P_n^\circ - P_n^\circ V \widetilde{P}_n \frac{1}{2}(I+M)\widetilde{L}_n(E-\gamma_0 G)P^\circ P_n^\circ \\ &= P_n^\circ - P_n^\circ V \frac{1}{2}(I+M)\widetilde{L}_n(E-\gamma_0 G)P^\circ P_n^\circ. \end{aligned}$$

Since $||V||_{L_2^m \mapsto \overset{\circ}{L}{}_2^+} \leq \sqrt{2}$ we have

$$\begin{aligned} &||P_n^\circ V \frac{1}{2}(I+M)\widetilde{L}_n(E-\gamma_0 G)P^\circ P_n^\circ x_n|| \\ &\qquad \leq ||V\frac{1}{2}(I+M)||_{L_2^m \mapsto \overset{\circ}{L}{}_2^+} ||\widetilde{L}_n(E-\gamma_0 G)P^\circ P_n^\circ|| \leq q||P_n x_n||, \end{aligned}$$

where $0 < q < 1$, and this inequality implies that the operator $V_n A_n$ and hence A_n are invertible. □

Theorem 3.4.11. *Suppose the coefficient $G \in L_\infty^{m\times m}$ admits the representation $G = G_0 h$ with $G_0 \in \widetilde{A}_{0,R}^{m\times m}$ and $h^* \in \mathbf{C} \cap \overset{\circ}{L}{}_2^{+,m\times m}$. The operator $K \in \Pi\{P_n^\circ, \widetilde{L}_n\}$ if and only if the operators K and $\widetilde{K} = \frac{1}{2}(I+M)(P^\circ + tQ)\widetilde{G}P^\circ$ are invertible.*

Proof. Necessity follows from Theorem 1.8.4 and Lemma 3.2.3. For sufficiency one has to establish the invertibility of the coset $(\widetilde{L}_n \frac{1}{2}(I+M)GP^\circ P_n^\circ)^\circ$ in the corresponding para-algebra. Considering the representation

$$(\widetilde{L}_n \frac{1}{2}(I+M)GP^\circ P_n^\circ)^\circ = (\widetilde{L}_n \frac{1}{2}(I+M)G_0 P^\circ P_n^\circ)^\circ (P_n^\circ P^\circ h_0 P^\circ P_n^\circ)^\circ,$$

we note that the invertibility of the coset $(P_n^\circ P^\circ h_0 P^\circ P_n^\circ)^\circ$ follows from the equation (cf. the proof of Lemma 3.3.2)

$$P_n^\circ P^\circ f P_n^\circ P^\circ f^{-1} P^\circ P_n^\circ = P_n^\circ - P_n^\circ P^\circ f Q^\circ f^{-1} P^\circ P_n^\circ - W_n^\circ P^\circ \widetilde{f} Q^\circ \widetilde{f}^{-1} P^\circ W_n^\circ,$$

while $(\widetilde{L}_n \frac{1}{2}(I+M)G_0 P^\circ P_n^\circ)^\circ$ is invertible by Lemma 3.4.10. To complete the proof one again applies Theorem 1.8.4. □

3.5 Approximate Solution of the Generalized Riemann-Hilbert-Poincaré Problem

Let Δ denote the Laplace operator

$$\Delta := \frac{\partial^2}{\partial x^2} + \frac{\partial^2}{\partial y^2} \tag{3.48}$$

and let A, B, C be real-valued $(m \times m)$-matrix functions defined on the domain $D = \{z \in \mathbb{C} : |z| < 1\}$. The Poincaré boundary value problem involves determination of the solution $u = u(x, y)$ of the elliptic system

$$\begin{aligned} &\Delta u + A\frac{\partial u}{\partial x} + B\frac{\partial u}{\partial y} + Cu = 0 \\ &(x, y) \in D \end{aligned} \tag{3.49}$$

under the boundary condition

$$p^{(1)}(t)\frac{\partial u}{\partial x}(t) + p^{(2)}(t)\frac{\partial u}{\partial y}(t) + r(t)u = f(t), \quad t \in \Gamma_0, \tag{3.50}$$

where $p^{(1)}, p^{(2)}, r$ are $(m \times m)$-matrices and f is an m-dimensional vector.

It is well-known that this problem finds various applications [78, 162, 221], and there are many papers devoted to its approximate solution. Usually, the corresponding approximation methods are based on finite difference or finite element schemes with high computational costs. At the same time, the Poincaré problem can be reduced to a Hilbert-Riemann boundary value problem with derivatives. This lessens the dimension of the problem and expands the variety of approximation methods to use. Thus previous considerations allow us to construct simple approximation methods for the problem (3.49) – (3.50).

Recall that generalized Riemann-Hilbert-Poincaré boundary value problem consists in finding a function Φ that is analytic in D and satisfies the conditions [162]

$$\begin{aligned} &\operatorname{Re}\{L\Phi\} = f(t), \quad t \in \Gamma_0, \\ &\operatorname{Im}\Phi^{(q)}(0) = 0, \\ &\int_{\Gamma_0} \Phi(\tau)\tau^{-k-1}\, d\tau = 0, \quad k = 0, 1, \ldots, q-1, \end{aligned} \tag{3.51}$$

where L is an integro-differential operator of the form

$$(L\Phi)(t) = \sum_{j=0}^{q} \left(a_j(t)\Phi^{(j)}(t) + \int_{\Gamma_0} h_j(t,\tau)\Phi^{(j)}(\tau)\, d\tau \right). \tag{3.52}$$

An approximate solution of this problem was considered in [125] for $q = 1$ and $m = 1$. The applicability of approximation methods to associated problems is studied in [126]. Both [125] and [126] deal with finite difference methods that have a high computational cost. In the present section we shall use projection operators $\widetilde{P}_n$ and $\widetilde{L}_n$ introduced in Sections 3.1–3.2, and seek the solution of the problem (3.51) in the class

$$\overset{\circ}{H}{}_2^{q+} := \{\Phi : \Phi \in H_2^q \cap L_2^+, \quad \operatorname{Im} \Phi^{(q)}(0) = 0\}.$$

Let $\widetilde{P}_n^q, n = 1, 2, \ldots$ denote the projection defined by

$$\begin{aligned}(\widetilde{P}_n^q f)(t) &= \widetilde{P}_n^q \left(\sum_{k=0}^{\infty} f_k t^k \right) \\ &= (\operatorname{Re} f_0)t^q + f_{q+1}t^{q+1} + \ldots + f_{n+q-1}t^{n+q-1} + (\operatorname{Re} f_{n+q})t^{n+q},\end{aligned}$$

and consider two sequences of approximate equations

$$\widetilde{P}_n \operatorname{Re}(L\widetilde{P}_n^q x_n) = \widetilde{P}_n f, \quad x_n \in \operatorname{im} \widetilde{P}_n^q, \tag{3.53}$$

$$\widetilde{L}_n \operatorname{Re}(L\widetilde{P}_n^q x_n) = \widetilde{P}_n f, \quad x_n \in \operatorname{im} \widetilde{P}_n^q. \tag{3.54}$$

Theorem 3.5.1. *Let $a_j \in \mathbf{PC}(\Gamma_0)$, $h_j \in L_\infty(\Gamma_0 \times \Gamma_0)$. Then:*

1. *If for any $\tau \in \Gamma_0$ the inequality*

$$\left| \arg \frac{a_q(\tau + 0)}{a_q(\tau - 0)} \right| < \frac{\pi}{2}$$

holds, then the sequence $(\widetilde{P}_n \operatorname{Re}(L\widetilde{P}_n^q))$ is stable if and only if the operators $K = \operatorname{Re}(LP^\circ) : \overset{\circ}{H}{}_2^{q+}(\Gamma_0) \mapsto \widetilde{L}_2(\Gamma_0)$ and $\widetilde{K} = \operatorname{Re}(P^\circ \widetilde{a}_q P^\circ) : \overset{\circ}{L}{}_2^+(\Gamma_0) \mapsto \widetilde{L}_2(\Gamma_0)$ are invertible.

2. *If $a_q \in \mathbf{C}(\Gamma_0)$, then the sequence $(\widetilde{L}_n \operatorname{Re}(L\widetilde{P}_n^q))$ is stable if and only if the operators $K : \overset{\circ}{H}{}_2^{q+}(\Gamma_0) \mapsto \widetilde{L}_2(\Gamma_0)$ and $\widetilde{K} = \operatorname{Re}((P^\circ + tQ)\widetilde{a}_q P^\circ) : \overset{\circ}{L}{}_2^+(\Gamma_0) \mapsto \widetilde{L}_2(\Gamma_0)$ are invertible.*

The proof of this result is similar to the proofs of Theorems 2.3.4 and 2.3.5 and is omitted here.

3.6 Comments and References

Sections 3.1 – 3.4: The initial approach to approximate solution of various problems associated with the Riemann-Hilbert problem is based on using difference approximation methods [125, 126]. The algebraic methods similar to those employed in the previous chapter are not working since the corresponding operators act in a pair of spaces, so the corresponding operator spaces do not have a multiplication operation. It was noted in [44] that these difficulties can be overcame if one uses the concept of para-algebra developed in [139, 184]. Paper [44] contains stability results for the Galerkin method whereas the interpolation method is studied in [45]. In [154] an approximate solution of the homogeneous problem was constructed by polylogarithms.

Section 3.5: Approximation methods for the generalized Riemann-Hilbert-Poincaré boundary value problem are studied in [48].

Let us recall that the Riemann-Hilbert problem is closely connected with the singular integral equations with Hilbert kernels [91, 162]

$$a(x)u(x) + \frac{b(x)}{2\pi}\int_0^{2\pi} u(y)\cot\frac{x-y}{2}\,dy = f(x). \tag{3.55}$$

Consequently, the approximation methods considered in Sections 3.1 – 3.4 can be used to construct and study approximation methods for equation (3.55), as well as for more general equations.

We also would like to note that there is a well-developed theory, where the Poincaré problem (3.49) – (3.50) is reduced to two-dimensional singular integral equations with conjugation [78, 221]. Projection methods for such equations are little developed. In [60] the authors study the stability of certain approximation methods for the equations

$$\begin{aligned}(A\phi)(z) &\equiv a(z)\phi(z) + b(z)(S\phi)(z) + c(z)(\overline{S}\phi)(z) + d(z)(B\phi)(z)e(z)(N\phi)(z)\\ &= f(z), \quad z \in G,\end{aligned} \tag{3.56}$$

where $G := \{z \in \mathbb{C} : |z| < 1\}$, the coefficients $a, b, c, d, e \in \mathbf{C}(G \cup \Gamma_0)$ and the operators $S, \overline{S}, B$ and N are respectively defined by the relations

$$\begin{aligned}(S\phi)(z) &= -\frac{1}{\pi}\int_G \frac{\phi(\zeta)dG_\zeta}{(\zeta - z)^2} = -\frac{1}{\pi}\int_G \frac{\phi(\zeta)d\xi d\eta}{(\zeta - z)^2},\\ (\overline{S}\phi)(z) &= -\frac{1}{\pi}\int_G \frac{\phi(\zeta)dG_\zeta}{(\overline{\zeta} - \overline{z})^2} = -\frac{1}{\pi}\int_G \frac{\phi(\zeta)d\xi d\eta}{(\overline{\zeta} - \overline{z})^2},\\ (B\phi)(z) &= \frac{1}{\pi}\int_G \frac{\phi(\zeta)dG_\zeta}{(1 - \overline{\zeta}z)^2} = \frac{1}{\pi}\int_G \frac{\phi(\zeta)d\xi d\eta}{(1 - \overline{\zeta}z)^2},\\ (N\phi)(z) &= \frac{1}{\pi}\int_G \frac{\phi(\zeta)dG_\zeta}{(1 - \zeta\overline{z})^2} = \frac{1}{\pi}\int_G \frac{\phi(\zeta)d\xi d\eta}{(1 - \zeta\overline{z})^2},\end{aligned}$$

with $\zeta = \xi + \mathrm{i}\eta$. The methods under consideration are based on polynomials derived from the system $1, z, \overline{z}, z\overline{z}, z^2, \overline{z}^2, \overline{z}^3, \dots$ by the orthonormalization process of Gram-Schmidt. T. Fink [83] considered the stability problem for various spline approximation methods for a particular class of equations (3.56). However, the theory of approximation methods for equation (3.56) still remains extremely scanty in comparison with the results available for the one-dimensional Cauchy singular integral equations.

Chapter 4

Piecewise Smooth and Open Contours

In this chapter, we study different problems connected with approximation methods for singular integral equations of the form

$$\begin{aligned}(Au)(t) \equiv a(t)u(t) + \frac{b(t)}{\pi i}\int_\Gamma \frac{u(s)ds}{s-t} + c(t)\overline{u(t)} + \frac{d(t)}{\pi i}\int_\Gamma \frac{\overline{u(s)ds}}{s-t} + \frac{e(t)}{\pi i}\overline{\int_\Gamma \frac{u(s)ds}{s-t}} \\ + \frac{f(t)}{\pi i}\overline{\int_\Gamma \frac{\overline{u(s)ds}}{s-t}} + \int_\Gamma k_1(t,s)u(s)ds + \int_\Gamma k_2(t,s)\overline{u(s)}ds = g(t),\ t\in\Gamma,\end{aligned} \tag{4.1}$$

where a, b, c, d, e, f, g, k_1, and k_2 are given functions, and Γ is either a simple open or closed piecewise smooth curve in the complex plane $\mathbb{C}$. A case of particular interest is the double layer potential equation

$$(Au)(t) \equiv a(t)u(t) + \frac{b(t)}{\pi}\int_\Gamma u(s)\frac{d}{dn_s}\log|t-s|d\Gamma_s + (Tu)(t) = g(t)\ ,\ t\in\Gamma \tag{4.2}$$

where n_s refers to the inner normal to Γ at s, and T stands for a compact operator.

The approach used here is based on the thorough use of localization techniques in combination with Mellin calculus.

4.1 Stability of Approximation Methods

In this section, we derive necessary and sufficient conditions for the stability of certain quadrature, collocation, and qualocation methods for equation (4.1) on piecewise smooth curves. With each concrete approximation method we associate a family A^τ, $\tau \in \Gamma$ of operators acting on l_2-spaces, and show that the method under consideration is stable if and only if the operator A, all operators $A^\tau, \tau \in \Gamma$, and possibly an additional operator $\widetilde{A}$, are invertible. The operators A^τ are referred to as local operators, and in a sense they hold all information about the associated approximation method. If τ is a corner point of Γ, the structure of the corresponding local operators A^τ is often rather complicated, and in some cases there seems to be no way to verify the invertibility of A^τ effectively. Then we restrict our study to the Fredholmness of the operators A^τ and evaluate the indices of these operators, provided that they are Fredholm.

4.1.1 Quadrature Methods for Singular Integral Equations with Conjugation

We study three simple quadrature methods, which are similar to the methods studied in Sections 2.6 and 2.9 for integral operators considered on smooth closed curves.

Let Γ be a simple closed curve and let $\gamma : \mathbb{R} \to \Gamma$ be a 1-periodic continuous parametrization of Γ. Suppose that γ is twice continuously differentiable on each open interval $(j/n_0, (j+1)/n_0)$ where n_0 is a fixed positive integer and $j = 0, \dots, n_0 - 1$, and that γ' and γ'' have finite one-sided limits $\gamma'(j/n_0 \pm 0)$ and $\gamma''(j/n_0 \pm 0)$ for any $j = 0, \dots, n_0 - 1$ such that

$$\left|\gamma'\left(\frac{j}{n_0} - 0\right)\right| = \left|\gamma'\left(\frac{j}{n_0} + 0\right)\right|$$

but

$$\arg\gamma'\left(\frac{j}{n_0} - 0\right) \neq \arg\gamma'\left(\frac{j}{n_0} + 0\right).$$

Throughout this section we work in the Hilbert space $L_2(\Gamma)$ with the scalar product

$$(f, g) = \int_0^1 f(\gamma(s))\, \overline{g(\gamma(s))}\, ds.$$

For each integer multiple n of n_0 and for each integer k we put

$$t_k^{(n)} = \gamma\left(\frac{k+\delta}{n}\right), \quad \tau_k^{(n)} = \gamma\left(\frac{k+\varepsilon}{n}\right), \quad 0 < \varepsilon,\ \delta < 1,\ \varepsilon \neq \delta, \tag{4.3}$$

and determine approximate values $\xi_k^{(n)}$ for the exact solution of (4.1) at the points $t_k^{(n)}$ by solving one of the following discrete systems:

$$
\begin{aligned}
&[a(\tau_k^{(n)}) + (f(\tau_k^{(n)}) - b(\tau_k^{(n)}))i\cot(\pi(\varepsilon-\delta))]\xi_k^{(n)} \\
&\quad + \frac{1}{\pi i}\sum_{j=0}^{n-1}\left(\frac{b(\tau_k^{(n)})\Delta t_j^{(n)}}{t_j^{(n)}-\tau_k^{(n)}} - \frac{f(\tau_k^{(n)})\overline{\Delta t_j^{(n)}}}{\overline{t_j^{(n)}}-\overline{\tau_k^{(n)}}}\right)\xi_j^{(n)} \\
&\quad + [c(\tau_k^{(n)}) + (e(\tau_k^{(n)}) - d(\tau_k^{(n)}))i\cot(\pi(\varepsilon-\delta))]\overline{\xi_k^{(n)}} \\
&\quad + \frac{1}{\pi i}\sum_{j=0}^{n-1}\left(\frac{d(\tau_k^{(n)})\Delta t_j^{(n)}}{t_j^{(n)}-\tau_k^{(n)}} - \frac{e(\tau_k^{(n)})\overline{\Delta t_j^{(n)}}}{\overline{t_j^{(n)}}-\overline{\tau_k^{(n)}}}\right)\overline{\xi_j^{(n)}} \\
&\quad + \sum_{j=0}^{n-1} k_1(t_k^{(n)}, t_j^{(n)})\Delta t_j^{(n)}\xi_j^{(n)} + \sum_{j=0}^{n-1} k_2(t_k^{(n)}, t_j^{(n)})\Delta t_j^{(n)}\overline{\xi_j^{(n)}} \\
&= g(\tau_k^{(n)}), \qquad k = 0,1,\dots,n-1,
\end{aligned}
\tag{4.4}
$$

$$
\begin{aligned}
&a(t_k^{(n)})\xi_k^{(n)} + \frac{1}{\pi i}\sum_{\substack{j=0\\ j\neq k}}^{n-1}\left(\frac{b(t_k^{(n)})\Delta t_j^{(n)}}{t_j^{(n)}-t_k^{(n)}} - \frac{f(t_k^{(n)})\overline{\Delta t_j^{(n)}}}{\overline{t_j^{(n)}}-\overline{t_k^{(n)}}}\right)\xi_j^{(n)} \\
&\quad + c(t_k^{(n)})\overline{\xi_k^{(n)}} + \frac{1}{\pi i}\sum_{\substack{j=0\\ j\neq k}}^{n-1}\left(\frac{d(t_k^{(n)})\Delta t_j^{(n)}}{t_j^{(n)}-t_k^{(n)}} - \frac{e(t_k^{(n)})\overline{\Delta t_j^{(n)}}}{\overline{t_j^{(n)}}-\overline{t_k^{(n)}}}\right)\overline{\xi_j^{(n)}} \\
&\quad + \sum_{j=0}^{n-1} k_1(t_k^{(n)}, t_j^{(n)})\Delta t_j^{(n)}\xi_j^{(n)} + \sum_{j=0}^{n-1} k_2(t_k^{(n)}, t_j^{(n)})\Delta t_j^{(n)}\overline{\xi_j^{(n)}} \\
&= g(t_k^{(n)}), \qquad k = 0,1,\dots,n-1,
\end{aligned}
\tag{4.5}
$$

$$
\begin{aligned}
&a(t_k^{(n)})\xi_k^{(n)} + \frac{1}{\pi i}\sum_{\substack{j=0\\ j\equiv k+1 (\mathrm{mod}\, 2)}}^{n-1}\left(\frac{b(t_k^{(n)})\Delta t_j^{(n)}}{t_j^{(n)}-t_k^{(n)}} - \frac{f(t_k^{(n)})\overline{\Delta t_j^{(n)}}}{\overline{t_j^{(n)}}-\overline{t_k^{(n)}}}\right)\xi_j^{(n)} \\
&\quad + c(t_k^{(n)})\overline{\xi_k^{(n)}} + \frac{1}{\pi i}\sum_{\substack{j=0\\ j\equiv k+1 (\mathrm{mod}\, 2)}}^{n-1}\left(\frac{d(t_k^{(n)})\Delta t_j^{(n)}}{t_j^{(n)}-t_k^{(n)}} - \frac{e(t_k^{(n)})\overline{\Delta t_j^{(n)}}}{\overline{t_j^{(n)}}-\overline{t_k^{(n)}}}\right)\overline{\xi_j^{(n)}} \\
&\quad + \sum_{\substack{j=0\\ j\equiv k+1 (\mathrm{mod}\, 2)}}^{n-1} k_1(t_k^{(n)}, t_j^{(n)})\Delta t_j^{(n)}\xi_j^{(n)} + \sum_{\substack{j=0\\ j\equiv k+1 (\mathrm{mod}\, 2)}}^{n-1} k_2(t_k^{(n)}, t_j^{(n)})\Delta t_j^{(n)}\overline{\xi_j^{(n)}} \\
&= g(t_k^{(n)}), \qquad k = 0,1,\dots,n-1,
\end{aligned}
\tag{4.6}
$$

where $\Delta t_j^{(n)} = \gamma((j+1)/n) - \gamma(j/n)$ for (4.4) and (4.5) and $\Delta t_j^{(n)} = \gamma((j+1)/n) - \gamma((j-1)/n)$ for (4.6). The number n in (4.6) is supposed to be even. If $\chi_j^{(n)}$ stands for the characteristic function of the arc joining the points $\gamma(j/n)$ and $\gamma((j+1)/n)$, then the approximate solution of the integral equation (4.1) is given by

$$u_n = \sum_{j=0}^{n-1} \xi_j^{(n)} \chi_j^{(n)}. \tag{4.7}$$

We study the methods (4.4) – (4.6) by means of a local principle, i.e., to each of these methods we associate at each point $\tau \in \Gamma$ a model method $A_n^\tau u_n^\tau = g_n^\tau$ for a model equation

$$A^\tau u^\tau = g^\tau \tag{4.8}$$

on a model curve Γ_τ, and show that the stability of the method under consideration depends on the stability of each of the corresponding model systems.

If $\tau \in \Gamma$ and $s_\tau \in [0,1)$ denote the number satisfying $\gamma(s_\tau) = \tau$, define

$$\omega_\tau := \arg(-\gamma'(s_\tau - 0)/\gamma'(s_\tau + 0)) \quad \in (0, 2\pi),$$

and write Γ_τ for the curve

$$\Gamma_\tau := \mathbb{R}^+ \cup e^{i\omega_\tau}\mathbb{R}^+$$

where the second and first rays are respectively directed to and away from the origin. The model operator

$$A^\tau = a(\tau)I + b(\tau)S_{\Gamma_\tau} + c(\tau)M + d(\tau)S_{\Gamma_\tau}M - e(\tau)MS_{\Gamma_\tau} - f(\tau)MS_{\Gamma_\tau}M, \tag{4.9}$$

where the operators $(Mu)(t) := \overline{u(t)}$ and

$$(S_{\Gamma_\tau}u)(t) := \frac{1}{\pi i}\int_{\Gamma_\tau} \frac{u(s)ds}{s-t},$$

are considered on the space $L_2(\Gamma_\tau)$. Further, let us introduce the set $\mathbf{R}_2(\Gamma_\tau)$ of Riemann integrable functions on Γ_τ, consisting of all functions which are Riemann integrable on each finite subcurve of Γ_τ, and for which the norm

$$\|f\|_{R_2(\Gamma_\tau)} = \|f\|_{L_2(\Gamma_\tau)} + \left(\sum_{k=0}^{\infty} \sup_{t\in[k,k+1]} |f(t)|^2\right)^{1/2} + \left(\sum_{k=0}^{\infty} \sup_{t\in e^{i\omega_\tau}[k,k+1]} |f(t)|^2\right)^{1/2}$$

is finite. It is easily seen that $\mathbf{R}_2(\Gamma_\tau)$ is a Banach space which is continuously embedded into $L_2(\Gamma_\tau)$. For the approximate solution of the equation $A^\tau u^\tau = g^\tau$ with $g^\tau \in \mathbf{R}_2(\Gamma_\tau)$ we choose the numbers

$$\tilde{t}_k^{(n)} = \begin{cases} \dfrac{k+\delta}{n} & \text{if } k \geq 0, \\ -\dfrac{k+\delta}{n} e^{i\omega_\tau} & \text{if } k < 0, \end{cases}$$

$$\tilde{\tau}_k^{(n)} := \begin{cases} \dfrac{k+\varepsilon}{n} & \text{if} \quad k \geq 0, \\ -\dfrac{k+\varepsilon}{n} e^{i\omega_\tau} & \text{if} \quad k < 0, \end{cases}$$

and determine approximative values $\xi_k^{(n)}$ of the exact solutions $u^\tau(t_k^{(n)})$, $k \in \mathbb{Z}$, by solving one of the following model systems:

$$\begin{aligned}
&[a(\tau) + (f(\tau) - b(\tau))i\cot(\pi(\varepsilon-\delta))]\xi_k^{(n)} \\
&\quad + \frac{1}{\pi i}\left\{\sum_{j=0}^{\infty}\left(\frac{b(\tau)}{\tilde{t}_j^{(n)} - \tilde{\tau}_k^{(n)}} - \frac{f(\tau)}{\overline{\tilde{t}_j^{(n)} - \tilde{\tau}_k^{(n)}}}\right)\frac{\xi_j^{(n)}}{n}\right. \\
&\quad + \left.\sum_{j=-\infty}^{-1}\left(\frac{b(\tau)}{\tilde{t}_j^{(n)} - \tilde{\tau}_k^{(n)}}\left(-\frac{e^{+i\omega_\tau}}{n}\right) - \frac{f(\tau)}{\overline{\tilde{t}_j^{(n)}} - \overline{\tilde{\tau}_k^{(n)}}}\left(-\frac{e^{-i\omega_\tau}}{n}\right)\right)\xi_j^{(n)}\right\} \\
&\quad + [c(\tau) + (e(\tau) - d(\tau))i\cot(\pi(\varepsilon-\delta))]\overline{\xi_k^{(n)}} \\
&\quad + \frac{1}{\pi i}\left\{\sum_{j=0}^{\infty}\left(\frac{d(\tau)}{\tilde{t}_j^{(n)} - \tilde{\tau}_k^{(n)}} - \frac{e(\tau)}{\overline{\tilde{t}_j^{(n)} - \tilde{\tau}_k^{(n)}}}\right)\frac{\overline{\xi_j^{(n)}}}{n}\right. \\
&\quad + \left.\sum_{j=-\infty}^{-1}\left(\frac{d(\tau)}{\tilde{t}_j^{(n)} - \tilde{\tau}_k^{(n)}}\left(-\frac{e^{+i\omega_\tau}}{n}\right) - \frac{e(\tau)}{\overline{\tilde{t}_j^{(n)}} - \overline{\tilde{\tau}_k^{(n)}}}\left(-\frac{e^{-i\omega_\tau}}{n}\right)\right)\overline{\xi_j^{(n)}}\right\} \\
&= g^\tau(\tilde{\tau}_k^{(n)}), \quad k \in \mathbb{Z},
\end{aligned} \tag{4.10}$$

$$\begin{aligned}
&a(\tau)\xi_k^{(n)} + \frac{1}{\pi i}\left\{\sum_{\substack{j=0 \\ j\neq k}}^{\infty}\left(\frac{b(\tau)}{\tilde{t}_j^{(n)} - \tilde{t}_k^{(n)}} - \frac{f(\tau)}{\overline{\tilde{t}_j^{(n)} - \tilde{t}_k^{(n)}}}\right)\frac{\xi_j^{(n)}}{n}\right. \\
&\quad + \left.\sum_{\substack{j=-\infty \\ j\neq k}}^{-1}\left(\frac{b(\tau)}{\tilde{t}_j^{(n)} - \tilde{t}_k^{(n)}}\left(-\frac{e^{+i\omega_\tau}}{n}\right) - \frac{f(\tau)}{\overline{\tilde{t}_j^{(n)}} - \overline{\tilde{t}_k^{(n)}}}\left(-\frac{e^{-i\omega_\tau}}{n}\right)\right)\xi_j^{(n)}\right\} \\
&\quad + c(\tau)\xi_k^{(n)} + \frac{1}{\pi i}\left\{\sum_{\substack{j=0 \\ j\neq k}}^{\infty}\left(\frac{d(\tau)}{\tilde{t}_j^{(n)} - \tilde{t}_k^{(n)}} - \frac{e(\tau)}{\overline{\tilde{t}_j^{(n)} - \tilde{t}_k^{(n)}}}\right)\frac{\overline{\xi_j^{(n)}}}{n}\right. \\
&\quad + \left.\sum_{\substack{j=-\infty \\ j\neq k}}^{-1}\left(\frac{d(\tau)}{\tilde{t}_j^{(n)} - \tilde{t}_k^{(n)}}\left(-\frac{e^{+i\omega_\tau}}{n}\right) - \frac{e(\tau)}{\overline{\tilde{t}_j^{(n)}} - \overline{\tilde{t}_k^{(n)}}}\left(-\frac{e^{-i\omega_\tau}}{n}\right)\right)\overline{\xi_j^{(n)}}\right\} \\
&= g^\tau(\tilde{t}_k^{(n)}), \quad k \in \mathbb{Z},
\end{aligned} \tag{4.11}$$

$$
\begin{aligned}
&a(\tau)\xi_k^{(n)} + \frac{1}{\pi i}\left\{ \sum_{\substack{j=1\\ j\equiv k+1 (\mathrm{mod}\, 2)}}^{\infty} \left(\frac{b(\tau)}{\tilde{t}_j^{(n)} - \tilde{t}_k^{(n)}} - \frac{f(\tau)}{\overline{\tilde{t}_j^{(n)} - \tilde{t}_k^{(n)}}} \right) \frac{2\xi_j^{(n)}}{n} \right.\\
&\quad + \sum_{\substack{j=-\infty\\ j\equiv k+1 (\mathrm{mod}\, 2)}}^{-1} \left(\frac{b(\tau)}{\tilde{t}_j^{(n)} - \tilde{t}_k^{(n)}} \left(-\frac{2e^{+i\omega_\tau}}{n}\right) - \frac{f(\tau)}{\overline{\tilde{t}_j^{(n)} - \tilde{t}_k^{(n)}}} \left(-\frac{2e^{-i\omega_\tau}}{n}\right) \right) \xi_j^{(n)}\\
&\quad \left. + \sum_{\substack{j=0\\ j\equiv k+1 (\mathrm{mod}\, 2))}} \left(\frac{b(\tau)}{\tilde{t}_j^{(n)} - \tilde{t}_k^{(n)}} \frac{1-e^{-i\omega_\tau}}{n} - \frac{f(\tau)}{\overline{\tilde{t}_j^{(n)} - \tilde{t}_k^{(n)}}} \frac{1-e^{-i\omega_\tau}}{n} \right) \xi_j^{(n)} \right\}\\
&\quad + c(\tau)\overline{\xi_k^{(n)}} + \frac{1}{\pi i}\left\{ \sum_{\substack{j=1\\ j\equiv k+1 (\mathrm{mod}\, 2)}}^{\infty} \left(\frac{d(\tau)}{\tilde{t}_j^{(n)} - \tilde{t}_k^{(n)}} - \frac{e(\tau)}{\overline{\tilde{t}_j^{(n)} - \tilde{t}_k^{(n)}}} \right) \frac{\overline{2\xi_j^{(n)}}}{n} \right.\\
&\quad + \sum_{\substack{j=-\infty\\ j\equiv k+1 (\mathrm{mod}\, 2)}}^{-1} \left(\frac{d(\tau)}{\tilde{t}_j^{(n)} - \tilde{t}_k^{(n)}} \left(-\frac{2e^{+i\omega_\tau}}{n}\right) - \frac{e(\tau)}{\overline{\tilde{t}_j^{(n)} - \tilde{t}_k^{(n)}}} \left(-\frac{2e^{-i\omega_\tau}}{n}\right) \right) \overline{\xi_j^{(n)}}\\
&\quad \left. + \sum_{\substack{j=0\\ j\equiv k+1 (\mathrm{mod}\, 2)}} \left(\frac{d(\tau)}{\tilde{t}_j^{(n)} - \tilde{t}_k^{(n)}} \frac{1-e^{+i\omega_\tau}}{n} - \frac{e(\tau)}{\overline{\tilde{t}_j^{(n)} - \tilde{t}_k^{(n)}}} \frac{1-e^{-i\omega_\tau}}{n} \right) \overline{\xi_j^{(n)}} \right\}\\
&= g^\tau(\tilde{t}_k^{(n)}), \quad k \in \mathbb{Z}. \qquad (4.12)
\end{aligned}
$$

Using the solutions of systems (4.10) – (4.12) one can construct the approximate solutions u_n^τ of the model equation (4.8) as

$$
u_n^\tau = \sum_{k\in\mathbb{Z}} \xi_k^{(n)} \tilde{\chi}_k^{(n)} \qquad (4.13)
$$

where

$$
\tilde{\chi}_k^{(n)}(t) = \begin{cases} \begin{cases} 1 & \text{if} \quad \dfrac{k}{n} \le t < \dfrac{k+1}{n} \\ 0 & \text{else} \end{cases} & \text{if} \quad k \ge 0\,, \\ \begin{cases} 1 & \text{if} \quad \dfrac{k}{n} \le -e^{-i\omega_\tau} t < \dfrac{k+1}{n} \\ 0 & \text{else} \end{cases} & \text{if} \quad k < 0\,. \end{cases}
$$

Recall that the model equation (4.8) is considered in $L_2(\Gamma_\tau)$, and note that for any element $u_n \in L_2(\Gamma_\tau)$ of the form (4.13) one has

$$
\|\sum_k \xi_k^{(n)} \tilde{\chi}_k^{(n)}\|_{L_2(\Gamma_\tau)} = n^{-1/2} \|(\xi_k^{(n)})_{k\in\mathbb{Z}}\|_{l_2(\mathbb{Z})}, \qquad (4.14)
$$

so the coefficient sequence $(\xi_k^{(n)})_{k\in\mathbb{Z}}$ belongs to the Hilbert space $l_2(\mathbb{Z})$ of all square summable two-sided sequences [102]. Moreover, it is well known that the interpolation projections which send each function $g^\tau \in \mathbf{R}_2(\Gamma_\tau)$ into the functions

$$g_n^\tau = \sum_{k\in\mathbb{Z}} g^\tau(\tilde{\tau}_k^{(n)})\tilde{\chi}_k^{(n)} \in L_2(\Gamma_\tau)$$

are uniformly bounded with respect to n. Consequently, the sequence $(g^\tau(\tilde{\tau}_k^{(n)}))_{k\in\mathbb{Z}}$ also belongs to $l_2(\mathbb{Z})$, and we can think of system (4.10) – (4.12) as of operator equation

$$B_n^\tau U_n = G_n \tag{4.15}$$

where $U_n = (\xi_k^{(n)}), G_n = (g^\tau(\tilde{\tau}_k^{(n)}))$ and the operator B_n^τ acts on $l_2(\mathbb{Z})$. The identity (4.14) implies that the sequences (A_n^τ) are stable if and only if the corresponding sequences (B_n^τ) in (4.15) are stable. But it is readily verified that the system matrices B_n^τ of the model methods are actually independent of n, so the (constant) sequence (B_n^τ) is stable if and only if one of its elements, say B_1^τ, is invertible.

Our main results concerning the stability of methods (4.4) – (4.6) follow.

Theorem 4.1.1. *Let $a, b, c, d, e, f \in \mathbf{C}(\Gamma)$ and $k_1,\ k_2 \in \mathbf{C}(\Gamma\times\Gamma)$. Then:*

(a) *the quadrature methods (4.4) and (4.5) are stable if and only if the operator*

$$A = aI + bS_\Gamma + cM + dS_\Gamma M - eMS_\Gamma - fMS_\Gamma M + T$$

with

$$(Tu)(t) = \int_\Gamma (k_1(t,s)u(s) + k_2(t,s)\overline{u(s)})ds,$$

and all corresponding operators B_1^τ, $\tau\in\Gamma$, are invertible;

(b) *the quadrature method (4.6) is stable if and only if the operator A, the operator*

$$\tilde{A} = aI - bS_\Gamma + cM - dS_\Gamma M + eMS_\Gamma + fMS_\Gamma M,$$

and all corresponding operators B_1^τ, $\tau\in\Gamma$, are invertible.

We prove Theorem 4.1.1 by translating the stability problem into an invertibility problem in the Banach algebras $\mathcal{S}_C/\mathcal{J}_1$ or $\mathcal{A}/\mathcal{J}$ via Propositions 1.6.4 and 1.6.5.

Let L_n denote the orthogonal projection of $L_2(\Gamma)$ onto the subspace

$$S_n = \text{span}\ \{\chi_j^{(n)}\ ,\ j=0,\dots,n-1\}.$$

If n is even, define the operators W_n by

$$W_n f = W_n L_n f = W_n\left(\sum_{j=0}^{n-1}\xi_j\chi_j^{(n)}\right) = \sum_{j=0}^{n-1}(-1)^j\xi_j\chi_j^{(n)}.$$

Lemma 4.1.2. *The projection L_n and operator W_n satisfy all the conditions imposed in Section 1.6 on the projection P_n and operator W_n. Moreover,*

$$ML_n = L_nM, \quad MW_n = W_nM, \quad n = 1, 2, \dots. \tag{4.16}$$

Proof. The equations

$$W_n^2 = L_n, \quad W_nL_n = L_nW_n, \quad n = 1, 2, \dots$$

as well as relations (4.16) can be verified by straightforward calculation, and the strong convergence of the sequence (L_n) to the identity operator is well known [183]. Let us demonstrate the weak convergence of the sequence (W_nL_n) to 0. Since this sequence is uniformly bounded, and since the continuous functions are dense in $L_2(\Gamma)$, it remains to show that

$$(W_nL_nf, g) \to 0$$

for all continuous functions f and g on Γ. Let n be even and set

$$c_j = n \int_{j/n}^{(j+1)/n} f(\gamma(s))ds.$$

Since

$$\begin{aligned}(W_nL_nf, g) &= \int_0^1 \sum_{j=0}^{n-1} (-1)^j c_j \chi_j^{(n)}(\gamma(s))\overline{g(\gamma(s))}ds = \sum_{j=0}^{n-1} (-1)^j c_j \int_{j/n}^{(j+1)/n} \overline{g(\gamma(s))}ds \\ &= \sum_{j=0}^{n/2-1} (c_{2j} - c_{2j+1}) \int_{2j/n}^{(2j+1)/n} \overline{g(\gamma(s))}ds \\ &\quad + \sum_{j=0}^{n/2-1} c_{2j+1} \int_{2j/n}^{(2j+1)/n} \overline{(g(\gamma(s)) - g(\gamma(s+1/n))}ds,\end{aligned}$$

we get

$$\begin{aligned}|(W_nL_nf, g)| \le &\sum_{j=0}^{n/2-1} |c_{2j} - c_{2j+1}| \int_{2j/n}^{(2j+1)/n} |g(\gamma(s))|ds \\ &+ \sum_{j=0}^{n/2-1} |c_{2j+1}| \int_{2j/n}^{(2j+1)/n} |g(\gamma(s)) - g(\gamma(s+1/n))|ds\end{aligned}$$

$$\leq \sum_{j=0}^{n/2-1} n \int_{2j/n}^{(2j+1)/n} |f(\gamma(s)) - f(\gamma(s+1/n))| ds \int_{2j/n}^{(2j+1)/n} |g(\gamma(s))| ds$$

$$+ \sum_{j=0}^{n/2-1} n \int_{2j/n}^{(2j+1)/n} |f(\gamma(s))| ds \int_{2j/n}^{(2j+1)/n} |g(\gamma(s)) - g(\gamma(s+1/n))| ds$$

$$\leq \omega(f \circ \gamma, 1/n) \|g\|_{\mathbf{C}(\Gamma)} + \omega(g \circ \gamma, 1/n) \|f\|_{\mathbf{C}(\Gamma)},$$

where $\omega(h, 1/n)$ is the modulus of continuity of the function h. This proves the assertion, since $\omega(h, 1/n) \to 0$ as $n \to \infty$. □

Thus we can identify these operators L_n and W_n as the operators P_n and W_n from Section 1.6, and let $\mathcal{S}_C, \mathcal{J}_1, \mathcal{A}$, and $\mathcal{J}$ refer to the associated specified algebras and ideals.

Further, following (1.17) we can represent any sequence (A_n) of additive continuous operators in the form

$$A_n = A_n^{(1)} + A_n^{(2)} M \tag{4.17}$$

where $(A_n^{(1)}), (A_n^{(2)})$ are the sequences of linear continuous operators. Note that the sequence (A_n) converges uniformly (strongly or weakly) to the operator $A = A^{(1)} + A^{(2)} M \in \mathcal{L}_{add}(L_2)$ if and only if the sequences $(A_n^{(1)})$ and $(A_n^{(2)})$ converge uniformly (strongly or weakly) to the operators $A^{(1)}$ and $A^{(2)}$, respectively.

Lemma 4.1.3. *If $(A_n), A_n \in \mathcal{L}_{add}(\operatorname{im} L_n)$ is one of the approximation sequences corresponding to methods (4.4), (4.5) or (4.6), then (A_n) belongs to both $\mathcal{S}_C$ and $\mathcal{A}$.*

From relations (4.16) and (4.17), it is sufficient to prove this for approximation sequences for singular integral operators without conjugation. This has already been done [180, 183].

Let us proceed to show the necessity of the conditions in Theorem 4.1.1.

Lemma 4.1.4. (a) *If (4.4) and (4.5) are stable methods, then the operator A and all associated operators B_1^τ with $\tau \in \Gamma$ are invertible.*

(b) *If (4.6) is stable, then the operators $A, \tilde{A}$, and B_1^τ with $\tau \in \Gamma$ are invertible.*

Proof. The invertibility of A and $\tilde{A}$ is a consequence of Proposition 1.6.5 and Lemma 4.1.3. For the computation of $\tilde{A}$, we refer the reader to [180].

Let $P_n \in \mathcal{L}(l_2(\mathbb{Z}))$ denote the projection $P_n : (x_k)_{k\in\mathbb{Z}} \mapsto (y_k)_{k\in\mathbb{Z}}$ with

$$y_k = \begin{cases} x_k & \text{if } -n/2 < k \leq n/2 \,, \\ 0 & \text{else} \,, \end{cases}$$

and for fixed $\tau \in \Gamma$ and $n \in \mathbb{Z}^+$ introduce the operator $E_n^\tau \in \mathcal{L}(\operatorname{im} P_n, \operatorname{im} L_n)$ by

$$E_n^\tau(\delta_{jk})_{k\in\mathbb{Z}} = \chi^{(n)}_{j+j(\tau,n)}$$

where $j(\tau,n) \in \{0,1,\dots,n-1\}$ is such that $\tau \in [\gamma(j(\tau,n)/n), \gamma((j(\tau,n)+1)/n))$. It is easy to see that E_n^τ is bijective, and its inverse may be denoted by E_{-n}^τ. Further, one has $\sup_n \|E_n^\tau\|\|E_{-n}^\tau\| < \infty$ (cf. (4.14)) and we claim that

$$E_{-n}^\tau A_n E_n^\tau \to B_1^\tau \ , \ (E_{-n}^\tau A_n E_n^\tau)^* \to (B_1^\tau)^* \quad \text{strongly.} \tag{4.18}$$

Indeed, if $\overline{M} \in \mathcal{L}_{add}(l_2(\mathbb{Z}))$ refers to the operator $\overline{M}(x_k) = (\overline{x}_k)$, then

$$E_n^\tau \overline{M} P_n = M E_n^\tau P_n \ , \ \overline{M} E_{-n}^\tau L_n = E_{-n}^\tau M L_n. \tag{4.19}$$

This observation reduces the verification of relations (4.18) to sequences (A_n) for operators without conjugation, and for this we refer to [180, Proof of Theorem 1.4] again.

From (4.18) and from the stability of the sequence (A_n) we conclude that

$$\|E_{-n}^\tau A_n E_n^\tau P_n x\|_{l_2(\mathbb{Z})} \ge C\|P_n x\|_{l_2(\mathbb{Z})}, \quad x \in l_2(\mathbb{Z}),$$

and

$$\|(E_{-n}^\tau A_n E_n^\tau)^* P_n x\|_{l_2(\mathbb{Z})} \ge C\|P_n x\|_{l_2(\mathbb{Z})}, \quad x \in l_2(\mathbb{Z})$$

and passing to the limits yields

$$\|B_1^\tau x\| \ge C\|x\| \ , \ \|(B_1^\tau)^* x\| \ge C\|x\|$$

whence the invertibility of B_1^τ follows.

For the converse it remains to show that the invertibility of all operators B_1^τ implies the invertibility of the coset $(A_n) + \mathcal{J}_1$ or $(A_n) + \mathcal{J}$ in the quotient algebras $\mathcal{S}_C/\mathcal{J}_1$ or $\mathcal{A}/\mathcal{J}$, respectively. Proposition 1.6.4 or 1.6.5 then applies to give the assertion. The invertibility of these cosets follows from the local principle of Gohberg and Krupnik of Section 1.9.1. □

By K_n^ε we denote the interpolation projection sending each Riemann integrable function g on Γ into the function

$$K_n^\varepsilon g = \sum_{j=0}^{n-1} g(\tau_j^{(n)})\chi_j^{(n)} \in L_2(\Gamma),$$

and given a function $h \in \mathbf{C}(\Gamma)$ we abbreviate the operator $K_n^\varepsilon h L_n$ by h_n. The set of all real-valued Lipschitz continuous functions on Γ which take values between 0 and 1 only may be denoted by $L(\Gamma)$.

Lemma 4.1.5. (a) *For $\tau \in \Gamma$ the set*

$$M_\tau^1 := \{(h_n) + \mathcal{J}_1,\ h \in L(\Gamma),\ h \equiv 1 \quad \text{in a neighbourhood of}\ \tau\}$$

is a localizing class in $\mathcal{S}_C/\mathcal{J}_1$ and the set

$$M_\tau^2 := \{(h_n) + \mathcal{J},\ h \in L(\Gamma),\ h \equiv 1 \quad \text{in a neighbourhood of}\ \tau\}$$

is a localizing class in $\mathcal{A}/\mathcal{J}$. Moreover, $\{M_\tau^1\}_{\tau\in\Gamma}$ and $\{M_\tau^2\}_{\tau\in\Gamma}$ form covering systems of localizing classes in algebras $\mathcal{S}_C/\mathcal{J}_1$ and $\mathcal{A}/\mathcal{J}$, respectively.

(b) *If (A_n) is the approximation sequence of method (4.4) or (4.5), then the elements of $\bigcup_{\tau\in\Gamma} M_\tau^1$ commute with the coset $(A_n) + \mathcal{J}_1$ in $\mathcal{S}_C/\mathcal{J}_1$. If (A_n) corresponds to (4.6), then the elements of $\bigcup_{\tau\in\Gamma} M_\tau^2$ commute with the coset $(A_n) + \mathcal{J}$ in $\mathcal{A}/\mathcal{J}$.*

The proof of (a) is almost evident. Thus because of the identity

$$Mh_n = h_n M \quad \text{for all} \quad h \in L(\Gamma),$$

one can consider approximation sequences for operators without conjugation, when it has been already shown [180, Theorem 1.3 (iii) and Theorem 1.4 (iii)′].

Lemma 4.1.6. (a) *Let (A_n) be the approximation sequence of (4.4) or of (4.5). If all associated operators $B_\tau^1, \tau \in \Gamma$ are invertible, then the coset $(A_n) + \mathcal{J}_1$ is invertible in $\mathcal{S}_C/\mathcal{J}_1$.*

(b) *Let (A_n) be the approximation sequence for (4.6). If all associated operators B_τ^1 are invertible, then the coset $(A_n) + \mathcal{J}$ is invertible in $\mathcal{A}/\mathcal{J}$.*

Proof. We only consider (4.4), since the proof for (4.5) and (4.6) is analogous. Let B_1^τ be invertible. In [180], Lemma 1.4 it has been shown that the sequences $(E_n^\tau B_1^\tau E_{-n}^\tau)$ and $(E_n^\tau (B_1^\tau)^{-1} E_{-n}^\tau)$ belong to $\mathcal{S}_C$ whenever B_1^τ is related to singular integral operators without conjugation. If the conjugation appears, then this case can be traced back to that without conjugation via (4.19).

We evidently have

$$(E_n^\tau B_1^\tau E_{-n}^\tau)\,(E_n^\tau (B_1^\tau)^{-1} E_{-n}^\tau) = (L_n),$$

i.e., all sequences $(E_n^\tau B_1^\tau E_{-n}^\tau)$ are invertible in $\mathcal{S}_C$, whence in particular it follows that all cosets $(E_n^\tau B_1^\tau E_{-n}^\tau) + \mathcal{J}_1$ are M_τ^1-invertible in the quotient algebra $\mathcal{S}_C/\mathcal{J}_1$.

It remains to verify the M_τ^1-equivalence of (A_n) and $(E_n^\tau B_1^\tau E_{-n}^\tau)$. Again this can be reduced to sequences without conjugation, and again we may invoke [180, Theorem 1.3 (iv)]. Observe that the notion of M_τ-equivalence used in [180] is equivalent to our definition since $\sup_n \|E_n^\tau\|\,\|E_{-n}^\tau\| < \infty$. □

We have seen in Lemma 4.1.3 that the approximation sequences (A_n) associated with (4.4), (4.5) and (4.6) belong both to $\mathcal{S}_C$ and to $\mathcal{A}$. Further, by Lemma 4.1.5, the sequences (4.4) and (4.5) commute with $(h_n), h \in L(\Gamma)$, modulo $\mathcal{J}_1$ (and, thus, modulo $\mathcal{J}$) whereas (4.6) commutes with (h_n) modulo $\mathcal{J}$ (but not modulo $\mathcal{J}_1$ as one can show). Hence, we could study (4.4) and (4.5) by either Proposition 1.6.4 or 1.6.5, but we prefer Proposition 1.6.4 for simplicity whereas (4.6) requires application of Proposition 1.6.5. □

4.1.2 The ε-Collocation Method

The applicability of the ε-collocation method to equation (4.1) is now considered. This method determines an approximate solution u_n in the form $u_n = \sum_{j=0}^{n-1} \xi_j^{(n)} \chi_j^{(n)}$ by solving the $n \times n$ algebraic system

$$(Au_n)(\tau_j^{n,\varepsilon}) = g(\tau_j^{n,\varepsilon}), \quad j = 0, 1, \dots, n-1 \tag{4.20}$$

where

$$\tau_j^{n,\varepsilon} = \gamma\left(\frac{j+\varepsilon}{n}\right), \quad 0 < \varepsilon < 1.$$

The system (4.20) can be rewritten as the operator equation

$$K_n^\varepsilon A u_n = K_n^\varepsilon g, \quad u_n \in \operatorname{im} L_n \subseteq L_2(\Gamma), \tag{4.21}$$

and so our main concern is the stability of the sequence $(K_n^\varepsilon A L_n)$ of approximation operators. As for the quadrature method, we associate to (4.21) a family of model operators as follows.

Let A^τ denote again the operator (4.9), and set

$$\tilde{\tau}_k^{n,\varepsilon} = \begin{cases} \dfrac{k+\varepsilon}{n} & \text{if } k \geq 0, \\ -\dfrac{k+\varepsilon}{n} e^{i\omega_\tau} & \text{if } k < 0. \end{cases}$$

For Riemann integrable functions $g \in \mathbf{R}_2(\Gamma_\tau)$, we seek an approximate solution of the model equation

$$A^\tau u^\tau = g^\tau, \quad u^\tau \in L_2(\Gamma_\tau) \tag{4.22}$$

in the form

$$u_n^\tau = \sum_{k \in \mathbb{Z}} \tilde{\xi}_k^{(n)} \tilde{\chi}_k^{(n)}, \tag{4.23}$$

where $\tilde{\chi}_k^{(n)}$ are the functions introduced in (4.13) and the coefficients $\tilde{\xi}_k^{(n)}$, $k \in \mathbb{Z}$, satisfy the infinite algebraic system

$$(A^\tau u_n^\tau)(\tilde{\tau}_k^{n,\varepsilon}) = g^\tau(\tilde{\tau}_k^{n,\varepsilon}), \ k \in \mathbb{Z}.$$

This system can be rewritten as

$$A_{n,\varepsilon}^\tau u_n^\tau = K_n^\tau g^\tau$$

where

$$A_{n,\varepsilon}^\tau = K_n^\tau A^\tau L_n^\tau \in \mathcal{L}_{add}(\operatorname{im} L_n^\tau),$$

with L_n^τ referring to the orthogonal projection of $L_2(\Gamma_\tau)$ onto its closed subspace spanned by all functions $\tilde{\chi}_k^{(n)}$, $k \in \mathbb{Z}$, and K_n^τ again denotes the interpolation operator

$$K_n^\tau h = \sum_{k\in\mathbb{Z}} h(\tilde{\tau}_k^{n,\varepsilon})\tilde{\chi}_k^{(n)}.$$

As in Section 4.1.1, we may identify the operators $A_{n,\varepsilon}^\tau \in \mathcal{L}_{add}(\operatorname{im} L_n^\tau)$ with the operators $B_{n,\varepsilon}^\tau \in \mathcal{L}_{add}(l_2(\mathbb{Z}))$ whose matrix representations with respect to the standard basis of $l_2(\mathbb{Z})$ are the same as those of $A_{n,\varepsilon}^\tau$ in the basis $(\tilde{\chi}_k^{(n)})_{k\in\mathbb{Z}}$ of $\operatorname{im} L_n^\tau$, and both sequences $(A_{n,\varepsilon}^\tau)$ and $(B_{n,\varepsilon}^\tau)$ are simultaneously either stable or not. Furthermore, it again turns out that the operators $B_{n,\varepsilon}^\tau$ are actually independent of n, so the sequence $(A_{n,\varepsilon}^\tau)$ is stable if and only if the operator $B_{1,\varepsilon}^\tau$ is invertible.

Employing the readily verifiable identities

$$\left((MS_{\Gamma_\tau})(\tilde{\xi}_j^{(n)}\tilde{\chi}_j^{(n)})\right)(\tilde{\tau}_k^{n,\varepsilon}) = -(S_{\Gamma_\tau^*}\tilde{\tilde{\chi}}_j^{(n)})(\overline{\tilde{\tau}_k^{n,\varepsilon}})\overline{\tilde{\xi}_j^{(n)}} \tag{4.24}$$

and

$$\left((MS_{\Gamma_\tau}M)(\tilde{\xi}_j^{(n)}\tilde{\chi}_j^{(n)})\right)(\tilde{\tau}_k^{n,\varepsilon}) = -(S_{\Gamma_\tau^*}\tilde{\tilde{\chi}}_j^{(n)})(\overline{\tilde{\tau}_k^{n,\varepsilon}})\tilde{\xi}_j^{(n)} \tag{4.25}$$

where

$$\Gamma_\tau^* = \left\{te^{(2\pi-\omega_\tau)i}, 0 \le t < \infty\right\} \cup \{t, 0 \le t < \infty\}$$

and $\tilde{\tilde{\chi}}_j^{(n)}$ is defined as in (4.13) but with ω_τ being replaced by $2\pi - \omega_\tau$, it is not hard to find the explicit matrix representation of the operator $B_{1,\varepsilon}^\tau$:

$$\begin{aligned} B_{1,\varepsilon}^\tau = a(\tau)I &+ \frac{b(\tau)}{\pi i}\left(\int_{\Gamma_\tau} \frac{\tilde{\chi}_j^{(1)}(s)ds}{s-\tilde{\tau}_k^{1,\varepsilon}}\right)_{k,j\in\mathbb{Z}} - \frac{f(\tau)}{\pi i}\left(\int_{\Gamma_\tau^*} \frac{\tilde{\tilde{\chi}}_j^{(1)}(s)ds}{s-\overline{\tilde{\tau}_k^{1,\varepsilon}}}\right)_{k,j\in\mathbb{Z}} \\ &+ \left[c(\tau)I + \frac{d(\tau)}{\pi i}\left(\int_{\Gamma_\tau} \frac{\tilde{\chi}_j^{(1)}(s)ds}{s-\tilde{\tau}_k^{1,\varepsilon}}\right)_{k,j\in\mathbb{Z}} - \frac{e(\tau)}{\pi i}\left(\int_{\Gamma_\tau^*} \frac{\tilde{\tilde{\chi}}_j^{(1)}(s)ds}{s-\overline{\tilde{\tau}_k^{1,\varepsilon}}}\right)_{k,j\in\mathbb{Z}}\right]\overline{M}. \end{aligned} \tag{4.26}$$

Theorem 4.1.7. *Let $a, b, c, d, e, f \in \mathbf{C}(\Gamma)$, $k_1, k_2 \in \mathbf{C}(\Gamma \times \Gamma)$, and $0 < \varepsilon < 1$.*

(a) *The ε-collocation method for the operator*

$$A = aI + bS_\Gamma + cM + dS_\Gamma M - eMS_\Gamma - fMS_\Gamma M$$

is stable if and only if the operator $A \in \mathcal{L}_{add}(L_2(\Gamma))$ and all operators $B^\tau_{1,\varepsilon} \in \mathcal{L}_{add}(l_2(\mathbb{Z}))$ with $\tau \in \Gamma$ are invertible.

(b) *If the ε-collocation method is stable and g is Riemann integrable, then the systems (4.20) are uniquely solvable for all sufficiently large n, and the sequence (u_n) of their solutions tends to the solution of equation (4.1) in the norm of $L_2(\Gamma)$.*

The proof of this result is analogous to the proof of Theorem 4.1.1. Thus it can be reduced again to the operators without conjugation, for which we again refer the reader to [180].

4.1.3 Qualocation Method

Choose real numbers ε_r, $0 < \varepsilon_0 < \cdots < \varepsilon_{m-1} < 1$ and numbers w_r, $r = 0, 1, \ldots, m-1$, such that $\sum_{r=0}^{m-1} w_r = 1$, and consider the quadrature rule

$$Q_n(g) = \sum_{k=0}^{n-1} \frac{1}{n} \sum_{r=0}^{m-1} w_r g(\tau_k^{n,\varepsilon_r})$$

with $\tau_k^{n,\varepsilon_r} = \gamma((k+\varepsilon_r)/n)$.

In qualocation method one seeks an approximate solution u_n of equation (4.1) in the form (4.7) and determines the unknown coefficients $\xi_j^{(n)}$, $j = 0, \ldots, n-1$, from the algebraic system

$$Q_n(\chi_j^{(n)} A u_n) = Q_n(g\chi_j^{(n)}),\ j = 0, \ldots, n-1$$

or equivalently

$$\sum_{r=0}^{m-1} w_r (Au_n)(\tau_j^{n,\varepsilon_r}) = \sum_{r=0}^{m-1} w_r g(\tau_j^{n,\varepsilon_r}),\ j = 0, \ldots, n-1. \tag{4.27}$$

Let us again emphasize that the interest in method (4.27) comes from the possibility to force higher convergence rates than for pure collocation methods, by special choices of the parameters ε_r and w_r, cf. [104, 215, 216].

Set

$$\tilde{\tau}_k^{n,\varepsilon_r} = \begin{cases} \dfrac{k+\varepsilon_r}{n} & \text{if } k \geq 0, \\ -\dfrac{k+\varepsilon_r}{n} e^{i\omega_\tau} & \text{if } k < 0, \end{cases}$$

and define a quadrature rule on the model curve Γ_τ by

$$\tilde{Q}_n(g) = \sum_{k=-\infty}^{\infty} \frac{1}{n} \sum_{r=0}^{m-1} w_r g(\tilde{\tau}_k^{n,\varepsilon_r}).$$

Our model methods arise from looking for an approximative solution u_n^τ of the form (4.23) of equation (4.22), by determining the unknown coefficients $\tilde{\xi}_k^{(n)}$ via the algebraic system

$$\sum_{r=0}^{m-1} w_r (A^\tau u_n^\tau)(\tilde{\tau}_j^{n,\varepsilon_r}) = \sum_{r=0}^{m-1} w_r g^\tau(\tilde{\tau}_j^{n,\varepsilon_r}), \quad j \in \mathbb{Z}. \tag{4.28}$$

Theorem 4.1.8. *Let $a, b, c, d, e, f \in \mathbf{C}(\Gamma)$, $k_1, k_2 \in \mathbf{C}(\Gamma \times \Gamma)$, and $\varepsilon_r, w_r \in \mathbb{R}^+$ be such that $0 < \varepsilon_0 < \varepsilon_1 < \cdots < \varepsilon_{m-1} < 1$ and $\sum_{r=0}^{m-1} w_r = 1$. Then:*

(a) *The qualocation method (4.27) is stable if and only if the operator A and the operators $B_{Q(\varepsilon)}^\tau = \sum_{r=0}^{m-1} w_r B_{1,\varepsilon_r}^\tau \in \mathcal{L}_{add}(l_2(\mathbb{Z}))$ are invertible for all $\tau \in \Gamma$.*

(b) *If the method (4.27) is stable and if g is Riemann integrable, then the systems (4.27) are uniquely solvable for all sufficiently large n, and the sequence (u_n) of their solutions converges in $L_2(\Gamma)$ to the solution of equation (4.1).*

Recall that the operators $B_{1,\varepsilon}^\tau$ are defined by (4.26).

The proof of Theorem 4.1.8 proceeds in the very same way as that of Theorem 4.1.7 since the qualocation method can be represented as a linear combination of ε_r-collocation methods. We omit the details here.

4.2 Fredholm Properties of Local Operators

The Fredholmness and invertibility of local operators B_1^τ and $B_{1,\varepsilon}^\tau$, associated with the above approximation methods for singular integral operators without conjugation, have been investigated by S. Prössdorf and A. Rathsfeld [180]. They pointed out that the Fredholmness of these operators is independent of the angles ω_τ, $\tau \in \Gamma$, and that the index of the local operator B_1^τ is zero whenever B_1^τ is Fredholm.

In contrast, we observe that the Fredholmness of the operator B_1^τ for operators with conjugation depends essentially on the value of the angle ω_τ. Moreover, the operator B_1^τ can be Fredholm but $\operatorname{ind} B_1^\tau \neq 0$. Nevertheless, we can show that possible values for $\operatorname{ind} B_1^\tau$ are rather restricted, e.g., if $e = f = 0$ and the operator B_1^τ is Fredholm, then the index of this operator takes only one of the values: $1, 0$, or -1.

4.2.1 Quadrature Methods

Assume now that the coefficients $e = 0$ and $f = 0$ everywhere on Γ and restrict ourselves to the method (4.4) and, correspondingly, to the local operators B_1^τ associated with the system (4.10). Further we identify the Hilbert space $l_2(\mathbb{Z})$ with the Cartesian product $l_2^2 := l_2 \times l_2$ of the two Hilbert spaces $l_2 := l_2(\mathbb{Z}^+)$ of one-sided sequences, and therefore consider operators on $l_2(\mathbb{Z})$ as 2×2 matrices of operators acting on l_2^2. By $\mathcal{T}$ we denote the smallest closed C^*-subalgebra of $\mathcal{L}(l_2)$ containing all Toeplitz operators $T(a)$ with piecewise constant generating functions a. Recall that the operator $T(a)$ acts on finitely supported sequences via

$$T(a)(x_k) = (y_k)\ ,\ y_k = \sum_{r=0}^{\infty} a_{k-r} x_r$$

where a_k is the kth Fourier coefficient of a. Finally, we write $\mathcal{T}^{2\times 2}$ for the subalgebra of $\mathcal{L}(l_2(\mathbb{Z})) \cong \mathcal{L}(l_2^2) \cong \mathcal{L}^{2\times 2}(l_2)$ consisting of all 2×2 matrices with entries in $\mathcal{T}$.

Lemma 4.2.1. *If $\overline{M} : l_2(\mathbb{Z}^+) \mapsto l_2(\mathbb{Z}^+)$ is the operator defined by*

$$\overline{M}((x_k)_{k\in\mathbb{Z}^+}) = (\overline{x}_k)_{k\in\mathbb{Z}^+},$$

then the operator $\overline{M}$ and the complex C^*-algebra $\mathcal{T}$ satisfy all axioms $(A_1) - (A_5)$ of Section 1.2.

Proof. Note that (A_1) and (A_5) are the only non-trivial axioms to check. A straightforward computation shows that for any $a \in \mathbf{PC}(\Gamma)$,

$$\overline{M}\,T(a)\overline{M} = T(a^*), \tag{4.29}$$

where $a^*(t) = \overline{a(\overline{t})}$, $t \in \Gamma_0$. Thus $\overline{M}\,T(a)\overline{M} \in \mathcal{T}$, and since $\overline{M}^2 = I$, the inclusion $\overline{M}A\overline{M} \in \mathcal{T}$ is valid for any operator $A \in \mathcal{T}$. The same relation (4.29) yields

$$(\overline{M}\,T(a)\overline{M})^* = T(\widetilde{a}),$$

and

$$\overline{M}\,T^*(a)\overline{M} = \overline{M}\,T(\overline{a})\overline{M} = T(\widetilde{a}),$$

where $\widetilde{a} = a(\overline{t})$, $t \in \Gamma_0$, and our claim follows. □

Lemma 4.2.2. *If B_1^τ, $\tau \in \Gamma$ is the operator determined by the left-hand side of system* (4.10)*, then it has the form*

$$B_1^\tau = B^{1,\tau} + B^{2,\tau}\overline{M} \tag{4.30}$$

where $B^{1,\tau}, B^{2,\tau} \in \mathcal{T}^{2\times 2}$. Moreover, the operator B_1^τ, $\tau \in \Gamma$ is invertible (Fredholm) if and only if the operator

$$\tilde{B}_1^\tau = \begin{pmatrix} B^{1,\tau} & B^{2,\tau} \\ \overline{M}B^{2,\tau}\overline{M} & \overline{M}B^{1,\tau}\overline{M} \end{pmatrix}, \qquad \tau \in \Gamma_0,$$

is invertible (Fredholm).

Proof. A straightforward computation shows that B_1^τ can be written in the form (4.30), where

$$B^{1,\tau} = \begin{pmatrix} B_{11}^{1,\tau} & B_{12}^{1,\tau} \\ B_{21}^{1,\tau} & B_{22}^{1,\tau} \end{pmatrix} \in \mathcal{L}^{2\times 2}(l_2), \qquad B^{2,\tau} = \begin{pmatrix} B_{11}^{2,\tau} & B_{12}^{2,\tau} \\ B_{21}^{2,\tau} & B_{22}^{2,\tau} \end{pmatrix} \in \mathcal{L}^{2\times 2}(l_2)$$

with

$$B_{11}^{1,\tau} = (a(\tau) - b(\tau)i\,\cot(\pi(\varepsilon-\delta)))\,I + \frac{b(\tau)}{\pi i}\left(\frac{1}{-(k-j)-(\varepsilon-\delta)}\right)_{k,j=0}^{\infty},$$

$$B_{21}^{1,\tau} = \frac{b(\tau)}{\pi i}\left(\frac{1}{(j+\delta)+(-k-1+\varepsilon)e^{i\omega_\tau}}\right)_{k,j=0}^{\infty},$$

$$B_{12}^{1,\tau} = \frac{b(\tau)}{\pi i}\left(\frac{e^{i\omega_\tau}}{(-j+1+\delta)e^{i\omega_\tau}+(k+\varepsilon)}\right)_{k,j=0}^{\infty},$$

$$B_{22}^{1,\tau} = (a(\tau) - b(\tau)i\,\cot(\pi(\varepsilon-\delta)))\,I + \frac{b(\tau)}{\pi i}\left(\frac{1}{(k-j)+(\delta-\varepsilon)}\right)_{k,j=0}^{\infty}.$$

The operators $B_{lr}^{2,\tau}$, $l,r = 1,2$, can be derived from the operators $B_{lr}^{1,\tau}$ by replacing a and b by c and d, respectively. But each of the operators $B_{lr}^{j,\tau}$, $j,l,r \in \{1,2\}$, belongs to $\mathcal{T}$, as was shown in [180, Lemma 1.1].

The assertion concerning the invertibility (Fredholmness) of the operators B_1^τ follows immediately from Lemmas 4.2.1 and 1.4.6. □

The Fredholmness of the operators $\tilde{B}_1^\tau$, $\tau \in \Gamma_0$ can be verified via the Gohberg – Krupnik symbol calculus for $\mathcal{T}^{2\times 2}$. Thus with each operator $\tilde{B}_1^\tau$ one can associate a 4×4 matrix-valued function $\mathcal{A}_{\tilde{B}_1^\tau}$ on $\Gamma_0 \times [0,1]$ with the property that $\tilde{B}_1^\tau$ is Fredholm if and only if $\det \mathcal{A}_{\tilde{B}_1^\tau}(t,\mu) \neq 0$ for all $t \in \Gamma_0$ and $\mu \in [0,1]$, and the index of $\tilde{B}_1^\tau$ coincides with the negative winding number[1] of the determinant $\det \mathcal{A}_{\tilde{B}_1^\tau}(t,\mu)$.

In accordance with [11, 29, 33, 180], when $t \neq 1$ and $\mu \in [0,1]$ the matrix $\mathcal{A}_{\tilde{B}_1^\tau}(t,\mu)$ is

$$\mathcal{A}_{\tilde{B}_1^\tau}(t,\mu) = \begin{pmatrix} a(\tau)+b(\tau)f^{(\varepsilon-\delta)}(t) & 0 & c(\tau)+d(\tau)f^{(\varepsilon-\delta)}(t) & 0 \\ 0 & a(\tau)-b(\tau)f^{(\delta-\varepsilon)}(t) & 0 & c(\tau)-d(\tau)f^{(\delta-\varepsilon)}(t) \\ \overline{c(\tau)}+\overline{d(\tau)}f^{(\varepsilon-\delta)*}(t) & 0 & \overline{a(\tau)}+\overline{b(\tau)}f^{(\varepsilon-\delta)*}(t) & 0 \\ 0 & \overline{c(\tau)}-\overline{d(\tau)}f^{(\delta-\varepsilon)*}(t) & 0 & \overline{a(\tau)}-\overline{b(\tau)}f^{(\delta-\varepsilon)*}(t) \end{pmatrix} \tag{4.31}$$

[1]For the definition of the winding number the reader can consult [11, Section 2.41].

with

$$f^{(\nu)}(e^{i2\pi s}) := 2e^{-i2\pi\nu(s-1)}\frac{\sin(-\pi\nu s)}{\sin(-\pi\nu)} - 1, \quad 0 \le s \le 1, \tag{4.32}$$

and $f^*(t) = \overline{f(\overline{t})}$, whereas in the case $t = 1$ and $\mu \in [0,1]$ this matrix is given by

$$\mathcal{A}_{\tilde{B}_1^\tau}(1,\mu)$$

$$= \begin{pmatrix} a(\tau)+b(\tau)(1-2\mu) & \frac{ib(\tau)}{\beta}e^{i(\omega_\tau-\pi)\alpha} & c(\tau)+d(\tau)(1-2\mu) & \frac{id(\tau)}{\beta}e^{i(\omega_\tau-\pi)\alpha} \\ \frac{ib(\tau)}{\beta}e^{i(\pi-\omega_\tau)\alpha} & a(\tau)-b(\tau)(1-2\mu) & \frac{-id(\tau)}{\beta}e^{i(\pi-\omega_\tau)\alpha} & c(\tau)-d(\tau)(1-2\mu) \\ \overline{c(\tau)}-\overline{d(\tau)}(1-2\mu) & \frac{-i\overline{d(\tau)}}{\beta}e^{i(\pi-\omega_\tau)\alpha} & \overline{a(\tau)}-\overline{b(\tau)}(1-2\mu) & \frac{-i\overline{b(\tau)}}{\beta}e^{i(\pi-\omega_\tau)\alpha} \\ \frac{i\overline{d(\tau)}}{\beta}e^{i(\omega_\tau-\pi)\alpha} & \overline{c(\tau)}+\overline{d(\tau)}(1-2\mu) & \frac{i\overline{d(\tau)}}{\beta}e^{i(\omega_\tau-\pi)\alpha} & \overline{a(\tau)}+\overline{b(\tau)}(1-2\mu) \end{pmatrix} \tag{4.33}$$

with $\beta = \beta(\mu) = \sin(\pi(1/2 + (i/2\pi)\log(\mu/(1-\mu)))$ and $\alpha = \alpha(\mu) = 1/2 + (i/2\pi)\log(\mu/(1-\mu))$.

Theorem 4.2.3. *Let B_1^τ, $\tau \in \Gamma$ be the local operator associated with the method (4.4) in case $e = f = 0$. If B_1^τ is Fredholm, then the index*

$$\operatorname{ind} B_1^\tau := \operatorname{ind}_{\mathbb{R}} B_1^\tau$$

of this operator can take only three values 0, 1 *and* -1.

For the proof of this statement we need an auxiliary result.

Lemma 4.2.4. *If a, b, c, and d are any fixed complex numbers, then the function*

$$\det \mathcal{A}_{\tilde{B}_1^\tau} : (\Gamma_0 \setminus \{1\}) \times [0,1] \mapsto \mathbb{C}$$

is real-valued and non-negative.

Proof. Put $\nu = \varepsilon - \delta$ and multiply the matrix (4.31) both from the left- and from the right-hand side by the matrix

$$J := \begin{pmatrix} 1 & 0 & 0 & 0 \\ 0 & 0 & 1 & 0 \\ 0 & 1 & 0 & 0 \\ 0 & 0 & 0 & 1 \end{pmatrix}.$$

This transformation does not influence the determinant, and the new matrix has the form

$$\begin{pmatrix} a + bf^\nu(t) & c + df^\nu(t) & 0 & 0 \\ \overline{c} + \overline{d}f^{\nu*}(t) & \overline{a} + \overline{b}f^{\nu*}(t) & 0 & 0 \\ 0 & 0 & a - bf^{-\nu}(t) & c - df^{-\nu}(t) \\ 0 & 0 & \overline{c} - \overline{d}f^{-\nu*}(t) & \overline{a} - \overline{b}f^{-\nu*}(t) \end{pmatrix}.$$

Its determinant is just

$$\begin{aligned}&[(a+bf^{\nu}(t))(\overline{a}+\overline{b}f^{\nu*}(t))-(c+df^{\nu}(t))(\overline{c}+\overline{d}f^{\nu*}(t))]\\&\quad\cdot[(a-bf^{-\nu}(t))(\overline{a}-\overline{b}f^{-\nu*}(t))-(c-df^{-\nu}(t))(\overline{c}-\overline{d}f^{-\nu*}(t))],\end{aligned}\tag{4.34}$$

and using the identities

$$f^{\nu*}(t)=-f^{\nu}(t),\quad \overline{f^{\nu}(t)}=f^{-\nu}(t),$$

we can rewrite expression (4.34) in the form

$$\begin{aligned}&[(a+bf^{\nu}(t))(\overline{a}-\overline{b}f^{\nu}(t))-(c+df^{\nu}(t))(\overline{c}-\overline{d}f^{\nu}(t))]\\&\quad\cdot[\overline{(a+bf^{\nu}(t))}(a-b\overline{f^{\nu}(t)})-\overline{(c+df^{\nu}(t))}(c+d\overline{f^{\nu}(t)})],\end{aligned}$$

which is obviously real and non-negative. □

Proof of Theorem 4.2.3. First of all, let us recall that $\operatorname{ind} B_1^{\tau}=\operatorname{ind}_{\mathbb{C}}\widetilde{B}_1^{\tau}$, cf. (2.11). Now if $t=1$, then

$$\begin{aligned}&\det\mathcal{A}_{\tilde{B}_1^{\tau}}(1,\mu)\\&=\left[(a+b(1-\mu))(a-b(1-\mu))-\frac{b^2}{\beta^2}\right]\left[(\overline{a}+\overline{b}(1-\mu))(\overline{a}-\overline{b}(1-\mu))-\frac{\overline{b^2}}{\beta^2}\right]\\&\quad+\frac{1}{\beta^2}(ad-bc)(\overline{ad}-\overline{bc})e^{i2(\pi-\omega_\tau)\alpha}+\frac{1}{\beta^2}(ad-bc)(\overline{ad}-\overline{bc})e^{i2(\omega_\tau-\pi)\alpha}\\&\quad+\left[(a+b(1-\mu))(c-d(1-\mu))-\frac{bd}{\beta^2}\right]\left[\frac{\overline{bd}}{\beta^2}-(\overline{a}-\overline{b}(1-\mu))(\overline{c}+\overline{d}(1-\mu))\right]\\&\quad+\left[(\overline{a}+\overline{b}(1-\mu)(\overline{c}-\overline{d}(1-\mu))-\frac{\overline{bd}}{\beta^2}\right]\left[\frac{bd}{\beta^2}(a-b(1-\mu))(c+d(1-\mu))\right]\\&\quad+\left[(c+d(1-\mu))(c-d(1-\mu))-\frac{d^2}{\beta^2})\right]\left[(\overline{c}+\overline{d}(1-\mu))(\overline{c}-\overline{d}(1-\mu))\frac{\overline{d^2}}{\beta^2}\right]=\\&=|a^2-b^2(1-2\mu)^2-\frac{b^2}{\beta^2}|^2+|c^2-d^2(1-2\mu)^2-\frac{d^2}{\beta^2}|^2\\&\quad+\frac{2}{\beta^2}|ad-bc|\cos[(\omega_\tau-\pi)2\alpha]\\&\quad+2\mathrm{Re}\left\{\left[(a+b(1-\mu))(c-d(1-\mu))-\frac{bd}{\beta^2}\right]\right.\\&\qquad\qquad\left.\times\left[\frac{\overline{bd}}{\beta^2}-(\overline{a}-\overline{b}(1-\mu))(\overline{c}+\overline{d}(1-\mu))\right]\right\}.\end{aligned}$$

Since

$$\beta^2=\frac{1}{4\mu(1-\mu)},$$

$$\left[(a+b(1-\mu))(c-d(1-\mu))-\frac{bd}{\beta^2}\right]\left[\frac{\overline{bd}}{\beta^2}-(\overline{a}-\overline{b}(1-\mu))(\overline{c}+\overline{d}(1-\mu))\right]$$
$$=-|ac-bd|^2+|bc-ad|^2(1-2\mu)^2-(\overline{ac}-\overline{bd})(bc-ad)(1-2\mu)$$
$$+(ac-bd)(\overline{bc}-\overline{ad})(1-2\mu),$$

and

$$\frac{1}{\beta^2}\cos((\omega_\tau-\pi)2\alpha)=-2\left(\mu^{\frac{\omega_\tau}{\pi}}(1-\mu)^{2-\frac{\omega_\tau}{\pi}}+(1-\mu)^{\frac{\omega_\tau}{\pi}}\mu^{2-\frac{\omega_\tau}{\pi}}\right)\cos\omega_\tau$$
$$+2i\left(\mu^{\frac{\omega_\tau}{\pi}}(1-\mu)^{2-\frac{\omega_\tau}{\pi}}-(1-\mu)^{\frac{\omega_\tau}{\pi}}\mu^{2-\frac{\omega_\tau}{\pi}}\right)\sin\omega_\tau,$$

we get

$$\det\mathcal{A}_{\tilde{B}_1^\tau}(1,\mu)=|a^2-b^2|^2+|c^2-d^2|^2-2|ac-bd|^2+2|bc-ad|^2(1-2\mu)^2$$
$$-4|ad-bc|^2\left\{(\mu^{\frac{\omega_\tau}{\pi}}(1-\mu)^{2-\frac{\omega_\tau}{\pi}}+(1-\mu)^{\frac{\omega_\tau}{\pi}}\mu^{2-\frac{\omega_\tau}{\pi}})\cos\omega_\tau\right.$$
$$\left.+i\,(\mu^{\frac{\omega_\tau}{\pi}}(1-\mu)^{2-\frac{\omega_\tau}{\pi}}-(1-\mu)^{\frac{\omega_\tau}{\pi}}\mu^{2-\frac{\omega_\tau}{\pi}})\sin\omega_\tau\right\}. \tag{4.35}$$

Consider now the imaginary part of expression (4.35), i.e.,

$$\operatorname{Im}\det\mathcal{A}_{\tilde{B}_1^\tau}(1,\mu)=4|ad-bc|^2\psi(\mu)\sin\omega_\tau \tag{4.36}$$

where

$$\psi(\mu)=\mu^x(1-\mu)^{2-x}-(1-\mu)^x\mu^{2-x}\ ,\ x=\omega_\tau/\pi\ ,\ 0<x<2.$$

Since

$$\psi(\mu)=(\mu(1-\mu)^x+(1-\mu)\mu^x)\left((1-\mu)^{1-x}-\mu^{1-x}\right),$$

the sign of the real-valued function ψ remains constant in each of the intervals $(0,1/2)$ and $(1/2,1)$. Together with the fact that $\det\mathcal{A}_{\tilde{B}_1^\tau}(t,\mu)$ is real-valued for $t\neq 0$, this shows that the winding number of the determinant function around the origin can only take the values $+1,0,-1$. This finishes the proof of Theorem 4.2.3. □

Remark 4.2.5. Observe that (4.36) actually expresses the dependence of the Fredholmness of the operator $\tilde{B}_1^\tau$ and thus of B_1^τ on the angle ω_τ.

4.2.2 The ε-Collocation Method

Now we turn our attention to the Fredholmness of the local operators $B_{1,\varepsilon}^\tau$ connected with the ε-collocation method. Write a, b, c, d, e, f in place of $a(\tau)$, $b(\tau)$,

$c(\tau)$, $d(\tau)$, $e(\tau)$, $f(\tau)$ respectively, and let $B^{1,\tau}_{1,\varepsilon}$ and $B^{2,\tau}_{1,\varepsilon}$ stand for the operators on $l_2(\mathbb{Z})$ with matrix representation

$$B^{1,\tau}_{1,\varepsilon} = aI + \frac{b}{\pi i}\left(\int\limits_{\Gamma_\tau} \frac{\tilde{\chi}_j^{(1)}(s)ds}{s-\tilde{\tau}_k^{1,\varepsilon}}\right)_{k,j\in\mathbb{Z}} - \frac{f}{\pi i}\left(\int\limits_{\Gamma^*_\tau} \frac{\overset{\approx}{\chi}{}_j^{(1)}(s)ds}{s-\overline{\tilde{\tau}_k^{1,\varepsilon}}}\right)_{k,j\in\mathbb{Z}} \tag{4.37}$$

and

$$B^{2,\tau}_{1,\varepsilon} = cI + \frac{d}{\pi i}\left(\int\limits_{\Gamma_\tau} \frac{\tilde{\chi}_j^{(1)}(s)ds}{s-\tilde{\tau}_k^{1,\varepsilon}}\right)_{k,j\in\mathbb{Z}} - \frac{e}{\pi i}\left(\int\limits_{\Gamma^*_\tau} \frac{\overset{\approx}{\chi}{}_j^{(1)}(s)ds}{s-\overline{\tilde{\tau}_k^{1,\varepsilon}}}\right)_{k,j\in\mathbb{Z}}. \tag{4.38}$$

Clearly, $B^{\tau}_{1,\varepsilon} = B^{1,\tau}_{1,\varepsilon} + B^{2,\tau}_{1,\varepsilon}\overline{M}$, so it follows as in Lemma 4.2.2 that $B^{\tau}_{1,\varepsilon}$ is Fredholm if and only if the operator

$$\begin{pmatrix} B^{1,\tau}_{1,\varepsilon} & B^{2,\tau}_{1,\varepsilon} \\ \overline{M}B^{2,\tau}_{1,\varepsilon}\overline{M} & \overline{M}B^{1,\tau}_{1,\varepsilon}\overline{M} \end{pmatrix} \in \mathcal{L}(l_2(\mathbb{Z}) \times l_2(\mathbb{Z}))$$

is Fredholm. By (4.24) and (4.25),

$$\overline{M}B^{1,\tau}_{1,\varepsilon}\overline{M} = \overline{a}I - \frac{\overline{b}}{\pi i}\left(\int\limits_{\Gamma^*_\tau} \frac{\overset{\approx}{\chi}{}_j^{(1)}(s)ds}{s-\overline{\tilde{\tau}_k^{1,\varepsilon}}}\right)_{k,j\in\mathbb{Z}} + \frac{\overline{f}}{\pi i}\left(\int\limits_{\Gamma_\tau} \frac{\tilde{\chi}_j^{(1)}(s)ds}{s-\tilde{\tau}_k^{1,\varepsilon}}\right)_{k,j\in\mathbb{Z}}$$

and

$$\overline{M}B^{2,\tau}_{1,\varepsilon}\overline{M} = \overline{c}I - \frac{\overline{d}}{\pi i}\left(\int\limits_{\Gamma^*_\tau} \frac{\overset{\approx}{\chi}{}_j^{(1)}(s)ds}{s-\overline{\tilde{\tau}_k^{1,\varepsilon}}}\right)_{k,j\in\mathbb{Z}} + \frac{\overline{e}}{\pi i}\left(\int\limits_{\Gamma_\tau} \frac{\tilde{\chi}_j^{(1)}(s)ds}{s-\tilde{\tau}_k^{1,\varepsilon}}\right)_{k,j\in\mathbb{Z}}.$$

In particular, all operators $B^{1,\tau}_{1,\varepsilon}, B^{2,\tau}_{1,\varepsilon}$, $\overline{M}B^{1,\tau}_{1,\varepsilon}\overline{M}$ and $\overline{M}B^{2,\tau}_{1,\varepsilon}\overline{M}$ are of the same structure, and we may consider only one of these operators (say $B^{1,\tau}_{1,\varepsilon}$).

Given $k \in \mathbb{Z}$, and $\delta \in (0,1)$, define t^δ_k by

$$t^\delta_k = \begin{cases} k+\delta & \text{if } k \geq 0, \\ -(k+\delta)e^{i\omega_\tau} & \text{if } k < 0, \end{cases}$$

and introduce the operator C^δ_ε as

$$C^\delta_\varepsilon = (a - i(b-f)ctg(\pi(\varepsilon-\delta)))I + \frac{b}{\pi i}\left(\frac{\Delta t_j^{(1)}}{t^\delta_j - \tilde{\tau}_k^{1,\varepsilon}}\right)_{k,j\in\mathbb{Z}} - \frac{f}{\pi i}\left(\frac{\overline{\Delta t_j^{(1)}}}{t^\delta_j - \tilde{\tau}_k^{1,\varepsilon}}\right)_{k,j\in\mathbb{Z}}$$

with

$$\Delta t_j^{(1)} = \begin{cases} 1 & \text{if } j \geq 0, \\ -e^{i\omega_\tau} & \text{if } j < 0. \end{cases}$$

Consider the operator function

$$\delta \mapsto C_\epsilon(\delta) = 2\varepsilon C_\varepsilon^{2\varepsilon\delta} + 2(\varepsilon - 1) C_\varepsilon^{1+2(\varepsilon-1)\delta} \quad \text{with} \quad \delta \in (0,1).$$

The results of the preceding section imply that $C_\varepsilon(\delta) \in \mathcal{T}^{2\times 2} \subseteq \mathcal{L}(l_2(\mathbb{Z}))$; and since this function is continuous on the interval (0,1/2), we conclude that the operator

$$B_{1,\varepsilon}^{1,\tau} = \int_0^{1/2} C_\varepsilon(\delta) d\delta$$

also belongs to the algebra $\mathcal{T}^{2\times 2}$. The Gohberg–Krupnik symbol of this operator can now be computed as

$$\begin{aligned} \mathcal{A}_{B_{1,\varepsilon}^{1,\tau}}(t,\mu) &= \int_0^{1/2} \mathcal{A}_{C_\varepsilon(\delta)}(t,\mu) d\delta \\ &= \int_0^{1/2} \Big[2\varepsilon (\mathcal{A}_{C_\varepsilon^{2\varepsilon\delta}}(t,\mu) + 2(\varepsilon-1)\mathcal{A}_{C_\varepsilon^{1+2(\varepsilon-1)\delta}}(t,\mu) \Big] d\delta. \end{aligned}$$

From (4.31) and (4.33), one obtains by straightforward calculation

$$\mathcal{A}_{B_{1,\varepsilon}^{1,\tau}}(t,\mu) = \begin{pmatrix} a + (b-f)\int_0^1 f^{(\varepsilon-\delta)}(t) d\delta & 0 \\ 0 & a - (b-f)\int_0^1 f^{(\delta-\varepsilon)}(t) d\delta \end{pmatrix} \tag{4.39}$$

for $t \in \Gamma_0 \backslash \{1\}, \mu \in [0,1]$, and

$$\mathcal{A}_{B_{1,\varepsilon}^{1,\tau}}(1,\mu) = \begin{pmatrix} a + (b-f)(1-2\mu) & i\frac{b}{\beta} e^{-i(\pi-\omega_\tau)\alpha} - i\frac{f}{\beta} e^{-i(\omega_\tau-\pi)\alpha} \\ i\frac{f}{\beta} e^{-i(\pi-\omega_\tau)\alpha} - i\frac{b}{\beta} e^{-i(\omega_\tau-\pi)\alpha} & a - (b-f)(1-2\mu) \end{pmatrix}, \tag{4.40}$$

for $\mu \in [0,1]$, where f^ν, α and β are as in the preceding section.

Now, the symbols of $B_{1,\varepsilon}^{2,\tau}, \overline{M} B_{1,\varepsilon}^{1,\tau} \overline{M}$ and $\overline{M} B_{1,\varepsilon}^{2,\tau} \overline{M}$ can be derived from (4.39) and (4.40) by an obvious substitution of the coefficients, so the symbol of the operator $B_{1,\varepsilon}^\tau$ is the 4×4 matrix function

$$\mathcal{A}_{B_{1,\varepsilon}^\tau} = \begin{pmatrix} \mathcal{A}_{B_{1,\varepsilon}^{1,\tau}} & \mathcal{A}_{B_{1,\varepsilon}^{2,\tau}} \\ \mathcal{A}_{\overline{M} B_{1,\varepsilon}^{2,\tau} \overline{M}} & \mathcal{A}_{\overline{M} B_{1,\varepsilon}^{1,\tau} \overline{M}} \end{pmatrix},$$

on the cylinder $\Gamma_0 \times [0,1]$.

Corollary 4.2.6. *The operator $B_{1,\varepsilon}^\tau$ is Fredholm if and only if the matrices $\mathcal{A}_{B_{1,\varepsilon}^\tau}(t,\mu)$ are invertible for all $(t,\mu) \in \mathbb{T} \times [0,1]$.*

Corollary 4.2.7. *If $t \neq 1$, then $\det \mathcal{A}_{B_{1,\varepsilon}^\tau}(t,\mu)$ takes real values only.*

Corollary 4.2.8. *If the operators B_{1,ε_1}^τ and B_{1,ε_2}^τ, $\varepsilon_1, \varepsilon_2 \in (0,1)$ are Fredholm, then*

$$\operatorname{ind} B_{1,\varepsilon_1}^\tau = \operatorname{ind} B_{1,\varepsilon_2}^\tau.$$

Corollary 4.2.9. *Let $e = f = 0$. If $B_{1,\varepsilon}^\tau$ is Fredholm, then $\operatorname{ind} B_{1,\varepsilon}^\tau \in \{-1,0,1\}$.*

4.2.3   Qualocation Method

We study some peculiarities of the local operators for the qualocation method, and in particular, point out that, if the local operators are Fredholm, then their indices are independent of the values of the parameters ε_r and w_r.

As in the preceding section, the corresponding local operators can be written as

$$B_{Q(\varepsilon)}^\tau = B_{Q(\varepsilon)}^{1,\tau} + B_{Q(\varepsilon)}^{2,\tau}\overline{M}$$

where

$$B_{Q(\varepsilon)}^{i,\tau} = \sum_{r=0}^{m-1} w_r B_{1,\varepsilon_r}^{i,\tau_r}, \quad i = 1,2$$

(compare (4.37), (4.38)), and their symbols are the 4×4 matrix functions on $\mathbb{T} \times [0,1]$,

$$\mathcal{A}_{B_{Q(\varepsilon)}^\tau} = \begin{pmatrix} \mathcal{A}_{B_{Q(\varepsilon)}^{1,\tau}} & \mathcal{A}_{B_{Q(\varepsilon)}^{2,\tau}} \\ \mathcal{A}_{\overline{M}B_{Q(\varepsilon)}^{2,\tau}\overline{M}} & \mathcal{A}_{\overline{M}B_{Q(\varepsilon)}^{1,\tau}\overline{M}} \end{pmatrix}.$$

The symbols of $B_{Q(\varepsilon)}^{i,\tau}$ and $\overline{M}B_{Q(\varepsilon)}^{i,\tau}\overline{M}$ ($i = 1,2$) can be computed via (4.39) and (4.40); for example,

$$\mathcal{A}_{B_{Q(\varepsilon)}^{1,\tau}}(t,\mu) = \begin{cases} \begin{pmatrix} a+(b-f)\sum_{r=0}^{m-1} w_r \int_0^1 f^{(\varepsilon_r-\delta)}(t)d\delta & 0 \\ 0 & a-(b-f)\sum_{r=0}^{m-1} w_r \int_0^1 f^{(\delta-\varepsilon_r)}(t)d\delta \end{pmatrix} \\ \qquad\qquad\qquad \text{if } t \neq 1, \ \mu \in [0,1] \\[2mm] \begin{pmatrix} a+(b-f)(1-2\mu) & i\frac{b}{\beta}e^{-i(\pi-\omega_\tau)\alpha} - i\frac{f}{\beta}e^{-i(\omega_\tau-\pi)\alpha} \\ i\frac{f}{\beta}e^{-i(\pi-\omega_\tau)\alpha} - i\frac{b}{\beta}e^{-i(\omega_\tau-\pi)\alpha} & a-(b-f)(1-2\mu) \end{pmatrix} \\ \qquad\qquad\qquad \text{if } t = 1, \ \mu \in [0,1]. \end{cases}$$

Corollary 4.2.10. *The operator $B^\tau_{Q(\varepsilon)}$ is Fredholm if and only if the matrix $A_{B^\tau_{Q(\varepsilon)}}(t,\mu)$ is invertible for all $(t,\mu) \in \mathbb{T} \times [0,1]$.*

The result on the index independence is as follows.

Theorem 4.2.11. *If $(\varepsilon^i_r)^{m-1}_{r=0}$ and $(w^i_r)^{m-1}_{r=0}$, $i = 1,2$ are two collections of numbers satisfying the hypotheses of Theorem 4.1.8, and if the associated local operators $B^\tau_{Q_1(\varepsilon)}$ and $B^\tau_{Q_2(\varepsilon)}$ are Fredholm, then $\operatorname{ind} B^\tau_{Q_1(\varepsilon)} = \operatorname{ind} B^\tau_{Q_2(\varepsilon)}$.*

The proof of this theorem, as well as the proof of the subsequent corollary, follows by repeating arguments from the proof of Theorem 4.2.3, so the details are omitted.

Corollary 4.2.12. *Let $e = f = 0$, and the operator $B^\tau_{Q(\varepsilon)}$ be Fredholm. Then $\operatorname{ind} B^\tau_{Q(\varepsilon)} \in \{-1,0,1\}$.*

4.3 Stability of Approximation methods in L_2-Spaces with Weight

In this section, we again consider a special case of equation (4.1), where we suppose the coefficients e and r are identically equal to zero on Γ, viz., the equation

$$(Au)(t) \equiv a(t)u(t) + \frac{b(t)}{\pi i} \int_\Gamma \frac{u(\tau)d\tau}{\tau - t} + c(t)\overline{u(t)} + \frac{d(t)}{\pi i} \int_\Gamma \frac{\overline{u(\tau)}d\tau}{\tau - t} \qquad (4.41)$$

$$+ \int_\Gamma k_1(t,\tau)u(\tau)d\tau + \int_\Gamma k_2(t,\tau)\overline{u(\tau)}\,d\tau = f(t), \quad t \in \Gamma.$$

4.3.1 A Quadrature Method and Its Stability

Let Γ be a simple closed curve and let $\gamma : \mathbb{R} \to \Gamma$ be a 1-periodic parametrization of Γ. We suppose that there are points $u_1 < u_2 < \cdots < u_{n_0} = u_1 + 1$ such that γ is twice continuously differentiable on each of the intervals (u_j, u_{j+1}), $j = 1, 2, \ldots, n_0 - 1$, and that the derivatives γ' and γ'' possess finite one-sided limits $\gamma'(u_j \pm 0)$ and $\gamma''(u_j \pm 0)$ such that $|\gamma'(u_j - 0)| = |\gamma'(u_j + 0)|$, but possibly $\arg\gamma'(u_j - 0) \neq \arg\gamma'(u_j + 0)$ at every point u_j, $j = 1, \ldots, n_0 - 1$. Consider the function

$$\omega(t) := \prod_{j=1}^{n_0-1} |t - \gamma(u_j)|^{-\rho_j}, \quad t \in \Gamma,$$

with $\rho_j \in (-1/2, 1/2)$ for all j. Under these conditions, the operator A of (4.41) acts boundedly on the Lebesgue space $L_{2,\omega}(\Gamma)$ of all complex-valued measurable functions x with

$$\|x\|_{L_{2,\omega}} := \left(\int_\Gamma |x(t)|^2 \omega^2(t)\mathrm{d}t \right)^{1/2} < \infty,$$

considered as a Banach space over the field $\mathbb{R}$.

To improve the convergence rate for the approximation method under consideration, we use discretizations based on graded meshes. Thus given $\sigma \geq 1$, choose a 1-periodic function $g : \mathbb{R} \to \mathbb{R}$ which satisfies the following conditions:

1) $g(j/n_0) = u_j$ for all $j = 1, 2, \ldots, n_0 - 1$.
2) $g(s) = g(j/n_0) + |s - j/n_0|^\sigma \operatorname{sign}(s - j/n_0)$ for all s in some neighbourhood U_{s_j} of the point $s_j = j/n_0$, with $U_{s_j} \cap U_{s_k} = \emptyset$ if $j \neq k$.
3) The function g is strictly monotonically increasing on $[0, 1]$, and its restriction onto the set $(s_j, s_{j+1}) \setminus (U_{s_j} \cup U_{s_{j+1}})$ is a twice differentiable function for every j.

Let $n = ln_0$ with $l \in \mathbb{N}$, choose numbers $\delta, \varepsilon \in (0,1)$ with $\delta \neq \varepsilon$, define $\tilde{\gamma}(s) = \gamma(g(s))$, and for each $k = 0, 1, \ldots, n-1$ put

$$t_k^{(n)} := \tilde{\gamma}\left(\frac{k+\delta}{n}\right), \quad \tau_k^{(n)} := \tilde{\gamma}\left(\frac{k+\varepsilon}{n}\right), \quad \Delta t_j^{(n)} = \frac{1}{n}\tilde{\gamma}\left(\frac{j+\delta}{n}\right).$$

Applying a shifted trapezoidal rule to the singular integrals in (4.41) and then collocating at the shifted points $\tau_k^{(n)}$ leads to the system of algebraic equations (4.4). If $\xi_k^{(n)}, k = 0, 1, \ldots, n-1$ are the solutions of (4.4) and if $\chi_j^{(n)}(t)$ denotes the characteristic function of the arc $[\tilde{\gamma}(j/n), \tilde{\gamma}((j+1)/n))$ of Γ, then the function

$$x_n(t) = \sum_{j=0}^{n-1} \xi_j^{(n)} \chi_j^{(n)}(t), \quad t \in \Gamma \tag{4.42}$$

can be viewed as an approximate solution of equation (4.4).

The stability of this quadrature method (4.4), (4.42) in the weighted space $L_{2,\omega} = L_{2,\omega}(\Gamma)$ can be studied by localizing techniques analogous to considerations of Section 4.1.1. Thus with every point $\tau \in \Gamma$ we again associate a model equation $A^\tau x^\tau = f^\tau$, which locally represents the equation (4.41), as well as a model approximation method for the model equation

$$A_n^\tau x_n^\tau = f_n^\tau,$$

which locally represents the equation (4.4). Recall that this association can be performed as follows: given $\tau \in \Gamma$, let s_τ refer to that point of $[0,1)$ with $\tilde{\gamma}(s_\tau) = \tau$. Denote by ω_τ the number

$$\omega_\tau := \arg(-\tilde{\gamma}'(s_\tau - 0)/\tilde{\gamma}'(s_\tau + 0)), \quad \omega_\tau \in [0, 2\pi),$$

and write Γ_τ for the curve $\Gamma_\tau := \mathrm{e}^{i\omega_\tau}\mathbb{R}^+ \cup \mathbb{R}^+$, where the ray $\mathrm{e}^{i\omega_\tau}\mathbb{R}^+$ is directed to the origin and the ray $\mathbb{R}^+$ is directed away. Then define $L_2(\Gamma_\tau, \rho_\tau)$ as the real

Hilbert space of all complex-valued Lebesgue measurable functions x on Γ_τ such that

$$\int_{\Gamma_\tau} |x(t)|^2 |t|^{-2\rho_\tau}(t)dt < +\infty,$$

where $\rho_\tau = \rho_j$ if $\tau = \gamma(u_j)$, $j = 1, 2, \ldots, n_0 - 1$ and $\rho_\tau = 0$ at all other points $\tau \in \Gamma$. Further, introduce the operators

$$(S_{\Gamma_\tau} x)(t) := \frac{1}{\pi i} \int_{\Gamma_\tau} \frac{x(u)du}{u - t}, \quad t \in \Gamma_\tau,$$

as well as

$$A^\tau := a(\tau)I + b(\tau)S_{\Gamma_\tau} + c(\tau)M + d(\tau)S_{\Gamma_\tau}M,$$

which act boundedly on $L_2(\Gamma_\tau, \rho_\tau)$. The model equation has the form

$$A^\tau x^\tau = f^\tau, \quad x^\tau \in L_2(\Gamma_\tau, \rho_\tau), \tag{4.43}$$

with a function f^τ belonging to the Banach space $\mathbf{R}_2(\Gamma_\tau, \rho_\tau)$ of all functions that are Riemann integrable on every finite subcurve of Γ_τ and have a finite Riemann norm

$$\begin{aligned} \|f\|_{\mathbf{R}_2(\Gamma_\tau, \rho_\tau)} &:= \|f\|_{L_2(\Gamma_\tau, \rho_\tau)} + \left(\sum_{k=0}^\infty \sup_{t \in [k, k+1]} |f(t)|^2 |t|^{-2\rho_\tau} \right)^{1/2} \\ &+ \left(\sum_{k=0}^\infty \sup_{t \in e^{i\omega_\tau}[k, k+1]} |f(t)|^2 |t|^{-2\rho_\tau} \right)^{1/2}. \end{aligned}$$

For the approximate solution of equation (4.43), introduce points $\tilde{t}_k^{(n)}$ and $\tilde{\tau}_k^{(n)}$, $k \in \mathbb{Z}$ by

$$\tilde{t}_k^{(n)} := \begin{cases} \left(\dfrac{k + \delta}{n} \right)^\sigma & \text{if } k \geq 0, \\ -\left(\dfrac{k + \delta}{n} \right)^\sigma e^{i\omega_\tau} & \text{if } k < 0, \end{cases}$$

$$\tilde{\tau}_k^{(n)} := \begin{cases} \left(\dfrac{k + \varepsilon}{n} \right)^\sigma & \text{if } k \geq 0, \\ -\left(\dfrac{k + \varepsilon}{n} \right)^\sigma e^{i\omega_\tau} & \text{if } k < 0, \end{cases}$$

and determine approximate values $\xi_k^{(n)}$ of the exact solution $x^\tau(\tilde{t}_k^{(n)})$, $k \in \mathbb{Z}$, of

(4.43) by solving the infinite model system of algebraic equations

$$[a(\tau) - b(\tau)i\cot(\pi(\varepsilon-\delta))]\xi_k^{(n)} + \frac{b(\tau)}{\pi i}\sum_{j=-\infty}^{+\infty}\frac{\Delta\tilde{t}_j^{(n)}}{\tilde{t}_j^{(n)} - \tilde{\tau}_k^{(n)}}\xi_j^{(n)}$$
$$+ [c(\tau) - d(\tau)i\cot(\pi(\varepsilon-\delta))]\overline{\xi_k^{(n)}} + \frac{d(\tau)}{\pi i}\sum_{j=-\infty}^{+\infty}\frac{\overline{\Delta\tilde{t}_j^{(n)}}}{\overline{\tilde{t}_j^{(n)} - \tilde{\tau}_k^{(n)}}}\overline{\xi_j^{(n)}}$$
$$= f(\tilde{\tau}_k^{(n)}) \qquad \text{for } k \in \mathbb{Z} \tag{4.44}$$

where

$$\Delta\tilde{t}_j^{(n)} = \begin{cases} \dfrac{\sigma}{n}\cdot\left(\dfrac{j+\delta}{n}\right)^{\sigma-1} & \text{if } j \geq 0,\\ -\dfrac{\sigma}{n}\cdot\left(\dfrac{j+\delta}{n}\right)^{\sigma-1} e^{i\omega_\tau} & \text{if } j < 0. \end{cases}$$

If $\widetilde{\chi}_k^{(n)}$ stands for the characteristic function of the interval $[(k/n)^\sigma, ((k+1)/n)^\sigma)$ when $k \geq 0$ and of the interval $[e^{i\omega_\tau}(k/n)^\sigma, e^{i\omega_\tau}((k+1)/n)^\sigma)$ when $k < 0$, the approximate solution of (4.43) on the whole curve Γ_τ is given by

$$x_n^\tau(t) = \sum_{k\in\mathbb{Z}} \xi_k^{(n)}\widetilde{\chi}_k^{(n)}(t), \quad t \in \Gamma_\tau. \tag{4.45}$$

One can think of the infinite model system (4.44) – (4.45) as an operator equation

$$A_n^\tau x_n^\tau = f_n^\tau, \tag{4.46}$$

with x_n^τ, $f_n^\tau \in L_2(\Gamma_\tau, \rho_\tau)$ and $A_n^\tau \in \mathcal{L}_{add}(L_2(\Gamma_\tau, \rho_\tau))$. Note that the boundedness of A_n^τ follows from the corresponding results of [102, Chapters 2 and 3] or of [180], but it is more convenient for our purposes to translate this system into an equation on an appropriately chosen weighted l_2-space.

Let $\tilde{l}_{2,\nu} = \tilde{l}_{2,\nu}(\mathbb{Z})$, $\nu \in \mathbb{R}$ be the Hilbert space of all two-sided sequences $\{\xi_k\}_{k\in\mathbb{Z}}$ of complex numbers such that

$$\|\{\xi_k\}_{k\in\mathbb{Z}}\|_{\tilde{l}_{2,\nu}} := \left(\sum_{k\in\mathbb{Z}} |\xi_k|^2(|k|+1)^{2\nu}\right)^{1/2} < \infty$$

and set $r_\tau := (\sigma-1)/2 - \rho_\tau\sigma$. It is well known (e.g., [102, 179]) that there is a constant $c > 0$ such that

$$\frac{1}{c}\|\sum_{k\in\mathbb{Z}}\xi_k\widetilde{\chi}_k^{(n)}\|_{L_2(\Gamma_\tau,\rho_\tau)} \leq n^{-r_\tau-1/2}\|\{\xi_k\}_{k\in\mathbb{Z}}\|_{\tilde{l}_{2,r_\tau}} \leq c\|\sum_{k\in\mathbb{Z}}\xi_k\widetilde{\chi}_k^{(n)}\|_{L_2(\Gamma_\tau,\rho_\tau)} \tag{4.47}$$

for every sequence $\{\xi_k\} \in \tilde{l}_{2,r_\tau}$. Then de Boor inequalities (4.47) allow us to identify the system (4.44) with an operator equation

$$\tilde{A}_n^\tau \xi_n = \eta_n \tag{4.48}$$

where $\xi_n = \{\xi_k^{(n)}\}_{k\in\mathbb{Z}}$, the sequence $\eta_n = \{f(\tilde{\tau}_k^{(n)})\}_{k\in\mathbb{Z}}$ belongs to the space $\tilde{l}_{2,r_\tau}$, and $\tilde{A}_n^\tau \in \mathcal{L}_{add}(\tilde{l}_{2,r_\tau})$. Moreover, again due to (4.47), the operator sequence $(A_n^\tau)_{n\in\mathbb{N}}$ in (4.46) is stable if and only if the sequence $(\tilde{A}_n^\tau)_{n\in\mathbb{N}}$ in (4.48) is stable. Now the advantage of considering the model problem on $\tilde{l}_{2,r_\tau}$ becomes obvious, viz., the operators $\tilde{A}_n^\tau, n \in \mathbb{N}$ turn out to be independent of n, so the sequence $(\tilde{A}_n^\tau)_{n\in\mathbb{N}}$ is actually constant, and this sequence is stable if and only if one of its members (say the operator $\tilde{A}_1^\tau$) is invertible.

Applying the local principle analogously to Section 4.1.1, one arrives at the following result.

Theorem 4.3.1. *Let a, b, c, $d \in \mathbf{C}(\Gamma)$, and let k_1, $k_2 \in \mathbf{C}(\Gamma \times \Gamma)$. The quadrature method* (4.4), (4.42) *is stable if and only if the operator $A \in \mathcal{L}_{add}(L_2(\Gamma_\tau, \rho_\tau))$ and all operators $\tilde{A}_1^\tau \in \mathcal{L}_{add}(\tilde{l}_{2,r_\tau})$, $\tau \in \Gamma$, are invertible.*

In contrast to Section 4.1.1, here we study operators defined on weighted spaces and associated with non-uniformly graded meshes, i.e., the situation where $\rho \neq 0$ and $\sigma \neq 0$. This modification does not seriously influence the proof of the corresponding stability results but the Fredholm properties of the local operators $\tilde{A}_1^\tau$ considered on such spaces differ from those for spaces without weight.

4.3.2 Local Operators and Their Indices

As mentioned above, both the Fredholm properties and the index of the local operators $\tilde{A}_1^\tau$ corresponding to pure singular integral equations without conjugation are independent of both the angles ω_τ and the weights ρ_τ (cf. [179, 180]). On the other hand, as observed in Section 4.2.1 where the weight function was supposed to be identically 1, the index of the operators $\tilde{A}_1^\tau$ corresponding to singular integral equations with conjugation can differ from zero, although it is restricted to the set $\{-1, 0, 1\}$. Let us now examine how the change of the weight influences the index.

We start with a matrix representation for the local operators $\tilde{A}_1^\tau$. Let us again identify the Hilbert space $\tilde{l}_{2,r_\tau}$ with the Cartesian product $l^2_{2,r_\tau} := l_{2,r_\tau} \times l_{2,r_\tau}$ of the corresponding Hilbert spaces of one-sided sequences. Then every operator on $\tilde{l}_{2,r_\tau}$ corresponds to a 2×2 matrix of operators acting on the space l_{2,r_τ}.

Further, given ν let $\mathcal{T}_\nu$ stand for the smallest closed subalgebra of $\mathcal{L}(l_{2,\nu})$ containing all Toeplitz operators generated by piecewise constant functions, and let $\mathcal{T}_\nu^{2\times 2}$ be the subalgebra of $\mathcal{L}(\tilde{l}_{2,\nu}) \cong \mathcal{L}(l^2_{2,\nu}) \cong \mathcal{L}^{2\times 2}(l_{2,\nu})$ which consists of all 2×2 matrices with entries in $\mathcal{T}_\nu$. Recall that the Toeplitz operator $T(a)$ with generating function $a \in L_\infty$ is defined via its matrix representation with respect to the standard basis of $l_{2,\nu}$ by $T(a) = (a_{j-k})_{j,k=0}^\infty$, with a_k referring to the kth Fourier coefficient of a. Moreover, for any piecewise smooth function a, the operator $T(a)$ is bounded on the space $l_{2,\nu}$ if $-1/2 < \nu < 1/2$.

Lemma 4.3.2. *Every operator $\tilde{A}_1^\tau$ as in* (4.48) *is isometrically isomorphic to the operator*

$$\widehat{A}_1^\tau = \widehat{A}^{1,\tau} + \widehat{A}^{2,\tau}\overline{M}, \tag{4.49}$$

where $\widehat{A}^{1,\tau}, \widehat{A}^{2,\tau} \subset \mathcal{T}^{2\times 2}$ *and* $\overline{M}$ *is the operator*

$$\overline{M}\{\{\xi_k\}_{k=0}^{\infty}, \quad \{\eta_k\}_{k=0}^{\infty}\} = \{\{\overline{\xi_k}\}_{k=0}^{\infty}, \{\overline{\eta_k}\}_{k=0}^{\infty}\}.$$

Proof. Fix a point $\tau \in \Gamma$ and let for brevity

$$a = a(\tau), \quad b = b(\tau), \quad c = c(\tau), \quad d = d(\tau), \quad \rho = \rho_\tau, \quad \omega = \omega_\tau.$$

Straightforward computation shows that the operator $\tilde{A}_1^\tau$ can be written in the form

$$\tilde{A}_1^\tau = \tilde{A}^{1,\tau} + \tilde{A}^{2,\tau}\overline{M},$$

with

$$\tilde{A}^{1,\tau} = \left(A_{k,r}^{1,\tau}\right)_{k,r=1}^2, \quad \tilde{A}^{2,\tau} = \left(A_{k,r}^{2,\tau}\right)_{k,r=1}^2,$$

where

$$A_{1,1}^{1,\tau} = (a - ib\cot(\pi(\varepsilon-\delta)))I + \frac{b}{\pi i}\left(\frac{\sigma(j+\delta)^{\sigma-1}}{(j+\delta)^\sigma - (k+\varepsilon)^\sigma}\right)_{k,j=0}^\infty,$$

$$A_{2,1}^{1,\tau} = \frac{b}{\pi i}\left(\frac{\sigma(j+\delta)^{\sigma-1}}{(j+\delta)^\sigma - (k+1-\varepsilon)^\sigma}e^{i\omega}\right)_{k,j=0}^\infty,$$

$$A_{1,2}^{1,\tau} = \frac{b}{\pi i}\left(\frac{\sigma(j+1-\delta)^{\sigma-1}e^{i\omega}}{(j+1-\delta)^\sigma e^{i\omega} - (k+\varepsilon)^\sigma}\right)_{k,j=0}^\infty,$$

$$A_{2,2}^{1,\tau} = (a - ib\cot(\pi(\delta-\varepsilon)))I - \frac{b}{\pi i}\left(\frac{\sigma(j+1-\delta)^{\sigma-1}}{(j+1-\delta)^\sigma - (k+1-\varepsilon)^\sigma}\right)_{k,j=0}^\infty.$$

Replacing a and b by c and d respectively, one gets analogous representations for the operators $A_{k,r}^{2,\tau}$, $k, r = 1, 2$.

Now let Λ^z, $z \in \mathbb{C}$ refer to the operator of multiplication by the diagonal matrix $\Lambda^z := ((j+1)^z\delta_{j,k})_{k,j=0}^{+\infty}$. Evidently, Λ^z is a linear isometry from $l_{2,z}$ onto l_2, where $l_2 := l_{2,0}$, so $\tilde{A}_1^\tau$ is a continuous additive operator on $l_{2,r_\tau} \times l_{2,r_\tau}$ if and only if the operator

$$\hat{A}_1^\tau := \begin{pmatrix} \Lambda^{r_\tau} & 0 \\ 0 & \Lambda^{r_\tau} \end{pmatrix} \tilde{A}_1^\tau \begin{pmatrix} \Lambda^{-r_\tau} & 0 \\ 0 & \Lambda^{-r_\tau} \end{pmatrix}$$

belongs to $\mathcal{L}_{add}^{2\times 2}(l_2)$. Moreover, $\mathcal{T}_z = \Lambda^{-z}\mathcal{T}_0\Lambda^z$. Since $\Lambda^z\overline{M} = \overline{M}\Lambda^z$ for every real z, the operator $\hat{A}_1^\tau$ can be rewritten in the form

$$\hat{A}_1^\tau = \hat{A}^{1,\tau} + \hat{A}^{2,\tau}\overline{M},$$

with $\hat{A}^{i,\tau} = \Lambda^{(\sigma-1)/2-\rho\sigma}\tilde{A}^{i,\tau}\Lambda^{(-\sigma+1)/2+\rho\sigma}$ for $i = 1, 2$. To complete the proof, one has to check that these operators belong to the algebra $\mathcal{T}^{2\times 2} = \mathcal{T}_0^{2\times 2}$. This has been done in [179, pp. 46–49] and [102, Sections 2.4 and 2.11]. □

Lemma 4.3.3. (a) *If $\hat{A} \in \mathcal{T}$, then $\overline{M}\hat{A}\overline{M}$ is also in $\mathcal{T}$.*

(b) *The operator $\widehat{A}_1^\tau$ in (4.49) is Fredholm on the real space $\tilde{l}_2$ if and only if the operator*

$$A_{1,\tau} := \begin{pmatrix} \hat{A}^{1,\tau} & \hat{A}^{2,\tau} \\ \overline{M}\hat{A}^{2,\tau}\overline{M} & \overline{M}\hat{A}^{1,\tau}\overline{M} \end{pmatrix} \tag{4.50}$$

is Fredholm on the corresponding complex space. In that case,

$$\operatorname{ind}_{\mathbb{R}} \tilde{A}_1^\tau = \operatorname{ind}_{\mathbb{C}} A_{1,\tau}.$$

Assertion (a) is proved in Lemma 4.2.1, and (b) is a consequence of Lemma 1.4.6 (see also (2.11)).

Let us fix a point $\tau \in \Gamma$, and drop the subscript τ in $A_{1,\tau}$ but indicate the dependence of that operator on σ, ρ and ω by writing $A_1^{\sigma,\rho,\omega}$ in place of $A_{1,\tau}$. The Fredholmness of the operator $A_1^{\sigma,\rho,\omega}$ can be studied via the Gohberg-Krupnik symbol calculus for the algebra $\mathcal{T}^{2\times 2}$. We again consider the function f^ν,

$$f^\nu(e^{i2\pi s}) = 2e^{i\pi\nu(s-1)}\frac{\sin(-\pi\nu s)}{\sin(-\pi\nu)} - 1 \quad \text{for } 0 \le s < 1,$$

first defined in (4.32). If we set $f^*(t) := \overline{f(\overline{t})}$, $\beta := \beta(\mu) = \sin(\pi\theta)$, $\gamma := \gamma(\mu) = -i\cot(\pi\theta)$, and

$$\theta := \theta(\mu) = \frac{1}{2} - \rho + \frac{i}{2\pi\sigma}\log\frac{\mu}{1-\mu}, \quad \mu \in [0,1], \tag{4.51}$$

then the symbol $\mathcal{A}_{A_1^{\sigma,\rho,\omega}}$ of the operator $A_1^{\sigma,\rho,\omega}$, which is a 4×4-matrix-valued function defined on $\mathbb{T} \times [0, 1]$, is given as follows (cf. Section 4.2.1 and [11, 94, 102, 179]):

If $t \ne 1$ and $\mu \in [0,1]$, then

$$\mathcal{A}_{A_1^{\sigma,\rho,\omega}}(t,\mu) = \begin{pmatrix} a + bf^{(\varepsilon-\delta)}(t) & 0 & c + df^{(\varepsilon-\delta)}(t) & 0 \\ 0 & a - bf^{(\delta-\varepsilon)}(t) & 0 & c - df^{(\delta-\varepsilon)}(t) \\ \bar{c} + \bar{d}f^{(\varepsilon-\delta)*}(t) & 0 & \bar{a} + \bar{b}f^{(\varepsilon-\delta)*}(t) & 0 \\ 0 & \bar{c} - \bar{d}f^{(\delta-\varepsilon)*}(t) & 0 & \bar{a} - \bar{b}f^{(\delta-\varepsilon)*}(t) \end{pmatrix},$$

and if $t = 1$ and $\mu \in [0,1]$, then

$$\mathcal{A}_{A_1^{\sigma,\rho,\omega}}(1,\mu) = \begin{pmatrix} a + b\gamma & i\frac{b}{\beta}e^{i(\omega-\pi)\theta} & c + d\gamma & i\frac{d}{\beta}e^{i(\omega-\pi)\theta} \\ -i\frac{b}{\beta}e^{i(\pi-\omega)\theta} & a - b\gamma & -i\frac{d}{\beta}e^{i(\pi-\omega)\theta} & c - d\gamma \\ \bar{c} - \bar{d}\gamma & -i\frac{\bar{d}}{\beta}e^{i(\pi-\omega)\theta} & \bar{a} - \bar{b}\gamma & -i\frac{\bar{b}}{\beta}e^{i(\pi-\omega)\theta} \\ i\frac{\bar{d}}{\beta}e^{i(\omega-\pi)\theta} & \bar{c} + \bar{d}\gamma & i\frac{\bar{b}}{\beta}e^{i(\omega-\pi)\theta} & \bar{a} + \bar{b}\gamma \end{pmatrix}. \tag{4.52}$$

We prepare for studying the Fredholmness of the local operator $A_1^{\sigma,\rho,\omega}$ by noting two of its important properties.

Lemma 4.3.4. *If a, b, c, and d are any fixed complex numbers, then the function*

$$\det \mathcal{A}_{A_1^{\sigma,\rho,\omega}} : (\Gamma_0 \setminus \{1\}) \times [0,1] \mapsto \mathbb{C}$$

is real-valued and non-negative.

Proof. If $t \neq 1$, then $\mathcal{A}_{A_1^{\sigma,\rho,\omega}}(t,\mu) = \mathcal{A}_{\tilde{B}_1^\tau}(t,\mu)$, where the operator $\tilde{B}_1^\tau(t,\mu)$ is defined by (4.30), so the result follows from Lemma 4.2.4. □

Lemma 4.3.5. *For any $\sigma_1, \sigma_2 \in \mathbb{R}$, $\sigma_1 \geq 1$, $\sigma_2 \geq 1$, the operators $A_1^{\sigma_1,\rho,\omega}$ and $A_1^{\sigma_2,\rho,\omega}$ are either simultaneously Fredholm or not. If they are Fredholm,*

$$\operatorname{ind} A_1^{\sigma_1,\rho,\omega} = \operatorname{ind} A_1^{\sigma_2,\rho,\omega}.$$

Proof. A closer look at the symbol of the operator $A_1^{\sigma,\rho,\omega}$ reveals that only the matrix (4.52) depends on the grading parameter σ via the function θ defined by (4.51), but the image of the function $\theta = \theta(\mu)$, $\mu \in [0,1]$, coincides with $1/2-\rho+i\mathbb{R}$, which is independent on σ. □

Thus we can suppose hereafter that $\sigma = 1$, and write $A_1^{\rho,\omega}$ in place of $A_1^{1,\rho,\omega}$. Recall that by Theorem 4.2.3, the index of the operator $A_1^{0,\omega}$ considered on the l_2-space without weight can only take the values -1, 0 or 1. Let us now verify that this still remains true for the weighted spaces if the weight ρ is in the interval $(0,1/2)$.

Lemma 4.3.6. *The determinant of $\mathcal{A}_{A_1^{\rho,\omega}}(1,\mu)$ can be represented as*

$$\det \mathcal{A}_{A_1^{\rho,\omega}}(1,\mu) = U + V\Phi^2(\mu),$$

with real numbers U, V, and with Φ denoting the function

$$\Phi(\mu) = \Phi_{\rho,\omega}(\mu) = \frac{\sin((\omega-\pi)\theta)}{\sin(\pi\theta)}, \quad \mu \in [0,\, 1],$$

where

$$\theta = \theta(\mu) = \frac{1}{2} - \rho + \frac{i}{2\pi}\log\frac{\mu}{1-\mu}, \quad \mu \in [0,1]$$

Proof. Expanding the determinant according to the second-order minors of the first two rows leads to the expression

$$\begin{aligned}
\det \mathcal{A}_{A_1^{\rho,\omega}}(1,\mu) &= [(a+b\gamma)(a-b\gamma) - \frac{b^2}{\beta^2}][(\overline{a}+\overline{b}\gamma)(\overline{a}-\overline{b}\gamma) - \frac{\overline{b^2}}{\beta^2}] \\
&+ [(c+d\gamma)(c-d\gamma) - \frac{d^2}{\beta^2}][(\overline{c}+\overline{d}\gamma)(\overline{c}-\overline{d}\gamma) - \frac{\overline{d^2}}{\beta^2}] \\
&+ [(a+b\gamma)(c-d\gamma) - \frac{bd}{\beta^2}][\frac{\overline{bd}}{\beta^2} - (\overline{a}-\overline{b}\gamma)(\overline{c}+\overline{d}\gamma)] \\
&+ [(\overline{a}+\overline{b}\gamma)(\overline{c}-\overline{d}\gamma) - \frac{\overline{bd}}{\beta^2}][\frac{bd}{\beta^2} - (a-b\gamma)(c+d\gamma)] \\
&+ \frac{1}{\beta^2}(ad-bc)(\overline{ad-bc})e^{i2(\pi-\omega)\theta} \\
&+ \frac{1}{\beta^2}(ad-bc)(\overline{ad-bc})e^{i2(\omega-\pi)\theta}. \qquad (4.53)
\end{aligned}$$

The first line on the right-hand side of (4.53) is equal to

$$\begin{aligned}
&[a^2 - b^2\gamma^2 - \frac{b^2}{\beta^2}][\overline{a}^2 - \overline{b}^2\gamma^2 - \frac{\overline{b^2}}{\beta^2}] \\
&\quad = [a^2 - b^2(-\cot^2(\pi\theta) + \frac{1}{\sin^2(\pi\theta)})][\overline{a}^2 - \overline{b}^2(-\cot^2(\pi\theta) + \frac{1}{\sin^2(\pi\theta)})] \\
&\quad = |a^2 - b^2|^2,
\end{aligned}$$

and the second coincides with $|c^2 - d^2|^2$. For the third line, abbreviate $ac - bd$ and $cb - ad$ by N and P respectively, and note that

$$(a+b\gamma)(c-d\gamma) - \frac{bd}{\beta^2} = N + P\gamma, \quad \frac{\overline{bd}}{\beta^2} - (\overline{a}-\overline{b}\gamma)(\overline{c}+\overline{d}\gamma) = -\overline{N} + \overline{P}\gamma,$$

which finally gives $-|N|^2 - 2i\,\mathrm{Im}(P\overline{N})\gamma + |P|^2\gamma^2$ for the third and $-|N|^2 + 2i\,\mathrm{Im}(P\overline{N})\gamma + |P|^2\gamma^2$ for the fourth line. Thus

$$\begin{aligned}
\det \mathcal{A}_{A_1^{\rho,\omega}(1,\mu)} &= |a^2 - b^2|^2 + |c^2 - d^2|^2 - 2|N|^2 + 2|P|^2\gamma^2 + 2|P|\frac{\cos(2(\omega-\pi)\theta)}{\sin^2(\pi\theta)} \\
&= |a^2 - b^2|^2 + |c^2 - d^2|^2 - 2|N|^2 + 2|P|^2\left(1 - 2\left(\frac{\sin((\omega-\pi)\theta)}{\sin(\pi\theta)}\right)^2\right) \\
&= U + V\Phi^2(\mu)
\end{aligned}$$

with

$$V = -4|P|^2, \quad U = |a^2 - b^2|^2 + |c^2 - d^2|^2 + 2(|P|^2 - |N|^2) \qquad (4.54)$$

as desired. □

Corollary 4.3.7. *If $cb - ad = 0$ and if the operator $A_1^{\rho,\omega}$ is Fredholm, then its index is zero.*

The condition $cb - ad = 0$ is, for example, satisfied for singular integral equations without conjugation, where $c = 0$, $d = 0$. The corresponding assertion for the index of $A_1^{\rho,\omega}$ is well known (cf. [180]). Lemma 4.3.6 provides another proof for this result.

Corollary 4.3.8. *If $\omega = \pi$ and if the operator $A_1^{\rho,\pi}$ is Fredholm, then its index is zero.*

It is obvious from Lemma 4.3.6 that the winding number of $\det \mathcal{A}_{A_1^{\rho,\omega}}(1,\mu)$ and thus the index of the operator $A_1^{\rho,\omega}$, depend essentially on the properties of the function

$$\Phi(\mu) = \frac{\sin((\omega-\pi)\theta)}{\sin(\pi\theta)}, \qquad \theta = \frac{1}{2} - \rho + \frac{i}{2\pi}\log\frac{\mu}{1-\mu}.$$

In connection with the index evaluation, the location of the zeros of the real and imaginary parts of the function Φ in the open interval $(0,\,1)$ is of a special interest. For the imaginary part the corresponding results are summarized in Lemma 4.3.12, whereas Lemma 4.3.18 contains results concerning the zeros of the real part.

Since $\theta = \alpha + i\beta$ with $\alpha = 1/2 - \rho$ and $\beta = (1/2\pi)\log(\mu/(1-\mu))$, one has

$$\begin{aligned}
\Phi(\mu) &= \frac{\sin((\omega-\pi)\theta)\overline{\sin(\pi\theta)}}{\sin(\pi\theta)\overline{\sin(\pi\theta)}} = \frac{\sin((\omega-\pi)\theta)\sin(\pi\bar{\theta})}{|\sin(\pi\theta)|^2} \\
&= \frac{\cos((\omega-2\pi\alpha)+i\omega\beta) - \cos(\omega\alpha + i(\omega-2\pi)\beta)}{2|\sin(\pi\theta)|^2} \\
&= \frac{\cos((\omega-2\pi)\alpha)\cosh(\omega\beta) - \cos(\omega\alpha)\cosh((\omega-2\pi)\beta)}{2|\sin(\pi\theta)|^2} \\
&\quad - i\,\frac{\sin((\omega-2\pi)\beta)\sinh(\omega\beta) - \sin(\omega\alpha)\sinh((\omega-2\pi)\beta)}{2|\sin(\pi\theta)|^2};
\end{aligned}$$

so the real and imaginary parts of the function Φ have the form

$$\operatorname{Re}\Phi(\mu) = \frac{\cos((\omega-2\pi)\alpha)\cosh(\omega\beta) - \cos(\omega\alpha)\cosh((\omega-2\pi)\beta)}{2|\sin(\pi\theta)|^2}, \tag{4.55}$$

$$\operatorname{Im}\Phi(\mu) = \frac{\sin(\omega\alpha)\sinh((\omega-2\pi)\beta) - \sin((\omega-2\pi)\beta)\sinh(\omega\beta)}{2|\sin(\pi\theta)|^2}. \tag{4.56}$$

The common denominator of (4.55) and (4.56) can be written as

$$
\begin{aligned}
|\sin(\pi\theta)|^2 = \sin(\pi\theta)\sin(\pi\overline{\theta}) &= \frac{\cos(\pi(\theta-\overline{\theta})) - \cos(\pi(\theta+\overline{\theta}))}{2} \\
&= \frac{\cosh(2\pi\beta) - \cos(2\pi\sigma)}{2} = \frac{1}{4}\left[\left(\frac{\mu}{1-\mu} + \frac{1-\mu}{\mu}\right) + 2\cos(2\pi\rho)\right) \\
&= \frac{(2\mu-1)^2 + 4\mu(1-\mu)\cos^2(\pi\rho)}{4\mu(1-\mu)},
\end{aligned}
$$

whence via straightforward calculation follows

$$
\begin{aligned}
\operatorname{Re}\Phi(\mu) = \quad & \frac{\mu(1-\mu)}{R(\mu)}\left[\left(\left(\frac{\mu}{1-\mu}\right)^{\delta} + \left(\frac{1-\mu}{\mu}\right)^{\delta}\right)\cos(2\pi(\delta-1)\alpha)\right. \\
& \left. - \left(\left(\frac{\mu}{1-\mu}\right)^{\delta-1} + \left(\frac{1-\mu}{\mu}\right)^{\delta-1}\right)\cos(2\pi\delta\alpha)\right), \qquad (4.57)
\end{aligned}
$$

$$
\begin{aligned}
\operatorname{Im}\Phi(\mu) = \quad & \frac{\mu(1-\mu)}{R(\mu)}\left[\left(\left(\frac{\mu}{1-\mu}\right)^{\delta-1} - \left(\frac{1-\mu}{\mu}\right)^{\delta-1}\right)\sin(2\pi\delta\alpha)\right. \\
& \left. - \left(\left(\frac{\mu}{1-\mu}\right)^{\delta} - \left(\frac{1-\mu}{\mu}\right)^{\delta}\right)\sin(2\pi(\delta-1)\alpha)\right) \qquad (4.58)
\end{aligned}
$$

with

$$R(\mu) := (2\mu-1)^2 + 4\mu(1-\mu)\cos^2(\pi\rho), \quad -1/2 < \rho < 1/2,$$

and $\delta := \omega/(2\pi)$, $\delta \in (0,1)$.

Let us start with study of the imaginary part of Φ. For brevity, we introduce a new variable x by $x = (1-\mu)/\mu$ and set $y = 2\pi\alpha$, i.e., x runs through $(0,+\infty)$, and y is in the interval $(0, 2\pi)$. Because the function $\mu \mapsto \mu(1-\mu)/R(\mu)$ has no zeros in the interval $(0,1)$, the zeros of the imaginary part of the function Φ are completely determined by those of the function

$$\psi_\delta(x) := (x^{1-\delta} - x^{\delta-1})\sin(\delta y) + (x^{-\delta} - x^{\delta})\sin((1-\delta)y).$$

The following property of the function ψ_δ is obvious.

Lemma 4.3.9. *If δ is in $(1/2, 1)$ and $y = \pi/\delta$, then the function ψ_δ has no zeros in $(1, \infty)$.*

Thus considering the function ψ_δ in case $\delta > 1/2$, we can restrict ourselves to the case when $y \neq \pi/\delta$.

Write the function ψ_δ as

$$\psi_\delta(x) = \sin(\delta y)(x^{-\delta} - x^{\delta})(\phi_\delta(x) - A_\delta(y))$$

with

$$A_\delta(y) := \frac{\sin((1-\delta)y)}{\sin(\delta y)} \quad \text{and} \quad \phi_\delta(x) := \frac{x^{1-\delta} - x^{\delta-1}}{x^{-\delta} - x^{\delta}}. \tag{4.59}$$

Further, given a function f, denote by $I_f(a, b)$ the image of the interval (a, b) under the mapping f.

Lemma 4.3.10. *If ϕ_δ is the function defined in (4.59), then*

$$I_{\phi_\delta}(1, \infty) = \begin{cases} ((1-\delta)/\delta, \infty) & \text{if } \delta \in (0, 1/2), \\ (0, (1-\delta)/\delta) & \text{if } \delta \in (1/2, 1). \end{cases}$$

Proof. The function ϕ_δ is continuous on $(1, \infty)$ with one-sided limits

$$\lim_{x\to 1+0} \phi_\delta(x) = (1-\delta)/\delta, \qquad \lim_{x\to +\infty} \phi_\delta(x) = \begin{cases} +\infty & \text{if} \quad \delta \in (0, 1/2), \\ 0 & \text{if} \quad \delta \in (1/2, 1), \end{cases}$$

and the elementary inequality

$$\frac{x^\alpha - x^{-\alpha}}{\alpha} < \frac{x^\beta - x^{-\beta}}{\beta}, \tag{4.60}$$

which is valid for $x > 1$ and $0 < \alpha < \beta < 1$, yields

$$\phi_\delta(x) < \frac{1-\delta}{\delta} \quad \text{if } x \in (1, \infty) \text{ and } \delta \in (0, 1/2),$$

and

$$\frac{1-\delta}{\delta} < \phi_\delta(x) \quad \text{if } x \in (1, \infty) \text{ and } \delta \in (1/2, 1).$$

(Put $\alpha = \delta$, $\beta = 1-\delta$ and $\alpha = 1-\delta$, $\beta = \delta$ in (4.60), respectively.) □

A similar discussion of the function A_δ leads to the following result.

Lemma 4.3.11. *Let A_δ be defined by (4.59) and let $y \in (0, 2\pi)$.*

(a) *If $\delta \in (0, 1/2)$, then $A_\delta(y) < (1-\delta)/\delta$ for all $y \in (0, 2\pi)$.*

(b) *If $\delta \in (1/2, 1)$, then $A_\delta(y) > (1-\delta)/\delta$ for all $y \in (0, \pi/\delta)$, and $A_\delta(y) < 0$ for all $y \in (\pi/\delta, 2\pi)$.*

Proof. For $0 < \alpha < \beta < \pi$ one has

$$\frac{\sin\beta}{\beta} < \frac{\sin\alpha}{\alpha}. \tag{4.61}$$

If $\delta \in (0, 1/2)$ and we set $\alpha = \delta y$ and $\beta = (1-\delta)y$ in (4.61), then this inequality yields

$$A_\delta(y) = \frac{\sin((1-\delta)y)}{\sin(\delta y)} < \frac{1-\delta}{\delta},$$

for all $y \in (0, \pi/(1-\delta))$, which shows that assertion (a) is correct in this case. In case $y \in (\pi/(1-\delta), 2\pi)$, one has $\sin((1-\delta)y) \le 0$, which verifies (a) for such y too.

If we similarly put $\alpha = (1-\delta)y$ and $\beta = \delta y$ with $y \in (0, \pi/\delta)$ in (4.61), then we get the first estimate in (b), and the second one is evident. □

Lemma 4.3.12. *If $\delta \in (0, 1/2) \cup (1/2, 1)$ and $\rho \in (-1/2, 1/2)$, then the function $\mu \mapsto \operatorname{Im}\Phi(\mu)$ has no zero in the intervals $(0, 1/2)$ and $(1/2, 1)$.*

Proof. The assertion concerning the interval $(0, 1/2)$ follows from Lemmas 4.3.9, 4.3.10 and 4.3.11, and for $(1/2, 1)$ one can make use of the obvious identity

$$\operatorname{Im}\Phi(\mu) = -\operatorname{Im}\Phi(1-\mu), \quad \mu \in (0, 1/2).$$ □

Let us now turn to the real part of the function Φ given by (4.57). As before, we introduce a new variable $x := (1-\mu)/\mu$, set $y := 2\pi\alpha$, and consider the function

$$\hat{\psi}_\delta(x) = (x^\delta + x^{-\delta})\cos((\delta-1)y) - (x^{\delta-1} + x^{1-\delta})\cos(\delta y), \quad x \in (1, +\infty).$$

For $\delta \in (1/4, 1/2) \cup (1/2, 3/4)$, there is only one number $y \in (0, 2\pi)$ such that $\cos(\delta y)$ vanishes, whereas in case $\delta \in (3/4, 1)$ there are exactly two different zeros $y_1, y_2 \in (0, 2\pi)$ of the equation $\cos(\delta y) = 0$. In any case, if $\cos(\delta y) = 0$, then the sign of the function $\hat{\psi}_\delta$ is constant on the interval $(1, \infty)$. Thus we suppose that $y \in (0, 2\pi)$ and $\delta \in (0, 1/2) \cup (1/2, 1)$ are chosen such that $\cos(\delta y) \neq 0$, whence the function $\hat{\psi}_\delta$ can be written as

$$\hat{\psi}_\delta(x) = (x^\delta + x^{-\delta})\cos(\delta y)(\hat{A}_\delta - \hat{\phi}_\delta(x))$$

where

$$\hat{\phi}_\delta(x) = \frac{x^{1-\delta} + x^{\delta-1}}{x^\delta + x^{-\delta}} \quad \text{and} \quad \hat{A}_\delta = \hat{A}_\delta(y) = \frac{\cos((\delta-1)y)}{\cos(\delta y)}. \tag{4.62}$$

Lemma 4.3.13. *If $\hat{\phi}_\delta(x)$ is the function defined in (4.62), then*

$$I_{\hat{\phi}_\delta}(1, \infty) = \begin{cases} (1, \infty) & \text{if} \quad \delta \in (0, 1/2), \\ (0, 1) & \text{if} \quad \delta \in (1/2, 1). \end{cases}$$

The proof is similar to that of Lemma 4.3.10, although the inequality

$$x^\beta + x^{-\beta} < x^\alpha + x^{-\alpha} \quad \text{for all } 0 < \beta < \alpha < 1 \text{ and } x > 1$$

must be used instead of (4.60). □

Let us now examine the function $\hat{A}_\delta$ given by (4.62). The behaviour of the derivative

$$\hat{A}'_\delta(y) = \frac{\sin((2\delta-1)y) + (2\delta-1)\sin y}{2\cos^2(\delta y)} \tag{4.63}$$

of the function $\hat{A}_\delta$ depends essentially on the parameter δ, so the cases $\delta \in (0, 1/4]$, $\delta \in (1/4, 1/2)$, $\delta \in (1/2, 3/4)$ and $\delta \in [3/4, 1)$ are considered separately in the subsequent four lemmas.

Lemma 4.3.14. *If $\delta \in (0, 1/4]$, then $I_{\hat{A}_\delta}(0,\, 2\pi) \cap I_{\hat{\phi}_\delta}(1,\, \infty) = \emptyset$.*

Proof. First suppose $\delta \in (0, 1/4)$. If $\hat{\eta}_\delta$ denotes the numerator of (4.63), its derivative

$$\hat{\eta}_\delta'(y) = 2(2\delta - 1)\cos(\delta y)\cos((1-\delta)y) \tag{4.64}$$

has the two zeros

$$y_1 := \frac{\pi}{2(1-\delta)} \in \left(\frac{\pi}{2}, \frac{2\pi}{3}\right), \quad y_2 := \frac{3\pi}{2(1-\delta)} \in \left(\frac{3\pi}{2}, 2\pi\right),$$

in the interval $(0,\, 2\pi)$, and $\hat{\eta}_\delta'(y)$ is negative for $y \in (0,\, y_1) \cup (y_2,\, 2\pi)$ and positive for $y \in (y_1,\, y_2)$. In particular, $\hat{\eta}_\delta$ is a monotonically decreasing function on both of the intervals $(0,\, y_1)$ and $(y_2,\, 2\pi)$. Since $\hat{\eta}_\delta(+0) = 0$ and $\hat{\eta}_\delta(2\pi - 0) = \sin(4\pi\delta) > 0$, one has

$$\hat{A}_\delta'(y) < 0 \text{ for all } y \in (0, y_1) \quad \text{and} \quad \hat{A}_\delta'(y) > 0 \text{ for all } y \in (y_2, 2\pi).$$

These inequalities imply that the function $\hat{A}_\delta$ is monotonically decreasing on $(0,\, y_1)$ and monotonically increasing on $(y_2,\, 2\pi)$. Since $\hat{A}_\delta(+0) = \hat{A}_\delta(2\pi - 0) = 1$, we henceforth obtain $\hat{A}_\delta(y) < 1$ for all $y \in (0,\, y_1) \cup (y_2, 2\pi)$. Moreover, it is easily seen that $\hat{A}_\delta(y) \le 0$ for all $y \in [y_1,\, y_2]$, so that $I_{\hat{A}_\delta}(0, 2\pi) \subseteq (-\infty, 1)$, which together with Lemma 4.3.13 proves the claim for $\delta \in (0,\, 1/4)$. Finally, if $\delta = 1/4$, then $y_2 = 2\pi$, and $\hat{A}_\delta(y)$ is negative everywhere on the interval $(y_1, 2\pi)$. □

Lemma 4.3.15. *Let $\delta \in (1/4,\, 1/2)$. Set $y_1 := \pi/(2\delta) \in (\pi,\, 2\pi)$ and $y_2 := \pi/(2(1-\delta)) \in (2\pi/3,\, \pi)$. Then*

(a) $I_{\hat{A}_\delta}(0,\, y_1) \cap I_{\hat{\phi}_\delta}(1,\, \infty) = \emptyset$.

(b) $I_{\hat{A}_\delta}(y_1,\, 2\pi) \subseteq I_{\hat{\phi}_\delta}(1,\, \infty)$. *Moreover, for every $y \in (y_1,\, 2\pi)$, there is only one $x \in (1,\, \infty)$ such that $\hat{\phi}_\delta(x) = \hat{A}_\delta(y)$.*

Proof. The points y_1 and y_2 are zeros of the derivative $\hat{\eta}_\delta'$ in (4.64), and $\hat{\eta}_\delta'(y) < 0$ for all $y \in (0, y_2) \cup (y_1, 2\pi)$ but $\hat{\eta}_\delta'(y) > 0$ if $y \in (y_2, y_1)$, hence the function $\hat{\eta}_\delta$ is monotonically decreasing on $(0,\, y_2) \cup (y_1,\, 2\pi)$ and monotonically increasing on $(y_2,\, y_1)$. Since $\hat{\eta}_\delta(+0) = 0$ and $\hat{\eta}_\delta(y_1) = 2\delta\sin(\pi/2\delta) < 0$, the derivative $\hat{A}_\delta'$ is negative on $(0,\, y_1) \cup (y_1,\, 2\pi)$, so the function $\hat{A}_\delta$ is monotonically decreasing on the intervals $(0,\, y_1)$ and $(y_1,\, 2\pi)$. Moreover, $\hat{A}_\delta(+0) = 1$. Combining these observations with Lemma 4.3.13, we obtain assertion (a).

Now consider the function $\hat{A}_\delta$ on the interval $(y_1,\, 2\pi)$. It is continuous there, and has the one-sided limits $\hat{A}_\delta(y_1 + 0) = \infty$ and $\hat{A}_\delta(2\pi - 0) = 1$. Again using Lemma 4.3.13, we find that $\hat{A}_\delta(y) \in I_{\hat{\phi}_\delta}(1,\, \infty)$ for every $y \in (y_1,\, 2\pi)$, and the uniqueness of an element x such that $\hat{\phi}_\delta(x) = \hat{A}_\delta(y)$ follows from the strict monotonicity of the function $\hat{\phi}_\delta$. □

Analogous discussions lead to the following two lemmas.

Lemma 4.3.16. *Let $\delta \in (1/2, 3/4)$, and set $y_1 := \pi/(2\delta) \in (2\pi/3, \pi)$ and $y_2 := \pi/(2(1-\delta)) \in (\pi, 2\pi)$. Then*

(a) $I_{\hat{A}_\delta}((0, y_1) \cup (y_1, y_2)) \cap I_{\hat{\phi}_\delta}(1, \infty) = \emptyset$.

(b) $I_{\hat{A}_\delta}(y_2, 2\pi) \subseteq I_{\hat{\phi}_\delta}(1, \infty)$. *Moreover, for every $y \in (y_2, 2\pi)$, there is only one $x \in (1, \infty)$ such that $\hat{\phi}_\delta(x) = \hat{A}_\delta(y)$.*

Lemma 4.3.17. *Let $\delta \in (3/4, 1)$ and set $y_1 := \pi/(2\delta)$ and $y_2 := 3\pi/(2\delta)$. Then*

$$I_{\hat{A}_\delta}((0, y_1) \cup (y_1, y_2) \cup (y_2, 2\pi)) \cap I_{\hat{\phi}_\delta}(1, \infty) = \emptyset.$$

Summarizing the preceding four lemmas and taking into account the identity

$$\operatorname{Re}\Phi(1-\mu) = \operatorname{Re}\Phi(\mu), \quad \text{for } \mu \in (0, 1/2),$$

we arrive at the following characterization of the zeros of the real part of the function Φ.

Lemma 4.3.18. *Let Φ be the function defined in Lemma 4.3.6.*

(a) *If $\delta \in (0, \frac{1}{4}] \cup [\frac{3}{4}, 1)$ and $\rho \in (-\frac{1}{2}, \frac{1}{2})$, then the real part $\operatorname{Re}\Phi$ of the function Φ has no zeros in $(0, \frac{1}{2}) \cup (\frac{1}{2}, 1)$.*

(b) *If $\delta \in (\frac{1}{4}, \frac{1}{2})$ and $\rho \in [\frac{1}{2} - \frac{1}{4\delta}, \frac{1}{2})$, then $\operatorname{Re}\Phi$ has no zeros on $(0, \frac{1}{2}) \cup (\frac{1}{2}, 1)$; whereas in case $\rho \in (-\frac{1}{2}, \frac{1}{2} - \frac{1}{4\delta})$ the function $\operatorname{Re}\Phi$ has exactly one zero in $(0, \frac{1}{2})$ and one in $(\frac{1}{2}, 1)$.*

(c) *If $\delta \in (\frac{1}{2}, \frac{3}{4})$ and $\rho \in [\frac{1}{2} - \frac{1}{4(1-\delta)}, \frac{1}{2})$, then $\operatorname{Re}\Phi$ has no zeros on $(0, \frac{1}{2}) \cup (\frac{1}{2}, 1)$; whereas in case $\rho \in (-\frac{1}{2}, \frac{1}{2} - \frac{1}{4(1-\delta)})$, the function $\operatorname{Re}\Phi$ has exactly one zero in $(0, \frac{1}{2})$ and one in $(\frac{1}{2}, 1)$.*

Corollary 4.3.19. *If $\delta \in (0, 1/2) \cup (1/2, 1)$ and $\rho \geq 0$, then the function $\operatorname{Re}\Phi$ has no zeros on $(0, 1/2) \cup (1/2, 1)$.*

Here is the main result of this section.

Theorem 4.3.20. *Let $\rho \in (-1/2, 1/2)$, $\omega \in (0, 2\pi)$, and let the operator $A_1^{\rho,\omega}$ be Fredholm. Then the index $\kappa := \kappa(A_1^{\rho,\omega})$ of $A_1^{\rho,\omega}$ satisfies the estimates summarized in the following table.*

Angle	*Weight*	*Index*
$0 < \omega \le \frac{\pi}{2}$	$-\frac{1}{2} < \rho < \frac{1}{2}$	$\lvert\kappa\rvert \le 1$
$\frac{\pi}{2} < \omega < \pi$	$-\frac{1}{2} < \rho < \frac{1}{2} - \frac{\pi}{2\omega}$	$\lvert\kappa\rvert \le 2$
$\frac{\pi}{2} < \omega < \pi$	$\frac{1}{2} - \frac{\pi}{2\omega} \le \rho < \frac{1}{2}$	$\lvert\kappa\rvert \le 1$
$\omega = \pi$	$-\frac{1}{2} < \rho < \frac{1}{2}$	$\lvert\kappa\rvert = 0$
$\pi < \omega < \frac{3\pi}{2}$	$-\frac{1}{2} < \rho < \frac{1}{2} - \frac{\pi}{2(2\pi-\omega)}$	$\lvert\kappa\rvert \le 2$
$\pi < \omega < \frac{3\pi}{2}$	$\frac{1}{2} - \frac{\pi}{2(2\pi-\omega)} \le \rho < \frac{1}{2}$	$\lvert\kappa\rvert \le 1$
$\frac{3\pi}{2} \le \omega < 2\pi$	$-\frac{1}{2} < \rho < \frac{1}{2}$	$\lvert\kappa\rvert \le 1$

The table has to be read as follows: If the angle ω and the weight ρ belong to the mentioned intervals, then the index κ is subject to the corresponding estimate in the last column.

Proof. Lemma 4.3.6 implies that the imaginary part of the determinant of the symbol $\mathcal{A}_{A_1^{\rho,\omega}}(1,\mu)$ of the operator $A_1^{\rho,\omega}$ has the form

$$\operatorname{Im}\det \mathcal{A}_{A_1^{\rho,\omega}}(1,\mu) = -8|N|^2 \operatorname{Im}\Phi(\mu)\operatorname{Re}\Phi(\mu).$$

From Lemmas 4.3.12 and 4.3.18, the function $\operatorname{Im}\Phi \cdot \operatorname{Re}\Phi$ can possess either three ($\mu = 0$, $\mu = 1/2$, $\mu = 1$) or five zeros on $[0, 1]$. Since $\det \mathcal{A}_{A_1^{\rho,\omega}}(t,\mu)$ is real for $t \ne 1$ and $\mu \in [0,1]$ (cf. Lemma 4.3.4), the curve

$$\{\det \mathcal{A}_{A_1^{\rho,\omega}}(1,\mu),\ 0 \le \mu \le 1\} \cup \{\det \mathcal{A}_{A_1^{\rho,\omega}}(t,0),\ t \ne 1\}$$

can perform at most one rotation around the origin when there are three zeros, or two when there are five zeros. □

Corollary 4.3.21. *Let $\omega \in (0, 2\pi)$, $\rho \in [0,1/2)$, and let the operator $A_1^{\rho,\omega}$ be Fredholm. Then $|\kappa(A_1^{\rho,\omega})| \le 1$.*

Consider an example where the index of the corresponding local operator can take value 2 or −2. Thus Figure 4.1 shows the image of the interval $[0,1]$ under the function Φ^2 in the case $\omega = 5\pi/4$, $\rho = -0.49$.

At first glance, Figure 4.1(a) does not give any hint that there might be a second coil of the curve. However, after "zooming" this picture one observes that in the neighbourhoods of the origin, the curve actually has two coils (cf., Figure 4.1(b)).

It is interesting to note that if the graph of the function Φ^2 consists of two coils, one of each is substantially smaller than other, so in most cases it cannot be spotted without zooming. One of the rare cases where a graph has a visible second coil is presented in Figure 4.2 on page 196.

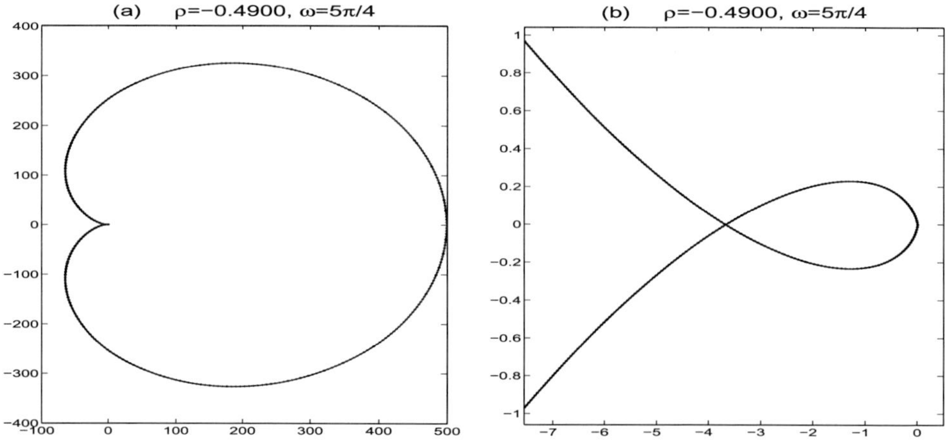

Figure 4.1: Graph of the function Φ^2 for $\rho = -0.49$, $\omega = 5\pi/4$.

4.3.3 The Vanishing of the Index of Local Operators

Vanishing of the index $\kappa(A_1^{\rho,\omega})$ of the operator $A_1^{\rho,\omega}$ is a necessary condition for applicability of a modified quadrature methods considered in Section 4.3.4. When

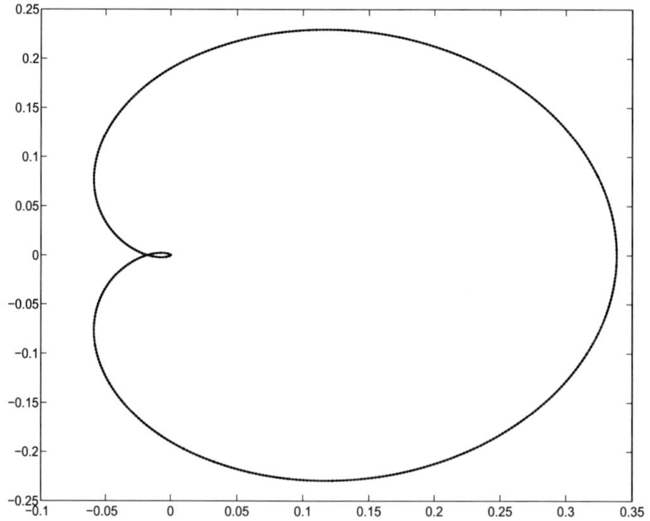

Figure 4.2: Graph of the function Φ^2 for $\rho = -0.35$, $\omega = 1.1\pi$.

the estimate $|\kappa(A_1^{\rho,\omega})| \leq 1$ can be guaranteed, it is possible to force the index $\kappa(A_1^{\rho,\omega})$ to vanish by an appropriately chosen weight function ρ. Recall that U and V are given by (4.54).

Theorem 4.3.22. *Let $\omega \in (0,\pi)\cup(\pi,2\pi)$, assume the operator $A_1^{0,\omega}$ to be Fredholm, and let one of the following conditions be satisfied:*

(i) $\omega \in (0,\,\pi] \cup [3\pi/2,\,2\pi)$ *and* $\rho \in (-1/2, 1/2)$,

(ii) $\omega \in (\pi/2,\,\pi)$ *and* $\rho \in [-1/2 - \pi/(2\omega),\,1/2)$,

(iii) $\omega \in (\pi,\,3/2\pi)$ *and* $\rho \in [-1/2 - \pi/(2(2\pi - \omega)),\,1/2)$.

Then the following assertions are true:

(a) *The operator $A_1^{\rho,\omega}$ is Fredholm if and only if*

$$t_2 = t_2(\rho) := t_1 - 4|cb - ad|^2 \frac{\sin^2((\omega-\pi)(1/2-\rho))}{\cos^2(\pi\rho)} \neq 0, \tag{4.65}$$

with $t_1 := U = |a^2 - b^2|^2 + |c^2 - d^2|^2 + 2(|cb - ad|^2 - |ac - bd|^2)$.

(b) *If the operator $A_1^{\rho,\omega}$ is Fredholm, then its index vanishes if and only if* $t_1 \cdot t_2 > 0$.

Proof. If $A_1^{0,\omega}$ is a Fredholm operator, then $t \mapsto \det \mathcal{A}_{A_1^{0,\omega}}(t,0)$, $t \neq 1$ is a real-valued function without a zero. Since

$$\det \mathcal{A}_{A_1^{\rho,\omega}}(t,0)|_{t\neq 1} = \det \mathcal{A}_{A_1^{0,\omega}}(t,0)|_{t\neq 1},$$

the operator $A_1^{\rho,\omega}$ is Fredholm if and only if the origin does not lie on the arc

$$\widetilde{\Gamma} := \{\det \mathcal{A}_{A_1^{\rho,\omega}}(1,\,\mu),\ \mu \in [0,\,1]\} = \{U + V\Phi^2(\mu),\ \mu \in [0,\,1]\}.$$

Each of the above conditions (i) – (iii) implies that the real axis $\mathbb{R}$ and the arc $\widetilde{\Gamma}$ have only two common points that correspond to the parameters $\mu = 0$ and $\mu = 1/2$. Let us denote these points by $\tilde{t}_1$ and $\tilde{t}_2$. Since

$$\Phi^2(\mu) = (\operatorname{Re}\Phi(\mu))^2 - (\operatorname{Im}\Phi(\mu))^2 + 2\mathrm{i}\operatorname{Re}\Phi(\mu)\operatorname{Im}\Phi(\mu)$$

with $\operatorname{Re}\Phi$ and $\operatorname{Im}\Phi$ given by (4.57) and (4.58) respectively, and since

$$\operatorname{Re}\Phi(0) = \operatorname{Im}\Phi(0) = \operatorname{Im}\Phi(1/2) = 0,$$

$$\operatorname{Re}\Phi(1/2) = \frac{\sin((\omega-\pi)(1/2-\rho))}{\cos(\pi\rho)},$$

we conclude that $\tilde{t}_1 = t_1$ and $\tilde{t}_2 = t_2$. However, t_1 cannot be zero since $A_1^{0,\omega}$ is a Fredholm operator. This finishes the proof of assertion (a), and that of (b) is now obvious. ☐

Thus, under each of the hypotheses (i) – (iii) of Theorem 4.3.22, the operator $A_1^{\rho,\omega}$ is Fredholm; and if $t_1 \cdot t_2 < 0$, then necessarily $\kappa(A_1^{\rho,\omega}) \neq 0$. Since the location of the point t_2 on the real line depends on the weight ρ, one might try to find a new weight $\tilde{\rho}$ such that $\kappa(A_1^{\tilde{\rho},\omega}) = 0$. The following corollary establishes conditions when such a choice is possible.

Corollary 4.3.23. *Let one of conditions* (i) – (iii) *of the previous theorem be satisfied, and suppose that* $t_1 \cdot t_2 < 0$. *Then there is a weight* $\rho \in (-1/2, 1/2)$ *such that* $\kappa(A_1^{\rho,\omega}) = 0$ *if and only if one of the following conditions is fulfilled:*

(a) $\omega \in (0, \pi/2] \cup [3\pi/2, 2\pi)$ *and*

$$0 \in (-\infty, U - 4|cb - ad|^2(1 - \pi/\omega)^2).$$

(b) $\omega \in (\pi/2, \pi)$ *and*

$$0 \in (U - 4|cb - ad|^2 \cot^2(\pi^2/2\omega), U - 4|cb - ad|^2(1 - \pi/\omega)^2).$$

(c) $\omega \in (\pi, 3\pi/2)$ *and*

$$0 \in (U - 4|cb - ad|^2 \cot^2(\pi^2/2(2\pi - \omega)), U - 4|cb - ad|^2(1 - \pi/\omega)^2).$$

In particular, if $U > 0$ but $U - 4|cb - ad|^2(1 - \pi/\omega)^2 < 0$ then, for every $\rho \in (-1/2, 1/2)$ that satisfies the conditions of Theorem 4.3.22, the index of $\kappa(A_1^{\rho,\omega})$ is not equal to zero. Conversely, if $U > 0$ and $U - 4|cb - ad|^2(1 - \pi/\omega)^2 > 0$, then there always exists a ρ such that all conditions of Theorem 4.3.22 are satisfied and that $\kappa(A_1^{\rho,\omega}) = 0$.

Proof of Corollary 4.3.23. To prove the claim one has to describe the range of the function

$$\rho \mapsto (\operatorname{Re}\Phi(1/2)(\rho))^2 = \left(\frac{\sin((\omega - \pi)(1/2 - \rho))}{\cos(\pi\rho)}\right)^2;$$

or equivalently, after substituting $y := \pi(1/2 - \rho) \in (0, \pi)$ and $\delta := \omega/2\pi$, that of the function

$$\psi_\delta^2(y) := \left(\frac{\sin((2\delta - 1)y)^2)}{\sin y}\right)^2.$$

The one-sided limits of ψ_δ at 0 and π are

$$\lim_{y\to+0} \psi_\delta(y) = 2\delta - 1 \quad \text{and} \quad \lim_{y\to\pi-0} \psi_\delta(y) = \begin{cases} -\infty & \text{if} \quad 2\delta - 1 < 0, \\ +\infty & \text{if} \quad 2\delta - 1 > 0, \end{cases} \tag{4.66}$$

respectively, the derivative of $\psi_\delta(y)$ is

$$\psi_\delta'(y) = \frac{(\delta - 1)\sin(2\delta y) + 2\delta(\sin(1 - \delta)y)}{\sin^2 y},$$

and the derivative of the numerator $\eta_\delta(y)$ of that fraction is given by

$$\eta'_\delta(y) = 4\delta(1-\delta)\sin y \sin((2\delta-1)y).$$

Other properties of the function ψ_δ depend on the location of the point δ in the interval $(0, 1)$.

If $\delta \in (0, 1/4)$, then $\eta'_\delta(y) < 0$ for all $y \in (0, \pi)$ and hence η_δ is monotonically decreasing on $[0, \pi]$. Since $\eta_\delta(0) = 0$, the function η_δ is negative on $(0, \pi)$, which implies that $\psi'_\delta(y) < 0$ for all $y \in (0, \pi)$. Thus ψ_δ is monotonically decreasing on this interval, which together with (4.66) shows that $I_{\psi_\delta}(0,\pi) = (-\infty, 2\delta-1)$ and

$$I_{\psi_\delta}(0,\pi) = ((2\delta-1)^2, \infty) \quad \text{if } \delta \in (0, 1/4). \tag{4.67}$$

If $\delta = 1/4$, then $\psi_{1/4}(y) = -1/(2\cos(y/2))$, which immediately gives

$$I_{\psi^2_{1/4}}(0,\pi) = (1/4, \infty). \tag{4.68}$$

When $\delta \in (1/4, 1/2)$, we again have $\eta'_\delta(y) < 0$ for all $y \in (0,\pi)$. However, we now have to restrict ourselves to those y for which the curve $\widetilde{\Gamma}$ has at most two common points with the real line, since otherwise, we have no information about the location of the zeros of the symbol. Thus according to Theorem 4.3.22, we get $y \in (0, \pi/(4\delta))$, and therefore $I_{\psi_\delta}(0, \pi/(4\delta))$ is the interval $(\cot\frac{\pi}{4\delta}, 2\delta-1)$, whence

$$I_{\psi^2_\delta}(0, \pi/(4\delta)) = ((2\delta-1)^2, \cot^2\frac{\pi}{4\delta}) \quad \text{if } \delta \in (1/4, 1/2). \tag{4.69}$$

Similarly we obtain

$$I_{\psi^2_\delta}(0, \pi/(4(1-\delta))) = ((2\delta-1)^2, \cot^2\frac{\pi}{4(1-\delta)}) \quad \text{if } \delta \in (1/2, 3/4), \tag{4.70}$$

$$I_{\psi^2_{3/4}}(0, \pi) = (1/4, \infty), \tag{4.71}$$

$$I_{\psi^2_\delta}(0, \pi) = ((2\delta-1)^2, \infty) \quad \text{if } \delta \in (3/4, 1). \tag{4.72}$$

A comparison of (4.67) – (4.72) with (4.65) indicates that the conditions of Corollary 4.3.23 allow us to choose the weight parameter ρ in such a way that the product $t_1 \cdot t_2(\rho)$ becomes positive, whence the index of $A_1^{\rho,\omega}$ becomes zero. □

4.3.4 Modification of the Quadrature Method by Cutting Off Corner Singularities

When the operator A is invertible and the local operators A_1^τ defined by (4.6) are not invertible but still Fredholm with index 0, the quadrature rule (4.4) can be modified in such a way that the stability conditions become essentially weaker. The modification under consideration, proposed by I.G. Graham and G.A. Chandler [98], is to cut off the quadrature rule in the neighbourhood of the corner points.

Thus choosing a non-negative integer i_0, one next defines a set of subscripts I_{n,i_0} as the set of all k in $\{0, 1, \ldots, n-1\}$ such that

$$|k/n - j/n_0| \geq i_0/n \quad \text{for every } j = 1, 2, \ldots, n_0 - 1,$$

and replace (4.4) by the system

$$\begin{aligned}
&[a(\tau_k^{(n)}) - b(\tau_k^{(n)}) i \cot(\pi(\varepsilon - \delta))]\xi_k^{(n)} + \frac{b(\tau_k^{(n)})}{\pi i} \sum_{j \in I_{n,i_0}} \frac{\Delta t_j^{(n)}}{t_j^{(n)} - \tau_k^{(n)}} \xi_j^{(n)} \\
&\quad + [c(\tau_k^{(n)}) - d(\tau_k^{(n)}) i \cot(\pi(\varepsilon - \delta))]\overline{\xi_k^{(n)}} + \frac{d(\tau_k^{(n)})}{\pi i} \sum_{j \in I_{n,i_0}} \frac{\Delta t_j^{(n)}}{t_j^{(n)} - \tau_k^{(n)}} \overline{\xi_j^{(n)}} \\
&\quad + \sum_{j \in I_{n,i_0}} k_1(t_k^{(n)}, t_j^{(n)}) \Delta t_j^{(n)} \xi_j^{(n)} + \sum_{j \in I_{n,i_0}} k_2(t_k^{(n)}, t_j^{(n)}) \Delta t_j^{(n)} \overline{\xi_j^{(n)}} \\
&\quad = f(\tau_k^{(n)}), \qquad k \in I_{n,i_0}. \qquad (4.73)
\end{aligned}$$

Observe that the choice $i_0 = 0$ yields the original system (4.4).

The modified method can be analyzed in a similar way to the original (cf. Sections 4.2.4 – 4.2.6 in [102] for details), and results are summarized in Theorem 4.3.24 below. Note that $r_\tau = (\alpha - 1)/2 - \rho_\tau \alpha$ again, and $P_i, i \geq 1$ stand for the projection operators

$$P_i : l_{2,r_\tau} \to l_{2,r_\tau}, \quad (x_0, x_1, \ldots) \mapsto (x_0, \ldots, x_{i-1}, 0, 0, \ldots),$$

and $Q_i := I - P_i$. The same notation is used for the diagonal operators diag (P_i, P_i, P_i, P_i) and diag (Q_i, Q_i, Q_i, Q_i), respectively.

Theorem 4.3.24. *The modified quadrature method* (4.73) *is stable if and only if:*

(a) *the operator A is invertible,*

(b) *the local operators A_1^τ are invertible for all $\tau \in \Gamma \setminus \{\gamma(u_1), \ldots, \gamma(u_{n_0})\}$, and*

(c) *the sequences $(P_i + Q_i A_1^\tau Q_i)_{i\geq 1}$ are stable for all $\tau \in \{\gamma(u_1), \ldots, \gamma(u_{n_0})\}$.*

If the sequence $(P_i + Q_i A_1^\tau Q_i)_{i\geq 1}$ is stable, then the operators $P_i + Q_i A_1^\tau Q_i$ are invertible for sufficiently large i, so A_1^τ is necessarily a Fredholm operator with index 0 (due to the compactness of the P_i). On the other hand, the invertibility of A_1^τ is not necessary for stability of the sequence, since the strong limit of the operators $P_i + Q_i A_1^\tau Q_i$ as $i \to \infty$ is the identity operator (which obviously is invertible). In that sense, $(P_i + Q_i A_1^\tau Q_i)_{i\geq 1}$ is an approximation method for the identity operator.

The operator A_1^τ has the form of a *Toeplitz operator + discretized Mellin convolution*; thus Theorem 4.4 in [102] applies and yields the following stability result for the sequence $(P_i + Q_i A_1^\tau Q_i)_{i\geq 1}$ (cf. also Example 4.7 in [102]). We again make use of the abbreviations $a := a(\tau)$, $b := b(\tau)$, $c := c(\tau)$, $d := d(\tau)$.

Theorem 4.3.25. *The sequence* $(P_i + Q_i A_1^\tau Q_i)_{i\geq 1}$ *is stable if and only if the following conditions are satisfied:*

(a) *the operator* A_1^τ *is Fredholm on the space* $l_{2,r_\tau} \times l_{2,r_\tau} \times l_{2,r_\tau} \times l_{2,r_\tau}$,

(b) *the* 4×4 *block Toeplitz operator* T *with generating function*

$$\begin{pmatrix} a+bf^{(\varepsilon-\delta)} & 0 & c+df^{(\varepsilon-\delta)} & 0 \\ 0 & a-bf^{(\delta-\varepsilon)} & 0 & c-df^{(\delta-\varepsilon)} \\ \bar{c}+\bar{d}(f^{(\varepsilon-\delta)})^* & 0 & \bar{a}+\bar{b}(f^{(\varepsilon-\delta)})^* & 0 \\ 0 & \bar{c}-\bar{d}(f^{(\delta-\varepsilon)})^* & 0 & \bar{a}-\bar{b}(f^{(\delta-\varepsilon)})^* \end{pmatrix}$$

is invertible on $l_{2,0} \times l_{2,0} \times l_{2,0} \times l_{2,0}$, *and*

(c) *the singular integral operator*

$$\chi_{\Gamma_\tau \setminus \Gamma_{\tau,1}}(aI + bS + (cI + dS)M)\chi_{\Gamma_\tau \setminus \Gamma_{\tau,1}} I$$

with $\Gamma_\tau := e^{i\omega_\tau}\mathbb{R} \cup \mathbb{R}$ *and* $\Gamma_{\tau,1} := e^{i\omega_\tau}[0,\,1) \cup [0,\,1)$, *is invertible on* $L_2(\Gamma_\tau \setminus \Gamma_{\tau,1},\, r_\tau) \subseteq L_2(\Gamma_\tau,\, r_\tau)$.

It may seem difficult to analyze condition (c) further, but it is actually equivalent to the invertibility of the Wiener-Hopf operator acting on the space $L_2([0,\infty),\, r_\tau) \oplus \ldots \oplus L_2([0,\infty),\, r_\tau$ with generating function

$$\begin{pmatrix} a+bs & -bn_{2\pi-\beta} & c+ds & -dn_{2\pi-\beta} \\ bn_\beta & a-bs & dn_\beta & c-ds \\ \bar{c}-\bar{d}s & \bar{d}n_\beta & \bar{a}-\bar{b}s & \bar{b}n_\beta \\ -\bar{d}n_{2\pi-\beta} & \bar{c}+\bar{d}s & -\bar{b}n_{2\pi-\beta} & \bar{a}+\bar{b}s \end{pmatrix}$$

where $\beta := \omega_\tau$ and where $s(z) := \coth \pi(z + i(1/2 + r_\tau))$ and $n_\beta(z) := e^{(z+i(1/2+r_\tau))(\pi-\beta)}/\sinh \pi(z + i(1/2 + r_\tau)))$. On the other hand, we have effective criteria to check the invertibility of the Toeplitz operator T in condition (b).

As in the proof of Lemma 4.3.4 one readily gets that the operator T is invertible if and only if the Toeplitz operator with generating function

$$\begin{pmatrix} a+bf^{(\varepsilon-\delta)} & c+df^{(\varepsilon-\delta)} & 0 & 0 \\ \bar{c}+\bar{d}(f^{(\varepsilon-\delta)})^* & \bar{a}+\bar{b}(f^{(\varepsilon-\delta)})^* & 0 & 0 \\ 0 & 0 & a-bf^{(\delta-\varepsilon)} & c-df^{(\delta-\varepsilon)} \\ 0 & 0 & \bar{c}-\bar{d}(f^{(\delta-\varepsilon)})^* & \bar{a}-\bar{b}(f^{(\delta-\varepsilon)})^* \end{pmatrix}$$

is invertible, and hence, if and only if the 2×2 block Toeplitz operators T_1 and T_2 with generating functions

$$\begin{pmatrix} a+bf^{(\varepsilon-\delta)} & c+df^{(\varepsilon-\delta)} \\ \bar{c}+\bar{d}(f^{(\varepsilon-\delta)})^* & \bar{a}+\bar{b}(f^{(\varepsilon-\delta)})^* \end{pmatrix}, \quad \begin{pmatrix} a-bf^{(\delta-\varepsilon)} & c-df^{(\delta-\varepsilon)} \\ \bar{c}-\bar{d}(f^{(\delta-\varepsilon)})^* & \bar{a}-\bar{b}(f^{(\delta-\varepsilon)})^* \end{pmatrix},$$

respectively are invertible.

Making use of the identity $(f^{(\nu)})^* = -f^{(\nu)}$ (again from the proof of Lemma 4.2.4), we can rewrite the operator T_1 as

$$\begin{pmatrix} a + bT(f^{(\varepsilon-\delta)}) & c + dT(f^{(\varepsilon-\delta)}) \\ \bar{c} - \bar{d}T(f^{(\varepsilon-\delta)}) & \bar{a} - \bar{b}T(f^{(\varepsilon-\delta)}) \end{pmatrix};$$

and since all entries in this operator matrix commute with each other, we conclude that the operator T_1 is invertible if and only if the operator

$$\begin{aligned}(a + bT(f^{(\varepsilon-\delta)}))(\bar{a} - \bar{b}T(f^{(\varepsilon-\delta)})) - (c + dT(f^{(\varepsilon-\delta)}))(\bar{c} - \bar{d}T(f^{(\varepsilon-\delta)}))\\ = (a\bar{a} - c\bar{c})I + (\bar{a}b - a\bar{b} - \bar{c}d + c\bar{d})T(f^{(\varepsilon-\delta)}) - (b\bar{b} - d\bar{d})(T(f^{(\varepsilon-\delta)}))^2\end{aligned}$$

is invertible. This clearly is true if and only if

$$(a\bar{a} - c\bar{c})I + (\bar{a}b - a\bar{b} - \bar{c}d + c\bar{d})\lambda - (b\bar{b} - d\bar{d})\lambda^2 \neq 0 \tag{4.74}$$

for all λ belonging to the spectrum of the Toeplitz operator $T(f^{(\varepsilon-\delta)})$.

We still need the spectrum of that Toeplitz operator, considered as acting on the Hilbert space l_2. Abbreviating $\varepsilon - \delta$ by ν and observing that for $\nu \neq 0$ the piecewise continuous function $f^{(\nu)}$ maps the unit circle into a circular arc $\mathcal{C}_\nu$ in the complex plane which joins -1 to 1 and runs through the point $i\tan(\pi\nu/2)$ (the center of that circular arc is $-i\cot\pi\nu$, and its radius is equal to $1/|\sin\pi\nu|$). The well-known criterion for the invertibility of Toeplitz operators with piecewise continuous generating function (e.g., Corollary 2.40 and Theorem 2.74 in [11]) states that the spectrum of $T(f^{(\nu)})$ just coincides with that compact region in the complex plane which is bounded by the circular arc $\mathcal{C}_\nu$ and the interval $[-1, 1]$. Thus denoting this region by $\mathcal{D}_\nu$, we get

Lemma 4.3.26. *The Toeplitz operator T_1 is invertible if and only if (4.74) is satisfied for all $\lambda \in \mathcal{D}_{\varepsilon-\delta}$.*

An analogous result holds for invertibility of the operator T_2.

4.4 Double Layer Potential Equation

As above, let Γ be a curve in the complex plane and write n_τ for the inner normal to Γ at $\tau \in \Gamma$, which exists for all points of Γ with exception of the corner points $\tau_0, \tau_1, \ldots, \tau_{n_0-1}$. By V_Γ we denote the double layer potential operator on Γ, viz.,

$$(V_\Gamma u)(t) = \frac{1}{\pi} \int_\Gamma u(\tau) \frac{d}{dn_\tau} \log|t - \tau| ds_\tau, \quad t \in \Gamma$$

where ds_τ refers to the arc length differential. In this notation, we can rewrite equation (4.2) in the form

$$Au = (aI + bV_\Gamma + T)u = g\,, \quad g \in L_2(\Gamma) \tag{4.75}$$

where I is the identity operator, T is compact and $a, b \in \mathbf{C}(\Gamma)$.

If Γ is a sufficiently smooth curve, then V_Γ is a compact operator and this essentially simplifies the question of stability of projection methods for (4.75). In this section, we proceed to stability investigation for collocation, qualocation, and quadrature methods for solving (4.75) when the curve Γ is not smooth, and we can rely upon results of preceding sections. Indeed, using the well-known identity [162]

$$2V_\Gamma = S_\Gamma + MS_\Gamma M$$

we can rewrite equation (4.75) as

$$Au = (aI + \frac{b}{2}S_\Gamma + \frac{b}{2}MS_\Gamma M + T)u = g, \tag{4.76}$$

which is a particular case of equation (4.1) and hence subject to Theorems 4.1.1, 4.1.7 and 4.1.8. For simplicity, throughout this section we suppose that $T = 0$ and consider the operator A on the space L_2 without weight.

4.4.1 Quadrature Methods

Let $t_k^{(n)}$ and $\tau_k^{(n)}$ be as in (4.3), and set $\Delta t_j^{(n)} = \gamma((j+1)/n) - \gamma(j/n)$. We determine approximate values $\xi_k^{(n)}$ for the exact values of the solution u of (4.76) at the points $t_k^{(n)}$ by solving the linear system

$$a(\tau_k^{(n)}) + \frac{b(\tau_k^{(n)})}{2\pi i}\sum_{j=0}^{n-1}\left(\frac{\Delta t_j^{(n)}}{t_j^{(n)} - \tau_k^{(n)}} - \frac{\overline{\Delta t_j^{(n)}}}{\overline{t_j^{(n)}} - \overline{\tau_k^{(n)}}}\right)\xi_j^{(n)} = g(\tau_k^{(n)}) \tag{4.77}$$

for $k = 0, 1, \ldots, n-1$. Fix $\tau \in \Gamma$ and abbreviate $a(\tau)$ and $b(\tau)$ by a and b. The local (model) system associated with (4.77) at $\tau \in \Gamma$ is given by

$$a\tilde{\xi}_k^{(n)} + \frac{b}{2\pi i n}\left[\sum_{j=0}^{\infty}\frac{2\operatorname{Im}\tilde{\tau}_k^{(n)}}{|\tilde{t}_j^{(n)} - \tilde{\tau}_k^{(n)}|^2} - \sum_{j=-\infty}^{-1}\left(\frac{e^{i\omega_\tau}}{\tilde{t}_j^{(n)} - \tilde{\tau}_k^{(n)}} - \frac{e^{-i\omega_\tau}}{\overline{\tilde{t}_j^{(n)}} - \overline{\tilde{\tau}_k^{(n)}}}\right)\right]\tilde{\xi}_j^{(n)} = g^\tau(\tilde{\tau}_k^{(n)})$$

where $k \in \mathbb{Z}$, and the corresponding operator $A_1^\tau \in \mathcal{L}(l_2^2)$ is of the form $(A_{rl}^\tau)_{r,l=1}^2$ where

$$A_{11}^\tau = A_{22}^\tau = aI,$$

$$A_{21}^\tau = \frac{b}{2\pi i}\left(\frac{1}{j+\delta+(\varepsilon-k-1)e^{i\omega_\tau}} - \frac{1}{j+\delta+(\varepsilon-k-1)e^{i(2\pi-\omega_\tau)}}\right)_{k,j=0}^{\infty},$$

$$A_{12}^\tau = \frac{b}{2\pi i}\left(\frac{e^{i\omega_\tau}}{(\delta-j+1)e^{i\omega_\tau}+k+\varepsilon} - \frac{e^{i(2\pi-\omega_\tau)}}{(\delta-j+1)e^{i(2\pi-\omega_\tau)}+k+\varepsilon}\right)_{k,j=0}^{\infty}.$$

Corollary 4.4.1. *If $a, b, \in \mathbf{C}(\Gamma)$, then the quadrature method (4.77) is stable if and only if the operator $A \in \mathcal{L}(L_2(\Gamma))$ and all operators $A_1^\tau \in \mathcal{L}(l_2^2)$, $\tau \in \Gamma$ are invertible.*

Our next goal is to consider the Fredholmness of the local operator A_1^τ. Computing its Gohberg–Krupnik symbol yields

$$\mathcal{A}_{A_1^\tau}(t,\mu) = \begin{cases} \begin{pmatrix} a & 0 \\ 0 & a \end{pmatrix} & \text{if } t \neq 1, \mu \in [0,1] \\ \begin{pmatrix} a & \frac{bi}{2\beta}\left(e^{-i(\pi-\omega_\tau)\alpha} - e^{-i(\omega_\tau-\pi)\alpha}\right) \\ -\frac{bi}{2\beta}\left(e^{-i(\omega_\tau-\pi)\alpha} - e^{-i(\pi-\omega_\tau)\alpha}\right) & a \end{pmatrix} & \text{if } t=1, \mu \in [0,1] \end{cases}$$

where α, β are as in Section 4.2.1.

Corollary 4.4.2. *Let $\theta_\tau := \omega_\tau/2\pi$. The operator A_1^τ is Fredholm if and only if the function*

$$\Phi_\tau(\mu) = a^2 - \frac{b^2}{2}\left[4\mu(1-\mu) + 2(\mu^{\theta_\tau}(1-\mu)^{2-\theta_\tau} + (1-\mu)^{\theta_\tau}\mu^{2-\theta_\tau})\cos\omega_\tau \right.$$
$$\left. -2i(\mu^{\theta_\tau}(1-\mu)^{2-\theta_\tau} - (1-\mu)^{\theta_\tau}\mu^{2-\theta_\tau})\sin\omega_\tau\right] \tag{4.78}$$

does not vanish on the interval $[0,1]$.

Proof. The determinant of the symbol $\mathcal{A}_{A_1^\tau}(t,\mu)$ of the operator A_1^τ is

$$\det \mathcal{A}_{A_1^\tau}(t,\mu) = a^2 - \frac{b^2}{\beta^2}\sin^2((\pi-\omega_\tau)\alpha) = a^2 - \frac{b^2}{2\beta^2}(1-\cos(2(\pi-\omega_\tau\alpha))).$$

Recalling the expressions for $1/\beta^2$ and $\cos(2(\pi-\omega_\tau)\alpha)/\beta^2$ quoted above, we obtain (4.78). □

Corollary 4.4.3. *If the quadrature method is stable, then*

$$a(t) \neq 0 \quad \textit{for} \quad \textit{all} \quad t \in \Gamma.$$

For proof, set $\mu = 0$ in (4.78).

Now let $\tau_0, \tau_1, \ldots, \tau_{n_0-1}$ denote the corner points of Γ, i.e., $\omega_{\tau_r} \neq \pi$ for $r = 0, 1, \ldots, n_0 - 1$ and $\omega_\tau = \pi$ for $\tau \in \Gamma \setminus \{\tau_0, \tau_1, \ldots, \tau_{n_0-1}\}$. Consider the curves

$$\mathcal{C}_r := \left\{ \frac{(1+s)^2}{4s+2(s^{\theta_r}+s^{2-\theta_r})\cos\omega_{\tau_r} - 2i(s^{\theta_r}-s^{2-\theta_r})\sin\omega_{\tau_r}},\ s \in (0,\infty),\ \theta_r := \frac{\omega_{\tau_r}}{2\pi} \right\}$$

for $r = 0, 1, 2, \ldots, n_0 - 1$, and

$$\mathcal{C}_{a,b} = \left\{ \frac{b^2(t)}{2a^2(t)},\quad t \in \Gamma \right\}.$$

Corollary 4.4.4. *Let $a(t) \neq 0$ for all $t \in \Gamma$.*

(a) *The local operators $A_1^\tau, \tau \in \Gamma$ are Fredholm if and only if*

$$\mathcal{C}_{a,b} \bigcap \left(\bigcup_{r=0}^{n_0-1} \mathcal{C}_r \right) = \emptyset. \tag{4.79}$$

(b) *If (4.79) is satisfied, then the index of $A_1^{\tau_r}, r = 0, \ldots, n_0 - 1$ is equal to the negative winding number of the curve $\mathcal{B}_r = \{\Phi_{\tau_r}(\mu), \mu \in [0,1]\}$.*

(c) *If $\omega_\tau = \pi$, then* $\operatorname{ind} A_1^\tau = 0$.

Proof. If $\tau \neq \tau_r$, $r = 0, \ldots, n_0 - 1$, then $\det \mathcal{A}_{A_1^\tau}(t, \mu) = a^2$ for all $[t, \mu] \in \mathbb{T} \times [0,1]$, i.e., A_1^τ is Fredholm with the index zero.

Let now $\tau = \tau_r$ for some $r = 0, \ldots, n_0 - 1$. Then the determinant of the symbol of A_1^τ does not vanish if and only if

$$\frac{b^2}{2a^2} \neq \Big[4\mu(1-\mu) + 2(\mu^{\theta_\tau}(1-\mu)^{2-\theta_\tau} + (1-\mu)^{\theta_\tau}\mu^{2-\theta_\tau}) \cos \omega_{\tau_r} \\ -2i(\mu^{\theta_r}(1-\mu)^{2-\theta_r} - (1-\mu)^{\theta_r}\mu^{2-\theta_r}) \sin \omega_{\tau_r}\Big]^{-1}.$$

Substituting $\mu/(1-\mu)$ by s one arrives at the assertion (4.79). □

Corollary 4.4.4 can now be reformulated in the following way:

Corollary 4.4.5. *Let $a, b \in \mathbf{C}(\Gamma)$ and $a(t) \neq 0$ for all $t \in \Gamma$. The quadrature method (4.77) is stable if and only if the following conditions are fulfilled:*

(a) *the operator A is invertible in $\mathcal{L}(L_2(\Gamma))$.*

(b) $\mathcal{C}_{a,b} \bigcap \left(\bigcup_{r=0}^{n_0-1} \mathcal{C}_r \right) = \emptyset.$

(c) *the winding number of each curve $\mathcal{B}_r, r = 0, \cdots, n_0 - 1$, is zero.*

(d) $\dim \ker A_1^\tau = 0$ *for all $\tau \in \Gamma$.*

Remark 4.4.6. The conditions (b) and (c) are easily verifiable, but the authors are not aware of any analytic method to evaluate the kernel dimension of the operators $A_1^\tau, \tau \in \Gamma$. On the other hand, there are results [209, 210] that allow us to study the kernel dimensions of Toeplitz operators by using singular values of certain sequences of matrices associated with the operator under consideration. This approach can be used to study the kernel dimensions of the operators A_1^τ.

It is interesting to compare the conditions (b) and (c), which are responsible for the Fredholmness of the local operators A_1^τ, with the corresponding conditions for quadrature methods for singular integral equations without conjugation. In our setting, the form of the curves $\mathcal{C}_r$ is independent of the parameters ε and δ, but it essentially depends on the angle ω_{τ_r}. In case of singular integral equations without conjugation, $\mathcal{C}_r$ is a semi-axis which forms the angle $\pi(\delta-\varepsilon)$ with the real axis (see [180]). There is also a decidable difference in the behaviour of the indices of the local operators: for singular integral operators without conjugation, the index of the local operators is always zero, whereas for local operators associated with double layer potential operators one has either $b(\tau_r) = 0$, which implies $\operatorname{ind} A_1^{\tau_r} = 0$, or $b(\tau_r) \neq 0$, which implies $\operatorname{ind} A_1^{\tau_r} = -\operatorname{wind} \mathcal{B}_r = -\operatorname{wind} \mathcal{B}'_r$ where

$$\mathcal{B}'_r = \left\{ \frac{2a^2(\tau_r)}{b^2(\tau_r)} - \left[4\mu(1-\mu) + 2(\mu^{\theta_r}(1-\mu)^{2-\theta_r} + (1-\mu)^{\theta_r}\mu^{2-\theta_r})\cos\omega_{\tau_r} \right.\right.$$
$$\left.\left. -2i(\mu^{\theta_r}(1-\mu)^{2-\theta_r} - (1-\mu)^{\theta_r}\mu^{2-\theta_r})\sin\omega_{\tau_r}\right], \quad \mu \in [0,1].\right\}.$$

Corollary 4.4.7. *Assume that the operator $A_1^{\tau_r}$ is Fredholm and $b(\tau_r) \neq 0$. Then:*

(a) *If* $\dfrac{a^2(\tau_r)}{b^2(\tau_r)} \in \mathbb{R}$ *and* $0 \notin \left(\dfrac{2a^2(\tau_r)}{b^2(\tau_r)} - 1 - \cos\omega_{\tau_r}, \dfrac{2a^2(\tau_r)}{b^2(\tau_r)}\right)$, *then*

$$\operatorname{ind} A_1^{\tau_r} = 0.$$

(b) *If* $\dfrac{a^2(\tau_r)}{b^2(\tau_r)} \in \mathbb{R}$ *and* $0 \in \left(\dfrac{2a^2(\tau_r)}{b^2(\tau_r)} - 1 - \cos\omega_{\tau_r}, \dfrac{2a^2(\tau_r)}{b^2(\tau_r)}\right)$, *then*

$$|\operatorname{ind} A_1^{\tau_r}| = 1.$$

4.4.2 The ε-Collocation Method

Consider an approximate solution of equation (4.76) in the form (4.7) and determine the unknown coefficients $\xi_j^{(n)}$, $j = 0, \dots, n-1$, by solving the linear algebraic system

$$(Au_n)(\tau_j^{n,\varepsilon}) = g(\tau_j^{n,\varepsilon}), \quad j = 0, \dots, n-1 \tag{4.80}$$

where $A = aI + \frac{b}{2}S_\Gamma + \frac{b}{2}MS_\Gamma M$ and $\tau_j^{n,\varepsilon}$ is given by (4.18). In accordance with (4.26), the local operators at $\tau \in \Gamma$ are of the form

$$A_{1,\varepsilon}^{\tau} = aI + \frac{b}{2\pi i}\left(\int_{\Gamma_\tau} \frac{\tilde{\chi}_j^{(1)}(t)dt}{t - \tilde{\tau}_k^{1,\varepsilon}}\right)_{k,j\in\mathbb{Z}} - \frac{b}{2\pi i}\left(\int_{\Gamma_\tau^*} \frac{\overset{\approx}{\chi}{}_j^{(1)}(t)dt}{t - \overline{\tilde{\tau}_k^{1,\varepsilon}}}\right)_{k,j\in\mathbb{Z}}. \tag{4.81}$$

Moreover, the symbols of the local operators $A_{1,\varepsilon}^{\tau}$ can be calculated by (4.38) and (4.39), so

$$\mathcal{A}_{A_{1,\varepsilon}^{\tau}}(t,\mu) = \mathcal{A}_{A_1^{\tau}}(t,\mu)$$

for all $(t,\mu) \in \Gamma \times [0,1]$, where A_1^{τ} is the local operator considered in the preceding subsection. Thus the results on Fredholmness of the local operators A_1^{τ} are similar to the corresponding results for the local operators $A_{1,\varepsilon}^{\tau}$ associated with the ε-collocation method.

Corollary 4.4.8. *Let $a, b \in \mathbf{C}(\Gamma)$ and $a(t) \neq 0$ on Γ. The ε-collocation method* (4.80) *is stable if and only if conditions* (a) – (c) *from Corollary* 4.4.5 *are satisfied and if* $\dim \ker A^{\tau}_{1,\varepsilon} = 0$ *for all* $\tau \in \Gamma$.

Corollary 4.4.9. *If $a, b \in \mathbb{C}$ and $a \neq 0$, then the Fredholmness of $A^{\tau}_{1,\varepsilon}$ is independent of the choice of $\varepsilon \in [0,1]$. If $A^{\tau}_{1,\varepsilon}$ is Fredholm, then its index is also independent of ε.*

From (4.39) one can conclude that the choice of $\varepsilon \in (0,1)$ essentially influences the Fredholmness of the local operators corresponding to equation (4.1), but it does not influence their indices. For the double layer potential operator, ε influences neither the indices nor Fredholmness of the local operators.

4.4.3 Qualocation Method

Given $r = 0, 1, \ldots, m-1$, choose real numbers ε_r and w_r as in Section 4.1.3. For the qualocation method, we seek an approximate solution u_n of equation (4.76) in the form (4.7), whence the unknown coefficients $\xi_j^{(n)}$ will be obtained by solving the algebraic systems

$$\sum_{r=0}^{m-1} w_r (Au_n)(\tau_j^{n,\varepsilon_r}) = \sum_{r=0}^{m-1} w_r g(\tau_j^{n,\varepsilon_r}), \quad j = 0, \ldots, n-1. \tag{4.82}$$

The local operators $A^{\tau}_{Q(\varepsilon)}$ associated with $\tau \in \Gamma$ are

$$A^{\tau}_{Q(\varepsilon)} = \sum_{r=0}^{m-1} w_r A^{\tau}_{1,\varepsilon_r}$$

where the operators $A^{\tau}_{1,\varepsilon_r}$ are now defined in (4.81).

Corollary 4.4.10. *Let $a, b \in \mathbf{C}(\Gamma)$ and $a(t) \neq 0$ on Γ, and suppose the numbers ε_r and w_r, $r = 0, \ldots, m-1$ are subject to the conditions of Theorem* 4.1.8. *The qualocation method* (4.82) *is stable if and only conditions* (a) – (c) *of Corollary* 4.4.5 *are satisfied and if* $\dim \ker A^{\tau}_{Q(\varepsilon)} = 0$ *for all* $\tau \in \Gamma$.

The analogue of Corollary 4.4.9 also remains valid.

Finally, let us point out once more a characteristic and, although at first glance unexpected, common property of all of these approximation methods for the double layer potential equations.

Corollary 4.4.11. *Let $\tau \in \Gamma$. The local operators $A^{\tau}_{1}, A^{\tau}_{1,\varepsilon}$, and $A^{\tau}_{Q(\varepsilon)}$ – corresponding to the quadrature method, the ε-collocation method, and the qualocation method, respectively – are either simultaneously Fredholm or not. Moreover, if these operators are Fredholm, their indices coincide.*

Consequently, if the index of at least one of these local operators is non-zero, then none of the above approximation methods is suitable to solve the double layer potential equation (4.75) in the space $L_2(\Gamma)$.

4.5 Approximation Methods for Mellin Operators with Conjugation

Let ρ_0, ρ_1 be fixed numbers in the interval $(-1/2, 1/2)$ and $J := [0,1]$. By $L_2(\rho_0,\rho_1) = L_2(J,\rho_0,\rho_1)$ we denote the set of all Lebesgue measurable functions ϕ such that

$$\|\phi\| = \|\phi\|_{2,\rho_0,\rho_1} = \left(\int_0^1 \tau^{2\rho_0}(1-\tau)^{2\rho_1}|\phi(t)|^2\, d\tau\right)^{1/2} < +\infty.$$

The set $L_2(\rho_0,\rho_1)$ with the scalar product

$$(\phi,\psi) = \int_0^1 \tau^{2\rho_0}(1-\tau)^{2\rho_1}\phi(t)\overline{\psi(\tau)}d\tau,$$

is a Hilbert space. In the space $L_2(\rho_0,\rho_1)$, let us consider an additive operator A defined by

$$\begin{aligned}(A\phi)(\tau) := &\frac{1}{2\pi i}\int_{Re(z)=1/2} \tau^{-z}\mathcal{A}_1(\tau,z)\tilde{\phi}(z)dz \\ &+\frac{1}{2\pi i}\int_{Re(z)=1/2} \tau^{-z}\mathcal{A}_2(\tau,z)(\widetilde{M\phi})(z)dz \\ &+\int_0^1 K_1(\tau,s)\phi(s)ds + \int_0^1 K_2(\tau,s)(M\phi)(s)ds, \qquad (4.83)\end{aligned}$$

where M is the operator of complex conjugation, and $\tilde{\varphi}$ denotes the Mellin transformation,

$$\tilde{\phi}(z) = \int_0^1 \tau^{z-1}\phi(\tau)d\tau$$

of the function φ, and the functions $K_1(\tau,s)$, $K_2(\tau,s)$ are continuous on $J\times J$. We assume that $\mathcal{A}_1(\tau,z)$, $\mathcal{A}_2(\tau,z)$ have the form

$$\mathcal{A}_1(t,z) = a(t) - b(t)\ i\ \cot(\pi z) + e(t,z), \qquad (4.84)$$

$$\mathcal{A}_2(t,z) = c(t) - d(t)\ i\ \cot(\pi z) + m(t,z), \qquad (4.85)$$

where:

- the functions a, b, c, d, are infinitely differentiable over $[0,1]$;
- there is a strip $\Pi_{\gamma,\beta} = \{z \in \mathbf{C}\colon \gamma < \operatorname{Re} z < \beta\ ,\ \gamma,\beta\in\mathbf{R},\ \gamma<\beta\}$ and complex numbers z_l and $\breve{z}_p$, $l = 1,2,\dots,k$, $p = 1,2,\dots,r$ such that $1/2$, $1/2$-ρ, $z_l-\rho$, $\breve{z}_p-\rho\in\Pi_{\gamma,\beta}$, but $z_l-\rho$, $\breve{z}_p-\rho\notin\Pi_{1/2,1/2-\rho}$ or $z_l-\rho$, $\breve{z}_p-\rho\notin\Pi_{1/2-\rho,1/2}$. Moreover, the function $e(t,z)$ (correspondingly $m(t,z)$) for every $t\in J$ and for every $z\in\Pi_{\gamma,\beta}\setminus\{\ z_l-\rho,\ l=1,2,\dots,k\}$ (correspondingly

for every $z \in \Pi_{\gamma,\beta} \setminus \{\breve{z}_p - \rho,\ p = 1, 2, \dots, r\}$) is infinitely differentiable in t and analytic in z. If $t \in J$ is fixed, then the points $z_l - \rho$, $l = 1, 2, \dots, k$ (correspondingly $\breve{z}_p - \rho$, $p = 1, 2, \dots, r$) are poles for the function $e(t, z)$ (correspondingly $m(t, z)$) with degree at most N, the same for all points t;

- for any $\epsilon > 0$ and for any natural numbers n, s the following inequalities hold:

$$\sup_{z \in \Pi_{\alpha,\beta} \setminus \bigcup_{l=1}^{k} B_\epsilon(z_l),\ t \in J} |(\partial/\partial t)^n (\partial/\partial z)^s e(t, z)(1 + |z|)^{1+s}| < +\infty,$$

$$\sup_{z \in \Pi_{\alpha,\beta} \setminus \bigcup_{p=1}^{r} B_\epsilon(\breve{z}_p),\ t \in J} |(\partial/\partial t)^n (\partial/\partial z)^s m(t, z)(1 + |z|)^{1+s}| < +\infty,$$

where $B_\epsilon(z_k) = \{z \in \mathbf{C} : |z - (z_k - \rho)| < \epsilon\}$.

It is notable that the above-defined Mellin operators interact well with the conjugation operator.

Lemma 4.5.1. *Let the function $\mathcal{A}(t, z)$ satisfy the above assumptions. If A is a Mellin operator generated by the function $\mathcal{A}(t, z)$, i.e.,*

$$(Ax)(\tau) = \frac{1}{2\pi i} \int_{Re(z)=1/2} \tau^{-z} \mathcal{A}(\tau, z) \widetilde{x(z)} dz, \tag{4.86}$$

then

$$MAM = \widehat{A} \tag{4.87}$$

where $\widehat{A}$ is the Mellin operator generated by the function $\widehat{\mathcal{A}}(t, z) = \overline{\mathcal{A}(t, \bar{z})}$.

Proof. Since $M\tilde{x}(z) = (\widetilde{Mx})(\bar{z})$,

$$(MAx)(\tau) = -\frac{1}{2\pi i} \int_{Re(z)=1/2} \tau^{-\bar{z}}\, \overline{\mathcal{A}(\tau, z)} (\widetilde{Mx})(\bar{z}) \overline{dz},$$

and $\overline{dz} = d\bar{z}$, from the substitution $\bar{z} = u$ one obtains the equation $MA = \widehat{A}M$ that implies (4.87). □

Due to Lemma 4.5.1 one can now use Corollary 1.4.7 to investigate the algebra generated by the Mellin operators (4.86) and the operator of conjugation.

Assumptions (4.84) and (4.85) and the requirements on the functions $e(t, z)$, $m(t, z)$ allow us to rewrite (4.83) in the form

$$(A\phi)(t) = a(t)\phi(t) + \frac{b(t)}{\pi i} \int_0^1 \frac{\phi(s)}{s - t} ds + c(t)\overline{\phi(t)} + \frac{d(t)}{\pi i} \int_0^1 \frac{\overline{\phi(s)}}{s - t} ds$$
$$+ \int_0^1 k_1(t, s)\phi(s) ds + \int_0^1 k_2(t, s)\overline{\phi(s)} ds = f(t),\ t \in [0, 1]. \tag{4.88}$$

where the integrals

$$\int_0^1 k_1(t,s)\phi(s)ds, \quad \int_0^1 k_2(t,s)\overline{\phi(s)}ds$$

are absolutely convergent and their kernels k_1 k_2 are continuous on the unit square except on the lines $t=s$ and $t,\ s=0$ [187] .

4.5.1 A Quadrature Method

Let us now apply a quadrature method to equation (4.88). We first choose real numbers $\delta_0, \delta_1 \in (0,1/2)$ and $\sigma_0, \sigma_1 \geq 1$ such that $\sigma_i(1-2\rho_i) < 2$, $i=0,1$. Secondly, we define a function g by

$$g(t) = \begin{cases} t^{\sigma_0} & \text{if} \quad t \in [0,\delta_0], \\ 1-(1-t)^{\sigma_1} & \text{if} \quad t \in [1-\delta_1, 1] \end{cases}$$

and extend g on the whole interval $(0,1)$ as a twice differentiable and monotonically increasing function.

Now we take an $n \in \mathbb{N}$, fix an $\varepsilon \in (-1/2, 1/2]$, $\varepsilon \neq 0$ and put

$$t_j^{(n)} := g\left(\frac{j+1/2}{n}\right), \quad \tau_j^{(n)} := g\left(\frac{j+1/2+\varepsilon}{n}\right), \quad j=0,1,\ldots,n-1.$$

Let ϕ denote the solution of equation (4.88). The approximate values $\xi_j^{(n)}$ of $\phi(t_j^{(n)}), j=0,1,\ldots,n-1$ are determined from the following system of algebraic equations:

$$\begin{aligned}
&[a(\tau_k^{(n)}) - b(\tau_k^{(n)})i\cot(\pi\varepsilon)]\xi_k^{(n)} + \frac{b(\tau_k^{(n)})}{\pi i}\sum_{j=0}^{n-1}\frac{\Delta t_j^{(n)}}{t_j^{(n)}-\tau_k^{(n)}}\xi_j^{(n)} \\
&+ [c(\tau_k^{(n)}) - d(\tau_k^{(n)})i\cot(\pi\varepsilon)]\overline{\xi_k^{(n)}} + \frac{d(\tau_k^{(n)})}{\pi i}\sum_{j=0}^{n-1}\frac{\Delta t_j^{(n)}}{t_j^{(n)}-\tau_k^{(n)}}\overline{\xi_j^{(n)}} \\
&+ \sum_{j=0}^{n-1} k_1(t_k^{(n)}, t_j^{(n)})\Delta t_j^{(n)}\xi_j^{(n)} + \sum_{j=0}^{n-1} k_2(t_k^{(n)}, t_j^{(n)})\Delta t_j^{(n)}\overline{\xi_j^{(n)}} \\
&= f(\tau_k^{(n)}), \qquad k=0,1,\ldots,n-1,
\end{aligned} \tag{4.89}$$

where

$$\Delta t_j^{(n)} := \frac{1}{n}g'\left(\frac{j+1/2}{n}\right).$$

Let $\chi_{[0,1)}$ represent the characteristic function of the interval [0,1) and let $n \in \mathbb{N}$ be as above. By $\phi_{k,n}$ we denote the functions

$$\phi_{k,n}(t) = \chi_{[0,1)}(nt-k), \quad k=0,1,\ldots,n-1,$$

and define the operator $P_n : L_2(\rho_0, \rho_1) \to L_2(\rho_0, \rho_1)$ by

$$(P_n x)(t) = \sum_{k=0}^{n-1} (x, \phi_{k,n}) \phi_{k,n}(t), \quad x \in L_2(\rho_0, \rho_1).$$

Lemma 4.5.2. (cf. [102, 183]) *The operators $P_n, n = 1, 2, \ldots$ are orthogonal projections onto the subspaces generated by the function systems $\{\phi_{k,n}(t),\ k = 0, 1, \ldots, n-1\}$, and the sequences (P_n) and (P_n^*) converge strongly to the identity operators as $n \to \infty$.*

Let us introduce the real Banach algebra $\mathcal{A}$ of all sequences (B_n) of additive continuous operators $B_n : \operatorname{im} P_n \to \operatorname{im} P_n$ for which there are operators $B \in \mathcal{L}_{add}(L_2(\rho_0, \rho_1))$ such that

$$s - \lim_{n\to\infty} B_n P_n = B, \quad s - \lim_{n\to\infty} (B_n^*) P_n^* = B^*.$$

Let $\mathcal{K}_{add}(L_2(\rho_0, \rho_1))$ denote the subset of all additive compact operators of $\mathcal{L}_{add}(L_2(\rho_0, \rho_1))$. It is easily seen that the set $\mathcal{J}$,

$$\mathcal{J} := \{(P_n T P_n + G_n) : T \in \mathcal{K}_{add}(L_2(\rho_0, \rho_1)),\ \|G_n\| \to 0 \text{ as } n \to \infty\},$$

forms an ideal in $\mathcal{A}$.

Let A be the operator defined by (4.88) and let $A_n \in \mathcal{L}_{add}(\operatorname{im} P_n)$ be the operator which corresponds to the approximation method (4.89). Recall that A_n has a special representation, viz.,

$$A_n = A_n^1 + A_n^2 M,$$

where A_n^1 and A_n^2 are the linear operators whose matrices with respect to the basis $\phi_{k,n}, k = 0, 1, \ldots, n-1$ are the matrices of the linear and antilinear parts of system (4.89), respectively.

Theorem 4.5.3. *The operator sequence (A_n) belongs to the algebra $\mathcal{A}$, and is stable if and only if the operator A is invertible in $\mathcal{L}_{add}(L_2(\rho_0, \rho_1))$ and if the coset $(A_n) + \mathcal{J}$ is invertible in the quotient algebra $\mathcal{A}/\mathcal{J}$.*

The proof of this result follows immediately from Proposition 1.6.4.

Thus the applicability of the approximation method (4.89) to equation (4.88) depends on the invertibility of the coset $(A_n) + \mathcal{J}$ in the quotient algebra $\mathcal{A}/\mathcal{J}$. As already mentioned, this problem can be studied via a localization technique. To proceed, we introduce additional operators.

For $\sigma_0 \geq 1, \varepsilon \in (-1/2, 1/2]$, $\varepsilon \neq 0$, we set

$$t_{jn}^{\sigma_0} := \left(\frac{j + 1/2}{n}\right)^{\sigma_0}, \quad \tau_{jn}^{\sigma_0} := \left(\frac{j + 1/2 + \varepsilon}{n}\right)^{\sigma_0}, \quad j = 0, 1, \ldots,$$

$$\Delta t_{jn}^{\sigma_0} := \frac{\sigma_0}{n} \left(\frac{j + 1/2}{n}\right)^{\sigma_0 - 1}, j = 0, 1, \ldots,$$

and depending on the location of the point τ, we introduce three kinds of systems of algebraic equations.

1. The first system corresponds to the point $\tau = 0$ and is

$$\begin{aligned}
&[a(0) - b(0)i\cot(\pi\varepsilon)]\xi_k^{(n)} + \frac{b(0)}{\pi i}\sum_{j=0}^{+\infty}\frac{\Delta t_{jn}^{\sigma_0}}{t_j^{(n)} - \tau_k^{(n)}}\xi_j^{(n)} \\
&+ [c(0) - d(0)i\cot(\pi\varepsilon)]\overline{\xi_k^{(n)}} + \frac{d(0)}{\pi i}\sum_{j=0}^{+\infty}\frac{\Delta t_{jn}^{\sigma_0}}{t_j^{(n)} - \tau_k^{(n)}}\overline{\xi_j^{(n)}} \\
&+ \sum_{j=1}^{\infty} k_1^{(0)}(\tau_{kn}^{\sigma_0}, t_{jn}^{\sigma_0})\Delta t_{jn}^{\sigma_0}\,\xi_j^{(n)} + \sum_{j=1}^{\infty} k_2^{(0)}(\tau_{kn}^{\sigma_0}, t_{jn}^{\sigma_0})\Delta t_{jn}^{\sigma_0}\,\overline{\xi_j^{(n)}} \\
&= f(\tau_{kn}^{\sigma_0}), \qquad k = 0, 1, \ldots,
\end{aligned} \tag{4.90}$$

where $k_1^{(0)}$, $k_2^{(0)}$ represent kernels of integral operators

$$\int_0^1 k_1^0(t,s)x(s)ds = \frac{1}{2\pi i}\int_{\mathrm{Re}\, z=1/2} t^{-z}e(0,z)\widetilde{x(z)}dz,$$

$$\int_0^1 k_2^0(t,s)\overline{x(s)}ds = \frac{1}{2\pi i}\int_{\mathrm{Re}\, z=1/2} t^{-z}m(0,z)\widetilde{(Mx)}(z)dz.$$

2. The second system corresponds to $\tau = 1$, and has just the same structure as system (4.90) except that $a(0), b(0), c(0), d(0), \sigma_0$ and ε should be replaced by $a(1), -b(1), c(1), -d(1), \sigma_1$ and $-\varepsilon$, respectively.

3. If τ is an arbitrary fixed point of (0,1), then

$$\begin{aligned}
&[a(\tau) - b(\tau)i\cot(\pi\varepsilon)]\xi_k^{(n)} + \frac{b(\tau)}{\pi i}\sum_{j=-\infty}^{+\infty}\frac{1}{j-k-\varepsilon}\xi_j^{(n)} \\
&+ [c(\tau) - d(\tau)i\cot(\pi\varepsilon)]\overline{\xi_k^{(n)}} + \frac{d(\tau)}{\pi i}\sum_{j=-\infty}^{+\infty}\frac{1}{j-k-\varepsilon}\overline{\xi_j^{(n)}} \\
&= f(\frac{k+1/2+\varepsilon}{n}), \quad k \in \mathbb{Z}.
\end{aligned} \tag{4.91}$$

Let $H_\tau, \tau \in [0,1]$ denote one of the following Hilbert spaces:

$$H_\tau := \begin{cases} l_{2,(\sigma_0-1)/2-\sigma_0\rho_0} & \text{if} \quad \tau = 0, \\ \tilde{l}_2 & \text{if} \quad \tau \in (0,1), \\ l_{2,(\sigma_1-1)/2-\sigma_1\rho_1} & \text{if} \quad \tau = 1, \end{cases}$$

where $l_{2,\gamma}$ and $\tilde{l}_2$ are respectively the spaces of one- and two-sided sequences $\{\xi_k\}_{k=0}^{+\infty}$ and $\{\xi_k\}_{k=-\infty}^{+\infty}$ of complex numbers such that

$$\|\{\xi_k\}_{k=0}^{+\infty}\|_{l_{2,\gamma}} = \left(\sum_{k=0}^{+\infty}|\xi_k|^2(1+k)^{2\gamma}\right)^{1/2} < +\infty,$$

and

$$\|\{\xi_k\}_{k=-\infty}^{+\infty}\|_{\tilde{l}_2} = \left(\sum_{k=-\infty}^{+\infty} |\xi_k|^2\right)^{1/2} < +\infty.$$

By $\overline{M}$ we again denote the operator defined on spaces of one- or two-sided sequences $\{\xi_k\}$ by the rule

$$\overline{M}\{\xi_k\} = \{\overline{\xi_k}\}.$$

Let $\tilde{A}_n^\tau \in \mathcal{L}_{add}(H_\tau), \tau \in [0,1]$ refer to the operator which corresponds to system (4.90) or (4.91). This operator has the form

$$\tilde{A}_n^\tau = \tilde{A}_{1,n}^\tau + \tilde{A}_{2,n}^\tau \overline{M}, \tag{4.92}$$

where $\tilde{A}_{1,n}^\tau, \tilde{A}_{2,n}^\tau$ are the linear operators whose matrices with respect to the basis $\phi_{k,n}, k = 0, 1, \dots$ are the matrices of the linear and antilinear parts of system (4.90) or (4.91). In the sequel we will identify the operators $\tilde{A}_{1,n}^\tau, \tilde{A}_{2,n}^\tau$ with their matrices.

Lemma 4.5.4. *Let $\tilde{A}_n^\tau, \tau \in [0,1]$ be the operator defined by (4.92). Then:*

1. *The operators $\tilde{A}_n^\tau, \tau \in [0,1]$ are independent of n, so all elements of the sequence $(\tilde{A}_n^\tau)_{n\in\mathbb{N}}$ coincide with the operator $\tilde{A}_1^\tau$.*
2. *The coset $(A_n) + \mathcal{J}$ is invertible in $\mathcal{A}/\mathcal{J}$ if and only if the operators $\tilde{A}_n^\tau \in \mathcal{L}_{add}(H_\tau)$, $\tau \in [0,1]$ are invertible.*

The first statement of the lemma is proved by straightforward calculation, and the second follows from the Gohberg-Krupnik local principle, (cf. Section 1.9.1).

4.5.2 Fredholm Properties of Local Operators

In order to use the approximation method (4.89), we need to know conditions for the invertibility of the operators $\tilde{A}_n^\tau, \tau \in [0,1]$. However, the complicated structure of the above operators does not allow us to treat the problem completely. Nevertheless, it turns out that these operators belong to a well-studied operator algebra, so conditions for their Fredholmness can be established. Referring again to Lemma 1.4.4 and Corollary 1.4.7, we note that the operator $\tilde{A}_n^\tau : H_\tau \mapsto H_\tau$ is invertible (Fredholm) if and only if the operator $A_n^\tau = \Psi(\tilde{A}_n^\tau) : H_\tau \times H_\tau \mapsto H_\tau \times H_\tau$ given by

$$A_n^\tau = \begin{pmatrix} \tilde{A}_{1,1}^\tau & \tilde{A}_{2,1}^\tau \\ \overline{M}\tilde{A}_{2,1}^\tau\overline{M} & \overline{M}\tilde{A}_{1,1}^\tau\overline{M} \end{pmatrix}$$

is invertible (Fredholm). Furthermore,

$$\operatorname{ind}_{\mathbb{R}}\tilde{A}_n^\tau = \operatorname{ind}_{\mathbb{C}} A_n^\tau \tag{4.93}$$

where the left- and right-hand sides of (4.93) denote the indices of $\tilde{A}_n^\tau$ and A_n^τ acting on the real and complex versions of H_τ and $H_\tau \times H_\tau$, respectively (see (2.11)). Let us start with the operators A_1^τ in the simplest case, $\tau \in (0,1)$. From the invertibility of the original operator A and [143], it follows that

$$e(\tau) = (a(\tau)+b(\tau))\overline{(a(\tau)-b(\tau))} - (c(\tau)+d(\tau))\overline{(c(\tau)-d(\tau))} \neq 0 \qquad (4.94)$$

for any $\tau \in (0,1)$. Let us introduce the curves

$$\Gamma^1_{a,b,c,d} = \{z \in \mathbf{C} : z(\tau) = \frac{q(\tau) + \sqrt{q(\tau)^2 - 4|e(\tau)|^2}}{-2\overline{e(\tau)}},\ \tau \in (0,1)\},$$

$$\Gamma^2_{a,b,c,d} = \{z \in \mathbf{C} : z(\tau) = \frac{q(\tau) - \sqrt{q(\tau)^2 - 4|e(\tau)|^2}}{-2\overline{e(\tau)}},\ \tau \in (0,1)\},$$

where

$$q(\tau) = |a(\tau)-b(\tau)|^2 + |a(\tau)+b(\tau)|^2 - |c(\tau)-d(\tau)|^2 - |c(\tau)+d(\tau)|^2. \qquad (4.95)$$

Note that whenever the square root sign is used, this means the branch defined by $\sqrt{1} = 1$. Consider now the ray

$$R_\varepsilon = \{z \in \mathbf{C} : z(t) = te^{-i\pi\varepsilon},\ t \in (0,+\infty)\}.$$

Lemma 4.5.5. *Let the operator $A \in L_{add}(L_2(\rho_0,\rho_1))$ be invertible. The operator $\tilde{A}_1^\tau$, $\tau \in (0,1)$ is invertible if and only if*

$$(\Gamma^1_{a,b,c,d} \cup \Gamma^2_{a,b,c,d}) \cap R_\varepsilon = \emptyset. \qquad (4.96)$$

Proof. Let the operator A be invertible and $\tau \in (0,1)$. Set

$$\Phi_\varepsilon(t) = \begin{cases} 1-2\mu,\ if\ \varepsilon = 0,\ t = e^{2\pi i\mu},\ \mu \in (0,1), \\ 2e^{-i\pi\varepsilon(\mu-1)}\dfrac{\sin(-\pi\mu\varepsilon)}{\sin(-\pi\varepsilon)} - 1,\ if\ \varepsilon \neq 0,\ t = e^{2\pi i\mu},\ \mu \in (0,1). \end{cases}$$

We study the operator $\tilde{A}_1^\tau$ by means of the operator A_1^τ, and observe that A_1^τ is the discrete Wiener-Hopf operator generated by the matrix

$$\mathcal{A}_{A_1^\tau}(t) = \begin{pmatrix} a(\tau)+b(\tau)\Phi_\varepsilon(t) & c(\tau)+d(\tau)\Phi_\varepsilon(t) \\ c(\tau)-d(\tau)\Phi_\varepsilon(t) & a(\tau)-b(\tau)\Phi_\varepsilon(t) \end{pmatrix},\quad t = e^{2\pi i\mu},\ \mu \in (0,1). \qquad (4.97)$$

The invertibility conditions for such operators are well known, and involve the non-vanishing of the determinant of the matrix $\mathcal{A}_{A_1^\tau}(t)$ (e.g., see [89, Chapter 8]). Now we show that this matrix is non-vanishing if and only if conditions (4.96) are satisfied.

Set

$$\Psi_\varepsilon(t) = e^{-i\pi\varepsilon(\mu-1)} \frac{\sin(-\pi\mu\varepsilon)}{\sin(-\pi\varepsilon)}, \quad t = e^{2\pi i\mu}, \ \mu \in (0,1)$$

and write the matrix (4.97), in the form

$$\mathcal{A}_{A_1^\tau}(t) = \begin{pmatrix} (a+b)\Psi_\varepsilon(t) + (a-b)(1-\Psi_\varepsilon(t)) & (c+d)\Psi_\varepsilon(t) + (c-d)(1-\Psi_\varepsilon(t)) \\ (\overline{c}-\overline{d})\Psi_\varepsilon(t) + (\overline{c}+\overline{d})(1-\Psi_\varepsilon(t)) & (\overline{a}-\overline{b})\Psi_\varepsilon(t) + (\overline{a}+\overline{b})(1-\Psi_\varepsilon(t)) \end{pmatrix},$$

where the coefficients a, b, c, d are evaluated at τ. Then

$$\det \mathcal{A}_{A_1^\tau}(t) = \Psi_\varepsilon^2(t)[e(\tau) + q(\tau)R_\varepsilon(t) + \overline{e(\tau)}R_\varepsilon^2(t)],$$

with $R_\varepsilon(t) = (1 - \Psi_\varepsilon(t))/\Psi_\varepsilon(t)$. Since A is invertible, the function $\overline{e(\tau)}$ does not vanish in the interval $(0,1)$, so the condition $\det \mathcal{A}_{A_1^\tau}(t) \neq 0$ is equivalent to (4.96). □

We now consider the case $\tau = 0$. Writing $a = a(0)$, $b = b(0)$, $c = c(0)$, $d = d(0)$ and passing from the operator $\tilde{A}_1^0$ to the operator A_1^0, we get

$$A_1^0 = \begin{pmatrix} A_{11}^0 & A_{12}^0 \\ A_{21}^0 & A_{22}^0 \end{pmatrix},$$

where the operators A_{rp}^0, $r, p = 1, 2$ are

$$A_{11}^0 = \left((a - ib\cot(\pi\varepsilon))\delta_{jk} + \frac{b}{\pi i} \frac{\Delta t_{jn}^{\sigma_0}}{t_{jn}^{\sigma_0} - \tau_{jn}^{\sigma_0}} + k_1^0(\tau_{jn}^{\sigma_0}, t_{jn}^{\sigma_0})\Delta t_{jn}^{\sigma_0} \right)_{k,j=1}^{+\infty},$$

$$A_{12}^0 = \left((c - id\cot(\pi\varepsilon))\delta_{jk} + \frac{d}{\pi i} \frac{\Delta t_{jn}^{\sigma_0}}{t_{jn}^{\sigma_0} - \tau_{jn}^{\sigma_0}} + k_2^0(\tau_{jn}^{\sigma_0}, t_{jn}^{\sigma_0})\Delta t_{jn}^{\sigma_0} \right)_{k,j=1}^{+\infty},$$

$$A_{21}^0 = \left((\bar{c} + i\bar{d}\cot(\pi\varepsilon))\delta_{jk} - \frac{\bar{d}}{\pi i} \frac{\Delta t_{jn}^{\sigma_0}}{t_{jn}^{\sigma_0} - \tau_{jn}^{\sigma_0}} + \overline{k_2^0(\tau_{jn}^{\sigma_0}, t_{jn}^{\sigma_0})}\Delta t_{jn}^{\sigma_0} \right)_{k,j=1}^{+\infty},$$

$$A_{22}^0 = \left((\bar{a} + i\bar{b}\cot(\pi\varepsilon))\delta_{jk} - \frac{\bar{b}}{\pi i} \frac{\Delta t_{jn}^{\sigma_0}}{t_{jn}^{\sigma_0} - \tau_{jn}^{\sigma_0}} + \overline{k_1^0(\tau_{jn}^{\sigma_0}, t_{jn}^{\sigma_0})}\Delta t_{jn}^{\sigma_0} \right)_{k,j=1}^{+\infty}.$$

From Lemma 4.5.1 and [187], all operators A_{rp}^0, $r, p = 1, 2$, belong to the smallest algebra containing all Toeplitz operators generated by piecewise continuous functions and acting in the Hilbert space $l_{2,(\sigma_0-1)/2-\sigma_0\rho_0}$. As before this algebra is denoted by $\mathcal{T}^{2\times 2}_{(\sigma_0-1)/2-\rho_0\sigma_0}$. Since for every $z \in \mathbf{C}$,

$$\cot(\pi\bar{z}) = \overline{\cot(\pi z)}, \tag{4.98}$$

for the symbols $\mathcal{A}_{A^0_{rp}}(t,\mu)$ of the operators A^0_{rp}, $r,p=1,2$, (again from Lemma 4.5.1 and [187]) we get

$$\mathcal{A}_{A^0_{11}}(t,\mu)=\begin{cases}\mathcal{A}_1(0,1/2-\rho_0+\frac{i}{2\pi\sigma_0}\log(\mu/(1-\mu)),\ if\ \ t=1,\ \mu\in[0,1],\\ \mathcal{A}_1(0,1/2-\rho_0-i\infty)\Psi_\varepsilon(t)+\mathcal{A}_1(0,1/2-\rho_0+i\infty)(1-\Psi_\varepsilon(t)),\\ \qquad if\ \ t=e^{2\pi is},\ s\in(0,1),\ \mu\in[0,1].\end{cases} \tag{4.99}$$

$$\mathcal{A}_{A^0_{12}}(t,\mu)=\begin{cases}\mathcal{A}_2(0,1/2-\rho_0+\frac{i}{2\pi\sigma_0}\log(\mu/(1-\mu)),\ if\ \ t=1,\ \mu\in[0,1],\\ \mathcal{A}_2(0,1/2-\rho_0-i\infty)\Psi_\varepsilon(t)+\mathcal{A}_2(0,1/2-\rho_0+i\infty)(1-\Psi_\varepsilon(t)),\\ \qquad if\ \ t=e^{2\pi is},\ s\in(0,1),\ \mu\in[0,1].\end{cases} \tag{4.100}$$

$$\mathcal{A}_{A^0_{21}}(t,\mu)=\begin{cases}\overline{\mathcal{A}_2(0,1/2-\rho_0-\frac{i}{2\pi\sigma_0}\log(\mu/(1-\mu))},\ if\ \ t=1,\ \mu\in[0,1],\\ \overline{\mathcal{A}_2(0,1/2-\rho_0+i\infty)}\Psi_\varepsilon(t)+\overline{\mathcal{A}_2(0,1/2-\rho_0-i\infty)}(1-\Psi_\varepsilon(t)),\\ \qquad if\ \ t=e^{2\pi is},\ s\in(0,1),\ \mu\in[0,1].\end{cases} \tag{4.101}$$

$$\mathcal{A}_{A^0_{22}}(t,\mu)=\begin{cases}\overline{\mathcal{A}_1(0,1/2-\rho_0-\frac{i}{2\pi\sigma_0}\log(\mu/(1-\mu))},\ if\ \ t=1,\ \mu\in[0,1],\\ \overline{\mathcal{A}_1(0,1/2-\rho_0+i\infty)}\Psi_\varepsilon(t)+\overline{\mathcal{A}_1(0,1/2-\rho_0-i\infty)}(1-\Psi_\varepsilon(t)),\\ \qquad if\ \ t=e^{2\pi is},\ s\in(0,1),\ \mu\in[0,1].\end{cases} \tag{4.102}$$

Now the symbol of the operator A^0_1 is

$$\mathcal{A}_{A^0_1}(1,\mu)=\begin{pmatrix}\mathcal{A}_{A^0_{11}}(1,\mu) & \mathcal{A}_{A^0_{12}}(1,\mu)\\ \mathcal{A}_{A^0_{21}}(1,\mu) & \mathcal{A}_{A^0_{22}}(1,\mu)\end{pmatrix},\quad \mu\in[0,1], \tag{4.103}$$

$$\mathcal{A}_{A^0_1}(e^{2\pi is},\mu)=\begin{pmatrix}\mathcal{A}_{A^0_{11}}(e^{2\pi is},\mu) & \mathcal{A}_{A^0_{12}}(e^{2\pi is},\mu)\\ \mathcal{A}_{A^0_{21}}(e^{2\pi is},\mu) & \mathcal{A}_{A^0_{22}}(e^{2\pi is},\mu)\end{pmatrix},\ s\in(0,1),\quad \mu\in[0,1], \tag{4.104}$$

where the elements of the matrix (4.103) and (4.104) have the form (4.99) – (4.102).

Henceforth let us define $e(\tau)$, $q(\tau)$ $\tau\in[0,1]$ by

$$\begin{aligned}e(\tau):=&\ \mathcal{A}_1(\tau,1/2-\rho_0-i\infty)\overline{\mathcal{A}_1(\tau,1/2-\rho_0+i\infty)}\\ &-\mathcal{A}_2(\tau,1/2-\rho_0-i\infty)\overline{\mathcal{A}_2(\tau,1/2-\rho_0+i\infty)},\end{aligned} \tag{4.105}$$

and by

$$\begin{aligned}q(\tau):=&\ \mathcal{A}_1(\tau,1/2-\rho_0+i\infty)\overline{\mathcal{A}_1(\tau,1/2-\rho_0-i\infty)}\\ &+\mathcal{A}_1(\tau,1/2-\rho_0-i\infty)\overline{\mathcal{A}_1(\tau,1/2-\rho_0+i\infty)}\\ &-[\mathcal{A}_2(\tau,1/2-\rho_0+i\infty)\overline{\mathcal{A}_2(\tau,1/2-\rho_0-i\infty)}\end{aligned}$$

$$+ \mathcal{A}_2(\tau, 1/2 - \rho_0 - i\infty)\overline{\mathcal{A}_2(\tau, 1/2 - \rho_0 + i\infty)}]. \qquad (4.106)$$

Observe that if $\tau \in (0,1)$ these functions are equal to the functions q and e considered earlier.

Using the invertibility criterion for elements of $\mathcal{T}_\rho^{2\times 2}$ [94], we get

Lemma 4.5.6. *Let the operator A be invertible. Then the operator A_1^0 is invertible if and only if:*

1. *The points*
$$z_{1,2} = \frac{q(0) \pm \sqrt{q^2(0) - 4|e(0)|^2}}{-2\overline{e(0)}}$$
do not belong to the ray R_ε;
2. $\det \mathcal{A}_{A_1^0}(1,\mu) \neq 0$ *for every* $\mu \in [0,1]$;
3. *the winding number for the curve*
$$\Gamma_0 = \{\det \mathcal{A}_{A_1^0}(1,\mu),\ \mu \in [0,1]\} \cup \{\det \mathcal{A}_{A_1^0}(e^{2\pi i s}, 0),\ s \in (0,1)\}$$
is equal to zero;
4. $\dim\ker \tilde{A}_1^0 = 0$, *i.e., the operator $\tilde{A}_1^0$ is injective.*

Condition 1 ensures the non-vanishing of $\det \mathcal{A}_{A_1^0}(t,\mu)$, $t \neq 1$ (its proof is similar to the proof of Lemma 4.5.5).

The invertibility of the operator $\tilde{A}_1^1$ can be considered analogously, so one again arrives at a matrix Toeplitz operator whose symbol can be obtained by obvious substitutions in the symbol of the operator $\tilde{A}_1^0$. We omit unnecessary details and just write down the corresponding invertibility conditions.

Lemma 4.5.7. *Let the operator A be invertible. The operator A_1^1 is invertible if and only if:*

1. *The points*
$$z_{1,2} = \frac{q(1) \pm \sqrt{q^2(1) - 4|e(1)|^2}}{-2\overline{e(1)}}$$
do not belong to the ray $R_{-\varepsilon}$;
2. $\det \mathcal{A}_{A_1^1}(1,\mu) \neq 0$ *for every* $\mu \in [0,1]$;
3. *the winding number for the curve*
$$\Gamma_1 = \{\det \mathcal{A}_{A_1^1}(1,\mu),\ \mu \in [0,1]\} \cup \{\det \mathcal{A}_{A_1^1}(e^{2\pi i s}, 0),\ s \in (0,1)\}$$
is equal to zero;
4. $\dim\ker \tilde{A}_1^1 = 0$, *i.e., the operator $\tilde{A}_1^1$ is injective.*

We now consider the curves

$$\mathcal{C}^1_{a,b,c,d} = \{z \in \mathbf{C} : z(\tau) = \frac{q(\tau) + \sqrt{q(\tau)^2 - 4|e(\tau)|^2}}{-2\overline{e(\tau)}}, \ \tau \in [0,1)\},$$
$$\mathcal{C}^2_{a,b,c,d} = \{z \in \mathbf{C} : z(\tau) = \frac{q(\tau) - \sqrt{q(\tau)^2 - 4|e(\tau)|^2}}{-2\overline{e(\tau)}}, \ \tau \in [0,1)\},$$

which differ from $\Gamma^1_{a,b,c,d}$ and $\Gamma^2_{a,b,c,d}$ since we now use the formulas (4.105) and (4.106) in place of (4.94) and (4.95) respectively. Combining Lemmas 4.5.4, 4.5.5 and 4.5.7 leads to the following stability condition of the approximation method (4.89).

Theorem 4.5.8. *Assume that the operator A is invertible. The approximation method (4.89) is stable if and only if the following conditions hold:*

1. $(\mathcal{C}^1_{a,b,c,d} \cup \mathcal{C}^2_{a,b,c,d}) \cap R_\varepsilon = \emptyset$;
2. *the points*
$$z_{1,2} = \frac{q(1) \pm \sqrt{q^2(1) - 4|e(1)|^2}}{-2\overline{e(1)}}$$
do not belong to the ray $R_{-\varepsilon}$;
3. $\det \mathcal{A}_{A^0_1}(1,\mu) \neq 0$ *and* $\det \mathcal{A}_{A^1_1}(1,\mu) \neq 0$ *for every* $\mu \in [0,1]$;
4. *the winding number for every curve Γ_0, Γ_1 is equal to zero.*
5. $\dim\ker \tilde{A}^1_1 = \dim\ker \tilde{A}^0_1 = 0$, *i.e., the operators $\tilde{A}^1_1$ and $\tilde{A}^0_1$ are injective.*

Remark 4.5.9. Conditions 1 – 4 of Theorem 4.5.8 have a geometrical nature and can be easily verified. For Condition 5 see Remark 4.4.6.

Remark 4.5.10. The condition $\det \mathcal{A}_{A^0_1}(1,\mu) \neq 0$ for the dominant singular integral equations with conjugation, i.e., if $m(t,z) = n(t,z) \equiv 0$, has a very simple form (cf. Lemma 4.5.14 below)

$$|a|^2 - |c|^2 - (|b|^2 - |d|^2) \left|\cot\left(\pi\left(\frac{1}{2} - \rho_0 + \frac{i}{2\sigma_0\pi}\log(s_{1,2})\right)\right)\right|^2 \neq 0,$$

where $s_{1,2}$ are non-negative roots of the quadratic equation

$$\mathrm{Im}\,(c\bar{d} - b\bar{a})s^2 - 2\mathrm{Re}\,(c\bar{d} - b\bar{a})\sin(2\pi\rho_0)s - \mathrm{Im}\,(c\bar{d} - b\bar{a}) = 0,$$

and all coefficients a, b, c, d are evaluated at $\tau = 0$.

4.5.3 Index of the Local Operators

As already mentioned (cf. Lemma 4.5.5), there are simple geometrical conditions of invertibility for the operators A^τ_1 for $\tau \in (0,1)$. For $\tau = 0$ or $\tau = 1$, the situation is more involved. Although both these operators belong to the above-mentioned Toeplitz algebra, there are no general invertibility conditions in such algebras. However, one can study the index of the operators A_1^0 and A_1^1, and in some cases find a Hilbert space where the corresponding operator has index 0. This is important in applications, because the associated approximation method can be modified to achieve stability. For definiteness, let us consider the operator A_1^0, and recall that this operator also depends on the parameters $\varepsilon \in [0,1)$ and $\sigma \in [1,\infty)$. To make this dependence more visible, let us now re-denote the operator A_1^0 by $A_0(\varepsilon,\sigma_0)$, where subscript 0 means that the operator is associated with the corresponding approximation method for the point 0. Note an important consequence of representations (4.99) – (4.104).

Corollary 4.5.11. *For all $\sigma_0^{(1)}, \sigma_0^{(2)} \geq 1$ such that $\sigma_0^{(i)}(1-2\rho_0) < 2$, $i=1,2$, the operators $A_0(\varepsilon,\sigma_0^{(1)})$ and $A_0(\varepsilon,\sigma_0^{(2)})$ are either simultaneously Fredholm or not, and*

$$\operatorname{ind} A_0(\varepsilon,\sigma_0^{(1)}) = \operatorname{ind} A_0(\varepsilon,\sigma_0^{(2)}).$$

Proof. Consider the symbol of the operator $A_0(\varepsilon,\sigma_0)$, $\sigma_0 \geq 1$. It is easily seen that the image of the function $(i/2\pi\sigma_0)\log(\mu/(1-\mu))$ is independent of σ_0. Hence, the determinant of the symbol does not depend on this parameter either, implying this result. □

In the following let us therefore assume that $\sigma_0 = 1$ and study the operator $A_0(\varepsilon) := A_0(\varepsilon,1)$ in the space $l_{2,-\rho} := l_{2,-\rho_0}$. Moreover, let us now consider the Cauchy singular integral equations with conjugation

$$(A\phi)(t) = a(t)\phi(t) + \frac{b(t)}{\pi i}\int_0^1 \frac{\phi(s)}{s-t}ds + c(t)\overline{\phi(t)} + \frac{d(t)}{\pi i}\int_0^1 \frac{\overline{\phi(s)}}{s-t}ds = f(t),\ t\in[0,1]. \tag{4.107}$$

Note that the operator A in (4.107) is a Mellin operator with conjugation generated by the functions

$$\mathcal{A}_1(t,z) = a(t) - b(t)\ i\ \cot(\pi z),$$
$$\mathcal{A}_2(t,z) = c(t) - d(t)\ i\ \cot(\pi z).$$

As before, we define a curve Γ_0 by

$$\Gamma_0 = \{\det \mathcal{A}_{A_0(\varepsilon)}(1,\mu),\ \mu\in[0,1]\} \cup \{\det \mathcal{A}_{A_0(\varepsilon)}(e^{i2\pi s},0),\ s\in(0,1)\}.$$

Lemma 4.5.12. *If $A_0(\varepsilon)$ is a Fredholm operator, then*

$$|\operatorname{ind} A_0(\varepsilon)| \leq \left[\frac{N}{2}\right],$$

where N is the number of zeros of the function $E(s,\mu)$ given by

$$E(s,\mu) = \begin{cases} \operatorname{Im}\{\det \mathcal{A}_{A_0(\varepsilon)}\}(1,\mu), & s=0,\ \mu\in[0,1],\\ \operatorname{Im}\{\det \mathcal{A}_{A_0(\varepsilon)}\}(e^{i2\pi s},0), & s\in(0,1),\ \mu\in[0,1], \end{cases}$$

and $[\cdot]$ *denotes the integral part.*

Proof. Since $A_0(\varepsilon) \in \mathcal{T}^{2\times 2}_{-\rho}$, it follows that $\operatorname{ind} A_0(\varepsilon)$ is equal to the winding number of the curve Γ_0 [94]. Let p denote the number of intersections of Γ_0 with the real axis. Thus the lemma is an obvious consequence of the estimate

$$|\operatorname{ind} A_0(\varepsilon)| \leq \left[\frac{p}{2}\right]$$

and the equality $p = N$. □

Consequently, we have to estimate the number of zeros of the function $E(s,\mu)$, and to do so we consider the two parts Γ_0^1 and Γ_0^2 of the curve Γ_0 defined by

$$\begin{aligned} \Gamma_0^1 &= \{\det \mathcal{A}_{A_0(\varepsilon)}(e^{i2\pi s}, 0) : s \in (0,1)\}, \\ \Gamma_0^2 &= \{\det \mathcal{A}_{A_0(\varepsilon)}(1, \mu) : \mu \in [0,1]\}. \end{aligned} \tag{4.108}$$

Lemma 4.5.13. *The equation*

$$\operatorname{Im} \det \mathcal{A}_{A_0(\varepsilon)}(e^{2\pi i s}, 0) = 0, \quad s \in (0,1) \tag{4.109}$$

has at most four solutions.

Proof. Let us consider the case $\varepsilon \neq 0$. The behaviour of the curve Γ_0^1 for $\varepsilon = 0$ is simpler, so it will be analyzed in the proof of Theorem 4.5.15. Thus we assume that $\varepsilon \neq 0$ and set

$$\begin{aligned} \Psi_\varepsilon(s) &= e^{-i\pi\varepsilon(s-1)} \frac{\sin(-\pi s\varepsilon)}{\sin(-\pi\varepsilon)}, \\ e &= (a(0)+b(0))\overline{(a(0)-b(0))} - (c(0)+d(0))\overline{(c(0)-d(0))}, \\ q &= 2(|a(0)|^2 + |b(0)|^2 - |c(0)|^2 - |d(0)|^2). \end{aligned}$$

Then

$$\det \mathcal{A}_{A_0(\varepsilon)}(e^{2\pi i s}, 0) = (2\mathrm{Re}(e) - q)\Psi_\varepsilon^2(s) + (q - 2\bar{e})\Psi_\varepsilon(s) + \bar{e},$$

so using the notation

$$A = \frac{\sin(\pi s\varepsilon)}{\sin(\pi\varepsilon)},$$

one obtains

$$\begin{aligned} \operatorname{Im} \det \mathcal{A}_{A_0(\varepsilon)}(e^{2\pi i s}, 0) = {} & (2\mathrm{Re}\,(e) - q)A^2 \sin(-2\pi\varepsilon(s-1)) \\ & + (q - 2\mathrm{Re}\,(\bar{e}))A \sin(-\pi\varepsilon(s-1)) \\ & + 2A(\operatorname{Im}(\bar{e}) \cos(\pi\varepsilon(s-1)) + \operatorname{Im}(\bar{e}). \end{aligned}$$

Now the imaginary part of $\det \mathcal{A}_{A_0(\varepsilon)}(e^{i2\pi s}, 0)$ can be transformed as follows:

$$\begin{aligned} &\operatorname{Im} \det \mathcal{A}_{A_0(\varepsilon)}(e^{i2\pi s}, 0) \\ &= (2\mathrm{Re}\,(e) - q)A^2 \sin(-2\pi\varepsilon(s-1)) + (q - 2\mathrm{Re}\,(e))A \sin(-\pi\varepsilon(s-1)) \\ &\quad + 2A \operatorname{Im}(e) \cos(\pi\varepsilon(s-1)) + \operatorname{Im}(\bar{e}) \end{aligned}$$

$$
\begin{aligned}
&= (2\mathrm{Re}\,(e) - q)A\,[-A\sin(2\pi\varepsilon(s-1)) + \sin(\pi\varepsilon(s-1))] \\
&\quad + \frac{\mathrm{Im}\,(e)}{\sin(\pi\varepsilon)}\sin(\pi s\varepsilon)\cos(\pi\varepsilon(s-1)) + \mathrm{Im}\,(\bar{e}) \\
&= (q - 2\mathrm{Re}\,(e))A\left[\frac{\sin(\pi s\varepsilon)}{\sin(\pi\varepsilon)}\sin(2\pi\varepsilon(s-1)) - \sin(\pi\varepsilon(s-1))\right] \\
&\quad + \frac{\mathrm{Im}\,(e)}{2\sin(\pi\varepsilon)}[\sin((\pi s\varepsilon) + \pi(s-1)\varepsilon) + \sin((\pi s\varepsilon) - \pi(s-1)\varepsilon)] + \mathrm{Im}\,(\bar{e}) \\
&= \frac{(q - 2\mathrm{Re}\,(e))}{\sin(\pi\varepsilon)}A[\sin(\pi s\varepsilon)\sin(2\pi\varepsilon(s-1)) - \sin(\pi\varepsilon)\sin(\pi\varepsilon(s-1))] \\
&\quad + \frac{\mathrm{Im}\,(e)}{\sin(\pi\varepsilon)}[\sin(\pi\varepsilon(2s-1)) + \sin(\pi\varepsilon)] + \mathrm{Im}\,(\bar{e}) \\
&= \frac{(q - 2\mathrm{Re}\,(e))}{2\sin(\pi\varepsilon)}A[\cos(\pi s\varepsilon) - \cos(\pi\varepsilon(3s-2))] + \frac{\mathrm{Im}\,(e)}{\sin(\pi\varepsilon)}\sin(\pi\varepsilon(2s-1)) \\
&= \frac{(q - 2\mathrm{Re}\,(e))}{2\sin^2(\pi\varepsilon)}[\cos(\pi s\varepsilon)\sin(\pi s\varepsilon) - \cos(\pi\varepsilon(3s-2))\sin(\pi s\varepsilon)] \\
&\quad + \frac{\mathrm{Im}\,(e)}{\sin(\pi\varepsilon)}\sin(\pi\varepsilon(2s-1)) \\
&= \frac{(q - 2\mathrm{Re}\,(e))}{4\sin^2(\pi\varepsilon)}[\sin(2\pi s\varepsilon) - \sin(2\pi\varepsilon(2s-1)) - \sin(2\pi\varepsilon(s-1))] \\
&\quad + \frac{\mathrm{Im}\,(e)}{\sin(\pi\varepsilon)}\sin(\pi\varepsilon(2s-1)) \\
&= \frac{(q - 2\mathrm{Re}\,(e))}{2\sin^2(\pi\varepsilon)}[\sin(\pi\varepsilon)\cos(\pi\varepsilon(2s-1)) - \sin(\pi\varepsilon(2s-1))\cos(\pi\varepsilon(2s-1))] \\
&\quad + \frac{\mathrm{Im}\,(e)}{\sin(\pi\varepsilon)}\sin(\pi\varepsilon(2s-1)).
\end{aligned}
$$

Thus we have

$$
\begin{aligned}
&\mathrm{Im}\,\det \mathcal{A}_{A_0(\varepsilon)}(e^{2\pi i s}, 0) \\
&= \frac{q - 2\mathrm{Re}\,(e)}{2\sin^2\pi\varepsilon}[\sin(\pi\varepsilon)\cos(\pi\varepsilon(2s-1)) - \sin(\pi\varepsilon(2s-1))\cos(\pi\varepsilon(2s-1))] \\
&\quad + \frac{\mathrm{Im}\,(e)}{\sin\pi\varepsilon}\sin(\pi\varepsilon(2s-1)).
\end{aligned}
\tag{4.110}
$$

Moreover, the number of intersections of the curve $\det \mathcal{A}_{A_0(\varepsilon)}(e^{2\pi i s}, 0)$ with the real axis can be replaced by the number of roots in the interval $[-1, 1]$ of the equation

$$
p_1\sin(\pi\varepsilon t)\cos(\pi\varepsilon t) = p_2\sin(\pi\varepsilon t) + p_3\cos(\pi\varepsilon t), \tag{4.111}
$$

where

$$
p_1 = \frac{2\mathrm{Re}\,(e) - q}{2\sin^2\pi\varepsilon},\quad p_2 = -\frac{\mathrm{Im}\,(e)}{\sin\pi\varepsilon},\quad p_3 = \frac{2\mathrm{Re}\,(e) - q}{2\sin\pi\varepsilon}.
$$

Squaring both sides of (4.111), dividing by $\cos^2(\pi\varepsilon t)$ and making the substitution $y = \tan(\pi\varepsilon t)$, we obtain the quartic equation in y,

$$p_1^2 y^2 = (1+y^2)\left(p_2^2 y^2 + 2p_2 p_3 y + p_3^2\right),$$

which can have at most four roots y_i, $i = 1, \dots, 4$. The equation

$$\tan(\pi\varepsilon t) = y_k, \ k = 1, \dots, 4, \tag{4.112}$$

has at most one solution t_j on the segment $[-1, 1]$. Indeed, the solution of (4.112) has the form

$$\pi\varepsilon t_j = \arctan y_k + \pi j, \ j \in \mathbb{Z},$$

so

$$t_j = \frac{\arctan y_k}{\pi\varepsilon} + \frac{j}{\varepsilon}$$

and since $\varepsilon \in (-1/2, 1/2]$, then $|j/\varepsilon| \geq 2$. Thus both (4.111) and equation (4.109) have at most four solutions. □

Let us now consider the behavior of the sub-curve Γ_0^2 defined by (4.108).

Lemma 4.5.14. *Set $a = a(0)$, $b = b(0)$, $c = c(0)$, $d = d(0)$. Then the real and imaginary parts of* $\det \mathcal{A}_{A_0(\varepsilon)}(1,\mu)$, $\mu \in [0,1]$ *can be represented in the form*

$$\begin{aligned}
&\operatorname{Re}\{\det \mathcal{A}_{A_0(\varepsilon)}\}(1,\mu) \\
&= |a|^2 - |c|^2 + 2\operatorname{Im}(c\bar{d} - a\bar{b})\frac{2\sin(2\pi\rho)\mu(1-\mu)}{(2\mu-1)^2 + 4\mu(1-\mu)\cos^2(\pi\rho)} \\
&\quad + (|b|^2 - |d|^2) \\
&\quad \times \left[\left(\frac{2\sin(2\pi\rho)\mu(1-\mu)}{(2\mu-1)^2 + 4\mu(1-\mu)\cos^2(\pi\rho)}\right)^2 - \left(\frac{2\mu-1}{(2\mu-1)^2 + 4\mu(1-\mu)\cos^2(\pi\rho)}\right)^2\right],
\end{aligned}$$

and

$$\begin{aligned}
&\operatorname{Im}\{\det \mathcal{A}_{A_0(\varepsilon)}\}(1,\mu) \\
&= -2\left[\operatorname{Im}(c\bar{d} - a\bar{b}) + (|b|^2 - |d|^2)\frac{2\sin(2\pi\rho)\mu(1-\mu)}{(2\mu-1)^2 + 4\mu(1-\mu)\cos^2(\pi\rho)}\right] \\
&\quad \times \frac{2\mu-1}{(2\mu-1)^2 + 4\mu(1-\mu)\cos^2(\pi\rho)}.
\end{aligned} \tag{4.113}$$

Proof. First of all we note the following well-known identities:

$$\overline{\cot(\pi z)} = \cot(\pi\bar{z}), \ \sin(iz) = i\sinh z, \ \cos(iz) = \cosh z, \quad z \in \mathbb{C}, \ i^2 = -1,$$

which are used without additional references.

Setting

$$\gamma = \cot\left(\pi\left(\frac{1}{2} - \rho + \frac{i}{2\pi}\log\frac{\mu}{1-\mu}\right)\right),$$

we have

$$\mathcal{A}_{A_0(\varepsilon)}(1,\mu) = \begin{pmatrix} a - ib\gamma & c - id\gamma \\ \bar{c} + i\bar{d}\gamma & \bar{a} + i\bar{b}\gamma \end{pmatrix}$$

with determinant

$$\det \mathcal{A}_{A_0(\varepsilon)}(1,\mu) = |a|^2 - |c|^2 + 2\mathrm{Im}\,(c\bar{d} - a\bar{b})\gamma + (|b|^2 - |d|^2)\gamma^2.$$

Hence the real and imaginary parts of the determinant $\det \mathcal{A}_{A_0(\varepsilon)}$ can be expressed in the form

$$\begin{aligned} &\mathrm{Re}\,\{\det \mathcal{A}_{A_0(\varepsilon)}\}(1,\mu) \\ &= |a|^2 - |c|^2 + 2\mathrm{Im}\,(c\bar{d} - a\bar{b})\mathrm{Re}\,\gamma + (|b|^2 - |d|^2)((\mathrm{Re}\,\gamma)^2 - (\mathrm{Im}\,\gamma)^2), \quad (4.114) \\ &\mathrm{Im}\{\det \mathcal{A}_{A_0(\varepsilon)}\}(1,\mu) = 2\left[\mathrm{Im}\,(c\bar{d} - a\bar{b}) + (|b|^2 - |d|^2)\mathrm{Re}\,\gamma\right]\mathrm{Im}\,\gamma. \quad (4.115) \end{aligned}$$

However,

$$\mathrm{Re}\,\gamma = \frac{2\sin(2\pi\rho)\mu(1-\mu)}{(2\mu-1)^2 + 4\mu(1-\mu)\cos^2(\pi\rho)}, \quad (4.116)$$

$$\mathrm{Im}\,\gamma = -\frac{2\mu-1}{(2\mu-1)^2 + 4\mu(1-\mu)\cos^2(\pi\rho)}, \quad (4.117)$$

so substituting $\mathrm{Re}\,\gamma$ and $\mathrm{Im}\,\gamma$ in (4.114), (4.115) by their expressions (4.116), (4.117) we obtain the claim. □

Theorem 4.5.15. *If $A_0(\varepsilon)$ is a Fredholm operator, then $|\kappa_0| \le 3$. Moreover, if $\varepsilon = 0$, then $|\kappa_0| \le 2$.*

Proof. Consider first the case $\varepsilon \neq 0$. By Lemma 4.5.13 the curve Γ_0^1 has at most four common points with the real axis $\mathbb{R}$, and equation (4.113) implies that the curve Γ_0^2 and $\mathbb{R}$ have at most three common points. Applying Lemma 4.5.12 one obtains the estimate $|\kappa_0| \le 3$. Consider now the case $\varepsilon = 0$. Then

$$\det \mathcal{A}_{A_0(0)}(e^{2\pi i s}, 0) = es^2 + (1-s)sq + \bar{e}(1-s)^2 = (2\mathrm{Re}\,(e) - q)s^2 + (q - 2\bar{e})s + \bar{e}.$$

It follows that

$$\mathrm{Im}\det \mathcal{A}_{A_0(0)}(e^{2\pi i s}, 0) = (1-2s)\mathrm{Im}(\bar{e}),$$

from which $\mathrm{Im}\det \mathcal{A}_{A_0(\varepsilon)}(e^{2\pi i s}, 0)$ is either identically zero or has one root, $s = 1/2$. Using Lemma 4.5.12 once more, one obtains $|\kappa_0| \le 2$. □

Thus the indices κ of the local operators, for the approximation methods under consideration for singular integral equations on open contours, are in the set $S = \{-3,-2,-1,0,1,2,3\}$. These estimates differ from those for singular operators considered on closed contours. Recall that for the space l_2 the corresponding set is $S = \{-1,0,1\}$, and for the weighted l_2 spaces, $S = \{-2,-1,0,1,2\}$. However, it is possible to choose parameters of the approximation method and the space where the local operators are considered such that the index of the corresponding local operator becomes zero. To show this we need a few auxiliary results.

Lemma 4.5.16. *If $a\bar{b} - c\bar{d} \notin \mathbb{R}$, then the sign of $\varepsilon \in (-1/2, 1/2]$, $\varepsilon \neq 0$ can be chosen in such a way that the curve Γ_0^1 has at most one intersection with the real axis.*

Proof. Considering the equation

$$\operatorname{Im}\{\det \mathcal{A}_{A_0(\varepsilon)}\}(e^{i2\pi s}, 0) = 0, \tag{4.118}$$

and taking into account relation (4.110), we can write (4.118) in the form

$$(q - 2\mathrm{Re}\,(e)) \left(1 - \frac{\sin(\pi\varepsilon\lambda)}{\sin(\pi\varepsilon)}\right) = -2\mathrm{Im}\,(e)\tan(\pi\varepsilon\lambda), \quad \lambda \in (-1, 1). \tag{4.119}$$

Note that $\operatorname{Im}(a\bar{b} - c\bar{d}) = -2\mathrm{Im}\,(e)$, hence $\mathrm{Im}\,(e) \neq 0$. Moreover, for each $\varepsilon \in (-1/2, 1/2], \varepsilon \neq 0$, the function

$$\Phi(\lambda) = 1 - \frac{\sin(\pi\varepsilon\lambda)}{\sin(\pi\varepsilon)}$$

is monotonically decreasing. We choose the sign of ε so that the function

$$\mathrm{Im}\,(e)\tan(\pi\varepsilon\lambda), \quad \lambda \in (-1, 1),$$

has the same character of monotonicity as the left-hand side of (4.119). Then equation (4.119) has at most one root, and the proof is complete. □

Corollary 4.5.17. *If the conditions of Theorem 4.5.8 hold, then there is an $\varepsilon \neq 0$ such that the index κ_0 of the corresponding local operator $A_0(\varepsilon)$ satisfies the inequality $|\kappa_0| \leq 2$.*

Proof. If one chooses an ε according to Lemma 4.5.16, then the curves Γ_0^1 and Γ_0^2 have at most four common points with the real axis $\mathbb{R}$, so the result again follows from Lemma 4.5.12. □

Lemma 4.5.18. *Set*

$$A = \mathrm{Im}\,(c\bar{d} - a\bar{b}), \quad B = |b|^2 - |d|^2 \tag{4.120}$$

and suppose $A \neq 0$. Then the following assertions are true:

1. *If $AB < 0$, then for all ρ,*

$$\rho \in \left(-\frac{1}{2}, -\frac{1}{2} + \frac{1}{2\pi}\operatorname{arccot}\left(-\frac{B}{2A}\right)\right) \bigcup \left(0, \frac{1}{2\pi}\operatorname{arccot}\left(-\frac{B}{2A}\right)\right),$$

the equation

$$\operatorname{Im}\{\det \mathcal{A}_{A_0(\varepsilon)}\}(1, \mu) = 0 \tag{4.121}$$

has only one root $\mu = 1/2$ in the interval $[0, 1]$.

2. *If $AB > 0$, then for all ρ,*

$$\rho \in \left(-\frac{1}{2} + \frac{1}{2\pi}\operatorname{arccot}\left(-\frac{B}{2A}\right), 0\right) \bigcup \left(\frac{1}{2\pi}\operatorname{arccot}\left(-\frac{B}{2A}\right), \frac{1}{2}\right),$$

the equation (4.121) has only one root $\mu = 1/2$ in the interval $[0, 1]$.

3. *If $B = 0$, then for all $\rho \in (-1/2, 1/2)$ the equation (4.121) has only one root $\mu = 1/2$ in the interval $[0, 1]$.*

Proof. Let us denote the expression in the square brackets of (4.113) by $R(\mu)$. Then

$$R(\mu) = \frac{(4A\sin^2(\pi\rho) - 2B\sin(2\pi\rho))\mu^2 - (4A\sin^2(\pi\rho) - 2B\sin(2\pi\rho))\mu + A}{(2\mu-1)^2 + 4\mu(1-\mu)\cos^2(\pi\rho)}, \tag{4.122}$$

and consider the behaviour of the numerator of the fraction (4.122). Let D denote the discriminant of the trinomial in this numerator. Since

$$4A\sin^2(\pi\rho) - 2B\sin(2\pi\rho) = 4\sin(\pi\rho)[A\sin(\pi\rho) - B\cos(\pi\rho)],$$

we obtain

$$\begin{aligned} D &= -8\sin(2\pi\rho)[A\sin(\pi\rho) - B\cos(\pi\rho)][A\cos(\pi\rho) + B\sin(\pi\rho)] \\ &= -8\sin(2\pi\rho)[(A^2 - B^2)\sin(\pi\rho)\cos(\pi\rho) + AB(\sin^2(\pi\rho) - \cos^2(\pi\rho))] \\ &= -4\sin^2(2\pi\rho)[A^2 - B(B + 2A\cot(2\pi\rho))]. \end{aligned}$$

Suppose $AB \neq 0$. Choosing ρ from the corresponding set, we have that $B + 2A\cot(2\pi\rho)$ has sign opposite to the sign of B. This implies that $D < 0$ and that the numerator of (4.122) does not have any roots on $[0, 1]$. □

Consequently, if we choose an ε^* as in Lemma 4.5.16 and a ρ^* as in Lemma 4.5.18, then the curve Γ_0 will have at most two common points with the real axis, denoted by P_1 and P_2. According to Lemma 4.5.12, the index of $A_1^0(\varepsilon^*) : l_{2,-\rho^*} \to l_{2,-\rho^*}$ can be estimated by

$$|\operatorname{ind} A_1^0(\varepsilon^*)| \leq 1.$$

However, if the points P_1 and P_2 are located both on the right or both on the left of the origin, then

$$\operatorname{ind} A_1^0(\varepsilon^*) = 0.$$

Let us consider the quadratic equation

$$Bz^2 + 2Az + C = 0, \tag{4.123}$$

where B and A are defined by (4.120) and $C = |a|^2 - |c|^2$. In addition, we put

$$\alpha_{AB} := \frac{1}{2}\operatorname{arccot}\left(-\frac{B}{2A}\right) - \frac{\pi}{2}, \quad \beta_{AB} := \frac{1}{2}\operatorname{arccot}\left(-\frac{B}{2A}\right).$$

Theorem 4.5.19. *Let*

$$(|a|^2 - |c|^2)(|b|^2 - |d|^2) \leq 0$$

and assume at least one of the following conditions is satisfied:

1. $AB < 0$ *and at least one root of equation* (4.123) *belongs to the set* E^1_{AB},

$$E^1_{AB} := (-\infty, \tan \alpha_{AB}) \cup (0, \tan \beta_{AB});$$

2. $AB > 0$ *and at least one root of equation* (4.123) *belongs to the set* E^2_{AB},

$$E^2_{AB} := (\tan \alpha_{AB}, 0) \cup (\tan \beta_{AB}, +\infty);$$

3. $B = 0$ *and* $A \neq 0$.

Then there exists $\rho^* \in (-1/2, 1/2)$ *such that the index of the operator* $A_0(\varepsilon^*)$: $l_{2,-\rho^*} \to l_{2,-\rho^*}$ *is equal to zero.*

Proof. We take an ε^* and a ρ such as in Lemmas 4.5.16 and 4.5.18, respectively. Then the curve Γ_0 has at most two intersections with the real axis $\mathbb{R}$. Let P_2 be the common point of Γ_0^2 and $\mathbb{R}$. Then

$$\begin{aligned} P_2 &= \operatorname{Re}\{\det \mathcal{A}_{A_0(\varepsilon^*)}\}(1, 1/2) \\ &= |a|^2 - |c|^2 + 2\operatorname{Im}(c\bar{d} - a\bar{b}) \tan(\pi\rho) + (|b|^2 - |d|^2) \tan^2(\pi\rho). \end{aligned}$$

It is not hard to prove that each of the conditions of Theorem 4.5.19 implies that the function $B(\rho) = \operatorname{Re}\{\det \mathcal{A}_{A_0(\varepsilon^*)}\}(1, 1/2)$ takes both positive and negative values, hence we can choose a ρ^* such that the points P_1 and P_2 are located on the same side of the origin. This finishes the proof. □

4.5.4 Examples of the Essential Spectrum of Local Operators

Let us illustrate the results of Sections 4.5.2 – 4.5.3 by a few examples. We present the essential spectrum $\sigma_{ess}(A_0(\varepsilon))$ of local operators $A_0(\varepsilon)$ considered on spaces $l_{2,-\rho}$ for various values of the parameters $\varepsilon \in (-1/2, 1/2)$ and $\rho \in (-1/2, 1/2)$. As was already mentioned, the set $\sigma_{ess}(A_0(\varepsilon))$ consists of two curves Γ_0^1 and Γ_0^2. Let us draw the graphs of these curves such that Γ_0^1 is represented by a dashed line and Γ_0^2 by a solid line. The values $a := a(0)$, $b := b(0)$, $c := c(0)$ and $d := d(0)$ of the coefficients of the singular integral equations (4.107) at the point $t = 0$ are displayed in the title of the corresponding figure. The values of the parameters ε and ρ are given on the top of each subgraph.

We start with a simple example illustrating Theorem 4.5.19. If the coefficients of (4.88) are $a = 3$, $b = 2i$, $c = 1$, $d = -2i$, then $B = 0$, $A = 8$ and $P_2 = P_2(\rho) = 8 + 16\tan(\pi\rho)$. The function $P_2(\rho)$ takes both positive and negative values, so it is possible to choose a $\rho' \in (-1/2, 1/2)$ such that the points P_1 and $P_2(\rho')$ are

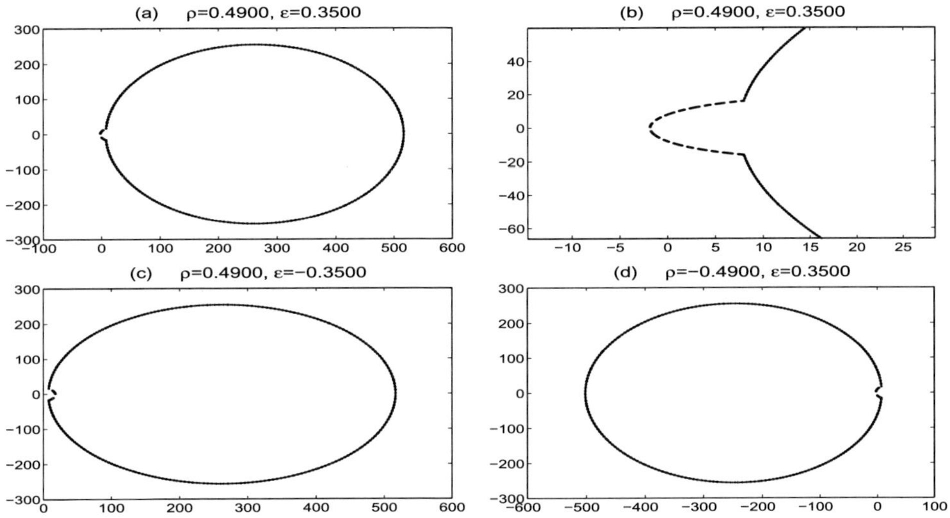

Figure 4.3: Essential spectrum of the operator $A_0(\varepsilon)$; $a = 3i$, $b = 2i$, $c = 1$, $d = -2i$.

located on the same side of the origin. Thus the index of the corresponding local operator will be equal to zero.

The graphs of the curve Γ_0 for some local operators of the approximation methods are presented in Figure 4.3 on page 227. Graphs 4.3(a) and 4.3(b) show that the local operator $A_0(0.35)$ on the space $l_{2,-0.49}$ has index $\kappa = -1$. To obtain an appropriate local operator with index zero one can change either the sign of ε (Graph 4.3(c)) or the sign of ρ (Graph 4.3(d)). In the first case, the local operator $A_0(-0.35)$ has index zero on the space $l_{2,-0.49}$, but in the second case the local operator $A_0(0.35)$ remains the same. However, it again has index zero but on the space $l_{2,0.49}$.

Let us now analyze a more complicated situation. The curve $\Gamma_0 = \Gamma_0^1 \cup \Gamma_0^2$ in Figure 4.4(a) has four intersections with the real axis $\mathbb{R}$; the operator $A_0(-0.45)$ on the space $l_{2,-0.30}$ is Fredholm and has index $\kappa = 2$. This operator $A_0(-0.45)$ is also Fredholm on the space $l_{2,0.30}$, but then its index is different, viz., $\kappa = 0$, cf. 4.4(b).

On the other hand, if one changes the sign of the parameter ε to $+$ and considers the corresponding local operator $A_0(0.45)$ on the space $l_{2,0.25}$, then the graphs 4.4(c) and 4.4(d) show that this local operator is again Fredholm; the corresponding curve Γ_0 has two coils, but $\kappa(A_0(0.45)) = 1$.

Note that Graph 4.4(d) is a "zoomed" image of the second coil, which is indiscernible in Graph 4.4(c).

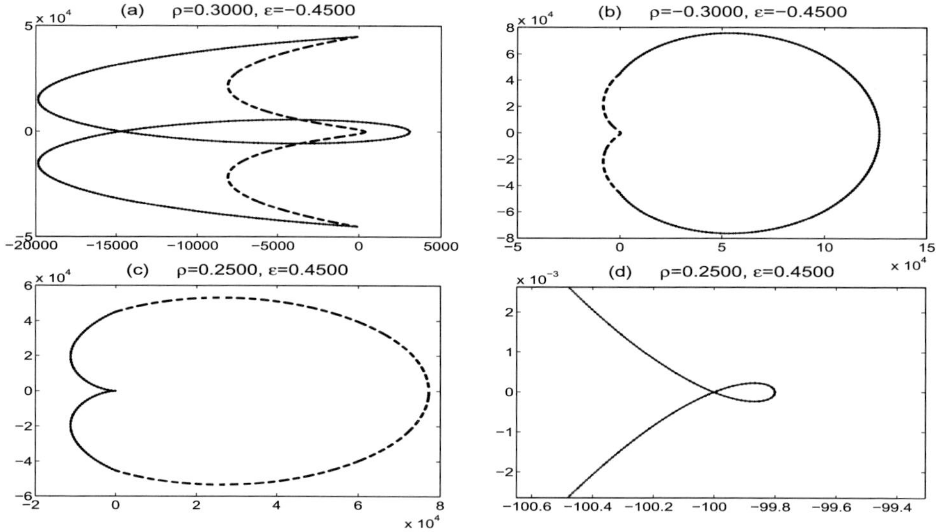

Figure 4.4: Essential spectrum of the operator $A_0(\varepsilon)$; $a = 150i$, $b = 150$, $c = 10i$, $d = 0.01$.

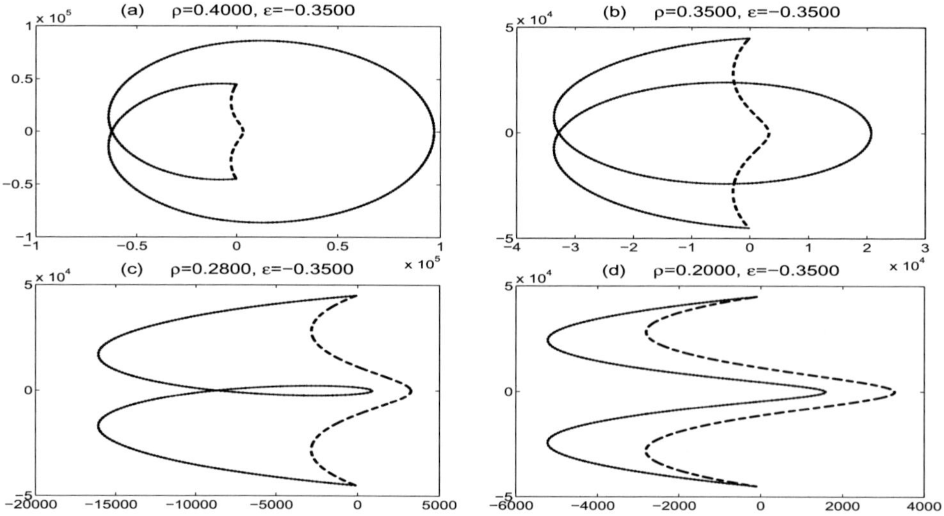

Figure 4.5: Essential spectrum of the operator $A_0(\varepsilon)$; $a = 150i$, $b = 150$, $c = 10i$, $d = 0.01$.

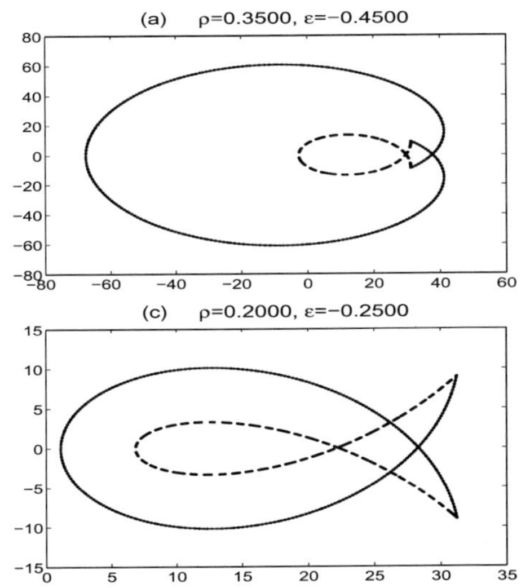

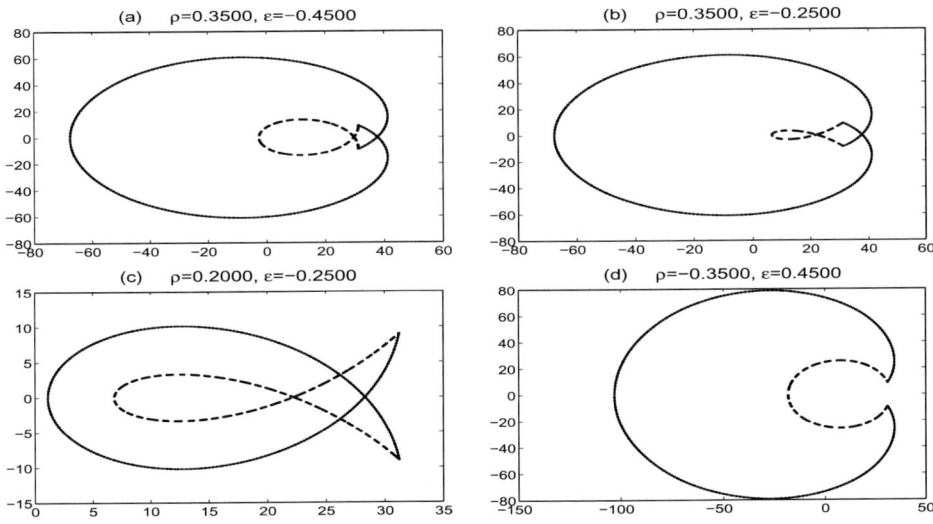

Figure 4.6: Essential spectrum of the operator $A_0(\varepsilon)$; $a = -3 - 0.5i$, $b = -1$, $c = -1 - i$, $d = 5i$.

Let us separately analyze the behaviour of the curves Γ_0^1 and Γ_0^2 by fixing one parameter (ε or ρ) and changing another. In the next example, the coefficients of the initial equation are the same as in Figure 4.4 and $\varepsilon = -0.35$; the essential spectrum of the operator $A_0(-0.35)$ in various spaces $l_{2,-\rho}$ is presented. Considered on $l_{2,-0.4}$ the index of this operator is equal to 2 (Graph 4.5(a)). If the parameter ρ decreases, then the shrinking coil of the curve Γ_0^2 moves to the left (Graphs 4.5(b) and 4.5(c)). As a result, the origin falls away from the set bounded by the curves Γ_0^1 and Γ_0^2, so the index of the operator under consideration in the space $l_{2,-0.2}$ is equal to zero (see Graph 4.5(d)).

Let us now consider how the essential spectrum of the local operators $A_0(\varepsilon)$ depends on the parameter ε. Starting with the Graph 4.6(a), one notes that the curves Γ_0^1 and Γ_0, respectively, have three and six common points with the real axis $\mathbb{R}$, and the local operator $A_0(-0.45)$ has index $\kappa = -2$ on $l_{2,-0.4}$. By changing the parameter ε to -0.35, one can reduce the index of the local operator to -1 (Graph 4.6(b)). However, in this special case it is not possible to obtain a local operator having index zero by varying the parameter ε only. To this end, local operators must be considered on appropriate spaces other than the space $l_{2,-0.35}$. One such case is established on Graph 4.6(c). The number of intersections of the curve Γ_0 with $\mathbb{R}$ is the same as before, but in comparison to the previous figure the left border of the graph has moved to the right, so the origin then falls outside of the curve Γ_0. Another approach is presented on the Graph 4.6(d). As was stated in Lemma 4.5.16, the sign of ε can be chosen so that Γ_0^1 will have at most one intersection point with $\mathbb{R}$. On the other side, Lemma 4.5.18 presents conditions when $\mathbb{R}$ and Γ_0^2 have only one common point; and Theorem 4.5.19 provides conditions when the index of the local operator can be made equal to zero. By changing the sign of parameters from the Graph 4.6(a) one notes that on the space $l_{2,0.35}$ the local operator $A_0(0.45)$ has the index $\kappa = 0$.

These examples and others not presented here, reflect the properties of the essential spectrum of the local operators $A_0(\varepsilon)$ very well. However, although the situation where the curve Γ_0 has three coils is not rare, we were unable to find local operators with the index 3 or -3. It seems that the modulus $\kappa(A_0(\varepsilon))$ cannot exceed 2, which is similar to the behaviour of local operators of approximation methods for singular integral equations with conjugation on closed contours with corner points. Thus a more detailed study of the symbol of local operators is needed.

4.6 Comments and References

Equations of type (4.1) containing the operator of complex conjugation often occur in elasticity theory, hydrodynamics, acoustics, and in deformation theory of solid surfaces (e.g., see [71, 70, 72, 73, 121, 123, 132, 133, 143, 150, 161, 168, 221, 222]). Fredholm properties of such equations on curves with corner points have been established in [74, 75, 76, 136, 137, 138, 164, 165, 213, 218]. In particular, it was shown that the operators arising in case of curves with angular points locally are equivalent to Mellin convolution operators. Moreover, there is a very well developed theory of approximation methods for Cauchy singular integral equations without conjugation on smooth and non-smooth open and closed contours and for various classes of Mellin operators [14, 19, 79, 80, 102, 104, 114, 115, 116, 180, 181, 186, 187, 179, 183, 189]. On the other hand, the methods of approximate solution of singular integral equations with conjugation based on replacement of the initial equation by a system of integral equations without conjugation meet additional difficulties because the operator $MS_\Gamma - S_\Gamma M$ is non-compact [166]. However, the algebraic approach to the stability investigation can be successfully applied in this situation too. One of the reasons for this is because in the case under consideration, all relevant operators A interact well with the operator of complex conjugation M in the sense that the operators A and MAM belong to the same operator algebras.

Sections 4.1–4.2: For quadrature method (4.4) the stability results were announced in [46]. Other results of these sections are taken from the subsequent paper [54]. The proof of Lemma 4.2.4 used here is from [55]. The initial proof given in [54] is more technical. Note that it was discovered in [54] that the local operators of approximation methods could have non-zero indices. This effect does not have a place for singular integral equations without conjugation where the indices of the corresponding operators are always zero [180].

The stability analysis of the approximation methods considered in Section 4.1 can also be based upon the fact that the related approximation sequences are indeed contained in the extension $\tilde{\mathcal{C}}$ of the complex algebra $\mathcal{C}$ defined on page 76 (cf. [183, Sections 11.1–11.17]).

Sections 4.3: The material of this section is taken from [55]. It turns out that consideration of approximation methods in spaces with weights leads to new effects, e.g., the indices of local operators of approximation methods can also take values 2 and -2. On the other hand, this gives more flexibility since one can vary the spaces where an approximation method is considered to find a situation when the local operators of the approximation method will possess the necessary properties. In particular, Section 4.3.4 uses the effect mentioned to modify quadrature methods in the situation when associated local operators are not invertible. Here we follow the approach of [188].

Section 4.4: When Γ is the boundary of a polygonal domain, and $a = b = 1$ on Γ, M. Costabel and E. Stephan [33, 35] established the stability of the 0–collocation and Galerkin methods for equation (4.2) and K.E. Atkinson, I.G. Graham, G.A. Chandler, and R. Kress studied the stability and convergence of quadrature methods [5, 98, 128]. The decisive observation widely used in these papers is that, under the above restrictions, the operator in (4.2) can be written as

$$A = I + R_1 + T_1 \tag{4.124}$$

where I denotes the identity operator, $\|R_1\| < 1$, and T_1 is compact. The representation (4.124) is no longer valid if the coefficients of A are not equal to 1.

Sections 4.5.1 – 4.5.3: Material of these sections is taken from [49, 63]. Initial study undertaken in [63] bounds the indices of local operators by numbers -3 and 3. However, the examples from Section 4.5.4 indicate that absolute value of the indices probably do not exceed 2, which is similar to the situation for closed contours with corner points. Numerical testing of the quadrature methods for singular integral equations with conjugation on open contours is undertaken in [53]

Note that the corresponding quadrature method for Mellin operators without conjugation has been studied in [187].

Chapter 5

Approximation Methods for the Muskhelishvili Equation

5.1 Boundary Problems for the Biharmonic Equation

Let $D \subset \mathbb{R}^2$ denote a domain bounded by a simple closed piecewise smooth contour Γ and let $\overline{D} := D \cup \Gamma$. It is well known that many problems in plane elasticity, radar imaging, and theory of slow viscous flows can be reduced to the biharmonic problem

$$\Delta^2 \mathbf{U}(x,y) = 0, \quad (x,y) \in D, \tag{5.1}$$

where Δ is the Laplace operator (3.48). We assume that the function $\mathbf{U}$ is from the space $W_p^1(\overline{D}) \cap W_p^4(D)$. The notation $W_p^k(X)$ is used for the Sobolev space of k-times differentiable functions on X, the derivatives of which belong to the corresponding space $L_p(X)$.

Equation (5.1) has been thoroughly studied [101]. Our aim now is to solve boundary problems for this equation using the theory of functions of complex variables. Thus let us identify sets from $\mathbb{R}^2$ and $\mathbb{C}$ via the correspondence $\mathbb{R}^2 \ni (x,y) \longleftrightarrow z = x + iy \in \mathbb{C}$, and denote the associated sets in $\mathbb{R}^2$ and $\mathbb{C}$ by the same symbol. Any function f biharmonic in a domain D can be represented as a combination of two functions analytic in D, viz., the following result is true.

Lemma 5.1.1 (Goursat Representation). *If D is a domain bounded by a simple closed contour, and if u is a biharmonic function in D, then there are two functions χ and φ that are analytic in D and such that*

$$u(x,y) = \operatorname{Re}\left(\overline{z}\varphi(z) + \chi(z)\right), \quad z = x + iy. \tag{5.2}$$

Proof. Since $\Delta^2 u = \Delta(\Delta u)$, the function Δu is harmonic in the domain D, so it is the real part of an analytic function $\psi = \psi(z)$, $z \in D$, i.e.,

$$\Delta u(x,y) = \operatorname{Re}\psi(z). \tag{5.3}$$

Choose a point $z_0 \in D$ and introduce a function

$$\varphi(z) := \frac{1}{4}\int_{z_0}^{z} \psi(\xi)\, d\xi, \quad z \in D.$$

Considering now the operators

$$\frac{\partial}{\partial z} = \frac{1}{2}\left(\frac{\partial}{\partial x} - i\frac{\partial}{\partial y}\right),$$
$$\frac{\partial}{\partial \overline{z}} = \frac{1}{2}\left(\frac{\partial}{\partial x} + i\frac{\partial}{\partial y}\right)$$

one notes that

$$\Delta = 4\frac{\partial^2}{\partial z \partial \overline{z}},$$

so

$$\Delta(\overline{z}\varphi(z)) = 4\frac{\partial^2}{\partial z\partial \overline{z}}(\overline{z}\varphi(z)) = 4\frac{\partial}{\partial z}(\varphi(z)) = \psi(z). \tag{5.4}$$

If we define a function v by

$$v := u - \operatorname{Re}(\overline{z}\varphi(z)),$$

then from (5.3), (5.4),

$$\Delta v = \Delta(u - \operatorname{Re}(\overline{z}\varphi(z)) = 0$$

everywhere in D. Thus, v is a harmonic function in D, so it is the real part of an analytic function χ, i.e.,

$$u - \operatorname{Re}(\overline{z}\varphi(z)) = \operatorname{Re}\chi(z),$$

or

$$u(x, y) = \operatorname{Re}(\overline{z}\varphi(z) + \chi(z)),$$

and the proof is complete. □

5.1.1 Reduction of Biharmonic Problems to a Boundary Problem for Two Analytic Functions

Let $c_1, c_2, \ldots, c_l$ be the corner points of Γ. By ω_j we denote the angle between the corresponding semi-tangents at the point c_j. Hence,

$$\omega_j \in (0, 2\pi), \quad \omega_j \neq \pi, \quad j = 1, 2, \ldots, l.$$

For real numbers p and α_j $j = 1, 2, \ldots, l$ satisfying the inequalities $p > 1$ and

$$0 < \alpha_j + \frac{1}{p} < 1, \quad j = 1, 2, \ldots, l, \tag{5.5}$$

we introduce the weight function

$$\rho = \rho(t) = \prod_{j=1}^{l} |t - c_j|^{\alpha_j}, \quad t \in \Gamma.$$

By $L_p(\Gamma, \rho)$ we denote the set of all Lebesgue measurable functions $f : \Gamma \mapsto \mathbb{C}$ such that

$$||f|| := \left(\int_\Gamma |f(t)\rho(t)|^p |dt| \right)^{1/p} < \infty.$$

Correspondingly, $W_p^1(\Gamma, \rho)$ refers to the Sobolev space of all Lebesgue measurable functions $f : \Gamma \mapsto \mathbb{C}$ such that

$$||f||_{W_p^1(\Gamma,\rho)} := \left(\int_\Gamma |f'(t)\rho(t)|^p |dt| + \int_\Gamma |f(t)\rho(t)|^p |dt| \right)^{1/p} < \infty.$$

We also consider the space $H_\mu = H_\mu(\Gamma)$ of Hölder continuous functions provided with the norm

$$||f||_{H_\mu} := \max_{t\in\Gamma} |f(t)| + \sup_{t_1,t_2\in\Gamma, t_1\neq t_2} \frac{|f(t_1) - f(t_2)|}{|t_1 - t_2|^\mu}.$$

It is known that the above-defined Sobolev space $W_p^1(\Gamma, \rho)$ is continuously embedded into a Hölder space, viz., the following result holds.

Lemma 5.1.2 ([72, 73]). *The Sobolev space $W_p^1(\Gamma, \rho)$ is continuously embedded into the Hölder space $H_{1-1/q}(\Gamma)$, where*

$$q = p \quad \textit{if} \quad \alpha_1 = \alpha_2 = \ldots = \alpha_l = 0,$$

and

$$1 < q < \min\left(p, \frac{p}{1+\alpha_1}, \ldots, \frac{p}{1+\alpha_l} \right) \quad \textit{if} \quad \sum_{j=1}^{l} \alpha_j^2 \neq 0.$$

Moreover, every function $f \in W_p^1(\Gamma, \rho)$ is Hölder continuous with the exponent $1 - 1/p$ everywhere on Γ except for the corner points $c_j, j = 1, \ldots, l$, in neighbourhoods of which it is Hölder continuous with exponents $1 - 1/q_j$, where $q_j \in (1, \min\{p/(1 + \alpha_j)\})$.

Let us now consider boundary problems for the biharmonic equation.

Lemma 5.1.3. *If $G_1, G_2 \in L_p(\Gamma, \rho)$, and if $\mathbf{U} = \mathbf{U}(x, y)$ is a solution of the biharmonic problem*

$$\begin{aligned} &\Delta^2 \mathbf{U}|_D = 0, \\ &\left.\frac{\partial \mathbf{U}}{\partial x}\right|_\Gamma = G_1, \quad \left.\frac{\partial \mathbf{U}}{\partial y}\right|_\Gamma = G_2, \end{aligned} \tag{5.6}$$

then $\mathbf{U}$ *can be represented in the form* (5.2), *where* φ *and* χ *are analytic functions in the domain* D *satisfying the boundary condition*

$$\varphi(t) + t\overline{\varphi'(t)} + \overline{\chi'(t)} = f(t), \quad t = x + iy \in \Gamma, \tag{5.7}$$

with $f(t) := G_1(t) + iG_2(t)$. *Moreover, if* φ *and* χ *are analytic functions satisfying* (5.7), *then the biharmonic function* $\mathbf{U}$ *constructed by* (5.2) *is a solution of the boundary value problem* (5.6).

Proof. Since for any analytic functions φ and χ function (5.2) is biharmonic, our task is to find φ and χ to satisfy boundary conditions (5.6).

Let $\varphi = u + iv$ and $z = x + iy$, with $u = \operatorname{Re}\varphi$, $v = \operatorname{Im}\varphi$. Then

$$\begin{aligned}\mathbf{U}(x,y) &= \operatorname{Re}\left[\bar{z}\cdot(u+iv)(z) + \chi(z)\right] \\ &= x\cdot u(x,y) + y\cdot v(x,y) + \operatorname{Re}\left[\chi(x,y)\right], \quad z \in D.\end{aligned}$$

From this immediately follows

$$\begin{aligned}\frac{\partial \mathbf{U}}{\partial x} &= u + x\frac{\partial u}{\partial x} + y\frac{\partial v}{\partial x} + \frac{\partial \operatorname{Re}[\chi]}{\partial x}, \\ \frac{\partial \mathbf{U}}{\partial y} &= x\frac{\partial u}{\partial y} + v + y\frac{\partial v}{\partial y} + \frac{\partial \operatorname{Re}[\chi]}{\partial y}.\end{aligned}$$

Letting z tend to Γ and taking into account the Cauchy-Riemann equations and boundary conditions (5.6) we obtain

$$\begin{aligned}\varphi(t) + t\overline{\varphi'(t)} + \overline{\chi'(t)} &= \varphi + x\left(\frac{\partial u}{\partial x} - i\frac{\partial v}{\partial x}\right) + iy\left(\frac{\partial v}{\partial y} - i\frac{\partial v}{\partial x}\right) \\ &\quad + \frac{\partial \operatorname{Re}[\chi]}{\partial x} - i\frac{\partial \operatorname{Im}[\chi]}{\partial x} \\ &= (u+iv) + x\left(\frac{\partial u}{\partial x} + i\frac{\partial u}{\partial y}\right) + y\left(\frac{\partial v}{\partial x} + i\frac{\partial v}{\partial y}\right) \\ &\quad + \frac{\partial \operatorname{Re}[\chi]}{\partial x} + i\frac{\partial \operatorname{Re}[\chi]}{\partial y} \\ &= \frac{\partial \mathbf{U}}{\partial x} + i\frac{\partial \mathbf{U}}{\partial y} = G_1(t) + iG_2(t), \quad t \in \Gamma,\end{aligned} \tag{5.8}$$

so the functions φ and χ satisfy the boundary problem (5.7). On the other hand, the converse follows from the first and last lines of transformation (5.8). □

Let $\alpha = \alpha(t), t \in \Gamma$ be the angle between the real axis and the outward normal n to Γ at the point t. By s we denote the unit vector such that the angle between s and the real axis is $\alpha - \pi/2$.

Lemma 5.1.4. *If* $(f_1, f_2) \in W_p^1(\Gamma, \rho) \times L_p(\Gamma, \rho)$ *and* $\mathbf{U} = \mathbf{U}(x, y)$ *is a solution of the biharmonic Dirichlet problem*

$$\begin{aligned} &\Delta^2 \mathbf{U}|_D = 0, \\ &\mathbf{U}|_\Gamma = f_1, \quad \left.\frac{\partial \mathbf{U}}{\partial \mathrm{n}}\right|_\Gamma = f_2, \end{aligned} \tag{5.9}$$

then it can be represented in the form (5.2), *where* φ *and* χ *are analytic functions in the domain* D *satisfying the boundary condition*

$$\overline{\varphi(t)} + \bar{t}\varphi'(t) + \chi'(t) = e^{-i\alpha}\left(f_2 + i\frac{\partial f_1}{\partial s}\right), \quad t \in \Gamma. \tag{5.10}$$

Moreover, if φ *and* χ *are analytic functions satisfying* (5.10), *and* $\mathbf{U}$ *is the corresponding biharmonic function* $\mathbf{U}$ *constructed by* (5.2), *then there is a constant* $c \in \mathbb{R}$ *such that* $\mathbf{U} + c$ *is a solution of the boundary value problem* (5.9).

Proof. Using representation (5.2) for the biharmonic function $\mathbf{U}$, let us first find appropriate functions φ and χ to satisfy the boundary conditions (5.9).

Since f_1 belongs to the Sobolev space $W_p^1(\Gamma, \rho)$ and $\mathbf{U}$ is differentiable on $\overline{D}$, one can write

$$\begin{aligned} f_2 + i\frac{\partial f_1}{\partial s} &= \frac{\partial \mathbf{U}}{\partial \mathrm{n}} + i\frac{\partial \mathbf{U}}{\partial s} \\ &= (\cos\alpha + i\sin\alpha)\left(\frac{\partial \mathbf{U}}{\partial x} - i\frac{\partial \mathbf{U}}{\partial y}\right) = e^{i\alpha}\overline{(\mathbf{U}_x + i\mathbf{U}_y)}. \end{aligned} \tag{5.11}$$

On the other hand, if the analytic function φ from the representation (5.2) is written in the form $\varphi = u + iv$, then

$$\mathbf{U}(x, y) = xu(x, y) + yv(x, y) + \operatorname{Re}\chi(x, y).$$

Immediate calculations and the Cauchy-Riemann equations lead to the formula

$$\begin{aligned} \mathbf{U}_x(z) + i\mathbf{U}_y(z) &= u(x, y) + xu_x(x, y) + yv_x(x, y) + \operatorname{Re}\chi_x(x, y) \\ &\quad + i(xu_y(x, y) + v(x, y) + yv_y(x, y) + \operatorname{Re}\chi_y(x, y)) \\ &= \varphi(z) + z\overline{\varphi'(z)} + \overline{\chi'(z)}. \end{aligned} \tag{5.12}$$

Comparing (5.11) and (5.12) we obtain that the functions φ and χ of (5.2) satisfy the boundary condition (5.10).

The converse also follows from relations (5.11) and (5.12). □

Thus the boundary value problem

$$\varphi(t) + t\overline{\varphi'(t)} + \overline{\psi(t)} = f(t), \quad t \in \Gamma \tag{5.13}$$

for two analytic functions φ and $\psi := \chi'$ can be used to find exact or approximate solutions of biharmonic problems. However, attempts to apply projection methods directly to the problem (5.13) meet an immediate difficulty, viz., the operator that corresponds to the left-hand side of (5.13) is not invertible in main functional spaces. Indeed, this is a consequence of the following result.

Lemma 5.1.5. *If Γ is a simple closed piecewise smooth contour, then the homogeneous problem*

$$\varphi(t) + t\overline{\varphi'(t)} + \overline{\psi(t)} = 0, \quad t \in \Gamma \tag{5.14}$$

has a non-trivial solution. Moreover, in the space of functions with boundary values from $W_p^1(\Gamma, \rho)$, any solution of (5.14) has the form

$$\varphi(z) = irz + c, \quad \psi(z) = -\overline{c}, \tag{5.15}$$

where $r \in \mathbb{R}$ and $c \in \mathbb{C}$.

Proof. It is easily seen that any pair of functions (5.15) is a solution for (5.14), so let us show that there are no other solutions of (5.14) with the boundary values from $W_p^1(\Gamma, \rho)$. For any analytic function φ we define the function q by

$$q(z) := \overline{\varphi(z)} + \overline{z}\varphi'(z), \quad z \in D.$$

Using the representation $\varphi(z) = u(x,y) + iv(x,y)$, $z = x + iy$ again, one obtains

$$q = U + iV = (u + xu_x + yv_x) - i(v + yu_x - xv_x).$$

Let us study whether q is an analytic function in D. The Cauchy-Riemann conditions for the function φ show that the equation

$$\frac{\partial U}{\partial y} = -\frac{\partial V}{\partial x}$$

holds everywhere in D. However, the equation

$$\frac{\partial U}{\partial x} = \frac{\partial V}{\partial y}$$

is satisfied if and only if

$$u_x(x,y) = 0, \quad (x,y) \in D.$$

It follows that

$$u(x,y) = m(y), \quad v(x,y) = n(x)$$

where m and n are differentiable functions, so using the Cauchy-Riemann condition $u_x = -v_y$, one obtains

$$n'(x) = -m'(y). \tag{5.16}$$

This is possible if and only if both functions in (5.16) are constants. Thus there is an $r \in \mathbb{R}$ such that

$$n(x) = rx + c_1, \quad m(y) = -ry + c_2,$$

where c_1 and c_2 are arbitrary real numbers. Henceforth,

$$\varphi(z) = -ry + irx + c = irz + c \tag{5.17}$$

with the constant $c = c_2 + ic_1 \in \mathbb{C}$. Thus q is an analytic function if and only if the function φ has the form (5.17), and $q = \bar{c}$ in this case. The function q also satisfies the boundary condition

$$\overline{\varphi(t)} + \bar{t}\varphi'(t) - q(t) = 0, \quad t \in \Gamma.$$

Hence if there is another function ψ that satisfies the boundary condition (5.14), is analytic in D, and such that $\psi \in W_p^1(\Gamma, \rho)$, then ψ is continuous on Γ, so the maximum modulus principle implies the equality

$$\psi(z) = -q(z) = -\bar{c}$$

for all $z \in D$. That completes the proof. □

An immediate consequence of Lemma 5.1.5 is the following result.

Proposition 5.1.6. *Let X denote one of the functional spaces $\mathbf{C}(\Gamma), L_p(\Gamma)$ or $W_p^1(\Gamma, \rho)$. If problem (5.13) has a solution φ_0, ψ_0 the boundary values of which belong to X, then for any $r \in \mathbb{R}$ and for any $c \in \mathbb{C}$ the pair*

$$\varphi(z) = \varphi_0(z) + irz + c,$$
$$\psi(z) = \psi_0(z) - \bar{c}$$

is also a solution of the problem (5.13).

Note that the non-uniqueness of a solution for problem (5.13) has a limited impact on solutions of the initial biharmonic problem, because the corresponding solutions differ by a constant only. For example, considering the biharmonic function constructed via functions φ and ψ one gets

$$\begin{aligned}\operatorname{Re}(\chi(z) + \bar{z}\varphi(z)) &= \operatorname{Re}(\chi_0 - \bar{c}z + c_1 + \bar{z}(\varphi_0(z) + irz + c)) \\ &= \operatorname{Re}(\chi_0 + \bar{z}\varphi_0(z) + c_1).\end{aligned}$$

Note that for the problem (5.9) the constant c_1 is chosen to satisfy the boundary condition

$$\mathbf{U}|_\Gamma = f_1,$$

whereas each of the above pairs φ, χ generates a solution of the boundary problem (5.6).

Thus, if problem (5.13) is solvable, its solution is not unique. It is also worth mentioning that this problem is not always solvable. More precisely, the following result holds.

Proposition 5.1.7. *If problem* (5.13) *has a solution* φ_0, ψ_0 *the boundary values of which belongs to* $W_p^1(\Gamma, \rho)$, *then*

$$\mathrm{Re} \int_\Gamma \overline{f(t)} dt = 0. \tag{5.18}$$

Proof. The function ψ is analytic in D and continuous on Γ, so by the Cauchy theorem

$$\int_\Gamma \psi(t)\, dt = 0,$$

and, correspondingly,

$$\mathrm{Re} \int_\Gamma \psi(t)\, dt = 0.$$

The last equation can be rewritten in the form

$$\mathrm{Re} \int_\Gamma (\overline{\varphi(t)}\, dt - \varphi(t)\, d\bar{t} + \psi(t)\, dt) = 0.$$

Since φ is also continuous on Γ, integration by parts gives

$$\int_\Gamma \bar{t}\varphi'(t)\, dt = - \int_\Gamma \varphi(t)\, d\bar{t},$$

so

$$\mathrm{Re} \int_\Gamma (\overline{\varphi(t)} + \bar{t}\varphi'(t) + \psi(t))\, dt = 0$$

that implies (5.18). □

In view of Proposition 5.1.7 we will always assume that the right-hand side f of equation (5.13) satisfies condition (5.18).

Let us now assume that boundary values $\varphi = \varphi(t)$, $\chi = \chi(t)$, $t \in \Gamma$ of the functions φ and χ have been found. Then using the Cauchy integral formula,

$$\varphi(z) = \frac{1}{2\pi i} \int_\Gamma \frac{\varphi(t)dt}{t - z}, \quad z \in D,$$
$$\chi(z) = \frac{1}{2\pi i} \int_\Gamma \frac{\chi(t)dt}{t - z}, \quad z \in D,$$

and (5.2), one can represent the biharmonic function $\mathbf{U}$ inside of the domain D. Later on we will see that the function $\varphi = \varphi(t)$, $t \in \Gamma$ is the solution of an integral equation, so if φ is known, then

$$\chi'(t) = f(t) - \overline{\varphi(t)} - \bar{t}\varphi'(t), \quad t \in \Gamma, \tag{5.19}$$

where f denotes the right-hand side of (5.7) or (5.10).

It is clear that equation (5.19) allows us to restore the function χ. Really, let $t = t(s)$ be a 1-periodic parametrization of Γ, and let $s_1, s_2, \dots, s_l$ be those points on [0,1) which correspond to the corner points of Γ, i.e.,,

$$t(s_j) = c_j, \quad j = 1, 2, \dots, l.$$

The function $t = t(s)$ is continuously differentiable on (s_j, s_{j+1}) because Γ was supposed to be a piecewise smooth curve. We set $s_{l+1} = s_1 + 1$, let Γ_k be the subarc of Γ which joins the points t_k and t_{k+1}, and let $s \in [s_k, s_{k+1}]$. Using (5.19) we can write

$$\chi'(t(s))t'(s) = f(t(s))t'(s) - \overline{\varphi(t(s))}t'(s) - \overline{t(s)}\varphi'(t(s))t'(s)$$

for any $s \in (s_k, s_{k+1})$ and hence introduce the functions

$$\zeta_k(t(s)) = \int_{s_k}^{s} f(t(s))t'(s)ds - \int_{s_k}^{s} \overline{\varphi(t(s))}t'(s)ds$$
$$+ \int_{s_k}^{s} \frac{d\overline{t(s)}}{ds}\varphi(t(s))ds - \overline{t(s)}\varphi(t(s)), \quad k = 1, 2, \dots, l.$$

Then the function χ may be represented in the form

$$\chi(t) = \begin{cases} \zeta_1(t) + C & \text{if } \ t \in \Gamma_1, \\ \zeta_2(t) + (\zeta_1(c_2) - \zeta_2(c_2)) + C & \text{if } \ t \in \Gamma_2, \\ \dots\dots\dots\dots\dots & \dots \\ \zeta_l(t) + (\zeta_{l-1}(c_l) - \zeta_l(c_l)) + \dots + (\zeta_1(c_2) - \zeta_2(c_2)) + C & \text{if } \ t \in \Gamma_l. \end{cases}$$

Let us recall that for the problem (5.9) the constant C has to be chosen to satisfy the first of the boundary conditions (5.9).

5.1.2 Reduction of Boundary Problems for Two Analytic Functions to Integral Equations

The aim of this section is to reduce the boundary problem for two analytic functions φ and ψ,

$$\psi(t) = \overline{f(t)} - k\overline{\varphi(t)} - \bar{t}\varphi'(t), \quad t \in \Gamma, \tag{5.20}$$

to an integral equation. If $k = 1$, we obtain equation (5.13), but boundary problems (5.20) with $k \neq 1$ also arise in plane elasticity [121, 161, 170]. Thus, in the case of the first fundamental boundary problem concerned with elastic equilibrium of a solid when boundary displacements are known, the constant k is defined by

$$k =: 3 - 4\nu \tag{5.21}$$

or by

$$k := \frac{3-\nu}{1+\nu} \tag{5.22}$$

for plane deformation and plane strain-state correspondingly. Note that the parameter $\nu \in (0, 1/2)$ is usually referred as the Poisson constant. On the other hand, the constant k is equal to 1 for the second fundamental boundary problem when a stress vector is given on the boundary Γ.

Assume that the functions $\varphi = \varphi(t)$, $\psi = \psi(t)$, $t \in \Gamma$ are Hölder continuous. Then they can be represented as the Cauchy integrals

$$\varphi(z) = \frac{1}{2\pi i} \int_\Gamma \frac{\varphi(\tau)d\tau}{\tau - z}, \quad \psi(z) = \frac{1}{2\pi i} \int_\Gamma \frac{\psi(\tau)d\tau}{\tau - z}, \quad z \in D.$$

Let D^- denote the complement of the set $D \cup \Gamma$. Then

$$\frac{1}{\pi i} \int_\Gamma \frac{\varphi(\tau)d\tau}{\tau - z} = 0, \quad \frac{1}{\pi i} \int_\Gamma \frac{\psi(\tau)d\tau}{\tau - z} = 0, \quad z \in D^-, \tag{5.23}$$

and taking into account equation (5.20) one can rewrite the first equation in (5.23) as

$$\frac{1}{\pi i} \int_\Gamma \frac{\overline{f(\tau)}d\tau}{\tau - z} - \frac{1}{\pi i} \int_\Gamma \frac{(\overline{k\varphi(\tau)} - \overline{\tau}\varphi'(\tau))d\tau}{\tau - z} = 0, \quad z \in D^-. \tag{5.24}$$

If $\Phi^+(t)$ and $\Phi^-(t)$ are the limit values of the Cauchy integral

$$\Phi(z) = \frac{1}{2\pi i} \int_\Gamma \frac{\varphi(\tau)d\tau}{\tau - z}$$

when $z \to t \in \Gamma$ by a non-tangential way from the domain D and D^-, respectively, then by the Sohotsky-Plemely formulas [162]

$$\Phi^\pm(t) = \left(\frac{1 \pm 1}{2} - \frac{\omega(t)}{2\pi}\right) \varphi(t) + \frac{1}{2\pi i} \int_\Gamma \frac{\varphi(\tau)d\tau}{\tau - t},$$

where $\omega(t) = \pi$ if $t \neq c_j$ and $\omega(t) = \omega_j$ if $t = c_j$, $j = 1, \ldots, l$. Now it follows from (5.24) that for $t \neq c_j$ the equation

$$-\frac{k\overline{\varphi(t)}}{2} + \frac{k}{2\pi i} \int_\Gamma \frac{\overline{\varphi(\tau)}d\tau}{\tau - t} + \frac{\overline{t}\varphi'(t)}{2} - \frac{1}{2\pi i} \int_\Gamma \frac{\overline{\tau}\varphi'(\tau)d\tau}{\tau - t} = f_0(t), \tag{5.25}$$

where

$$f_0(t) = -\frac{\overline{f(t)}}{2} + \frac{1}{2\pi i} \int_\Gamma \frac{\overline{f(\tau)}d\tau}{\tau - t},$$

holds.

Since φ and φ' are analytic in D, then

$$\frac{1}{2\pi i} \int_\Gamma \frac{\varphi(\tau)d\tau}{\tau - z} = 0, \quad \frac{1}{2\pi i} \int_\Gamma \frac{\varphi'(\tau)d\tau}{\tau - z} = 0$$

for $z \in D^-$, and another application of the Sohotsky-Plemely formulas leads to the relations

$$-\frac{\varphi(t)}{2} + \frac{1}{2\pi i} \int_\Gamma \frac{\varphi(\tau)\, d\tau}{\tau - t} = 0, \quad t \in \Gamma, \tag{5.26}$$

$$-\frac{\varphi'(t)}{2} + \frac{1}{2\pi i} \int_\Gamma \frac{\varphi'(\tau) d\tau}{\tau - t} = 0 \quad t \in \Gamma. \tag{5.27}$$

Using (5.26), (5.27) one can rewrite equation (5.25) in the form

$$-k\overline{\varphi(t)} - \frac{k}{2\pi i} \int_\Gamma \overline{\varphi(\tau)} \left(\frac{d\overline{\tau}}{\overline{\tau} - \overline{t}} - \frac{d\tau}{\tau - t} \right) - \frac{1}{2\pi i} \int_\Gamma \varphi'(\tau) \frac{\overline{\tau} - \overline{t}}{\tau - t}\, d\tau = f_0(t). \tag{5.28}$$

Since

$$\int_\Gamma \varphi'(\tau) \frac{\overline{\tau} - \overline{t}}{\tau - t}\, d\tau = \int_\Gamma \varphi(\tau) d\, \frac{\overline{\tau} - \overline{t}}{\tau - t}$$

and

$$\frac{d\overline{\tau}}{\overline{\tau} - \overline{t}} - \frac{d\tau}{\tau - t} = d \ln \frac{\overline{\tau} - \overline{t}}{\tau - t},$$

where the symbol d is associated with differentiation in the variable τ, it is customary to write equation (5.28) as

$$R\varphi(t) \equiv -k\overline{\varphi(t)} - \frac{k}{2\pi i} \int_\Gamma \overline{\varphi(\tau)} d \log \frac{\overline{\tau} - \overline{t}}{\tau - t} - \frac{1}{2\pi i} \int_\Gamma \varphi(\tau) d \frac{\overline{\tau} - \overline{t}}{\tau - t} = f_0(t), \; t \in \Gamma, \tag{5.29}$$

and call it the Muskhelishvili equation.

Thus the function φ is a solution of the Muskhelishvili equation (5.29). Therefore, the above-described procedure can be used to determine solutions of the boundary value problem (5.20) and, correspondingly, allows us to obtain solutions of biharmonic problems and associated problems of plane elasticity. Let us mention that, in general, equation (5.29) is not solvable in a closed form, so the applicability of approximation methods acquires a great importance. However, in the case $k = 1$ the operator R defined by the left-hand side of (5.29) is not invertible in main functional spaces since Proposition 5.1.6 yields that $\dim \ker R > 0$. A similar result can also be established for other relevant values of k. This means that neither of the projection methods considered earlier can be directly applied to the equation (5.29). One of the ways to overcome this difficulty consists in modifying the spaces where the operator R is considered. Such an approach also requires modification of the corresponding approximating subspaces and, normally, this leads to additional problems during implementation of the corresponding approximation methods. Therefore, here we prefer another approach. Instead of modifying the corresponding spaces where the operator R is considered, we will modify the operator R itself in a very special way. This allows us to apply most of the projection methods studied to the modified operator and obtain approximate solutions of the initial Muskhelishvili equation (5.29). For this, we need to study the operator R in more detail.

5.2 Muskhelishvili Operator. Fredholm Properties and Invertibility in $L_p(\Gamma, \rho)$

This section is devoted to investigating the Fredholm properties of the operator R in the spaces $L_p(\Gamma, \rho)$, where Γ is a piecewise Lyapunov curve. However, we start with studying the operator R on special curves: thus, let $\Gamma_{\beta,\omega}$, $\beta \in [0, 2\pi)$, $\omega \in (0, 2\pi)$ refer to the curve

$$\Gamma_{\beta,\omega} = \Gamma_1 \bigcup \Gamma_2$$

where

$$\Gamma_1 := e^{i(\beta+\omega)}\mathbb{R}^+, \quad \Gamma_2 := e^{i\beta}\mathbb{R}^+,$$

and Γ_1 is directed to 0 but Γ_2 is directed away from 0. Given $p \in (1, +\infty)$ and α satisfying the inequality (5.5), we consider the weighted Lebesgue spaces $L_p(\Gamma_{\beta,\omega}, \alpha)$ and $L_p(\mathbb{R}^+, \alpha)$ of measurable functions endowed with the norms

$$||f||_{p,\alpha,\omega} = \left(\int_{\Gamma_{\beta,\omega}} |f(t)|^p |t|^{\alpha p} |dt|\right)^{1/p}$$

and

$$||f||_{p,\alpha} := \left(\int_{\mathbb{R}^+} |f(t)|^p t^{\alpha p} dt\right)^{1/p},$$

respectively.

Let us further introduce the set $L_p^2(\mathbb{R}^+, \alpha)$ of all pairs $(f_1, f_2)^T$, $f_1, f_2 \in L_p(\mathbb{R}^+, \alpha)$ and the norm

$$||(f_1, f_2)^T|| := (||f_1||_{p,\alpha}^p + ||f_2||_{p,\alpha}^p)^{1/p},$$

together with a mapping $\eta : L_p(\Gamma_{\beta,\omega}, \alpha) \to L_p^2(\mathbb{R}^+, \alpha)$ defined as follows:

$$\eta(f) = (\eta_1(f), \eta_2(f))^T$$

with

$$\begin{aligned} \eta_1(f)(s) &= f(se^{i(\beta+\omega)}), \\ \eta_2(f)(s) &= f(se^{i\beta}) \end{aligned}$$

for all $s \in \mathbb{R}^+$. It is clear that η is a linear isometry from $L_p(\Gamma_{\beta,\omega}, \alpha)$ onto $L_p^2(\mathbb{R}^+, \alpha)$. In addition, the mapping $A \to \eta A \eta^{-1}$ is an isometric algebra isomorphism of $\mathcal{L}_{add}(L_p(\Gamma_{\beta,\omega}, \alpha))$ onto $\mathcal{L}_{add}(L_p^2(\mathbb{R}^+, \alpha))$.

In the space $L_p(\Gamma_{\beta,\omega}, \alpha)$ we consider the corresponding Muskhelishvili operator

$$R_\omega x(t) \equiv -k\overline{x(t)} - \frac{k}{2\pi i} \int_{\Gamma_{\beta,\omega}} \overline{x(\tau)} d \log \frac{\overline{\tau} - \overline{t}}{\tau - t} - \frac{1}{2\pi i} \int_{\Gamma_{\beta,\omega}} x(\tau) d \frac{\overline{\tau} - \overline{t}}{\tau - t}. \quad (5.30)$$

Now we are going to use Theorem 1.9.5 to investigate Fredholm properties of the Muskhelishvili operator in the space $L_p(\Gamma, \rho)$.

Let z be a point of Γ, and $\hat{\Gamma}_z$ denote the curve which consists of the semi-tangents to Γ at the point z with the orientation inherited from the orientation of Γ. By β_z we denote the angle between the real axis and that semi-tangent which is directed away from z. Then, $\Gamma_z := \hat{\Gamma}_z - z$. For the sake of simplicity, we will write β_j for the angle at the point $z = c_j$, $j = 1, 2, \dots, l$ where c_j are the corner points of Γ.

Theorem 5.2.1. *Let Γ be a simple closed piecewise smooth curve in the complex plane $\mathbb{C}$. Then the operator R is Fredholm if and only if the operators $R_{\omega_j} : L_p(\Gamma_{\beta_j,\omega_j}, \alpha_j) \to L_p(\Gamma_{\beta_j,\omega_j}, \alpha_j)$ are invertible for all $j = 1, 2, \dots, l$.*

Let us point out important steps of the proof. By $\mathcal{B}_p(\Gamma, \rho)$ we denote the smallest closed real subalgebra of $\mathcal{L}_{add}(L_p(\Gamma, \rho))$ which contains the operators

$$(K_{1,\Gamma}\varphi)(t) = \frac{1}{2\pi i} \int_\Gamma \varphi(\tau) d\frac{\overline{\tau} - \overline{t}}{\tau - t},$$

$$(K_{2,\Gamma}\varphi)(t) = \frac{1}{2\pi i} \int_\Gamma \varphi(\tau) d\ln \frac{\overline{\tau} - \overline{t}}{\tau - t},$$

the operator of the complex conjugation

$$(M_\Gamma\varphi)(t) := \overline{\varphi(t)}, \quad t \in \Gamma$$

and all operators of multiplication by continuous bounded functions. The algebra $\mathcal{B}_p(\Gamma_z, \alpha_z)$, where

$$\alpha_z = \begin{cases} \alpha_j, & \text{if} \quad z = c_j, \\ 0, & \text{if} \quad z \neq c_j,\ j = 1, 2 \dots, l \end{cases}$$

is defined analogously.

We mention two useful properties of the algebra $\mathcal{B}_p(\Gamma, \rho)$:

1. The algebra $\mathcal{B}_p(\Gamma, \rho)$ contains the ideal $K_{add}(L_p(\Gamma, \rho))$ of all compact operators on $L_p(\Gamma, \rho)$.
2. For any continuous real-valued function f, the operator $fA - AfI$, $A \in \mathcal{B}_p(\Gamma, \rho)$ is compact.

Due to these properties one can consider the algebra

$$\mathcal{B}_p(\Gamma, \rho)^\pi := \mathcal{B}_p(\Gamma, \rho)/K_{add}(L_p(\Gamma, \rho))$$

and localize it over Γ via Allan's local principle. This leads to the local algebras $\mathcal{B}_p(\Gamma, \rho)_z^\pi, z \in \Gamma$ with canonical real algebra homomorphisms $\Phi_z : \mathcal{B}_p(\Gamma, \rho)^\pi \to \mathcal{B}_p(\Gamma, \rho)_z^\pi, z \in \Gamma$. Simultaneously, we localize the algebra $\mathcal{B}_p(\Gamma_z, \alpha_z)^\pi$ over Γ_z

that leads to local algebras $\mathcal{B}_p(\Gamma_z,\alpha_z)^\pi_w$ with canonical homomorphisms $\Psi_w : \mathcal{B}_p(\Gamma_z,\alpha_z)^\pi \to \mathcal{B}_p(\Gamma_z,\alpha_z)^\pi_w$, $w \in \Gamma_z$. The algebras $\mathcal{B}_p(\Gamma,\rho)^\pi_z$ and $\mathcal{B}_p(\Gamma_z,\alpha_z)^\pi_0$ are topologically isomorphic, and there is an isomorphism which sends $\Phi_z(\pi(K_{j,\Gamma}))$ into $\Psi_0(\pi(K_{j,\Gamma_z})), j = 1,2$; $\Phi_z(\pi(M_\Gamma))$ into $\Psi_0(\pi(M_{\Gamma_z}))$, and $\Phi_z(\pi(fI))$ into $\Psi_0(\pi(f(z)I))$ for each continuous function f. However, the algebra $\mathcal{B}_p(\Gamma_z,\alpha_z)^\pi_0$ possesses the so-called homogenization property [102]. This property allows us to obtain a locally equivalent representation of the algebra $\mathcal{B}_p(\Gamma_z,\alpha_z)^\pi_0$ and, consequently, conditions of invertibility of the element $\pi(R)_z$ in the algebra $\mathcal{B}_p(\Gamma,\rho)^\pi_z$. Note that a description of the corresponding homomorphism between the algebras $\mathcal{B}_p(\Gamma,\rho)^\pi_z$ and $\mathcal{B}_p(\Gamma_z,\alpha_z)^\pi_0$ can be found in [102, pp. 286–289], although algebras considered there differ from those used above. □

It turns out that the integral operators in (5.30) are isometrically isomorphic to Mellin convolution operators. More precisely, let $\widetilde{K}_1,\widetilde{K}_2 : L_p(\Gamma_{\beta,\omega},\alpha) \to L_p(\Gamma_{\beta,\omega},\alpha)$ be the integral operators of the left-hand side of (5.30), viz.,

$$(\widetilde{K}_1\varphi)(t) = -\frac{1}{2\pi i}\int_{\Gamma_{\beta,\omega}} \varphi(\tau)d\frac{\overline{\tau}-\overline{t}}{\tau-t},$$

$$(\widetilde{K}_2\varphi)(t) = -\frac{1}{2\pi i}\int_{\Gamma_{\beta,\omega}} \varphi(\tau)d\ln\frac{\overline{\tau}-\overline{t}}{\tau-t}.$$

Given $\nu \in (0,2\pi)$ we consider the Mellin convolution operators $\mathcal{M}_\nu,\mathcal{N}_\nu : L_p(\mathbb{R}^+,\alpha) \to L_p(\mathbb{R}^+,\alpha)$ defined by

$$(\mathcal{M}_\nu(\varphi))(\sigma) = \frac{1}{\pi}\int_0^{+\infty}\left(\frac{\sigma}{s}\right)\frac{\sin\nu}{(1-(\sigma/s)e^{i\nu})^2}\frac{\varphi(s)}{s}ds \tag{5.31}$$

and

$$(\mathcal{N}_\nu(\varphi))(\sigma) = \frac{1}{2\pi i}\int_0^{+\infty}\frac{\varphi(s)ds}{s-\sigma e^{i\nu}}. \tag{5.32}$$

Lemma 5.2.2. *The operators $\widetilde{K}_1,\widetilde{K}_2 : L_p(\Gamma_{\beta,\omega},\alpha) \to L_p(\Gamma_{\beta,\omega},\alpha)$ are isometrically isomorphic to the block Mellin operators*

$$\hat{K}_1 = \begin{pmatrix} 0 & e^{-i2\beta}\mathcal{M}_\omega \\ -e^{-i2(\beta+\omega)}\mathcal{M}_{2\pi-\omega} & 0 \end{pmatrix} \in \mathcal{L}(L^2_p(\mathbb{R}^+,\alpha)), \tag{5.33}$$

$$\hat{K}_2 = \begin{pmatrix} 0 & \frac{1}{2}\left[\mathcal{N}_\omega - \mathcal{N}_{2\pi-\omega}\right] \\ \frac{1}{2}\left[\mathcal{N}_\omega - \mathcal{N}_{2\pi-\omega}\right] & 0 \end{pmatrix} \in \mathcal{L}(L^2_p(\mathbb{R}^+,\alpha)), \tag{5.34}$$

respectively.

Proof. Let $(f_1,f_2)^T$ be a vector of $L^2_p(\mathbb{R}^+,\alpha)$, and let $\widetilde{f}_1(t),\widetilde{f}_2(t)$, $t \in \Gamma_{\beta,\omega}$ denote the functions

$$\widetilde{f}_1(t) = \begin{cases} f_1(s), & \text{if} \quad t = se^{i(\beta+\omega)}, \\ 0, & \text{if} \quad t = se^{i\beta}, \end{cases}$$

$$\widetilde{f_2}(t) = \begin{cases} 0, & \text{if} \quad t = se^{i(\beta+\omega)}, \\ f_2(s), & \text{if} \quad t = se^{i\beta}. \end{cases}$$

Then $\eta^{-1} : L_p^2(\mathbb{R}^+, \alpha) \to L_p(\Gamma_{\beta,\omega}, \alpha)$ can be written in the form

$$\left(\eta^{-1}(f_1, f_2)^T\right)(t) = \widetilde{f_1}(t)\chi_{\Gamma_1}(t) + \widetilde{f_2}(t)\chi_{\Gamma_2}(t),$$

where χ_{Γ_j}, $j = 1, 2$ is the characteristic function of the curve Γ_j, $j = 1, 2$. Let K be an integral operator on $L_p(\Gamma_{\beta,\omega}, \alpha)$, i.e.,

$$(Kf)(t) = \int_{\Gamma_{\beta,\omega}} f(\tau)K(t,\tau)d\tau.$$

Then the operator $\eta K \eta^{-1} : L_p^2(\mathbb{R}^+, \alpha) \to L_p(\Gamma_{\beta,\omega}, \alpha)$ is a matrix operator of the form

$$\eta K \eta^{-1} = \begin{pmatrix} \widetilde{K}_{11} & \widetilde{K}_{12} \\ \widetilde{K}_{21} & \widetilde{K}_{22} \end{pmatrix} \tag{5.35}$$

with

$$\widetilde{K}_{jl} f_l = \eta_j \left(\int_{\Gamma_l} \widetilde{f_l}(\tau) K(t,\tau) d\tau \right), \quad j, l = 1, 2. \tag{5.36}$$

Now we set $K = \widetilde{K}_1$ and find the entries of the corresponding matrix (5.35). For instance, we have

$$\begin{aligned} \widetilde{K}_{12}^1 f_2(\sigma) &= \eta_1 \left(\int_{\Gamma_2} \widetilde{f_2}(\tau) \left(\frac{\overline{\tau} - \overline{t}}{(\tau - t)^2} d\tau - \frac{1}{\tau - t} d\overline{\tau} \right) \right) \\ &= \frac{1}{2\pi i} \int_0^{+\infty} f_2(s) \left(\frac{(se^{-i\beta} - \sigma e^{-i(\beta+\omega)})e^{i\beta}}{(se^{i\beta} - \sigma e^{i(\beta+\omega)})^2} - \frac{e^{-i\beta}}{se^{i\beta} - \sigma e^{i(\beta+\omega)}} \right) ds \\ &= \frac{e^{-i2\beta}}{\pi} \int_0^{+\infty} \left(\frac{\sigma}{s} \right) \frac{\sin\omega}{(1 - (\sigma/s)e^{i\omega})^2} f_2(s) \frac{ds}{s}. \end{aligned}$$

Thus, $\widetilde{K}_{12}^1$ is a Mellin convolution operator $\mathcal{M}_\omega$ defined by (5.31). Analogously,

$$\widetilde{K}_{11}^1 = 0, \quad \widetilde{K}_{22}^1 = 0, \quad \widetilde{K}_{21}^1 = -e^{-i2(\beta+\omega)} \mathcal{M}_{2\pi-\omega},$$

and we arrive at representation (5.33). Representation (5.34) for the operator $\widetilde{K}_2$ can be obtained by using the same formulas (5.35) and (5.36), completing the proof. □

Let M be the operator of complex conjugation on the space $L_p(\mathbb{R}^+, \alpha)$, and let $\widetilde{M} : L_p^2(\mathbb{R}^+, \alpha) \to L_p^2(\mathbb{R}^+, \alpha)$ stand for the diagonal operator $\operatorname{diag}(M, M)$. Then from Lemma 5.2.2 we immediately obtain the following result

Proposition 5.2.3. *The operator R_ω is isometrically isomorphic to the operator $A_\omega \in \mathcal{L}_{add}(L_p^2(\mathbb{R}^+, \alpha))$,*

$$A_\omega = \mathcal{A} + \mathcal{B}\widetilde{M}$$

with

$$\mathcal{A} = \begin{pmatrix} 0 & e^{-i2\beta}\mathcal{M}_\omega \\ -e^{-i2(\beta+\omega)}\mathcal{M}_{2\pi-\omega} & 0 \end{pmatrix}$$

and

$$\mathcal{B} = \begin{pmatrix} -kI & \frac{k}{2}\left[\mathcal{N}_\omega - \mathcal{N}_{2\pi-\omega}\right] \\ \frac{k}{2}\left[\mathcal{N}_\omega - \mathcal{N}_{2\pi-\omega}\right] & -kI \end{pmatrix}.$$

Thus A_ω is a block Mellin operator with conjugation.

Lemma 5.2.4. *Let $\mathcal{M}_\nu$ and $\mathcal{N}_\nu, \nu \in (0, 2\pi)$ be the operators defined by (5.31) and (5.32), respectively. Then*

$$M\mathcal{M}_\nu M = -\mathcal{M}_{2\pi-\nu}, \quad M\mathcal{N}_\nu M = -\mathcal{N}_{2\pi-\nu}. \tag{5.37}$$

The proof of relations (5.37) follows immediately from the definition of the operators $M, \mathcal{M}_\nu$ and $\mathcal{N}_\nu$.

Lemma 5.2.5. *The operator $R_\omega : L_p(\Gamma_{\beta,\omega}, \alpha) \to L_p(\Gamma_{\beta,\omega}, \alpha)$ is invertible if and only if the block Mellin operator*

$$\widetilde{A}_\omega = \begin{pmatrix} 0 & e^{-i2\beta}\mathcal{M}_\omega & -kI & \frac{k}{2}[\mathcal{N}_\omega - \mathcal{N}_{2\pi-\omega}] \\ -e^{-i2(\beta+\omega)}\mathcal{M}_{2\pi-\omega} & 0 & \frac{k}{2}[\mathcal{N}_\omega - \mathcal{N}_{2\pi-\omega}] & -kI \\ -kI & \frac{k}{2}[\mathcal{N}_\omega - \mathcal{N}_{2\pi-\omega}] & 0 & -e^{i2\beta}\mathcal{M}_{2\pi-\omega} \\ \frac{k}{2}[\mathcal{N}_\omega - \mathcal{N}_{2\pi-\omega}] & -kI & e^{i2(\beta+\omega)}\mathcal{M}_\omega & 0 \end{pmatrix} \tag{5.38}$$

is so.

The proof follows from Lemmas 1.4.16 and 5.2.4.

Proposition 5.2.6. *The operator $R_\omega : L_p(\Gamma_{\beta,\omega}, \alpha) \to L_p(\Gamma_{\beta,\omega}, \alpha)$ is invertible if and only if the function*

$$F(y) = \frac{[y^2 \sin^2\omega - k^2 \sinh^2(2\pi - \omega)y][y^2 \sin^2\omega - k^2 \sinh^2 \omega y]}{\sinh^4 \pi y} \tag{5.39}$$

does not vanish for all $y \in \mathbb{R} + i(1/p + \alpha)$.

Proof. The operators R_ω and $\widetilde{A}_\omega$ are simultaneously invertible or not. However, $\widetilde{A}_\omega$ is a matrix Mellin operator. The conditions of invertibility for such operators are well known, and consist in the invertibility of their symbols (see Section 1.10.2). To find the symbol of the operator $\widetilde{A}_\omega$, we compute the symbols of the Mellin operators $\mathcal{M}_\nu$ and $\mathcal{N}_\nu, \nu \in (0, 2\pi)$. From Sections 1.10.2 and 1.10.3, the symbol $m_\nu = m_\nu(z)$ of the operator $\mathcal{M}_\nu$ is

$$m_\nu(z) = \frac{1}{\pi} \int_0^{+\infty} x^{1/p+\alpha-zi-1} \frac{x \sin\nu}{(1 - xe^{i\nu})^2} dx.$$

Therefore, via formula (3.194.6) of [97] we get

$$\begin{aligned} m_\nu(z) &= \frac{\sin\nu}{\pi}\int_0^{+\infty}\frac{x^{(1+1/p+\alpha-zi)-1}}{(1-xe^{i\nu})^2}dx \\ &= -e^{i\nu}\sin\nu\frac{z+i(1/p+\alpha)}{\sinh\pi(z+i(1/p+\alpha))}e^{-(\nu-\pi)(z+i(1/p+\alpha))}, \quad z\in\mathbb{R}. \end{aligned}$$

Setting $\nu=\omega$ and $y=z+i(1/p+\alpha)$ we find

$$m_\omega(y) = -e^{-i\omega}\sin\omega\frac{y}{\sinh\pi y}e^{-(\omega-\pi)y}, \quad y=z+i(1/p+\alpha),\ z\in\mathbb{R}.$$

Analogously, the symbol n_ω of the operator $(\mathcal{N}_\omega-\mathcal{N}_{2\pi-\omega})/2$ is

$$n_\omega(y) = \frac{\sinh(\pi-\omega)y}{\sinh\pi y}, \quad y=z+i(1/p+\alpha),\ z\in\mathbb{R}.$$

Thus the symbol of $\widetilde{A}_\omega$ can be represented in the form

$$\mathcal{A}_{\widetilde{A}_\omega}(y) = \begin{pmatrix} 0 & e^{-i2\beta}m_\omega(y) & -k & kn_\omega(y) \\ -e^{-i2(\beta+\omega)}m_{2\pi-\omega}(y) & 0 & kn_\omega(y) & -k \\ -k & kn_\omega(y) & 0 & -e^{i2\beta}m_{2\pi-\omega}(y) \\ kn_\omega(y) & -k & e^{i2(\beta+\omega)}m_\omega(y) & 0 \end{pmatrix}. \tag{5.40}$$

Let us calculate the determinant of the matrix (5.40). Expanding it by the first two rows yields

$$\begin{aligned} \det\mathcal{A}_{\widetilde{A}_\omega}(y) &= (m_\omega(y)m_{2\pi-\omega}(y))^2 - k^2e^{-i2\omega}m^2_{2\pi-\omega}(y) + k^2m_{2\pi-\omega}(y)m_\omega(y)n^2_\omega(y) \\ &\quad + k^2m_\omega(y)m_{2\pi-\omega}(y)n^2_\omega(y) - k^2e^{i2\omega}m^2_\omega(y) + k^4(1-n^2_\omega(y))^2 \\ &= \sin^4\omega\frac{y^4}{\sinh^4\pi y} - k^2\sin^2\omega\frac{y^2}{\sinh^2\pi y}e^{-2(\pi-\omega)y} \\ &\quad - 2k^2\sin^2\omega\frac{y^2}{\sinh^2\pi y}\frac{\sinh^2(\pi-\omega)y}{\sinh^2\pi y} \\ &\quad - k^2\sin^2\omega\frac{y^2}{\sinh^2\pi y}e^{-2(\omega-\pi)y} + k^4\left(1-\frac{\sinh(\pi-\omega)y}{\sinh^2\pi y}\right)^2 \\ &= k^4\frac{\sinh^2(2\pi-\omega)y\sinh^2\omega y}{\sinh^4\pi y} + \frac{y^4\sin^4\omega}{\sinh^4\pi y} \\ &\quad - 2k^2\frac{y^2\sin^2\omega}{\sinh^4\pi y}(\sinh^2(\pi-\omega)y\cosh^2\pi y + \sinh^2\pi y\cosh^2(\pi-\omega)y) \\ &= \frac{[y^2\sin^2\omega - k^2\sinh^2(2\pi-\omega)y][y^2\sin^2\omega - k^2\sinh^2\omega y]}{\sinh^4\pi y}. \end{aligned}$$

Thus the right-hand side of (5.39) represents the determinant of the symbol of $\widetilde{A}_\omega$, so the operator R_ω is invertible if and only if the function F does not vanish on the line $y \in \mathbb{R} + i(1/p + \alpha)$. □

Consequently, invertibility of the Muskhelishvili operator R_ω in $\mathcal{L}_{add}(L_p(\Gamma_{\beta,\omega}, \alpha))$ depends on whether the function

$$g_{\gamma,\delta}(z) = k^2 \sinh^2 \gamma(z + i\delta) - (z + i\delta)^2 \sin^2 \gamma, \quad z \in \mathbb{R}$$

has any zeros when $\delta = 1/p + \alpha$ and $\gamma = \omega$ or $\gamma = 2\pi - \omega$. The behaviour of this function was studied by R. Duduchava [73], who inter alia proved a result which for convenience is formulated here as a lemma.

Lemma 5.2.7. (R.V. Duduchava, [73]). *Let k be the coefficient of the boundary problem* (5.20). *If $\pi < \gamma < 2\pi$, then the transcendental equation*

$$k \sin \gamma\delta = \delta |\sin \gamma| \tag{5.41}$$

has only one solution $\delta = \delta'_\gamma$ in the interval $(0, 1)$ such that

$$\frac{1}{2} < \delta'_\gamma < \frac{\pi}{\gamma}. \tag{5.42}$$

If $0 < \gamma < \pi$, then equation (5.41) *does not have any solutions in the interval* $(0, 1)$.

Remark 5.2.8. A result analogous to Lemma 5.2.7 has been also presented by J.E. Lewis [140].

From Proposition 5.2.6 and inequality (5.42) of Lemma 5.2.7 we immediately obtain the following result.

Corollary 5.2.9. *There exists $\delta'_\omega > 1/2$ such that the operator $R_\omega : L_p(\Gamma_{\beta,\omega}, \alpha) \to L_p(\Gamma_{\beta,\omega}, \alpha)$ is invertible for all p and α satisfying the inequality*

$$0 < \frac{1}{p} + \alpha < \delta'_\omega.$$

The principal significance of this corollary is that it provides us with conditions of invertibility of the local operators R_{ω_j}, $j = 1, 2, \ldots, l$ from Theorem 5.2.1.

With each operator R of the form (5.29) we can now associate the symbol of R, viz.,

$$\mathcal{A}_R(t, z) = \begin{cases} \mathcal{A}_{\widetilde{A}_{\omega_j}}(z),\ z \in \mathbb{R} & \text{if} \quad t = c_j, \quad j = 1, 2, \ldots l, \\ \begin{pmatrix} -k & 0 \\ 0 & -k \end{pmatrix} & \text{otherwise.} \end{cases}$$

Theorem 5.2.10. *Let $\delta'_{\omega_j}, j = 1, 2, \ldots, l$ be defined as in Corollary 5.2.9 for every corner point c_j, $j = 1, 2, \ldots, l$, and let*

$$\delta' = \min_{1 \le j \le l} \{\delta'_{\omega_j}\}.$$

If

$$0 < \max_{1 \le j \le l} \left\{ \frac{1}{p} + \alpha_j \right\} < \delta', \tag{5.43}$$

then the Muskhelishvili operator

$$R : L_p(\Gamma, \rho) \to L_p(\Gamma, \rho)$$

is Fredholm and

$$\operatorname{ind}_{L_p(\Gamma,\rho)} R = 0.$$

Proof. The first assertion follows from Theorem 5.2.1 and Corollary 5.2.9. For the second assertion we mention that the index of the operator R is equal to the winding number of the curve $L := \det \mathcal{A}_R(t, z)$, $t \in \Gamma$, $z \in \mathbb{R}$ around the origin. Homotopy arguments shows that in the case where the operator R is Fredholm, the winding number of the curve L is zero. □

Theorem 5.2.11. *There exist δ and δ', $\delta < 1/2 < \delta'$ such that, if*

$$\delta < \min_{1 \le j \le l} \left\{ \frac{1}{p} + \alpha_j \right\} \le \max_{1 \le j \le l} \left\{ \frac{1}{p} + \alpha_j \right\} < \delta', \tag{5.44}$$

then the operator R is Fredholm in both $L_p(\Gamma, \rho)$ and $W_p^1(\Gamma, \rho)$ and has the same index $\kappa = 0$ in each of these spaces.

Proof. As we have seen from Theorem 5.2.10, R is a Fredholm operator with index $\kappa = 0$ in all spaces $L_p(\Gamma, \rho)$ with p and $\rho = \rho(\alpha_1, \alpha_1, \ldots, \alpha_l)$ satisfying inequality (5.43). On the other hand there exists a $\delta \in (0, 1/2)$ such that R is a Fredholm operator with index $\kappa = 0$ in all spaces $W_p^1(\Gamma, \rho)$ with p and ρ satisfying the inequality [72, 73]:

$$\delta < \min_{1 \le j \le l} \left\{ \frac{1}{p} + \alpha_j \right\} < 1. \tag{5.45}$$

Comparing inequalities (5.43) and (5.45) completes the proof. □

Thus the operator R is Fredholm in both families of the Banach spaces, and its index vanishes. Nevertheless, this operator is not invertible in each of the spaces mentioned, so none of the approximation methods can be applied to R immediately. However, the operator R can be corrected in such a way that the consequent operator is already invertible. Moreover, some additional conditions ensure correspondence between solutions of the respective equations. To be more precise, we consider the operators

$$R_j = R + T_j, \quad j = 1, 2 \tag{5.46}$$

where $T = T_1$ if k is defined by (5.21) or (5.22), and $T = T_2$ if $k = 1$, with

$$(T_1\varphi)(t) = \frac{1}{2\pi i}\int_\Gamma \frac{\varphi(\tau)d\tau}{\tau},$$

and

$$(T_2\varphi)(t) = \frac{1}{2\pi i}\int_\Gamma \frac{\varphi(\tau)d\tau}{\tau} + \frac{1}{t}\frac{1}{2\pi i}\int_\Gamma \left(\frac{\varphi(\tau)}{\tau^2}d\tau + \frac{\overline{\varphi(\tau)}}{\overline{\tau}^2}\overline{d\tau}\right). \tag{5.47}$$

Let us study properties of the operators $R_j, j = 1, 2$ and their connections with various properties of the Muskhelishvili operators. Note that the operator R_2 will be studied in more detail, whereas relevant properties of the operator R_1 will only be mentioned in the end of this section. We start with considering an auxiliary boundary value problem for a pair of analytic functions.

Lemma 5.2.12. *If domain D is bounded by simple closed piecewise smooth curve Γ, then any solution of the exterior boundary value problem*

$$\overline{\varphi_-(t)} + \overline{t}\varphi'_-(t) + \psi_-(t) = 0, \quad t \in \Gamma$$

where $\varphi_-, \psi_- \in W^1_p(\Gamma, \rho)$ are the boundary values of functions φ and ψ, analytic in the exterior domain $D^- := \mathbb{C} \setminus \overline{D}$, has the form

$$\varphi(z) = irz + c, \quad \psi(z) = -\overline{c}$$

with an $r \in \mathbb{R}$ and a $c \in \mathbb{C}$.

The proof of Lemma 5.2.12 mainly follows the proof of lemma 5.1.5. However, it is not possible to use the maximum modulus principle now. Instead, one has to consider a function $\psi_1 : D \to \mathbb{C}$ defined by $\psi_1(z) = -\overline{c}$ for any $z \in D$ and use the uniqueness of the analytic continuation.

Lemma 5.2.13. *Let Γ be a simple closed piecewise smooth contour and let $f \in W^1_p(\Gamma, \rho)$. If the Muskhelishvili equation (5.29) is solvable in the space $W^1_p(\Gamma, \rho)$, then any of its solutions is a boundary value of a function analytic in the domain D.*

Proof. Assume that $\varphi_0 \in W^1_p(\Gamma, \rho)$ is a solution of equation (5.29) and define two functions $\Phi, \Psi : D^- \to \mathbb{C}$ by

$$\Phi(z) := -\frac{1}{2\pi}\int_\Gamma \frac{\varphi_0(\tau)}{\tau - z}\,d\tau, \tag{5.48}$$

$$\Psi(z) := \frac{1}{2\pi}\int_\Gamma \frac{\overline{\varphi_0(\tau)} + \overline{\tau}\varphi_0'(\tau) - \overline{f(\tau)}}{\tau - z}\,d\tau.$$

The functions Φ and Ψ are analytic in the domain D^-, so using the Sohotsky-Plemely formulas one can rewrite equation (5.29) as

$$\overline{\Phi(t)} + \overline{t}\Phi'(t) + \Psi(t) = 0, \quad t \in \Gamma, \tag{5.49}$$

where $\Phi(t)$ and $\Psi(t)$ are the boundary values of the functions $\Phi(z)$ and $\Psi(z)$ as $z \to t \in \Gamma$ from the domain D^-. By Lemma 5.2.12 any solution of the boundary problem (5.49) has the form

$$\Phi(z) = ir_0 z + c_0, \quad \Psi(z) = -\overline{c}_0$$

with $r_0 \in \mathbb{R}$ and $c_0 \in \mathbb{C}$. On the other hand, the function Φ is represented via a Cauchy type integral, so it vanishes at infinity. This is possible if and only if $r_0 = 0$ and $c_0 = 0$. Thus, $\Phi(z) = 0$ for any $z \in D^-$, and our assertion follows from (5.48). In passing note that $\Psi(z) = -\overline{c}_0 = 0$ everywhere in D^-. □

Lemma 5.2.14. *If Γ is a simple closed piecewise smooth curve, then*

$$\ker_{W_p^1(\Gamma,\rho)} R = \{irt + c : r \in \mathbb{R},\, c \in \mathbb{C}\}.$$

Proof. Let $\varphi_0 \in W_p^1(\Gamma, \rho)$ be a solution of the homogeneous Muskhelishvili equation

$$R\varphi = 0. \tag{5.50}$$

It follows from the proof of Lemma 5.2.13 that

$$\frac{1}{2\pi} \int_\Gamma \frac{\overline{\varphi_0(\tau)} + \overline{\tau}\varphi_0'(\tau)}{\tau - z}\, d\tau = 0, \quad z \in D^-,$$

so the function $\overline{\varphi_0(t)} + \overline{t}\varphi_0'(t)$ is the boundary value of a function ψ_0 analytic in D. Thus functions ψ_0 and φ_0 satisfy the boundary condition

$$\psi_0(t) = \overline{\varphi_0(t)} + \overline{t}\varphi_0'(t), \quad t \in \Gamma,$$

and Lemma 5.2.12 yields that $\varphi_0(t) = ir_0 t + c_0$ with $r_0 \in \mathbb{R}$ and $c_0 \in \mathbb{C}$. □

Lemma 5.2.15. *Let Γ be a simple closed piecewise smooth curve, and let $f \in W_p^1(\Gamma, \rho)$ and satisfy the condition* (5.18). *If equation*

$$(R + T_2)\varphi = f \tag{5.51}$$

has a solution $\varphi_0 \in W_p^1(\Gamma, \rho)$, then φ_0 is also a solution of the Muskhelishvili equation (5.29).

Proof. Let us show that any solution $\varphi_0 \in W_p^1(\Gamma, \rho)$ of equation (5.51) has the property

$$T_2\varphi_0 = 0.$$

Analogously to the proof of Lemma 5.2.13, one can introduce two functions $\Phi, \Psi : D \to \mathbb{C}$ by

$$\Phi(z) := -\frac{1}{2\pi}\int_\Gamma \frac{\varphi_0(\tau)}{\tau - z}\,d\tau,$$
$$\Psi(z) := \frac{1}{2\pi}\int_\Gamma \frac{\varphi_0(\tau)}{\tau}d\tau + \frac{1}{z}\frac{1}{2\pi}\int_\Gamma \left(\frac{\varphi_0(\tau)}{\tau^2}\,d\tau + \frac{\overline{\varphi_0(\tau)}}{\overline{\tau}^2}d\overline{\tau}\right)$$
$$+ \frac{1}{2\pi}\int_\Gamma \frac{\overline{\varphi_0(\tau)} + \overline{\tau}\varphi_0'(\tau) - \overline{f(\tau)}}{\tau - z}\,d\tau,$$

and rewrite the equation (5.51) as an homogeneous exterior boundary problem which again leads to the equation

$$\Psi(z) = 0 \tag{5.52}$$

for all $z \in D^-$. Consider the Laurent series expansion of the function Ψ around the point $z_0 = \infty$ in the domain $|z| > \max_{t\in\Gamma}|t|$. It is

$$\frac{1}{2\pi}\int_\Gamma \frac{\varphi_0(\tau)}{\tau}\,d\tau + \frac{1}{2\pi}\int_\Gamma \left(\frac{\varphi_0(\tau)}{\tau^2}d\tau + \frac{\overline{\varphi_0(\tau)}}{\overline{\tau}^2}d\overline{\tau}\right)\cdot\frac{1}{z}$$
$$+ \frac{1}{2\pi}\int_\Gamma (\overline{\varphi_0(\tau)} + \overline{\tau}\varphi_0'(\tau) - \overline{f(\tau)})d\tau \cdot \frac{1}{z} + \dots,$$

and from (5.52) one obtains that all Laurent coefficients in the above expansion must be equal to zero. In particular,

$$\frac{1}{2\pi}\int_\Gamma \frac{\varphi_0(\tau)}{\tau}\,d\tau = 0 \tag{5.53}$$

and

$$\frac{1}{2\pi}\int_\Gamma \left(\frac{\varphi_0(\tau)}{\tau^2}\,d\tau + \frac{\overline{\varphi_0(\tau)}}{\overline{\tau}^2}d\overline{\tau}\right) + \frac{1}{2\pi}\int_\Gamma (\overline{\varphi_0(\tau)} + \overline{\tau}\varphi_0'(\tau) - \overline{f(\tau)})d\tau = 0. \tag{5.54}$$

Using integration by parts, we rewrite equation (5.54) as

$$\begin{aligned}&\frac{1}{2\pi}\int_\Gamma \left(\frac{\varphi_0(\tau)}{\tau^2}d\tau + \frac{\overline{\varphi_0(\tau)}}{\overline{\tau}^2}d\overline{\tau}\right)\\ &- \frac{1}{2\pi}\int_\Gamma \left(\overline{\varphi_0(\tau)}d\tau - \varphi_0(\tau)d\overline{\tau}\right) + \frac{1}{2\pi}\int_\Gamma \overline{f(\tau)})d\tau = 0,\end{aligned} \tag{5.55}$$

and notice that the first integral in the left-hand side of (5.55) is a real number, whereas the second and third are pure imaginary (cf. (5.18)). Therefore

$$\frac{1}{2\pi}\int_\Gamma \left(\frac{\varphi_0(\tau)}{\tau^2}\,d\tau + \frac{\overline{\varphi_0(\tau)}}{\overline{\tau}^2}d\overline{\tau}\right) = 0 \tag{5.56}$$

and combining (5.53) and (5.56) completes the proof. □

Lemma 5.2.16. *Let* Γ *be a simple closed piecewise smooth curve. Then*

$$\ker_{W_p^1(\Gamma,\rho)}(R+T_2)=0.$$

Proof. Let $\varphi_0 \in \ker_{W_p^1(\Gamma,\rho)}(R+T_2)$. Since for the corresponding homogeneous equation condition (5.18) obviously holds, by Lemma 5.2.15 $\varphi_0 \in \ker_{W_p^1(\Gamma,\rho)}(R)$, so Lemma 5.2.14 yields that $\varphi_0(t) = ir_0t + c_0$ with $r_0 \in \mathbb{R}$ and $c_0 \in \mathbb{C}$. On the other hand, φ_0 must satisfy relations (5.53) and (5.56). Hence

$$\frac{1}{2\pi}\int_\Gamma \frac{\varphi_0(\tau)}{\tau}\,d\tau = \frac{1}{2\pi}\int_\Gamma \frac{ir_0\tau + c_0}{\tau}\,d\tau = ic_0 = 0,$$

which gives us $c_0 = 0$ and also $r_0 = 0$ because of the identity

$$\frac{1}{2\pi}\int_\Gamma\left(\frac{\varphi_0(\tau)}{\tau^2}\,d\tau + \frac{\overline{\varphi_0(\tau)}}{\overline{\tau}^2}\,d\overline{\tau}\right) = \frac{1}{2\pi}\int_\Gamma\left(\frac{ir_0\tau}{\tau^2}\,d\tau - i\frac{r_0\overline{\tau}}{\overline{\tau}^2}\,d\overline{\tau}\right) = -2r_0 = 0.$$

Thus the homogeneous equation $(R+T_2)\varphi = 0$ has the only solution $\varphi_0 = 0$. □

Theorem 5.2.17. *Let* Γ *be a piecewise smooth curve, and let* $\alpha_j, j = 1,2,\dots,l$ *satisfy conditions* (5.44). *Then the operator*

$$R_j : L_p(\Gamma,\rho) \to L_p(\Gamma,\rho), \quad j=1,2$$

is invertible. Moreover, if $f \in W_p^1(\Gamma,\rho)$, *then:*

1. *If* k *is defined by* (5.21) *or* (5.22), *then the solution of the equation*

$$R_1\varphi = f_0$$

is simultaneously a solution of the corresponding Muskhelishvili equation (5.29).

2. *If* $k=1$ *and if* f *satisfies the condition*

$$\operatorname{Re}\int_\Gamma \overline{f(\tau)}\,d\tau = 0,$$

then the solution of the equation

$$R_2\varphi = f_0 \tag{5.57}$$

is simultaneously a solution of the corresponding Muskhelishvili equation (5.29).

Proof. Let us consider the operator R_2. As was established in Lemma 5.2.16, in the space $W_p^1(\Gamma,\rho)$ the homogeneous equation

$$R_2\varphi = 0$$

has only the trivial solution. Since T_2 is a compact operator on $W_p^1(\Gamma,\rho)$, we have [89]

$$\operatorname{ind}_{W_p^1(\Gamma,\rho)} R_2 = \operatorname{ind}_{W_p^1(\Gamma,\rho)} R,$$

so $\operatorname{ind}_{W_p^1(\Gamma,\rho)} R_2 = 0$ and the operator $R_2 : W_p^1(\Gamma,\rho) \to W_p^1(\Gamma,\rho)$ is invertible. Because of condition (5.44) the operator $R_2 : L_p(\Gamma,\rho) \to L_p(\Gamma,\rho)$ is Fredholm and has the same index zero. However, the space $W_p^1(\Gamma,\rho)$ is dense in $L_p(\Gamma,\rho)$. Since the indices of R_2 are the same in both $W_p^1(\Gamma,\rho)$ and $L_p(\Gamma,\rho)$ the dimensions of null spaces of R_2 are equal as well [89], so $\dim\ker R_2|_{L_p(\Gamma,\rho)} = 0$ and the operator R_2 is invertible in $L_p(\Gamma,\rho)$. The application of Lemma 5.2.15 completes the proof. □

Corollary 5.2.18. *The operator $R_j, j = 1,2$ is invertible in the space $L_2(\Gamma)$.*

The last results allow us to apply and investigate different approximation methods for the operator R_j, and simultaneously obtain approximate solutions of the corresponding Muskhelishvili equation (5.29).

5.3 Approximation Methods for Equations on Smooth Contours

We start with a study of the stability of various projection methods for the Muskhelishvili equation on simple smooth contours. The assumption on the smoothness of the contour leads to a very nice result. Namely, all approximation methods under consideration converge without any additional conditions.

Let Γ be a simple closed Lyapunov curve in the complex plane $\mathbb{C}$ and let γ be a 1-periodic parametrization of Γ, which maps the interval $[0,1)$ one-to-one and onto Γ and $\gamma'(s) \neq 0$ for every $s \in [0,1)$. This curve divides the complex plane into two domains. As before, the interior domain is denoted by D and we also assume that $0 \in D$.

For definiteness, let us assume that the coefficient k in the Muskhelishvili equation (5.29) is equal to 1, so

$$Rx(t) \equiv -\overline{x(t)} - \frac{1}{2\pi i}\int_\Gamma \overline{x(\tau)} d\log\frac{\bar\tau-\bar t}{\tau - t} - \frac{1}{2\pi i}\int_\Gamma x(\tau)\, d\frac{\bar\tau-\bar t}{\tau-t} = f_0(t), \ t\in\Gamma. \tag{5.58}$$

Recall that for $k = 1$ the auxiliary operator T_2 in (5.57) is defined by (5.47), and consider the stability of approximation methods based on splines. Let $(f*g)$ denote the convolution of the functions f and g,

$$(f*g)(s) = \int_{\mathbb{R}} f(s-x)\, g(x)\, dx.$$

Consider the characteristic function $\chi = \chi(s)$, $s \in \mathbb{R}$ of the interval $[0, 1)$, i.e.,

$$\chi(s) = \begin{cases} 1 & s \in [0,1), \\ 0 & \text{otherwise.} \end{cases}$$

For any $d \in \mathbb{N}$ introduce the function ϕ^d by

$$\phi^d(s) = \left(\phi^0 * \phi^{d-1}\right)(s)$$

where $\phi^0(s) = \chi(s)$, $s \in \mathbb{R}$. The functions ϕ^d generate spline spaces on $\mathbb{R}$. To be more precise, we fix $d \in \mathbb{N}$ and set

$$\tilde{\phi}(s) = \phi^d(s), \quad s \in \mathbb{R}.$$

Also, fix a number $n \in \mathbb{N}$ and for each $j \in \mathbb{Z}$ define the function $\tilde{\phi}_{jn} = \tilde{\phi}_{jn}(s)$ by

$$\tilde{\phi}_{jn}(s) = \tilde{\phi}(ns - j), \quad s \in \mathbb{R}.$$

Then the set of all linear combinations of $\tilde{\phi}_{jn}$, $j \in \mathbb{Z}$ is a spline space on $\mathbb{R}$.

Using the above construction one can introduce the corresponding spline spaces on Γ. Thus, if γ is a 1-periodic parametrization of the curve Γ, then for any $t \in \Gamma$ we set

$$\phi_{jn}(t) = \tilde{\phi}_{jn}(s), \quad t = \gamma(s), \quad s \in \mathbb{R}.$$

Let $S_n^d = S_n^d(\Gamma)$ denote the corresponding spline space on Γ.

5.3.1 Galerkin Method

An approximate solution $x_n = x_n(t) \in S_n^d$ of equation (5.58) is sought in the form

$$x_n(t) = \sum_{j=0}^{n-1} c_j \phi_{jn}(t), \quad t \in \Gamma. \tag{5.59}$$

By $(\cdot, \cdot)$ we denote the usual inner product on Γ, i.e.,

$$(f, g) = \int_\Gamma f(t)\, \overline{g(t)}\, |dt|, \quad f, g \in L_2(\Gamma).$$

The unknown coefficients c_j, $j = 0, 1, \ldots, n-1$ of the approximate solution (5.59) are defined by the system of algebraic equations

$$(R_2 x_n, \phi_{kn}) = (f_0, \phi_{kn}), \quad k = 0, 1, \ldots, n-1 \tag{5.60}$$

where R_2 is defined by (5.46) and (5.47).

Theorem 5.3.1. *Let Γ be a simple closed Lyapunov curve, and let $f \in W_2^1(\Gamma)$. Then there exists n_0 such that for all $n \geq n_0$ the systems of algebraic equations (5.60) are solvable and the sequence $(x_n)_{n \geq n_0}$ of approximate solutions (5.59) of equation (5.57) converges to the exact solution of (5.58) in the space $L_2(\Gamma)$.*

Proof. Denote by P_n the orthogonal projections onto the spline subspaces $S_n^d(\Gamma)$. Then the systems of algebraic equations (5.60) are equivalent to the operator equations

$$P_n R_2 P_n x_n = P_n f_0, \quad n = 1, 2, \ldots. \tag{5.61}$$

Therefore, if we establish the stability of the sequence $\{P_n R_2 P_n\}$, our claim will follow from Theorem 5.2.17, estimate (1.30) and from the corresponding results of approximation theory. Let M refer to the operator of complex conjugation and let K be the operator defined by

$$\begin{aligned} Kx(t) = &-\frac{1}{2\pi i}\int_\Gamma \overline{x(\tau)} d\log\frac{\bar{\tau}-\bar{t}}{\tau - t} - \frac{1}{2\pi i}\int_\Gamma x(\tau)\, d\frac{\bar{\tau}-\bar{t}}{\tau - t} \\ &+ \frac{1}{2\pi i}\int_\Gamma \frac{x(\tau)\, d\tau}{\tau} + \frac{1}{t}\frac{1}{2\pi i}\int_\Gamma \left(\frac{x(\tau)\, d\tau}{\tau^2} + \frac{\overline{x(\tau)}}{\bar{\tau}^2} d\bar{\tau}\right). \end{aligned} \tag{5.62}$$

Then the operator $P_n R_2 P_n$ can be written as $P_n R_2 P_n = -P_n M P_n + P_n K P_n$. Using the easily verified equality $MP_n = P_n M$, we immediately obtain

$$P_n R_2 P_n (-P_n M P_n) = P_n - P_n K M P_n. \tag{5.63}$$

By Corollary 5.2.18 the operator R_2 is invertible on the space $L_2(\Gamma)$. We can thus introduce operators $B_n : \text{im } P_n \longrightarrow \text{im } P_n$ by

$$B_n = -P_n M P_n + P_n R_2^{-1} K M P_n.$$

Then by (5.63),

$$\begin{aligned} (P_n R_2 P_n) B_n &= (P_n R_2 P_n)(-P_n M P_n) + (P_n R_2 P_n)\left(P_n R_2^{-1} K M P_n\right) \\ &= P_n - P_n K M P_n + P_n R_2 P_n R_2^{-1} K M P_n \\ &= P_n - P_n R_2 (I - P_n)\left(R_2^{-1} K M\right) P_n. \end{aligned} \tag{5.64}$$

But by Lemma 5.21 of [183] the projections P_n converge strongly to the identity operator. Since Γ is a Lyapunov curve, the operator K is compact on $L_2(\Gamma)$, [162]. Therefore, the sequence $\left\{P_n R_2 (I - P_n) R_2^{-1} K M P_n\right\}$ converges uniformly to 0 as $n \longrightarrow \infty$. Hence, the operators in the right-hand side of (5.64) are invertible on im P_n for all n sufficiently large and the norms of their inverses are bounded, for example by 2.

This implies the invertibility from the right of the operators $(P_n R_2 P_n)$ and the inequality

$$\|\left(P_n R_2 P_n\right)^{-1} P_n\| \leq 2\|B_n\| \leq 2 + 2\|R_2^{-1} K\|$$

for the norms of the right inverses. The left invertibility of $P_n R_2 P_n$ can be proved analogously. Thus, the sequence $\{P_n R_2 P_n\}$ is stable and the proof of Theorem 5.3.1 follows from Theorem 5.2.17 and estimate (1.30). □

Taking into account the stability of the method, Theorem 5.2.17, and inequality (1.30), one easily obtains error estimates. More precisely, if the right-hand side f in (5.47) belongs to $W_2^1(\Gamma)$, then $f_0 \in W_2^1(\Gamma)$ [91], and the solution x of (5.58) belongs to $W_2^1(\Gamma)$ as well, [73]. Therefore, by [183, p. 44]

$$\|x - P_n x\|_{L_2(\Gamma)} \le \frac{d_1}{n} \|x\|_{W_2^1(\Gamma)}$$

and

$$\|f_0 - P_n f_0\|_{L_2(\Gamma)} \le \frac{d_2}{n} \|f_0\|_{W_2^1(\Gamma)},$$

where d_1, d_2 are constants independent of x, f_0 and n. Hence

$$\begin{aligned}\|x - x_n\|_{L_2(\Gamma)} \le \|x - P_n x\|_{L_2(\Gamma)} + \|A_n^{-1} P_n\| \big(\|P_n A P_n x - A P_n x\|_{L_2(\Gamma)} \\ + \|A P_n x - Ax\|_{L_2(\Gamma)} + \|f_0 - P_n f_0\|_{L_2(\Gamma)}\big) \le \frac{d_3}{n},\end{aligned}$$

where the constant d_3 is independent of n.

5.3.2 The ε-Collocation Method

Let $0 \le \varepsilon < 1$ be a real number and let $t_j^{(n)} \in \Gamma$ be defined as

$$t_j^{(n)} = \gamma\left(\frac{j+\varepsilon}{n}\right), \quad j = 0, 1, \dots, n-1.$$

The approximate solution of equation (5.58) is again sought in the form (5.59) but the unknown coefficients c_j, $j = 0, 1, \dots, n-1$ will be obtained from the system

$$R_2 x_n \left(t_j^{(n)}\right) = f_0\left(t_j^{(n)}\right), \quad j = 0, 1, \dots, n-1. \tag{5.65}$$

The collocation method is considered in the context of the space $\mathbf{C}(\Gamma)$. It follows from [183, p. 64], that if d is odd and $\varepsilon \ne 1/2$ or if d is even and $\varepsilon \ne 0$, then there exists a uniquely determined interpolation projection L_n onto the space $S_n^d(\Gamma)$ such that

$$L_n f\left(t_j^{(n)}\right) = f\left(t_j^{(n)}\right).$$

Using this notation and recalling the remark after (5.61), we see that the system (5.65) is equivalent to the operator equation

$$L_n R_2 P_n x_n = L_n f_0, \quad n \in \mathbb{N}. \tag{5.66}$$

However, the projections L_n are not defined on the space $L_2(\Gamma)$. Hence, to be able to study the collocation method we have to consider the operator R_2 on a more appropriate space.

Theorem 5.3.2. *Let Γ be a simple closed curve and let its parametrization γ be twice continuously differentiable on $[0,1]$. Assume that $f \in W_2^1(\Gamma)$. Then there exists an integer n_0 such that equations (5.66) are solvable for all $n \geq n_0$, and the sequence $\{x_n\}_{n\geq n_0}$ converges to a solution of equation (5.58) in the norm of $\mathbf{C}(\Gamma)$. More precisely, there is a constant d_4 such that for all $n \geq n_0$ the estimate*

$$||x - x_n||_{\mathbf{C}(\Gamma)} \leq \frac{d_4}{n} \tag{5.67}$$

holds.

Proof. We now consider the operator R_2 on the space $\mathbf{C}(\Gamma)$. First of all we mention that if γ is twice continuously differentiable, then the operator K of (5.62), is compact on $C(\Gamma)$. Really, this claim is obvious for the operator T because the kernels of the corresponding integral operators are continuous. Therefore consider for instance the operator T_1,

$$T_1 x(t) = \frac{1}{2\pi i}\int_\Gamma x(\tau)\, d\frac{\overline{\tau} - \overline{t}}{\tau - t} = \int_\Gamma K_1(t,\tau)\, x(\tau)\, d\tau.$$

The kernel $K_1(t,\tau)$ of this integral operator has the form

$$K_1(t,\tau) = \frac{1}{2\pi i}\left[\frac{(\tau - t)\frac{d\overline{\tau}}{d\tau} - (\overline{\tau} - \overline{t})}{(\tau - t)^2}\right].$$

Since the curve Γ does not have any intersections with itself, the function K_1 is obviously continuous for $\tau \neq t$. Let us study the behaviour of the expression

$$\Phi_1(t,\tau) = \frac{(\tau - t)\, d\overline{\tau} - (\overline{\tau} - \overline{t})\, d\tau}{(\tau - t)^2}$$

when τ tends to t. Setting $\tau = \gamma(\sigma)$, $t = \gamma(s)$, $\sigma, s \in [0,1)$, $\sigma \neq s$ and using the twice continuous differentiability of the function γ, we get as $\sigma \longrightarrow s$,

$$\Phi_1(t,\tau) = \frac{i\mathrm{Im}\left(\overline{\gamma''(\sigma)}\gamma'(\sigma)\right) + o(1)}{[\gamma'(\sigma)]^2}.$$

Hence

$$\lim_{\tau \longrightarrow t} \Phi_1(t,\tau) = i\frac{\mathrm{Im}\left(\overline{\gamma''(s)}\gamma'(s)\right)}{[\gamma'(s)]^2},$$

thus the function $\Phi_1(t,\tau)$ is continuous for all t, τ on Γ, and the operator $T_1 : \mathbf{C}(\Gamma) \longrightarrow \mathbf{C}(\Gamma)$ is compact. The compactness of the remaining integral operator T_2,

$$T_2 x(t) = -\frac{1}{2\pi i}\int_\Gamma \overline{x(\tau)}\, d\log\frac{\overline{\tau} - \overline{t}}{\tau - t} = \int_\Gamma K_2(t,\tau)\, \overline{x(\tau)}\, d\tau,$$

can be shown in the same way, thus obtaining

$$\lim_{\tau\longrightarrow t} \Phi_2(t,\tau) = i\frac{\operatorname{Im}\left(\overline{\gamma''(s)}\gamma'(s)\right)}{|\gamma'(s)|^2}.$$

Now we can show the invertibility of the operator R_2 on the space $\mathbf{C}(\Gamma)$. Since K is compact, the standard Fredholm theory implies that the index of the operator R_2 considered on $\mathbf{C}(\Gamma)$, is equal to zero. The space $\mathbf{C}(\Gamma)$ is dense in $L_2(\Gamma)$ and by Theorem 5.2.10 the index of R_2 on $L_2(\Gamma)$ is equal to zero. Therefore, by [89] the dimensions of the kernels of the operator R_2 on the spaces $L_2(\Gamma)$ and $\mathbf{C}(\Gamma)$ coincide. However the operator $R_2 : L_2(\Gamma) \longrightarrow L_2(\Gamma)$ is invertible, hence $\dim\ker R_2|_{L_2(\Gamma)} = 0$. This implies $\dim\ker R_2|_{\mathbf{C}(\Gamma)} = 0$. Taking into account that the index of the operator $R_2 : \mathbf{C}(\Gamma) \longrightarrow \mathbf{C}(\Gamma)$ is also zero, one obtains the invertibility of the operator R_2 on the Banach space $C(\Gamma)$.

The next steps mainly follow the Proof of Theorem 5.3.1. Representing the operator $L_nR_2P_n$ in the form

$$L_nR_2P_n = -L_nMP_n + L_nKP_n,$$

and multiplying the latter expression by the operator

$$\tilde{B}_n = -L_nMP_n + L_nR_n^{-1}KMP_n,$$

one obtains

$$(L_nR_2P_n)\tilde{B}_n = P_n - L_nR_2(I - L_n)R_2^{-1}KMP_n.$$

Since by Lemma 5.28 of [183] the sequence $\{I - L_n\}_{n\in\mathbb{N}}$ is uniformly bounded, the approximation properties of the splines guarantee that the sequence converges strongly to zero on $\mathbf{C}(\Gamma)$. Also observing that the operator $R^{-1}KM$ is compact, we get that for all n large enough the operators $L_nR_2P_n : \operatorname{im} P_n \longrightarrow \operatorname{im} P_n$ are right invertible and their right inverses are uniformly bounded. Since these operators are finite-dimensional their invertibility on im P_n follows. Having proved the stability of the sequence $\{L_nR_2P_n\}$ we again can use the inequality (1.30) to establish the estimate (5.67), and hence, convergence of the collocation method. □

5.3.3 Qualocation Method

Here we consider the stability of the following version of the qualocation method for the Muskhelishvili equation.

Let $0 < \varepsilon_1 < \varepsilon_2 < \ldots < \varepsilon_m < 1$ and $\omega_1, \omega_2, \ldots, \omega_m$, be positive numbers such that $\omega_1 + \omega_2 + \ldots + \omega_m = 1$ and let $t_{j,\varepsilon_k}^{(n)} = \gamma((j+\varepsilon_k)/n)$, $j \in \mathbb{Z}$, $k = 1, 2, \ldots, m$. By Q_n we denote the quadrature formula

$$Q_n(g) = \sum_{j=0}^{n-1}\sum_{r=1}^{m} \omega_r g\left(t_{j,\varepsilon_r}^{(n)}\right).$$

The approximate solution of equation (5.58) is sought in the form (5.59) but in contrast to (5.60) the coefficients c_j, $j = 0, 1, \ldots, n-1$ are determined from the system of algebraic equations

$$Q_n(R_2 x_n, v) = Q_n(f_0, v), \quad x_n \in S_n^d(\Gamma) \tag{5.68}$$

for all $v \in S_n^0(\Gamma)$, where $S_n^0(\Gamma)$ denotes the space of piecewise constant splines on Γ. Recall that such a spline qualocation method, though with the more general spline space $S_n^\mu(\Gamma)$ in place of $S_n^0(\Gamma)$, was studied in Section 2.9 for the Cauchy singular integral equations on the unit circle Γ.

Theorem 5.3.3. . *Let Γ be a simple closed curve such that its parametrization γ is a twice continuously differentiable function on $[0,1]$. Assume also that $f \in W_2^1(\Gamma)$, and let $\varepsilon_r \in (0,1)$ be real numbers such that $\varepsilon_r \neq 1/2$, $r = 1, 2, \ldots, m$ if d is odd. Then there exists an integer n_0 such that for all $n \geq n_0$ equations (5.68) are solvable and the corresponding approximate solutions x_n converge to an exact solution of equation (5.58) in the norm of $\mathbf{C}(\Gamma)$.*

Proof. The operators $A_n : S_n^d(\Gamma) \longrightarrow S_n^d(\Gamma)$ corresponding to the left-hand side of (5.68) can be represented in the form

$$A_n = \sum_{r=1}^{m} \omega_r L_n^{\varepsilon_r} R_2 P_n \tag{5.69}$$

where $L_n^{\varepsilon_r}$ denotes the interpolation projection onto the spline space $S_n^d(\Gamma)$ satisfying the property

$$L_n^{\varepsilon_r} u(t_{j,\varepsilon_r}) = u(t_{j,\varepsilon_r}), \quad j = 0, 1, \ldots, n-1.$$

Since $R_2 = -M + K$ and $L_n^{\varepsilon_r} M P_n = M P_n = L_n^{\varepsilon_1} M P_n$ for any $r = 1, 2, \ldots, m$ we can rewrite (5.69) as

$$A_n = -L_n^{\varepsilon_1} M P_n + L_n^{\varepsilon_1} K P_n + \sum_{r=2}^{m} \omega_r (L_n^{\varepsilon_r} - L_n^{\varepsilon_1}) K P_n.$$

The sequence $\{L_n^{\varepsilon_r} - L_n^{\varepsilon_1}\}$ converges strongly to 0 on the space $\mathbf{C}(\Gamma)$ as $n \longrightarrow \infty$. Taking into account the compactness of K we deduce that $\| \sum_{r=2}^{m} \omega_r (L_n^{\varepsilon_r} - L_n^{\varepsilon_1}) K P_n \| \longrightarrow 0$ as $n \longrightarrow \infty$. Now we can proceed as in the proofs of Theorems 5.3.1 and 5.3.2. □

5.3.4 Biharmonic Problem

Let us now study constructions connected with approximate solution of the biharmonic problem. For definiteness we assume that the boundary Γ of a domain D satisfies conditions of Section 5.3, and consider the boundary value problem

$$\begin{gathered} \Delta^2 \mathbf{U}|_D = 0, \\ \left.\frac{\partial \mathbf{U}}{\partial x}\right|_\Gamma = G_1, \quad \left.\frac{\partial \mathbf{U}}{\partial y}\right|_\Gamma = G_2, \end{gathered} \tag{5.70}$$

under the assumption $G_1, G_2 \in W_2^1(\Gamma)$. By Lemma 5.1.3, any solution of (5.70) can be represented in the form

$$\operatorname{Re}\left[\overline{z}\varphi(z) + \chi(z)\right], \quad z \in D \tag{5.71}$$

where φ and χ are analytic functions in the domain D satisfying the boundary condition

$$\varphi(t) + t\overline{\varphi'(t)} + \overline{\chi'(t)} = f(t), \quad t \in \Gamma, \tag{5.72}$$

with the right-hand side $f(t) := G_1(t) + iG_2(t)$. As was already mentioned, the function φ can be obtained as a solution of the Muskhelishvili equation (5.58). On the other hand, relation (5.72) allows us to express the boundary value $\chi'(t)$, $t \in \Gamma$, for the analytic function χ', viz.,

$$\chi'(t) = \overline{f(t)} - \overline{\varphi(t)} - \overline{t}\varphi'(t). \tag{5.73}$$

Making the substitution $t = \gamma(\sigma)$ in (5.73) and multiplying the resulting expression by $\gamma'(\sigma)$, we obtain

$$\chi'(\gamma(\sigma))\gamma'(\sigma) = \overline{f(\gamma(\sigma))}\gamma'(\sigma) - \overline{\varphi(\gamma(\sigma))}\gamma'(\sigma) - \overline{\gamma(\sigma)}\varphi'(\gamma(\sigma))\gamma'(\sigma). \tag{5.74}$$

Taking integrals of both sides of (5.74) and using integration by parts for the last integral in (5.74) we get

$$\begin{aligned}\chi(\gamma(s)) = &\int_0^s \overline{f(\gamma(\sigma))}\gamma'(\sigma)\,d\sigma - \int_0^s \overline{\varphi(\gamma(\sigma))}\gamma'(\sigma)\,d\sigma\\ &+ \int_0^s \overline{\gamma'(\sigma)}\varphi(\gamma(\sigma))\,d\overline{\sigma} \; - \overline{\gamma(s)}\varphi(\gamma(s)) + \overline{\gamma(0)}\varphi(\gamma(0)) + c,\end{aligned} \tag{5.75}$$

where $c \in \mathbb{C}$ is an arbitrary constant. If t denotes the point $\gamma(s)$ and if Γ_t denotes the arc of Γ joining the points $t_0 = \gamma(0)$ and $t = \gamma(s)$, then representation (5.75) takes the form

$$\chi(t) = \int_{\Gamma_t} \overline{f(\tau)}d\tau - \int_{\Gamma_t} \overline{\varphi(\tau)}d\tau + \int_{\Gamma_t} \varphi(\tau)\,d\overline{\tau} - \overline{t}\varphi(t) + \overline{t}_0\varphi(t_0) + C.$$

Thus having obtained the boundary representations $\varphi(t)$ and $\chi(t)$, $t \in \Gamma$ of the analytic functions φ and χ one can retrieve these functions via the Cauchy integrals, viz.,

$$\varphi(z) = \frac{1}{2\pi i}\int_\Gamma \frac{\varphi(\tau)\,d\tau}{\tau - z}, \quad \chi(z) = \frac{1}{2\pi i}\int_\Gamma \frac{\chi(\tau)\,d\tau}{\tau - z}, \quad z \in D.$$

Then using the Goursat representation (5.71) one can get a solution of the biharmonic problem (5.6).

Assume now that we have available an approximate solution φ_n of the Muskhelishvili equation (5.58) and let

$$\|\varphi - \varphi_n\|_{\mathbf{C}} \le \epsilon_n, \quad n \ge n_0. \tag{5.76}$$

From formulas (5.74), (5.75) we can also obtain an approximation for the function χ, viz.,

$$\chi_n(t) = \int_{\Gamma_t} \overline{f(\tau)} d\tau - \int_{\Gamma} \overline{\varphi_n(\tau)} d\tau + \int_{\Gamma} \varphi_n(\tau)\, d\bar{\tau} - \bar{t}\varphi_n(t) + \bar{t}_0\varphi_n(t_0) + C.$$

Now taking into account (5.76) one can easily find

$$\|\chi_n - \chi\|_{\mathbf{C}} \le d_1 \epsilon_n \tag{5.77}$$

where d_1 is a constant independent of n. Then

$$\tilde{\varphi}_n(z) \equiv \frac{1}{2\pi i} \int_{\Gamma} \frac{\varphi_n(\tau)\, d\tau}{\tau - z}, \quad \tilde{\chi}_n(z) \equiv \frac{1}{2\pi i} \int_{\Gamma} \frac{\chi_n(\tau)\, d\tau}{\tau - z}, \quad z \in D \tag{5.78}$$

and using the Goursat representation (5.71) once more, one obtains

$$U_n(x,y) \equiv \operatorname{Re}\left[\bar{z}\tilde{\varphi}_n(z) + \tilde{\chi}_n(z)\right], \quad z = x + iy \in D. \tag{5.79}$$

Theorem 5.3.4. *Let the approximate solution φ_n, $n \ge n_0$ of the Muskhelishvili equation (5.58) satisfy inequality (5.76). Then for any compact subset K of D the approximate solution (5.79) of the biharmonic problem satisfies the estimate*

$$\sup_{(x,y)\in K} |U_n(x,y) - U(x,y)| \le d_2 \epsilon_n, \quad n \ge n_0 \tag{5.80}$$

where d_2 is independent of the point $(x,y) \in K$, parameter $n \in \mathbb{N}$, and where U is the exact solution of the biharmonic problem (5.70).

Proof. Using the above construction we have

$$\begin{aligned} &|U_n(x,y) - U(x,y)| \\ &\le \frac{|z|}{2\pi} \int_{\Gamma} \frac{|\varphi_n(\tau) - \varphi(\tau)|}{|\tau - z|} |d\tau| + \frac{1}{2\pi} \int_{\Gamma} \frac{|\chi_n(\tau) - \chi(\tau)|}{|\tau - z|} |d\tau| \\ &\le \frac{|\Gamma|}{2\pi \operatorname{dist}(K,\Gamma)} \left[\max_{z\in\Gamma} |z| + d_1\right] \epsilon_n \end{aligned}$$

where $|\Gamma|$ stands for the length of Γ and the constant d_1 is taken from (5.77). □

Remark 5.3.5. Estimate (5.80) contains the constant $[\operatorname{dist}(K,\Gamma)]^{-1}$ which grows if the boundary of K tends to Γ. The estimate can be improved if it is known that the functions $\varphi_n(t)$, $\chi_n(t)$, $n \ge n_0$ belong to a subspace $W(\Gamma)$ of $\mathbf{C}(\Gamma)$ such that the Cauchy integral operator S,

$$Sx(t) = \frac{1}{\pi i} \int_{\Gamma} \frac{x(\tau)\, d\tau}{\tau - t}, \quad t \in \Gamma,$$

is bounded on $W(\Gamma)$, i.e., if there exists a constant d_3 such that

$$\|Sx\|_{\mathbf{C}} \leq d_3 \|x\|_{\mathbf{C}}$$

for any $x \in W(\Gamma)$. Then instead of φ_n and χ_n one can use in (5.78) the functions

$$\hat{\varphi}_n(t) = (P\varphi_n)(t), \quad \hat{\chi}_n(t) = (P\chi_n)(t)$$

with $P = \frac{1}{2}(I+S)$. Since φ and χ are boundary values of analytic functions in D we have

$$\|\varphi - \hat{\varphi}_n\|_{\mathbf{C}} = \|P(\varphi - \varphi_n)\|_{\mathbf{C}} \leq d_3 \|\varphi - \varphi_n\|_{\mathbf{C}},$$
$$\|\chi - \hat{\chi}_n\|_{\mathbf{C}} = \|P(\chi - \chi_n)\|_{\mathbf{C}} \leq d_3 \|\chi - \chi_n\|_{\mathbf{C}}.$$

An approximate solution for the biharmonic problem can now be constructed as

$$\hat{U}_n(x,y) = \operatorname{Re}\left[\bar{z}\widetilde{\hat{\varphi}_n}(z) + \widetilde{\hat{\chi}_n}(z)\right].$$

However, to estimate the error now one can use the maximum principle for analytic functions. This gives us

$$\sup_{(x,y)\in D} \left|\hat{U}_n(x,y) - U(x,y)\right| \leq d_4 \epsilon_n.$$

5.4 Approximation Methods for the Muskhelishvili Equation on Special Contours

Let $\Gamma_{\beta,\omega}$, $\beta \in [0, 2\pi)$, $\omega \in (0, 2\pi)$, $\omega \neq \pi$ be as above. In this section we examine the stability of approximation methods based on piecewise constant splines for the Muskhelishvili equation

$$R_\omega x = y, \quad x, y \in L_p(\Gamma_{\beta,\omega}, \alpha)$$

considered on the curve $\Gamma_{\beta,\omega}$. The corresponding methods represent the so-called local models, which are of great importance in studying approximation methods for equations on piecewise smooth contours. More precisely, the stability conditions of approximation methods for equation (5.29) can be formulated in terms of the local models mentioned.

First of all, we construct spline spaces on $\Gamma_{\beta,\omega}$. Let $\chi_{[0,1)} = \chi_{[0,1)}(s)$, $s \in \mathbb{R}$, denote the characteristic function of the interval [0,1). For any positive integer m, let us introduce the function

$$\psi^m(s) := (\psi^0 * \psi^{m-1})(s), \quad s \in \mathbb{R},$$

with

$$\psi^0(s) = \chi_{[0,1)}(s), \quad s \in \mathbb{R},$$

where $(f * g)$ denotes the convolution of the functions f and g, i.e.,

$$(f * g)(s) := \int_{\mathbb{R}} f(s-x)g(x)dx.$$

From now on we fix $m \in \mathbb{N}$ and set

$$\psi(s) = \psi^m(s), \quad s \in \mathbb{R}. \tag{5.81}$$

As a next step we also fix a positive integer n and, for each $k \in \mathbb{N}$, we define the function $\psi_{kn} = \psi_{kn}(s)$ by

$$\psi_{kn}(s) := \psi(ns - k), \quad s \in \mathbb{R}. \tag{5.82}$$

Lemma 5.4.1. *Let ψ denote the function defined by (5.81). Then*

1. $\operatorname{supp}\{\psi\} \subset [0, m+1]$;
2. *for every $s \geq 0$ one has*

$$\psi(-s+m+1) = \psi(s). \tag{5.83}$$

Proof. Both assertions of this lemma can be proved by induction. We show the second one only. Thus, if $m = 1$, then

$$\psi(s) = \begin{cases} s & \text{if } 0 \leq s < 1, \\ 2-s & \text{if } 1 \leq s < 2, \\ 0 & \text{otherwise,} \end{cases}$$

and (5.83) is obvious. Suppose that equality (5.83) is satisfied for $k = m$ and consider the case $k = m+1$. One has

$$\begin{aligned} \psi^{m+1}(-s+m+2) &= \int_{\mathbb{R}} \chi_{[0,1)}(-s+m+2-x)\psi^m(x)dx \\ &= \int_{\mathbb{R}} \chi_{[0,1)}(-s+1+u)\psi^m(-u+m+1)du \\ &= \int_{\mathbb{R}} \chi_{[0,1)}(-s+1+u)\psi^m(u)du \\ &= \int_{\mathbb{R}} \chi_{[0,1)}(s-u)\psi^m(u)du = \psi^{m+1}(s), \end{aligned}$$

which completes the proof. □

Now we are able to introduce spline spaces on $\Gamma_{\beta,\omega}$. Namely, we denote by $S_n^{\beta,\omega}$ the smallest closed subspace of $L_p(\Gamma_{\beta,\omega}, \alpha)$ which contains all the functions

$$\widetilde{\phi}_k^{(n)}(t) := \begin{cases} \begin{cases} \psi_{kn}(s) & \text{if } \ t = e^{i\beta}s \\ 0 & \text{otherwise} \end{cases} & k \geq 0, \\ \\ \begin{cases} \psi_{k-m,n}(s) & \text{if } \ t = e^{i(\beta+\omega)}s \\ 0 & \text{otherwise} \end{cases} & k < 0. \end{cases}$$

Let us consider the semi-linear form

$$\langle f, g\rangle := \int_{\Gamma_{\beta,\omega}} f(t)\overline{g(t)}|dt|,$$

where $f \in L_p(\Gamma_{\beta,\omega}, \alpha)$, $g \in L_q(\Gamma_{\beta,\omega}, -\alpha)$ and $1/p + 1/q = 1$. The Galerkin projection operators $\widetilde{L}_n$ from $L_p(\Gamma_{\beta,\omega}, \alpha)$ onto $S_n^{\beta,\omega}$ can be defined by the relations

$$\langle \widetilde{L}_n f, \phi_{kn}\rangle := \langle f, \phi_{kn}\rangle \quad \text{for} \quad f \in L_p(\Gamma_{\beta,\omega}, \alpha) \quad \text{and} \quad \phi_{kn} \in S_n^{\beta,\omega}.$$

It is known [102] that the operators $\widetilde{L}_n$, $n = 1, 2, \dots$ are well defined and the sequence $(\widetilde{L}_n)$ converges strongly to the identity operator I as n tends to ∞.

Our task now is to establish stability conditions for the sequence of Galerkin operators $(\widetilde{L}_n R_\omega \widetilde{L}_n)$ in the case where R_ω is the Muskhelishvili operator (5.30). To proceed with this problem we have to recall some notions. Following Section 5.2 consider the mapping $\eta : L_p(\Gamma_{\beta,\omega}, \alpha) \to L_p^2(\mathbb{R}^+, \alpha)$ defined by

$$\eta(f) = (\eta_1(f), \eta_2(f))^T$$

where

$$\begin{aligned} \eta_1(f)(s) &= f(se^{i(\beta+\omega)}), \quad s \in \mathbb{R}^+, \\ \eta_2(f)(s) &= f(se^{i\beta}), \quad s \in \mathbb{R}^+. \end{aligned}$$

Recall that the mapping $\eta : L_p(\Gamma_{\beta,\omega}, \alpha) \to L_p^2(\mathbb{R}^+, \alpha)$ is invertible and $A \to \eta A \eta^{-1}$ is an isometric algebra isomorphism of $\mathcal{L}(L_p(\Gamma_{\beta,\omega}, \alpha))$ onto $\mathcal{L}(L_p^2(\mathbb{R}^+, \alpha))$.

Lemma 5.4.2. *Let S_n be the smallest closed subspace of $L_p(\mathbb{R}^+, \alpha)$ which contains all functions $\psi_{kn} = \psi_{kn}(s)$, $s \in \mathbb{R}^+$ of (5.82) with $k \geq 0$ and let $L_n : L_p(\mathbb{R}^+, \alpha) \to S_n$ denote the Galerkin projection onto S_n. Then for every $n \in \mathbb{N}$ the operator $\widetilde{L}_n \in \mathcal{L}(L_p(\Gamma_{\beta,\omega}, \alpha))$ is isometrically isomorphic to the operator* $\operatorname{diag}(L_n, L_n) \in$ *$\mathcal{L}(L_p^2(\mathbb{R}^+, \alpha))$.*

Proof. Immediate calculations and Lemma 5.4.1 show that

$$\eta(\widetilde{L}_n)\eta^{-1} = \operatorname{diag}(L_n, L_n).$$

□

Note that in the sequel any diagonal operator of the form $\operatorname{diag}(T, T, \dots, T)$ will be written as T, so we write L_n instead of $\operatorname{diag}(L_n, L_n)$.

The next result immediately follows from Lemma 5.4.2 and Proposition 5.2.3.

Proposition 5.4.3. *The sequence $(\widetilde{L}_n R_\omega \widetilde{L}_n)$ is stable if and only if so is the sequence $(L_n A_\omega L_n)$.*

Recall that the operator A_ω is defined in Proposition 5.2.3. Now we can make further simplification. For, we consider the space $l_{p,\alpha}$ of all sequences $\{\xi_j\}_{j=0}^{+\infty}$ of complex numbers ξ_j such that

$$||\{\xi_j\}||_{p,\alpha}^p = \sum_{j=0}^{+\infty} (1 + j)^{\alpha p} |\xi_j|^p < \infty,$$

and we define the operators $E_n : l_{p,\alpha} \to S_n$ and $E_{-n} : S_n \to l_{p,\alpha}$ by

$$E_n : \{\xi_j\} \to \sum_{j=0}^{+\infty} \xi_j \psi_{jn}(t),$$

$$E_{-n} : \sum_{j=0}^{+\infty} \xi_j \psi_{jn}(t) \to \{\xi_j\}.$$

Proposition 5.4.4. ([39]) *The mappings $E_n : l_{p,\alpha} \to S_n$ and $E_{-n} : S_n \to l_{p,\alpha}$ are bounded linear operators, and there are constants $C_1 > 0$ and $C_2 > 0$ such that*

$$\|\sum_{j=0}^{+\infty} \xi_j \psi_{jn}\|_{L_p(\mathbb{R}^+,\alpha)} \le C_1 n^{-(1/p+\alpha)} \|\{\xi_j\}\|_{l_{p,\alpha}},$$

$$\|\{\xi_j\}\| \le C_2 n^{(1/p+\alpha)} \|\sum_{j=0}^{+\infty} \xi_j \psi_{jn}\|_{L_p(\mathbb{R}^+,\alpha)}.$$

Let $l^2_{p,\alpha}$ denote the Cartesian product of two copies of $l_{p,\alpha}$.

Lemma 5.4.5. *The sequence $(\widetilde{L}_n R_\omega \widetilde{L}_n)$ is stable if and only if the operator $B^1_\omega := E_{-1} L_1 A_\omega L_1 E_1 : l^2_{p,\alpha} \to l^2_{p,\alpha}$ is invertible.*

Proof. Let M and M^{-1} be the direct and inverse Mellin transforms introduced in Section 1.10.2, $\mathcal{M}(b)$ be the Mellin convolution operator with the symbol b, $k := M^{-1}(b)$, and let $F(b) = (a_{lq})_{l,q=0}^{\infty}$ be the matrix of the operator $E_{-n} L_n (M^{-1} b M) L_n E_n : l_{p,\alpha} \to l_{p,\alpha}$. Then

$$\begin{aligned} a_{lq} &= \int_{\mathbb{R}} (\mathcal{M}(b)\psi_{qn})(s)\psi_{ln}(\sigma)d\sigma \\ &= \int_{\mathbb{R}}\int_{\mathbb{R}^+} k\left(\frac{\sigma}{s}\right)\psi(ns-q)\frac{ds}{s}\psi(\sigma n - l)du \\ &= \int_{\mathbb{R}}\int_{\mathbb{R}^+} k\left(\frac{u+l}{t+q}\right)\psi(t)\frac{dt}{t}\psi(u)du. \end{aligned}$$

Thus, the entries of the matrix $F(b)$ are independent of n. Taking into account that the operator $B_\omega = B^1_\omega := E_{-1} L_1 A_\omega L_1 E_1$ admits the representation

$$B_\omega = \begin{pmatrix} 0 & F(b_1) \\ F(b_2) & 0 \end{pmatrix} + \begin{pmatrix} -I & F(b_3) \\ F(b_3) & -I \end{pmatrix} \begin{pmatrix} \overline{M} & 0 \\ 0 & \overline{M} \end{pmatrix} \tag{5.84}$$

where $\overline{N}(\{\xi_j\}) = \{\overline{\xi}_j\}$ and $b_1 = e^{-2i\beta} m_\omega(y)$, $b_2 = e^{-2i(\beta+\omega)} m_{2\pi-\omega}(y)$, $b_3 = k n_\omega(y)$, $y = z + i(1/p+\alpha)$, $z \in \mathbb{R}$, one obtains the claim. □

It is worth mentioning that the study of the invertibility of the operator B_ω is a very difficult problem. However, the conditions for invertibility of the operator A_ω are simultaneously the conditions for Fredholmness of the operator B_ω. Thus the following corollary holds.

Corollary 5.4.6. *For every $\omega \in (0, 2\pi)$ there exists a real number $\delta'_\omega > 1/2$ such that for every $p > 1$ and every α satisfying the inequality*

$$0 < \frac{1}{p} + \alpha < \delta'_\omega$$

the operator $B^1_\omega : l^2_{p,\alpha} \to l^2_{p,\alpha}$ is Fredholm of index zero.

Proof. Indeed, let us consider the operator $F(b)$ again, and let R and S denote the real and imaginary parts of the function $k = M^{-1}b$, respectively. Then

$$a_{lq} = \int_{\mathbb{R}} \int_{\mathbb{R}+} R\left(\frac{u+l}{t+q}\right) \psi(t)\frac{dt}{t}\psi(u)du + i \int_{\mathbb{R}} \int_{\mathbb{R}+} S\left(\frac{u+l}{t+q}\right) \psi(t)\frac{dt}{t}\psi(u)du.$$

Using the generalized mean value theorem we find that there are points $u^1_{lq}, u^2_{lq} \in [0, m]$ and $t^1_{lq}, t^2_{lq} \in [0, m]$ such that

$$a_{lq} = \left(\int_{\mathbb{R}+} \psi(t)dt\right)^2 \left(R\left(\frac{u^1_{lq}+l}{t^1_{lq}+q}\right)\frac{1}{t^1_{lq}} + iS\left(\frac{u^2_{lq}+l}{t^2_{lq}+q}\right)\frac{1}{t^2_{lq}}\right).$$

Now by [102, Corollary 2.1 and Proposition 2.11 on p. 65] there exists a compact operator K_1 such that

$$F(b) = \left(\int_{\mathbb{R}} \psi(t)dt\right)^2 \left(\int_l^{l+1}\int_q^{q+1} k\left(\frac{t}{s}\right)\frac{ds}{s}dt\right)_{l,q=0}^{+\infty} + K_1. \tag{5.85}$$

Recall that $k = M^{-1}(b)$ and define the operator $G(b)$ by

$$G(b) := \left(\int_l^{l+1}\int_q^{q+1} k\left(\frac{t}{s}\right)\frac{ds}{s}dt\right)_{l,q=0}^{+\infty}.$$

Considering now the operators $G(b_j)$ for the above defined symbols b_j, $j = 1, 2, 3$, we note that the functions b_1, b_2, b_3 are continuous on $\mathbb{R}$, possess a total finite variation and vanish at infinity. Consequently, by [102, Theorem 2.1, p. 69] the operators $G(b_j)$, $j = 1, 2, 3$ belong to the algebra of Toeplitz operators $alg\, T(PC_{p,\alpha})$ and the symbols of $G(b_j)$ are

$$\mathcal{A}_{G(b_j)} = \mathcal{M}(b_j), \quad j = 1, 2, 3,$$

and relation (5.85) implies

$$\mathcal{A}_{F(b_j)} = d\,\mathcal{M}(b_j), \quad j = 1, 2, 3, \tag{5.86}$$

where the constant

$$d = \left(\int_{\mathbb{R}} \psi(t)dt \right)^2 .$$

Since the operator B_ω has the form (5.84), one can use Lemma 1.4.16 to study its Fredholm properties. Thus the operator $B_\omega = \mathcal{F}_1 + \mathcal{F}_2 \overline{M}$ is Fredholm if and only if the operator

$$\widetilde{B}_\omega = \begin{pmatrix} \mathcal{F}_1 & \mathcal{F}_2 \\ \overline{M}\mathcal{F}_2\overline{F} & \overline{M}\mathcal{F}_1\overline{M} \end{pmatrix}$$

is Fredholm, so (5.86) and the equalities

$$\overline{M}G(b_j)\overline{M} = G(\widetilde{b}_j), \quad j = 1, 2, 3$$

where $\widetilde{b}_j(z) = \overline{b_j(-z)}$, $z \in \mathbb{R}$, imply that

$$\mathcal{A}_{B_\omega} = d\,\widetilde{A}_\omega .$$

Recall that the operator $\widetilde{A}_\omega$ is defined by (5.38). The operator B_ω is Fredholm if and only if its symbol is invertible. However, conditions of invertibility of the operator $\widetilde{A}_\omega$ have been established in Corollary 5.2.9. □

Corollary 5.4.7. *The operator B_ω considered on the space $l^2 := l^2_0$ is Fredholm for any $\omega \in (0, 2\pi)$, and its index vanishes.*

5.5 Galerkin Method. Piecewise Smooth Contour

This section contains some results concerning stability and convergence of the Galerkin method based on splines of degree $m \geq 1$.

Let γ be a 1-periodic parametrization of Γ such that the corner points c_j, $j = 0, 1, \ldots, l-1$, are represented as follows:

$$c_j = \gamma(j/l), \quad j = 0, 1, \ldots, l-1.$$

We also assume that γ is twice continuously differentiable on each of the intervals $(j/l, (j+1)/l)$, $j = 0, 1, \ldots, l-1$, and that there exist one-sided limits $\gamma'(j/l \pm 0)$ and $\gamma''(j/l \pm 0)$ and

$$|\gamma'(j/l + 0)| = |\gamma'(j/l - 0)|, \quad j = 0, 1, \ldots, l-1.$$

Choose $n = lr$, $r \in \mathbb{N}$ and set

$$\widetilde{\psi}_{kn}(t) := \psi_{kn}(s), \quad t = \gamma(s), \quad s \in [0, 1),$$

with ψ_{kn} defined by (5.82).

An approximate solution φ_n of equation (5.58) is sought in the form

$$\varphi_n = {\sum}' c_k \widetilde{\psi}_{kn}, \tag{5.87}$$

where the sum $\sum'$ includes only those functions $\widetilde{\psi}_{kn}$, the support of which is entirely contained in one of the arcs $[\gamma(j/l), \gamma((j+1)/l))$, $j = 0, 1, \ldots, l-1$. The smallest subspace of $L_p(\Gamma, \rho)$ containing all such functions $\widetilde{\psi}_{kn}$ will be referred to as $S_n(\Gamma)$. The corresponding subset of indices k $(0 \le k \le n-1)$ such that $\widetilde{\psi}_{kn} \in S_n(\Gamma)$ is denoted by A'.

To find the coefficients c_k, $k \in A'$ on the right-hand side of (5.87) we use the following system of algebraic equations:

$$\left(R_j \varphi_n, \widetilde{\psi}_{kn}\right)_\Gamma = \left(f_0, \widetilde{\psi}_{kn}\right)_\Gamma, \quad k \in A', \tag{5.88}$$

where the form of the operator R_j, $j = 1, 2$ depends on the coefficient k in the Muskhelishvili equation (cf. (5.46)), and

$$(x, y)_\Gamma = \sum_{j=0}^{l-1} \int_{j/l}^{(j+1)/l} x(\gamma(s))\overline{y(\gamma(s))}ds, \quad x \in L_p(\Gamma, \rho), \ y \in L_q(\Gamma, \rho^{-1}),$$

$1/p + 1/q = 1$.

With each corner point c_r $(r = 0, 1, \ldots, l-1)$ of Γ we associate an operator $B_{\omega_r} = B_{\beta_r, \omega_r}$. These operators are defined similarly to the operator B_ω in (5.84), but the parameters ω and β are replaced by ω_r and β_r, respectively. We recall that ω_r is the angle between corresponding semi-tangents to Γ at the point c_r and that β_r is the angle between the right semi-tangent to Γ at the point c_r and the real axis.

Theorem 5.5.1. *Let α_r, $r = 0, 1, \ldots, l-1$, and $p \in (1, \infty)$ satisfy inequality* (5.44). *Then:*

1. *The Galerkin method* (5.88) *for the operator R_j in $L_p(\Gamma, \rho)$ is stable if and only if the operators $B_{\omega_r} \in \mathcal{L}_{add}(l^2_{p,\alpha_r})$, $r = 0, 1, 2, \ldots, l-1$, are invertible.*
2. *Let the operators $B_{\omega_r} \in \mathcal{L}_{add}(l^2_{p,\alpha_r})$, $r = 0, 1, 2, \ldots, l-1$, be invertible. If, in addition, the right-hand side $f \in W^1_p(\Gamma, \rho)$ and satisfies other conditions of Theorem* 5.2.17, *then the approximate solutions* (5.87) *converge to a solution of Muskhelishvili equation* (5.58) *in the norm of $L_p(\Gamma, \rho)$.*

The proof of the first assertion follows from Theorems 1.6.6, 1.9.5 and from the first part of Theorem 5.2.17. Afterwards, the second assertion follows from the second part of Theorem 5.2.17 and estimates (1.29), (1.30).

Thus, the sequence $\{\varphi_n\}$ constructed by Galerkin method (5.88) converges to an exact solution of (5.58) in the norm $L_p(\Gamma, \rho)$. What is more important is that in the case $p = 2$ one can guarantee convergence of the sequence in the space $W^1_2(\Gamma)$.

Corollary 5.5.2. *Let the operators B_{ω_r}, $r = 0, 1, \ldots, l-1$, be invertible in l_2^2. Then the Galerkin method* (5.58) *is stable in $W_2^1(\Gamma)$, and if f satisfies condition* (5.18), *then the sequence $\{\varphi_n\}$ converges to a solution of equation* (5.58) *in the norm of $W_2^1(\Gamma)$.*

Proof. It follows from Corollary 5.2.18 and from the proof of Theorem 5.2.17 that the operator R_j is simultaneously invertible in the spaces $L_2(\Gamma)$ and $W_2^1(\Gamma)$. Hence, to prove the stability of the method (5.88) in $W_2^1(\Gamma)$ one can use Theorem 1.37 of [183]. Let L_n denote the corresponding Galerkin projection onto the subspace $S_n(\Gamma)$. A slight modification of the proof of Theorem 2.7 of [183] shows that there exists a constant $d_1 > 0$ such that

$$||g - L_n g||_{L_2(\Gamma)} < d_1 n^{-1} ||g||_{W_2^1(\Gamma)}, \quad g \in W_2^1(\Gamma).$$

The latter inequality yields the estimate

$$||Ag - L_n A L_n g||_{L_2(\Gamma)} < d_2 n^{-1} ||g||_{W_2^1(\Gamma)}, \quad g \in W_2^1(\Gamma).$$

Note that the positive constants d_1, d_2 are independent of $g \in W_2^1(\Gamma)$ and n.

Taking into account the inverse properties of the splines φ_n (recall that $n = lr$, $r \in \mathbb{N}$ and that the support of $\widetilde{\psi}_{kn}$ is entirely contained in one of the arcs $(\gamma(j/l), \gamma((j+1)/l))$, $j = 0, 1, \ldots, l-1$) and applying Theorem 1.37 of [183] one obtains stability of the Galerkin method (5.88). This implies convergence of $\{\varphi_n\}$ in $W_2^1(\Gamma)$. Using Theorem 5.2.17 once more we get the result. □

5.6 Comments and References

The biharmonic problem arises in different branches of applied mathematics, e.g., the behaviour of plane "slow" viscous flows, the deflection of plates, the elastic equilibrium of solids etc can be modelled by means of the biharmonic equation [25, 26, 34, 121, 134, 148, 153, 160, 161, 150, 167, 170, 217, 219]. It is therefore no wonder that this problem has been attracting great attention from numerical analysts. Suffice it to say that one of the most powerful approximation procedures, viz., the Galerkin method[1], was discovered while considering a special case of problem (5.1) [12, 13, 87, 211]. Among the variety of approaches to numerical treatment of problem (5.1), one can distinguish the so-called boundary element methods [16, 21, 31, 32, 109, 110]. They allow us to reduce the dimension of the initial problem, and, correspondingly, reduce the computation cost drastically. The authors of the aforementioned papers usually use various modifications of the integral equation proposed by S. Christiansen and P. Hougaard [22, 23, 84], or an integral equation of the first kind [111]. Although such approaches are widely used, they do have some drawbacks. Thus for some boundaries called *critical* the corresponding integral operators become non-invertible and corrections are necessary

[1] In Russian-speaking literature this method is often called the Bubnov-Galerkin method.

before one can start with the construction of approximation methods. Note that the analysis of stability of the approximation methods has mainly been conducted for smooth boundaries, although for piecewise smooth contours, invertibility of the corresponding integral operators is also studied [30].

On the other hand, there is a very nice alternative *complex* approach to the problem (5.1), which originates from the work of N.I. Muskhelishvili [160] and leads to integral equations without critical geometry. However, in a strange way the Muskhelishvili equation remains a little-known quantity in numerical analysis. The most common approach to the approximate solution of this equation is based on trigonometric Fourier expansions and was proposed by N.I. Muskhelishvili himself in the mid-1930s (cf. [17, 18, 24, 121]). Since then many new powerful approximation methods have been developed, but they have not been implemented and/or studied in the case of the Muskhelishvili equation. For example, the papers [99, 100, 107, 108, 129, 220, 130, 152, 172, 173] deal with variety of approximation methods for this and similar equations on smooth and non-smooth contours but do not contain a complete stability analysis.

It is also notable that the integral operators in the Muskhelishvili equation can be "locally" represented as elements of an algebra of Mellin operators with conjugation. The stability of approximation methods for Mellin convolutions was studied in [102, 179], so one can apply some of these results to the operators connected with the Muskhelishvili equation. It should also be noted that, from the practical point of view, convergence of the methods considered has to be proven in spaces of differentiable functions. However, since the technique used here is well adapted to the norms of L_p spaces, we first show stability in the spaces L_p, and then we use [183] to obtain some results for Sobolev norms.

Section 5.1: Material presented here can be found in the elasticity theory literature (cf. for example [121, 147, 161, 170, 219]) although some proofs may be new.

Section 5.2: In case of smooth boundaries, the equations of Muskhelishvili and Sherman-Laurichella possess weakly-singular kernels. Various properties and solvability of these equations is established in [160, 161, 162, 204, 205, 206]. For the non-smooth contours, these equations have considerably different properties. Therefore, a lot of effort has been spent and various methods have been used to investigate the Fredholmness of the corresponding operators in functional spaces $\mathbf{C}, L_p(\Gamma)$ or $L_p(\Gamma, \rho)$, [37, 145, 146, 149, 156, 169, 201, 202, 203]. However, since the initial boundary problems (5.13), (5.20) contain the derivative of the solution, the results mentioned do not allow one to show the equivalence of these boundary problems with the associated integral equations.

To overcome this difficulty, R.V. Duduchava studied these integral equations in the weighted Sobolev spaces $W_p^1(\Gamma, \rho)$ [72, 73]. His results were used in [61] to show the invertibility of the relevant integral operators in the space $L_p(\Gamma, \rho)$. If Ω is a plane sector, the spectra of the hydrostatic K_{Stoks} and elastostatic K_{Lame} layer potential operators in spaces $L_p(\Omega)$ are studied [158]. Note that the determinant of the symbol of these operators obtained in [158] coincide with the determinant of the symbol of the local Muskhelishvili operators (5.39) when the Muskhelishvili operators are also considered on spaces without weight, i.e., when $\alpha = 0$. Thus in the aforementioned case, the conditions of Fredholmness of all these three operators coincide.

Section 5.3: Results on approximate solution of the Muskhelishvili equation on smooth contours are established in [65]; approximate solution of the biharmonic problem in smooth domains is considered in [64].

Sections 5.4–5.5: Most of the material presented here is taken from [61, 62]. There are other approaches to approximate solution of biharmonic problems based on integral equations. Thus in [31] the authors construct approximate solutions of (5.9) via spline approximation of strongly elliptic integral equations obtained by [36]. Approximation methods based on integral equations from [22, 23, 84] are investigated in [32]. Another integral equation [111] is employed in [109, 110]. In [81], the authors apply spline collocation methods to a double layer potential equation to find solutions of a fundamental problem of the elasticity theory.

Note an excellent survey of earlier works on biharmonic problems presented in [153].

Concluding Remarks. There are various papers where elastostatic and hydrostatic problems are considered on multi-connected domains [100, 129, 130, 152, 220]. These problems are reduced to the so-called Sherman-Laurichella equation that differs from the Muskhelishvili equation by compact operators only. However, since such operators do not influence the properties of the local operators, the methods used here can also be applied to study the stability of the corresponding approximation methods for multi-connected domains as well.

Chapter 6

Numerical Examples

6.1 Muskhelishvili Equation

The Galerkin and collocation methods were implemented on several examples, using splines of order $m = d + 1$. We also performed an extensive investigation on the performance of the proposed schemes, examining in particular the behaviour of the code as a function of the various available parameters.

Convergence of the numerical schemes follows, comparing runs with an increasing number of grid points, n. As the analytic solution for these examples is not available, we show the graphs of the solutions $x_n(t)$, $t \in \Gamma$ for various increasing values of n. The figures below show convergence of the algorithms, as the former become closer the higher n is. This happens both with different choices of the order of the splines as well as by using different methods for their calculations. When the analytic solution is not available, these calculations also validate the code since always the same solution is reproduced.

Note that the major computational effort lies in solving systems of algebraic equations and thus it is the same for each example and the same approximation method if the matrices have the same size. The estimates on conditioning of the system are obtained from the standard Matlab function cond (M, p) with $p = 2$ and $p = \infty$.

Below we provide graphs that show collocation and Galerkin solutions of the Muskhelishvili equation (5.29) with the same right-hand side

$$f_0(t) = \frac{\cos(t) - it^2}{\sin(t)}$$

on a family of curves. Thus we start with the circular domain of radius 1 and proceed with equations formulated on ellipses with increasing eccentricity. It is obvious that solutions differ for different values of this parameter, but the algorithms are still stable and pick up the fine details of each solution, provided that a sufficient number of nodes is used. In addition, one can note a remarkable transformation of the solution caused by changing of the contour. As a final check, also the Galerkin method has been used, running it with different m but with the same eccentricity, to compare the results of the collocation method. The reader should compare Figures 6.3 and 6.4, 6.5 and 6.6, 6.7 and 6.8 respectively. We also remark that the condition numbers for the larger values of n in these cases are essentially the same, provided ϵ is not close to the value 0.5, independently of the number of nodes, up to $n = 512$ and of the order of the splines used. Specifically keeping $b = 2$, the condition number is evaluated in Table 6.1 for both methods.

Table 6.1: Conditioning in elliptic case for $b = 2$.

		$a = 6$	$a = 10$	$a = 18$
collocation	$m = 2$	5.1×10^2	8.3×10^2	1.5×10^3
$\|M\|_\infty \|M^{-1}\|_\infty$	$m = 3$	5.6×10^2	9.2×10^2	1.9×10^3
$n = 128$	$m = 4$	7.1×10^2	1.3×10^3	3.3×10^3
$\epsilon = .25$	$m = 5$	8.4×10^2	1.6×10^3	4.8×10^3
Galerkin $\|M\|_2 \|M^{-1}\|_2$	$m = 2$	1.2×10^2	3.0×10^2	9.9×10^2
$n = 129$	$m = 4$	8.0×10^2	2.2×10^3	7.5×10^3

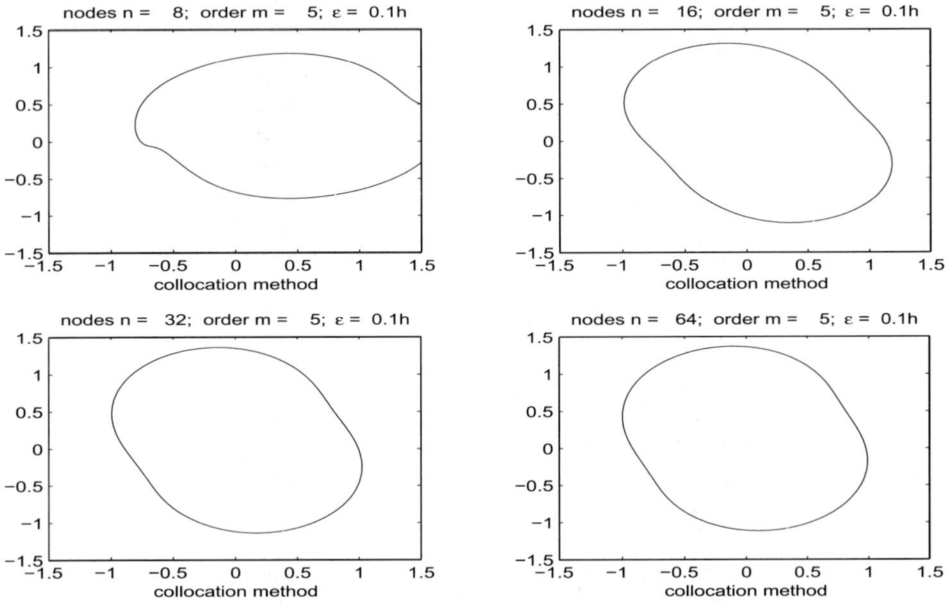

Figure 6.1: Collocation solution $x_n(t)$, $t \in \Gamma$, the unit circle.

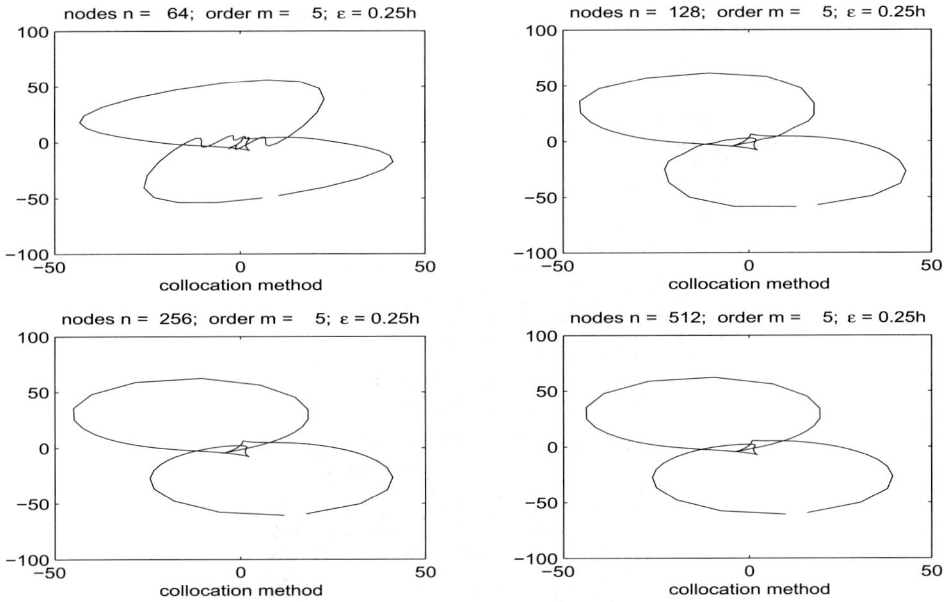

Figure 6.2: Collocation solution $x_n(t)$, $t \in \Gamma$ where Γ is now the ellipse with semiaxes $a = 3$, $b = 2$.

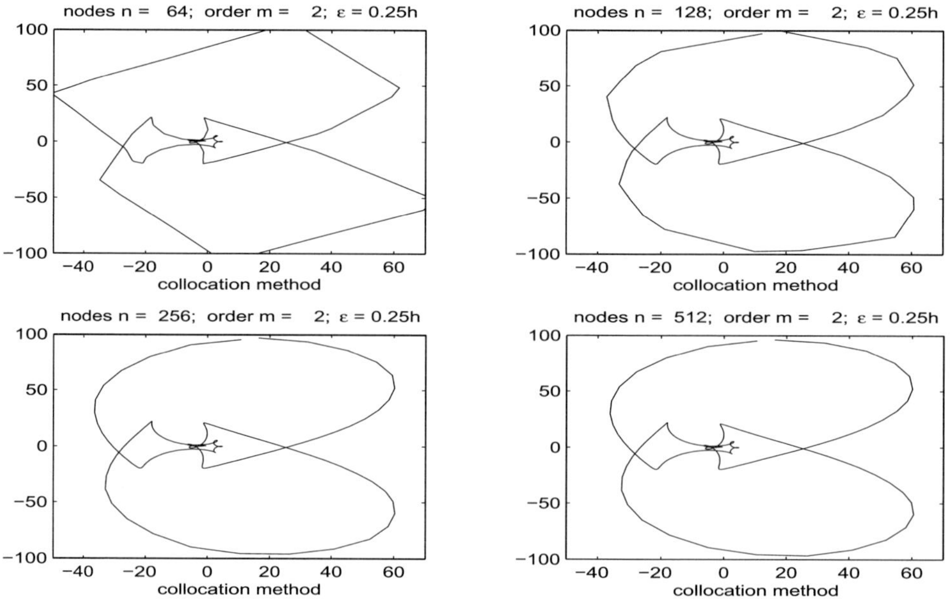

Figure 6.3: Collocation solution $x_n(t)$, $t \in \Gamma$ where Γ is now the ellipse with semiaxes $a = 6$, $b = 2$.

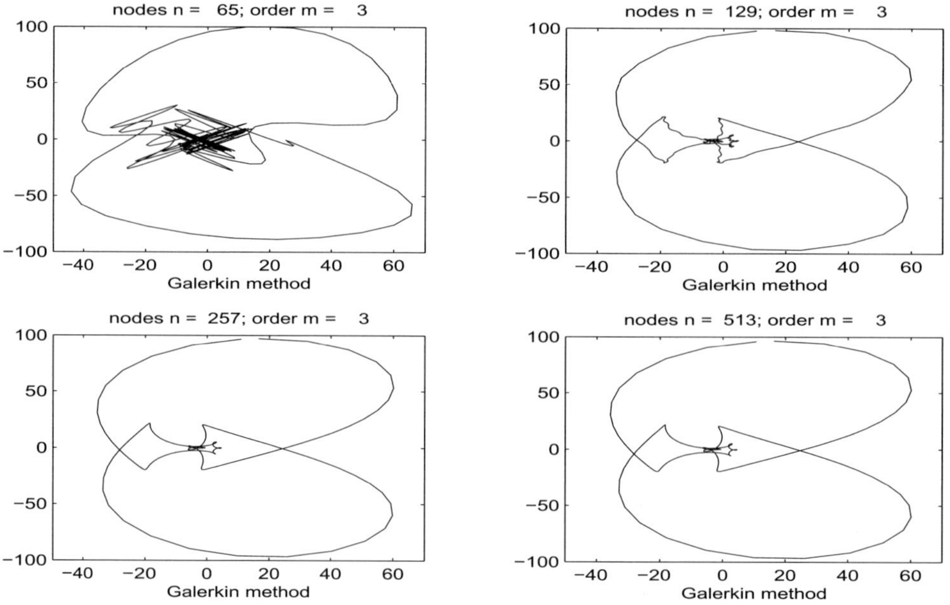

Figure 6.4: Galerkin solution $x_n(t)$, $t \in \Gamma$ where Γ is now the ellipse with semiaxes $a = 6$, $b = 2$.

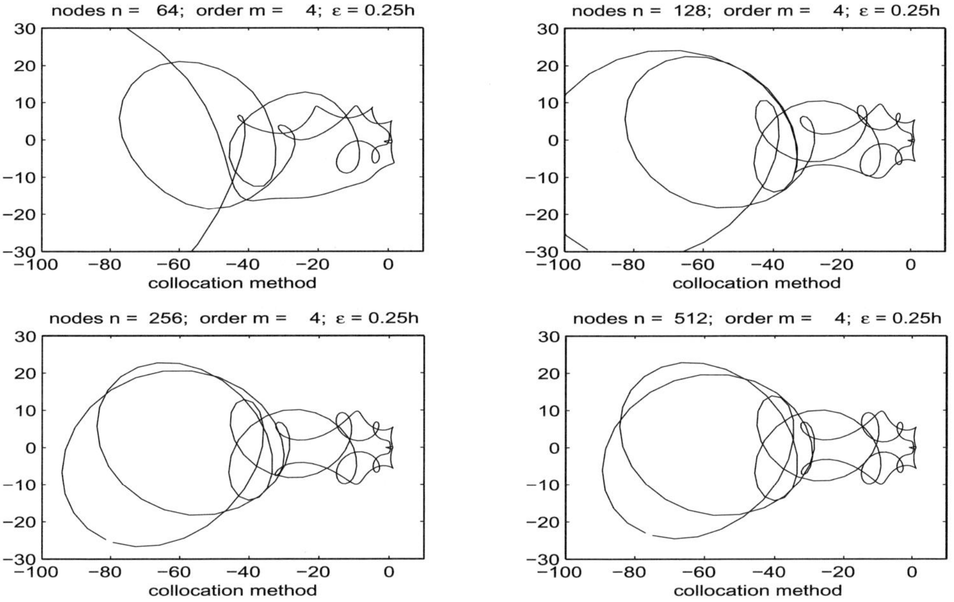

Figure 6.5: Collocation solution $x_n(t)$, $t \in \Gamma$ where Γ is now the ellipse with semiaxes $a = 10$, $b = 2$.

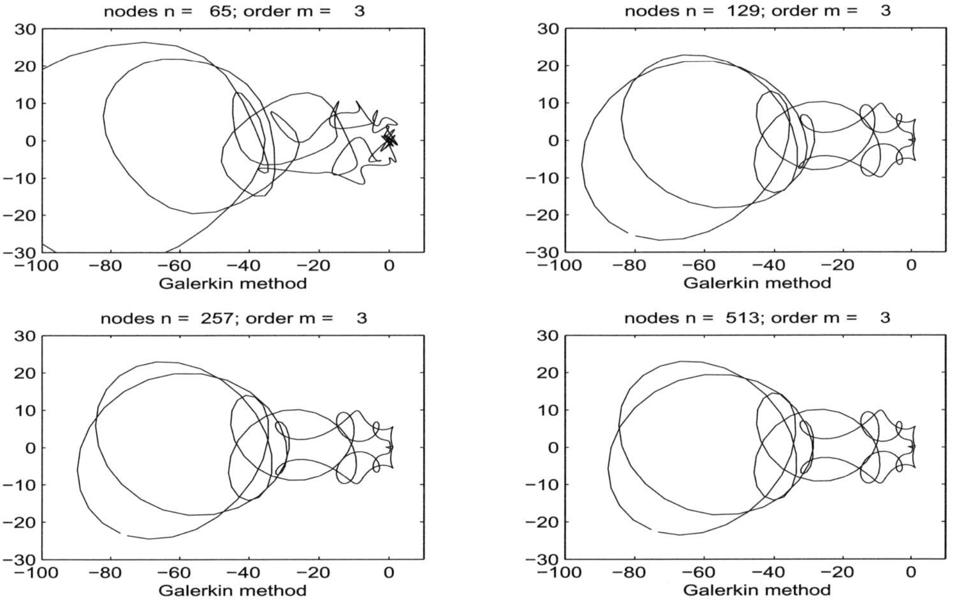

Figure 6.6: Galerkin solution $x_n(t)$, $t \in \Gamma$ where Γ is now the ellipse with semiaxes $a = 10$, $b = 2$.

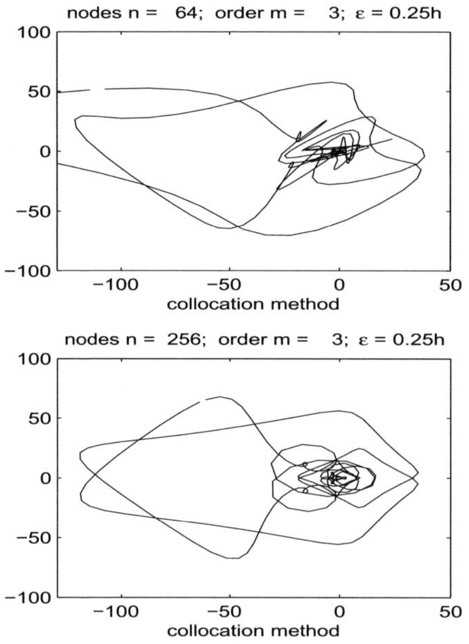

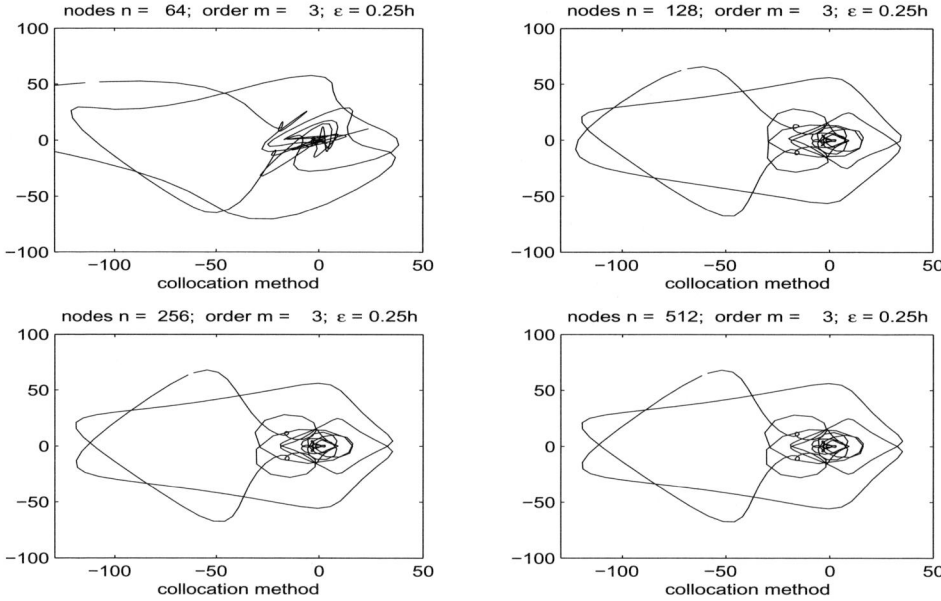

Figure 6.7: Collocation solution $x_n(t)$, $t \in \Gamma$ where Γ is now the ellipse with semiaxes $a = 18$, $b = 2$.

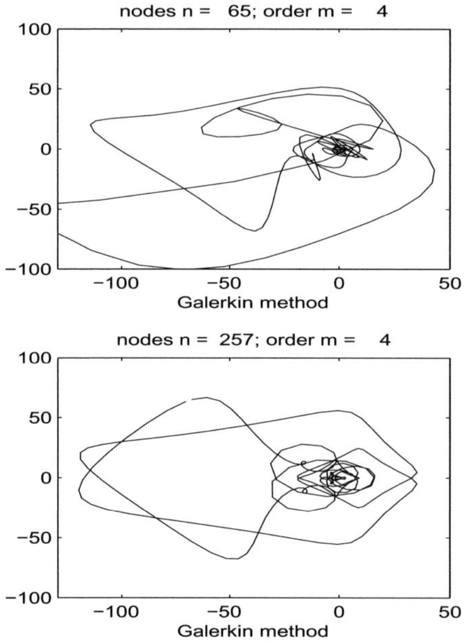

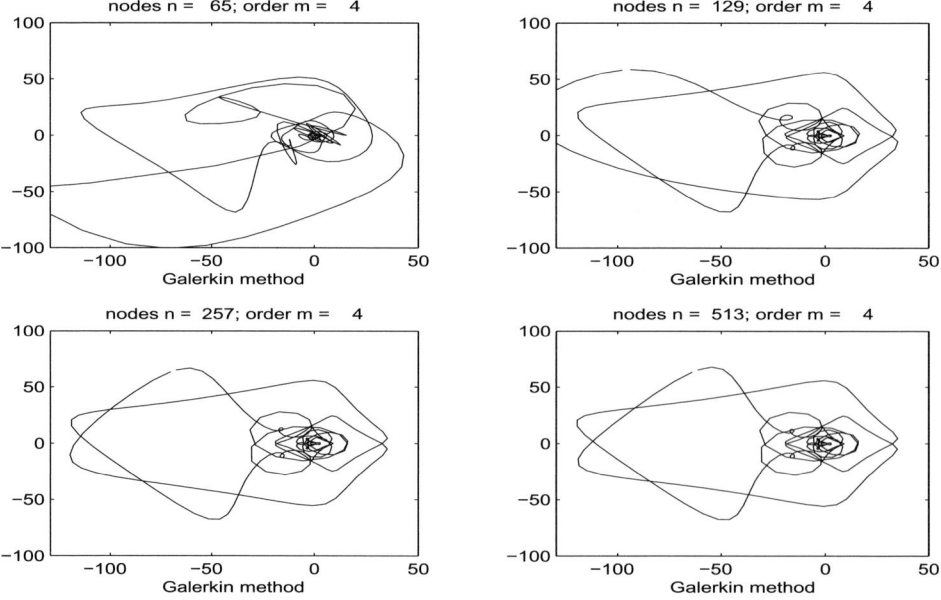

Figure 6.8: Galerkin solution $x_n(t)$, $t \in \Gamma$ where Γ is now the ellipse with semiaxes $a = 18$, $b = 2$.

6.2 Biharmonic Problem

Consider a few examples of numerical solution of the biharmonic problem studied in Section 5.3.4. These examples as well as the related figures empirically illustrate the numerical performance of the algorithm. In all examples the order of the splines used is $d = 2$. Also, the curve Γ is always the unit circle. Approximations of the function φ are found from the Muskhelishvili equation using the collocation method of Section 5.3.2 with the collocation points

$$t_j^{(n)} \equiv \gamma\left(\frac{j+\delta}{n}\right), \quad j = 0, 1, \ldots, n-1$$

where δ is a real number $0 \leq \delta < 1$. The collocation points are chosen with different values of δ to demonstrate that the choice of this parameter does not influence the conditioning of the resulting system away from the forbidden value $\delta = 0.5$. The number of basis elements n ranges from 16 to 128. The rows in the figures correspond respectively to the analytic solution, to the computed solution and to the contour plots of the absolute error. The tables show excellent conditioning of the algorithm. To study convergence, in the polar coordinate plane we use a rectangular grid G, with 25 points both in the radial as well as in the angular directions. The convergence is empirically determined from the absolute error $\|U - U_n\|_{G,\infty}$ calculated at grid points of the grid G. The results of the tables show it to be in line with the theory.

Example 6.2.1. Here we solve the problem with boundary functions $G_1 = 2x$, $G_2 = 2y$, with an analytic solution (up to an arbitrary constant) given by $U = x^2 - y^2 + 1$. The behaviour of the numerically evaluated solution is illustrated in Figure 6.9. Table 6.2 contains the conditioning and the numerical values of the error.

Example 6.2.2. The relevant functions in this case are $G_1 = 4x^3 - 12xy^2$, $G_2 = 4y^3 - 12x^2y$, with an analytic solution given by $U = x^4 - 6x^2y^2 + y^4$ and results plotted in Figure 6.10. Conditioning and behaviour of the error are found in Table 6.3.

Table 6.2: Results of Example 6.2.1, $\delta = 0.25$

n	conditioning	$\|U - U_n\|_{G,\infty}$
8	15.85	0.14291
16	15.90	0.04293
32	15.92	0.00898
64	15.92	0.00370

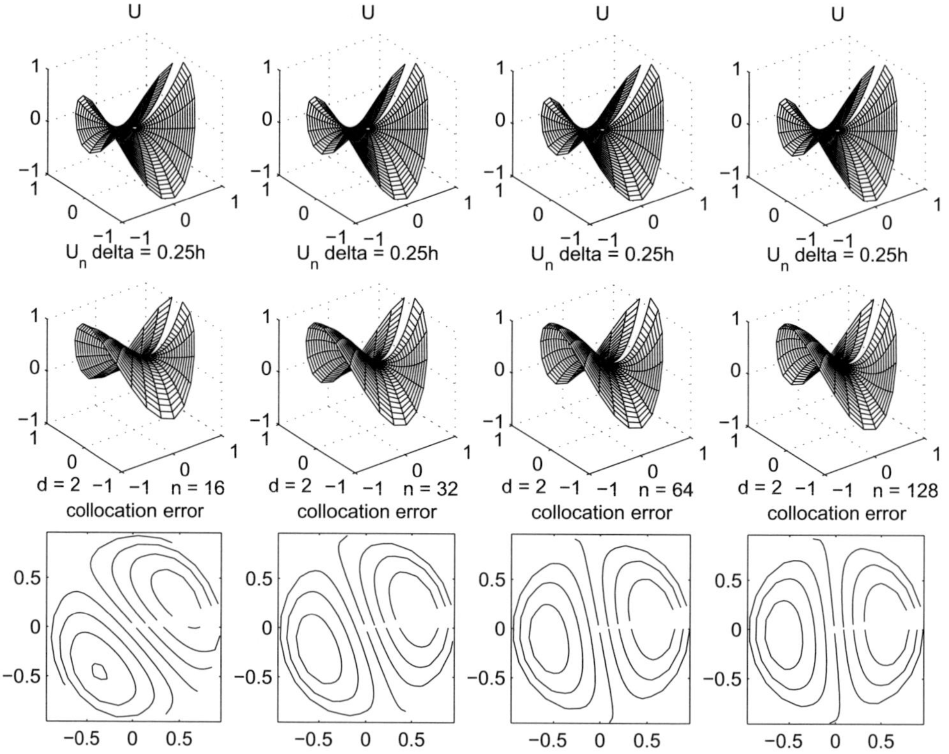

Figure 6.9: Analytic solution (first row) and numerical solution (second row) and contour plot of absolute error (third row) of Example 6.2.1.

Table 6.3: Results of Example 6.2.1, $\delta = 0.3$

n	conditioning	$\|U - U_n\|_{G,\infty}$
16	19.88	7.2196 E-2
32	19.89	4.1258 E-2
64	19.90	1.3493 E-2
128	19.90	5.8315 E-3

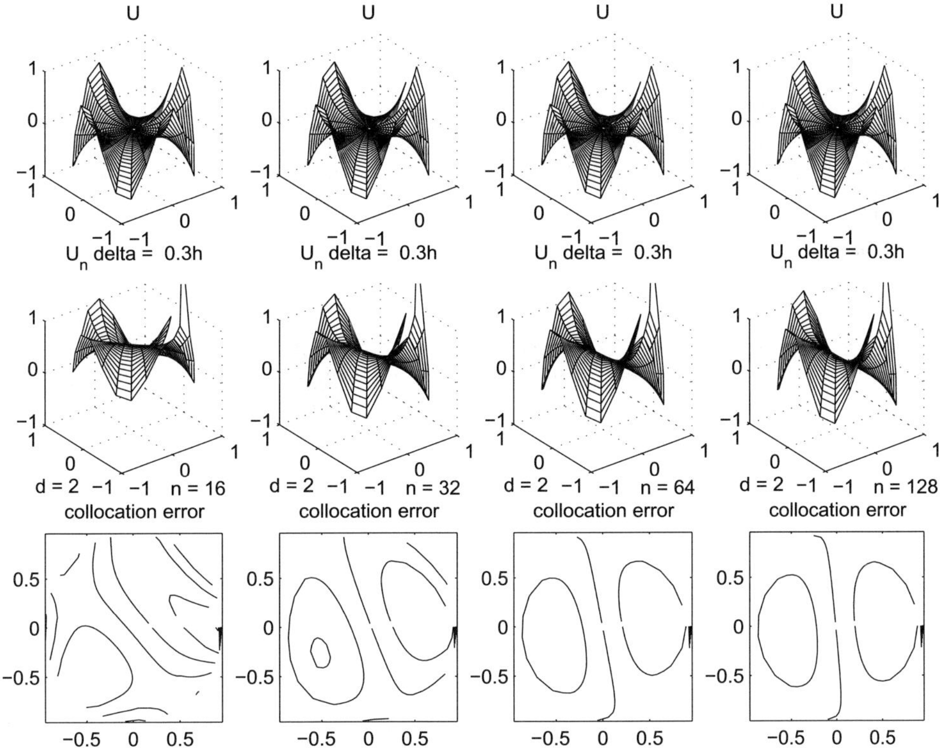

Figure 6.10: Analytic solution (first row) and numerical solution (second row) and contour plot of absolute error (third row) of Example 6.2.2.

Bibliography

[1] G. R. Allan. Ideals of vector-valued functions. *Proc. London Math. Soc.*, 18(3):193–216, 1968.

[2] D. N. Arnold and W. L. Wendland. On the asymptotic convergence of collocation methods. *Math. Comp.*, 41(164):349–381, 1983.

[3] D. N. Arnold and W. L. Wendland. The convergence of spline-collocation for strongly elliptic equations on curves. *Numer. Math.*, 47:317–341, 1985.

[4] W. Arveson. *An invitation to C^*-algebras*. Springer-Verlag, New York, 1976. Graduate Texts in Mathematics, No. 39.

[5] K. E. Atkinson and I. G. Graham. An iterative variant of the Nyström method for boundary integral equations on nonsmooth boundaries. In *The mathematics of finite elements and applications, VI (Uxbridge, 1987)*, pages 297–303. Academic Press, London, 1988.

[6] N. K. Bari. *Trigonometric series.* Gosudarstv. Izdat. Fiz.-Mat. Lit., Moscow, 1961.

[7] S. M. Belotserkovsky and I. K. Lifanov. *Method of discrete vortices.* CRC Press, Boca Raton, FL, 1993.

[8] B. Blackadar. *Operator algebras*, volume 122 of *Encyclopaedia of Mathematical Sciences.* Springer-Verlag, Berlin, 2006. Theory of C^*-algebras and von Neumann algebras, Operator Algebras and Non-commutative Geometry, III.

[9] A. Böttcher and Y. I. Karlovich. *Carleson curves, Muckenhoupt weights, and Toeplitz operators*, volume 154 of *Progress in Mathematics.* Birkhäuser Verlag, Basel, 1997.

[10] A. Böttcher, N. Krupnik, and B. Silbermann. A general look at local principles with special emphasis on the norm computation aspect. *Integral Equations Operator Theory*, 11(4):455–479, 1988.

[11] A. Böttcher and B. Silbermann. *Analysis of Toeplitz operators.* Springer-Verlag, Berlin, 1990.

[12] I. G. Bubnov. Referee report on the memoir by Professor Timoshenko "On the stability of classic systems" crowned by D.I. Zhuravskii prize. *Sb. Inst. Inzh. Put. Soobsch.*, 81:33–36, 1913. See also: *Selected Papers*, Sudpromgiz, Leningrad (1956), 136–139.

[13] I. G. Bubnov. *Structural Mechanics of a Ship.* Naval Academy Press, St. Peterburg, 1914.

[14] K. Bühring. A quadrature method for singular integral equation on curves with corners. *Math. Nachr.*, 167:43–81, 1994.

[15] P. L. Butzer and R. J. Nessel. *Fourier analysis and approximation.* Academic Press, New York, 1971. Volume 1: One-dimensional theory, Pure and Applied Mathematics, Vol. 40.

[16] C. V. Camp and G. S. Gipson. A boundary element method for viscous flows at low reynold numbers. *Eng. Analysis with Boundary Elements*, 6:144–151, 1989.

[17] R. H. Chan, T. K. Delillo, and M. A. Horn. The numerical solution of the biharmonic equation by conformal mapping. *SIAM J. Sci. Comput.*, 18(6):1571–1582, 1997.

[18] R. H. Chan, T. K. DeLilo, and M. A. Horn. Superlinear convergence estimates for a conjugate gradient method for the biharmonic equation. *SIAM J. Sci.Comput.*, 19(1):139–147, 1998.

[19] G. A. Chandler and I. G. Graham. Uniform convergence of Galerkin solutions to noncompact integral operator equations. *IMA J. Numer. Anal.*, 7(3):327–334, 1987.

[20] G. A. Chandler and I. H. Sloan. Spline qualocation methods for boundary integral equations. *Numer. Math.*, 58(5):537–567, 1990.

[21] C. Chang-jun and W. Rong. Boundary integral equations and the boundary element method for buckling analysis of perforated plates. *Eng. Analysis with Boundary Elements*, 17:54–68, 1996.

[22] S. Christiansen. Derivation and analytical investigation of three direct boundary integral equations for the fundamental biharmonic problem. *J. Comput. Appl. Math.*, 91:231–247, 1998.

[23] S. Christiansen and P. Hougaard. An investigation of a pair of integral equations for the biharmonic problem. *J. Inst. Mat. Appl.*, 22:15–27, 1978.

[24] J. M. Chuang and S. Z. Hu. Numerical computation of Muskhelishvili's integral equation in plane elasticity. *J. Comput. Appl. Math.*, 66:123 – 138, 1996.

[25] C. Constanda. Sur le probléme de Dirichlet dans deformation plane'. *Comptes Rendus de l'Academie de Sciences Paris, Serie I, Mathematique*, 316:1107–1109, 1993.

[26] C. Constanda. The boundary integral equations method in the plane elasticity. *Proc. Am. Math. Soc.*, 123(11):3385–3396, 1995.

[27] M. Costabel. Singular integral operators on curves with corners. *Integral Equations Operator Theory*, 3(3):323–349, 1980.

[28] M. Costabel. An inverse for the Gohberg-Krupnik symbol map. *Proc. Roy. Soc. Edinburgh Sect. A*, 87(1-2):153–165, 1980/81.

[29] M. Costabel. Boundary integral operators on curved polygons. *Ann. Mat. Pura Appl. (4)*, 133:305–326, 1983.

[30] M. Costabel and M. Dauge. Invertibility of the biharmonic single layer potential operator. *Integr. Equat. Oper. Th.*, 24:46–67, 1996.

[31] M. Costabel, I. Lusikka, and J. Saranen. Comparison of three boundary element approaches for the solution of the clamped plate problem. In *Boundary elements IX, Vol. 2 (Stuttgart, 1987)*, pages 19–34. Comput. Mech., Southampton, 1987.

[32] M. Costabel and J. Saranen. Boundary element analysis of a direct method for the biharmonic Dirichlet problem. In *The Gohberg anniversary collection, Vol. II (Calgary, AB, 1988)*, volume 41 of *Oper. Theory Adv. Appl.*, pages 77–95. Birkhäuser, Basel, 1989.

[33] M. Costabel and E. Stephan. Boundary integral equations for mixed boundary value problems in polygonal domains and Galerkin approximation. In *Mathematical models and methods in mechanics*, volume 15 of *Banach Center Publ.*, pages 175–251. PWN, Warsaw, 1985.

[34] M. Costabel, E. Stephan, and W. L. Wendland. On boundary integral equations of the first kind for the bi-Laplacian in a polygonal plane domain. *Ann. Scuola Norm. Sup. Pisa Cl. Sci. (4)*, 10(2):197–241, 1983.

[35] M. Costabel and E. P. Stephan. On the convergence of collocation methods for boundary integral equations on polygons. *Math. Comp.*, 49(180):461–478, 1987.

[36] M. Costabel and W. L. Wendland. Strong ellipticity of boundary integral operators. *J. Reine Angew. Math.*, 372:34–63, 1986.

[37] I. I. Danilyuk. *Irregular boundary value problems on the plane.* Nauka, Moscow, 1975.

[38] K. R. Davidson. *C^*-algebras by example*, volume 6 of *Fields Institute Monographs.* American Mathematical Society, Providence, RI, 1996.

[39] C. De Boor. *A practical guide to splines.* Springer Verlag, New-York-Heidelberg-Berlin, 1978.

[40] V. D. Didenko. On an approximate solution of singular integral equations with Carleman shift and complex conjugate values of the unknown function. *Ukrain. Mat. Zh.*, 32(3):378–382, 1980.

[41] V. D. Didenko. On the approximate solution of the generalized Riemann boundary value problem. *Ukrain. Mat. Zh.*, 35(1):85–89, 135, 1983.

[42] V. D. Didenko. Approximate solution of certain systems of singular integral equations with shift. *Differentsial′nye Uravneniya*, 20(6):1055–1060, 1984.

[43] V. D. Didenko. Some methods of approximate solution of a four-element generalized Riemann boundary value problem. *Differentsial′nye Uravneniya*, 20(8):1439–1441, 1984.

[44] V. D. Didenko. A direct method for solving the Hilbert boundary value problem. *Math. Nachr.*, 133:317–341, 1987.

[45] V. D. Didenko. The interpolation method for solving the Hilbert boundary value problem. *Math. Nachr.*, 142:337–348, 1989.

[46] V. D. Didenko. A method for the approximate solution of singular integral equations with conjugation on piecewise-smooth contours. *Dokl. Akad. Nauk SSSR*, 318(6):1298–1301, 1991.

[47] V. D. Didenko. Para-algebras and approximate solution of singular integro-differential equations. *Dokl. Akad. Nauk SSSR*, 316(6):1293–1297, 1991.

[48] V. D. Didenko. Stability of some operator sequences and approximate solution of integral equations with conjugation. Dr.Sc. Thesis, Odessa I.I. Mechnikov State University, October 1994, 312 pp.

[49] V. D. Didenko. On parameters influencing the stability of quadrature methods for singular integral equations with conjugation. *Numer. Funct. Anal. Optim.*, 21(1-2):107–119, 2000.

[50] V. D. Didenko and V. N. Matskul. The reduction method for the solution of singular integral equations with conjugation. *Dokl. Akad. Nauk Ukrain. SSR Ser. A*, (7):40–43, 86, 1987.

[51] V. D. Didenko and V. N. Matskul. On the approximate solution of singular integral equations with conjugation. *Zh. Vychisl. Mat. i Mat. Fiz.*, 29(3):392–404, 478, 1989.

[52] V. D. Didenko and G. L. Pel′ts. On the stability of the spline-qualocation method for singular integral equations with conjugation. *Differentsial′nye Uravneniya*, 29(9):1593–1601, 1654, 1993.

[53] V. D. Didenko, G. Pittaluga, L. Sacripante, and E. Venturino. Numerical methods on graded meshes for singular integral equations with conjugation. *J. Computat. Methods Science and Engineering*, 4:209–233, 2004.

[54] V. D. Didenko, S. Roch, and B. Silbermann. Approximation methods for singular integral equations with conjugation on curves with corners. *SIAM J. Numer. Anal.*, 32(6):1910–1939, 1995.

[55] V. D. Didenko, S. Roch, and B. Silbermann. Some peculiarities of approximation methods for singular integral equations with conjugation. *Methods Appl. Anal.*, 7(4):663–685, 2000.

[56] V. D. Didenko and B. Silberman. Symbols of some operator sequences and quadrature methods for solving singular integral equations with conjugation. *Funkcional. Anal. i Priložen.*, 26(4):67–70, 1992.

[57] V. D. Didenko and B. Silbermann. On the stability of some operator sequences and the approximate solution of singular integral equations with conjugation. *Integral Equations Operator Theory*, 16(2):224–243, 1993.

[58] V. D. Didenko and B. Silbermann. Extension of C^*-algebras and Moore-Penrose stability of sequences of additive operators. *Linear Algebra Appl.*, 275/276:121–140, 1998.

[59] V. D. Didenko and B. Silbermann. On real and complex spectra in some real C^*-algebras and applications. *Z. Anal. Anwendungen*, 18(3):669–686, 1999.

[60] V. D. Didenko and B. Silbermann. On the approximate solution of some two-dimensional singular integral equations. *Math. Methods Appl. Sci.*, 24(15):1125–1138, 2001.

[61] V. D. Didenko and B. Silbermann. On stability of approximation methods for the Muskhelishvili equation. *J. Comput. Appl. Math.*, 146(2):419–441, 2002.

[62] V. D. Didenko and B. Silbermann. *Spline approximation methods for the biharmonic Dirichlet problem on non-smooth domains*, volume 135 of *Oper. Theory Adv. Appl.*, pages 145–160. Birkhäuser, Basel, 2002.

[63] V. D. Didenko and E. Venturino. Approximate solutions of some Mellin equations with conjugation. *Integral Equations Operator Theory*, 25(2):163–181, 1996.

[64] V. D. Didenko and E. Venturino. Approximate solution of the biharmonic problem in smooth domains. *J. Integral Equations Appl.*, 18(3):399–413, 2006.

[65] V. D. Didenko and E. Venturino. Approximation methods for the Muskhelishvili equation on smooth curves. *Math. Comp.*, 76(259):1317–1339, 2007.

[66] J. Dixmier. *C^*–Algebras.* North–Holland, Amsterdam, 1977.

[67] R. G. Douglas. *Banach algebra techniques in operator theory.* Academic Press, New York, 1972. Pure and Applied Mathematics, Vol. 49.

[68] R. G. Douglas. *Banach algebra techniques in the theory of Toeplitz operators.* Amer. Math. Soc., Providence, R.I., 1973. Lecture Notes, Vol. 15.

[69] R. G. Douglas and R. Howe. On the C^*–algebra of Toeplitz operators on the quarter plane. *Indiana Univ. Math. J.*, 21:1031–1035, 1972.

[70] R. V. Dudučava. Convolution integral equations with discontinuous presymbols, singular integral equations with fixed singularities, and their applications to problems in mechanics. *Trudy Tbiliss. Mat. Inst. Razmadze Akad. Nauk Gruzin. SSR*, 60:136, 1979.

[71] R. V. Duduchava. *Integral equations in convolution with discontinuous presymbols, singular integral equations with fixed singularities, and their applications to some problems of mechanics.* BSB B. G. Teubner Verlagsgesellschaft, Leipzig, 1979.

[72] R. V. Duduchava. On general singular integral operators of the plane theory of elasticity. *Rend. Sem. Mat. Univ. Politec. Torino*, 42(3):15–41, 1984.

[73] R. V. Duduchava. General singular integral equations and fundamental problems of the plane theory of elasticity. *Trudy Tbiliss. Mat. Inst. Razmadze Akad. Nauk Gruzin. SSR*, 82:45–89, 1986.

[74] R. V. Duduchava and T. I. Latsabidze. The index of singular integral equations with complex-conjugate functions on piecewise smooth lines. *Soobshch. Akad. Nauk Gruzin. SSR*, 115(1):29–32, 1984.

[75] R. V. Duduchava and T. I. Latsabidze. The index of singular integral equations with complex-conjugate functions on piecewise-smooth lines. *Trudy Tbiliss. Mat. Inst. Razmadze Akad. Nauk Gruzin. SSR*, 76:40–59, 1985.

[76] R. V. Duduchava, T. I. Latsabidze, and A. I. Saginashvili. Singular integral operators with complex conjugation on piecewise-smooth lines. *Soobshch. Akad. Nauk Gruzii*, 146(1):21–24 (1993), 1992.

[77] N. Dunford and J. T. Schwartz. *Linear operators. Part I:General Theory.* Wiley, New York–Chichester–Brisbane-Toronto–Singapore, 1988.

[78] A. D. Dzhuraev. *Method of singular integral equations.* Nauka, Moscow, 1987.

[79] J. Elschner. Asymptotics of solutions to pseudodifferential equations of Mellin type. *Math. Nachr.*, 130:267–305, 1987.

[80] J. Elschner and I. G. Graham. Numerical methods for integral equations of Mellin type. *J. Comput. Appl. Math.*, 125:423–437, 2000.

[81] J. Elschner and O. Hansen. Collocation method for the solution of the first boundary value problem of elasticity in a polygonal domain in $\mathbf{R}^2$. *J. Integ. Equat. Appl.*, 11(2):141–196, 1999.

[82] J. Elschner, S. Prössdorf, A. Rathsfeld, and G. Schmidt. Spline approximation of singular integral equations. *Demonstratio Math.*, 18(3):661–672, 1985.

[83] T. Fink. Splineapproximationsverfahren für singuläre Integralgleichungen auf ebenen, glattberandeten Gebieten. Ph.D. Dissertation, Fakultät für Mathematik, TU Chemnitz, December 1998, 124 pp.

[84] B. Fuglege. On a direct method of integral equations for solving the biharmonic dirichlet problem. *ZAMM*, 61:449–459, 1981.

[85] B. G. Gabdulhaev. Approximate solution of singular integral equations by the method of mechanical quadratures. *Dokl. Akad. Nauk SSSR*, 179:260–263, 1968.

[86] B. G. Gabdulhaev. *Optimal approximations of solutions of linear problems.* Kazann University, Kazann, 1980.

[87] B. G. Galerkin. Rods and plates. *Vestn. Inzh.*, 1(19):897–908, 1915. See also: *Colected Papers*, Vol. 1, Akad. Sci. SSSR (1952), 168–195.

[88] I. M. Gelfand, D. A. Raikov, and G. E. Shilov. *Commutative normed rings.* Chelsea Publishing Co., New York, 1964.

[89] I. C. Gohberg and I. A. Fel′dman. *Convolution equations and projection methods for their solution.* American Mathematical Society, Providence, R.I., 1974.

[90] I. C. Gohberg and N. J. Krupnik. The algebra generated by the one-dimensional singular integral operators with piecewise continuous coefficients. *Funkcional. Anal. i Priložen.*, 4(3):26–36, 1970.

[91] I. C. Gohberg and N. J. Krupnik. *One-dimensional linear singular integral equations. I*, volume 53 of *Operator Theory: Advances and Applications.* Birkhäuser Verlag, Basel, 1992.

[92] I. C. Gohberg and N. J. Krupnik. *One-dimensional linear singular integral equations. Vol. II*, volume 54 of *Operator Theory: Advances and Applications.* Birkhäuser Verlag, Basel, 1992.

[93] I. C. Gohberg and N. J. Krupnik. The symbol of singular integral operators on a composite contour. In *Proceedings of the Symposium on Continuum Mechanics and Related Problems of Analysis (Tbilisi, 1971), Vol. 1 (Russian)*, pages 46–59. Izdat. "Mecniereba", Tbilisi, 1973.

[94] I. C. Gohberg and N. Y. Krupnik. On the algebra generated by Toeplitz matrices. *Funkcional. Anal. i Priložen.*, 3(2):46–56, 1969.

[95] M. A. Golberg and C. S. Chen. *Discrete projection methods for integral equations.* Computational Mechanics Publications, Southampton, 1997.

[96] K. R. Goordearl. *Notes on real and complex C^*-algebras.* Shiva Publ. Lim., Cheshire, 1982.

[97] I. S. Gradshteyn and I. M. Ryzhik. *Table of integrals, series, and products.* Academic Press Inc., Boston, MA, fifth edition, 1994.

[98] I. G. Graham and G. A. Chandler. High–order methods for linear functionals of solutions of second kind integral equations. *SIAM J. Num. Anal.*, 25:1118–1179, 1988.

[99] A. Greenbaum, L. Greengard, and A. Mayo. On the numerical solution of the biharmonic equation in the plane. *Phys. D*, 60(1-4):216–225, 1992. Experimental mathematics: computational issues in nonlinear science (Los Alamos, NM, 1991).

[100] L. Greengard, M. C. Kropinski, and A. Mayo. Integral equation methods for Stokes flow and isotropic elasticity in the plane. *J. Comput. Phys.*, 125(2):403–414, 1996.

[101] P. Grisvard. *Boundary value problems in non-smooth domains*, volume 24 of *Monographs and Studies in Mathematics.* Pitman, Boston-London-Melbourne, 1985.

[102] R. Hagen, S. Roch, and B. Silbermann. *Spectral theory of approximation methods for convolution equations*, volume 74 of *Operator Theory: Advances and Applications.* Birkhäuser Verlag, Basel, 1995.

[103] R. Hagen, S. Roch, and B. Silbermann. *C^*-algebras and numerical analysis*, volume 236 of *Monographs and Textbooks in Pure and Applied Mathematics.* Marcel Dekker Inc., New York, 2001.

[104] R. Hagen and B. Silbermann. On the stability of the qualocation method. In *Seminar Analysis (Berlin, 1987/1988)*, pages 43–52. Akademie-Verlag, Berlin, 1988.

[105] R. Hagen and B. Silbermann. A Banach algebra approach to the stability of projection methods for singular integral equations. *Math. Nachr.*, 140:285–297, 1989.

[106] P. R. Halmos. Two subspaces. *Trans. Amer. Math. Soc.*, 144:381–389, 1969.

[107] J. Helsing. On the interior stress problem for elastic bodies. *ASME J. Appl. Mech.*, 67:658–662, 2000.

[108] J. Helsing and A. Jonsson. Stress calculations on multiply connected domains. *J. Comput. Phys.*, 176:256–482, 2002.

[109] G. C. Hsiao, P. Kopp, and W. L. Wendland. A Galerkin collocation method for some integral equations of the first kind. *Computing*, 25:89–130, 1980.

[110] G. C. Hsiao, P. Kopp, and W. L. Wendland. Some applications of a Galerkin-collocation method for boundary integral equations of the first kind. *Math. Methods Appl. Sci.*, 6(2):280–325, 1984.

[111] G. C. Hsiao and R. MacCamy. Solution of boundary value problems by integral equations of the first kind. *SIAM Rev.*, 15:687–705, 1973.

[112] V. V. Ivanov. *Theory of approximation methods and their application to the numerical solution of singular integral equations.* Noordhof, Leyden, 1976.

[113] P. Junghanns. Kollokationsverfahren zur näherungsweisen Lösung singulärer Integralgleichungen mit unstetigen Koeffizienten. *Math. Nachr.*, 102:17–24, 1981.

[114] P. Junghanns and A. Rathsfeld. On polynomial collocation for Cauchy singular integral equations with fixed singularities. *Integral Equations Operator Theory*, 43(2):155–176, 2002.

[115] P. Junghanns, S. Roch, and B. Silbermann. Collocation methods for systems of Cauchy singular integral equations on an interval. *Vychisl. Tekhnol.*, 6(1):88–124, 2001.

[116] P. Junghanns and B. Silbermann. Zur Theorie der Näherungsverfahren für singuläre Integralgleichungen auf Intervallen. *Math. Nachr.*, 103:199–244, 1981.

[117] P. Junghanns and B. Silbermann. Local Theory of collocation method for the approximate solution of singular integral equations. *Integral Equations and Operator Theory*, 7:791–807, 1984.

[118] V. P. Kadušin. Approximate solution of singular integral equations with complex conjugate unknowns. *Izv. Vysš. Učebn. Zaved. Matematika*, (6):109–113, 1976.

[119] V. P. Kadušin. On direct methods for the solution of a certain class of singular integral equations. *Izv. Vysš. Učebn. Zaved. Matematika*, (11):109–111, 1976.

[120] V. P. Kadušin. The quadrature-interpolation method of solution of singular integro-differential equations with complex conjugate unknowns. *Izv. Vyssh. Uchebn. Zaved. Mat.*, (12):60–66, 1978.

[121] A. I. Kalandiya. *Mathematical methods of two-dimensional elasticity.* Mir Publishers, Moscow, 1975.

[122] L. V. Kantorovich and G. P. Akilov. *Funktsional analysis.* Nauka, Moscow, third edition, 1984.

[123] N. Karapetiants and S. Samko. *Equations with involutive operators.* Birkhäuser Boston Inc., Boston, MA, 2001.

[124] B. V. Khvedelidze. Linear discontinuous boundary problems of the function theory, singular integral equations and applications. *Trudy Tbilissk. Matem. Inst. AN Gruz. SSR*, 23:3–158, 1957. (Russian).

[125] L. S. Klabukova. Approximate method of solution for the problems of Hilbert and Poincaré. *Vyčisl. Mat.*, 3:34–87, 1958.

[126] L. S. Klabukova. An approximate method of solving the Riemann-Hilbert problem in a multiply connected domain. *Vyčisl. Mat.*, 7:115–132, 1961.

[127] A. V. Kozak. A local principle in the theory of projection methods. *Dokl. Akad. Nauk SSSR*, 212:1287–1289, 1973.

[128] R. Kress. A Nyström method for boundary integral equations in domains with corners. *Numer. Math.*, 58(2):145–161, 1990.

[129] M. C. A. Kropinski. Integral equation methods for particle simulations in creeping flows. *Comput. Math. Appl.*, 38(5-6):67–87, 1999.

[130] M. C. A. Kropinski. Numerical methods for multiple inviscid interfaces in creeping flows. *J. Comput. Phys.*, 180(1):1–24, 2002.

[131] S. H. Kulkarni and B. V. Limaye. *Real function algebras*, volume 168 of *Monographs and Textbooks in Pure and Applied Mathematics.* Marcel Dekker Inc., New York, 1992.

[132] V. D. Kupradze. *Fundamental problems in the mathematical theory of diffraction (steady state processes).* NBS Rep. 2008. U. S. Department of Commerce National Bureau of Standards, Los Angeles, Calif., 1952.

[133] V. D. Kupradze. *Potential methods in the theory of elasticity.* Israel Program for Scientific Translations, Jerusalem, 1965.

[134] V. D. Kupradze, T. G. Gegelia, M. O. Basheleĭshvili, and T. V. Burchuladze. *Three-dimensional problems of the mathematical theory of elasticity and thermoelasticity*, volume 25 of *North-Holland Series in Applied Mathematics and Mechanics.* North-Holland Publishing Co., Amsterdam, 1979.

[135] S. Lang. *Introduction to differentiable manifolds.* Columbia University, New York–London, 1962.

[136] T. I. Latsabidze. Singular integral operators with complex conjugation on piecewise smooth lines. *Soobshch. Akad. Nauk Gruzin. SSR*, 112(3):481–484, 1983.

[137] T. I. Latsabidze. A Banach algebra generated by singular integral operators and operators of complex conjugation. *Soobshch. Akad. Nauk Gruzin. SSR*, 114(2):265–268, 1984.

[138] T. I. Latsabidze. Singular integral operators with complex conjugation on piecewise-smooth lines. *Trudy Tbiliss. Mat. Inst. Razmadze Akad. Nauk Gruzin. SSR*, 76:107–122, 1985.

[139] H. Lausch and D. Przeworska-Rolewicz. Pseudocategories, para-algebras and linear operators. *Math. Nachr.*, 138:67–82, 1988.

[140] J. E. Lewis. Layer potentials for elastostatic and hydrostatic in curvilinear polygonal domains. *Trans. Amer. Math. Soc.*, 320:53–76, 1990.

[141] B. Li. *Real operator algebras.* World Scientific Publishing Co. Inc., River Edge, NJ, 2003.

[142] I. K. Lifanov and I. E. Polonskii. Proof of the numerical method of "discrete vortices" for solving singular integral equations. *Prikl. Mat. Meh.*, 39(4):742–746, 1975.

[143] G. S. Litvinchuk. *Boundary value problems and singular integral equations with shift.* Nauka, Moscow, 1977.

[144] G. S. Litvinchuk and I. M. Spitkovskii. *Factorization of measurable matrix functions*, volume 25 of *Operator Theory: Advances and Applications.* Birkhäuser Verlag, Basel, 1987.

[145] Y. B. Lopatinskiĭ. On a method of reducing boundary problems for a system of differential equations of elliptic type to regular integral equations. *Ukrain. Mat. Ž.*, 5:123–151, 1953.

[146] Y. B. Lopatinskiĭ. *Introduction to modern theory of partial differential equations.* Naukova Dumka, Kiev, 1980.

[147] J. K. Lu. *Complex variable methods in plane elasticity*, volume 22 of *Series in Pure and Applied Mathematics.* World Scientific, Singapore, 1995.

[148] S. A. Lurie and V. V. Vasiliev. *The biharmonic problem in the theory of elasticity.* Gordon and Breach Publishers, Luxembourg, 1995.

[149] L. G. Magnaradze. On a general theorem of I. I. Privalov and its applications to certain linear boundary problems of the theory of functions and to singular integral equations. *Doklady Akad. Nauk SSSR (N.S.)*, 68:657–660, 1949.

[150] G. F. Mandžavidze. On a singular integral equation with discontinuous coefficients and its applications in the theory of elasticity. *Akad. Nauk SSSR. Prikl. Mat. Meh.*, 15:279–296, 1951.

[151] M. Marcus and H. Minc. *A survey of matrix theory and matrix inequalities.* Dover Publications Inc., New York, 1992.

[152] A. Mayo and A. Greenbaum. Fast parallel iterative solution of Poisson's and the biharmonic equations on irregular regions. *SIAM J. Sci. Statist. Comput.*, 13(1):101–118, 1992.

[153] V. V. Meleshko. Biharmonic problem in a rectangle. *Appl. Sci. Res.*, 1–4:217–249, 1998.

[154] A. N. Meleško. Approximate solution of a homogeneous Hilbert boundary value problem for a disk using polylogarithms. *Zh. Vychisl. Mat. i Mat. Fiz.*, 21(3):678–684, 811, 1981.

[155] C. Meyer. Reduktionsverfahren für singuläre Integralgleichungen mit speziellen Carlemanschen Verschiebungen. *Wiss. Z. Tech. Hochsch. Karl-Marx-Stadt*, 25(3):381–387, 1983.

[156] L. G. Mikhailov. *A new class of singular integral equations and its application to differential equations with singular coefficients.* Wolters-Noordhoff Publishing, Groningen, 1970.

[157] S. G. Mikhlin and S. Prössdorf. *Singular integral operators.* Springer-Verlag, Berlin, 1986.

[158] I. Mitrea. On the spectra of elastostatic and hydrostatic layer potentials on curvilinear polygons. *J. Fourier Anal. Appl.*, 8(5):443–487, 2002.

[159] G. J. Murphy. *C^*-algebras and operator theory.* Academic Press Inc., Boston, MA, 1990.

[160] N. I. Muskhelishvili. A new general method of solution of the fundamental boundary problems of the plane theory of elasticity. *Dokl. Akad. Nauk SSSR*, 3(100):7–11, 1934.

[161] N. I. Muskhelishvili. *Fundamental problems in the theory of elasticity.* Nauka, Moscow, 1966.

[162] N. I. Muskhelishvili. *Singular integral equations.* Nauka, Moscow, 1968.

[163] M. A. Naĭmark. *Normed algebras.* Wolters-Noordhoff Publishing, Groningen, third edition, 1972.

[164] V. I. Nyaga. A singular integral operator with conjugation along a contour with corner points. *Mat. Issled.*, (73):47–50, 1983. Studies in differential equations.

[165] V. I. Nyaga. Symbol of singular integral operators with conjugation in the case of a piecewise Lyapunov contour. *Dokl. Akad. Nauk SSSR*, 268(4):806–808, 1983.

[166] V. I. Nyaga and M. N. Yanioglo. A counterexample in the theory of singular integral equations with conjugation. In *Operations research and programming*, pages 104–107. "Shtiintsa", Kishinev, 1982.

[167] H. Ockendon and J. R. Ockendon. *Viscous flow.* Cambridge Texts in Applied Mathematics. Cambridge University Press, Cambridge, 1995.

[168] V. V. Panasyuk, M. P. Savruk, and Z. T. Nazarchuk. *The method of singular integral equations in two-dimensional problems of diffraction.* Naukova Dumka, Kiev, 1984.

[169] E. N. Parasjuk. On the index of the integral operator corresponding to the second fundamental problem of the plane theory of elasticity. *Ukrain. Mat. Ž.*, 16:250–253, 1964.

[170] V. Z. Parton and P. I. Perlin. *Integral equations of elasticity theory.* Nauka, Moscow, 1977.

[171] G. K. Pedersen. *C^*-algebras and their automorphism groups*, volume 14 of *London Mathematical Society Monographs.* Academic Press Inc., London, 1979.

[172] P. I. Perlin and Y. N. Shalyukhin. On numerical solution of integral equations of elasticity theory. *Izv. AN Kaz. SSR, Ser. fiz.-mat.*, 1:86–88, 1976.

[173] P. I. Perlin and Y. N. Shalyukhin. On numerical solution of some plane problems of elasticity theory. *Prikl. Mekh.*, 15(4):83–86, 1977.

[174] S. Prössdorf. *Some classes of singular equations*, volume 17 of *North-Holland Mathematical Library.* North-Holland Publishing Co., Amsterdam, 1978.

[175] S. Prössdorf. Ein Lokalisierungsprinzip in der Theorie der Spline-Approximationen und einige Anwendungen. *Math. Nachr.*, 119:239–255, 1984.

[176] S. Prössdorf and A. Rathsfeld. A spline collocation method for singular integral equations with piecewise continuous coefficients. *Integral Equations Operator Theory*, 7(4):536–560, 1984.

[177] S. Prössdorf and A. Rathsfeld. On spline Galerkin methods for singular integral equations with piecewise continuous coefficients. *Numer. Math.*, 48(1):99–118, 1986.

[178] S. Prössdorf and A. Rathsfeld. Stabilitätskriterien für Näherungsverfahren bei singulären Integralgleichungen in L^p. *Z. Anal. Anwendungen*, 6(6):539–558, 1987.

[179] S. Prössdorf and A. Rathsfeld. *Mellin techniques in the numerical analysis for one-dimensional singular integral equations*, volume 88 of *Report MATH.* Akademie der Wissenschaften der DDR Karl-Weierstrass-Institut für Mathematik, Berlin, 1988.

[180] S. Prössdorf and A. Rathsfeld. Quadrature and collocation methods for singular integral equations on curves with corners. *Z. Anal. Anwendungen*, 8(3):197–220, 1989.

[181] S. Prössdorf and A. Rathsfeld. Quadrature methods for strongly elliptic Cauchy singular integral equations on an interval. In *The Gohberg anniversary collection, Vol. II (Calgary, AB, 1988)*, volume 41 of *Oper. Theory Adv. Appl.*, pages 435–471. Birkhäuser, Basel, 1989.

[182] S. Prössdorf and G. Schmidt. A finite element collocation method for singular integral equations. *Math. Nachr.*, 100:33–60, 1981.

[183] S. Prössdorf and B. Silbermann. *Numerical analysis for integral and related operator equations.* Birkhäuser Verlag, Berlin–Basel, 1991.

[184] D. Przeworska-Rolewicz and S. Rolewicz. *Equations in linear spaces.* PWN—Polish Scientific Publishers, Warsaw, 1968. Monografie Matematyczne, Tom 47.

[185] A. Rathsfeld. Quadraturformelverfahren für eindimensionale singuläre Integralgleichungen. In *Seminar analysis*, pages 147–186. Akad. Wiss. DDR, Berlin, 1986.

[186] A. Rathsfeld. Reduktionsverfahren für singuläre Integralgleichungen mit stückweise stetigen Koeffizienten. *Math. Nachr.*, 127:125–143, 1986.

[187] A. Rathsfeld. Eine Quadraturformelmethode für Mellin-Operatoren nullter Ordnung. *Math. Nachr.*, 137:321–354, 1988.

[188] S. Roch. Spline approximation method cutting off singularities. *Z. Anal. Anwendungen*, 13(2):329–345, 1994.

[189] S. Roch, P. A. Santos, and B. Silbermann. *Non-commutative Gelfand theory. A tool-kit for operator theoretists and numerical analysts.* In Preparation.

[190] S. Roch and B. Silbermann. The Calkin image of algebras of singular integral operators. *Integral Equations Operator Theory*, 12(6):855–897, 1989.

[191] S. Roch and B. Silbermann. *Algebras of convolution operators and their image in the Calkin algebra*, volume 90 of *Report MATH.* Akademie der Wissenschaften der DDR Karl-Weierstrass-Institut für Mathematik, Berlin, 1990.

[192] S. Roch and B. Silbermann. Asymptotic Moore-Penrose invertibility of singular integral operators. *Integral Equations Operator Theory*, 26(1):81–101, 1996.

[193] S. Roch and B. Silbermann. C^*-algebra techniques in numerical analysis. *J. Operator Theory*, 35(2):241–280, 1996.

[194] S. Roch and B. Silbermann. Index calculus for approximation methods and singular value decomposition. *J. Math. Anal. Appl.*, 225(2):401–426, 1998.

[195] R. S. Saks. *Boundary problems for elliptic differential equations.* University of Novosibirsk, Novosibirsk, 1975.

[196] J. Saranen and G. Vainikko. *Periodic integral and pseudodifferential equations with numerical approximation.* Springer Monographs in Mathematics. Springer-Verlag, Berlin, 2002.

[197] J. Saranen and W. L. Wendland. On the asymptotic convergence of collocation methods with spline functions of even degree. *Math. Comp.*, 45(171):91–108, 1985.

[198] A. H. Schatz, V. Thomée, and W. L. Wendland. *Mathematical theory of finite and boundary element methods*, volume 15 of *DMV Seminar*. Birkhäuser Verlag, Basel, 1990.

[199] G. Schmidt. On spline collocation for singular integral equations. *Math. Nachr.*, 111:177–196, 1983.

[200] G. Schmidt. On spline collocation methods for boundary integral equations in the plane. *Math. Methods Appl. Sci.*, 7(1):74–89, 1985.

[201] V. Y. Shelepov. The index of an integral operator of potential type in the space L_p. *Dokl. Akad. Nauk SSSR*, 186:1266–1268, 1969.

[202] V. Y. Shelepov. Investigation by the method of Ya. B. Lopatinskiĭ of matrix integral equations in a space of continuous functions. In *General theory of boundary value problems*, pages 220–226. Naukova Dumka, Kiev, 1983.

[203] V. Y. Shelepov. The Noethericity of integral equations of the plane theory of elasticity in the spaces L_p, $p > 1$. *Dokl. Akad. Nauk Ukrain. SSR Ser. A*, (8):27–31, 86, 1988.

[204] D. I. Sherman. On the solution of the second fundamental problem of the theory of elasticity for plane multiconnected regions. *Doklady AN SSSR*, 4(3):119–122, 1935.

[205] D. I. Sherman. The theory of elasticity of static plane problems. *Trudy Tbil. Mat. Inst.*, 2:163–225, 1937.

[206] D. I. Sherman. On the solution of the plane static problem of the theory of elasticity for displacements given on the boundary. *Doklady AN SSSR*, 27(9):911–913, 1940.

[207] B. Silbermann. Locale Theorie des Reduktionsverfahrens für Toeplitzoperatoren. *Math. Nachr.*, 105:137–146, 1981.

[208] B. Silbermann. Locale Theorie des Reduktionsverfahrens für singuläre Integral Gleichungen. *Z. Anal. Anwendungen*, 1(6):45–56, 1982.

[209] B. Silbermann. How to compute the partial indices of a regular and smooth matrix-valued function? In *Factorization, singular operators and related problems (Funchal, 2002)*, pages 291–300. Kluwer Acad. Publ., Dordrecht, 2003.

[210] B. Silbermann. Modified finite sections for Toeplitz operators and their singular values. *SIAM J. Matrix Anal. Appl.*, 24(3):678–692 (electronic), 2003.

[211] J. Simič. Ein Beitrag zur Berechnung der rechteckigen Platten. *Z. Östereich. Ing. Architekt. Verein*, 60:709–714, 1908.

[212] I. B. Simonenko. The Riemann boundary-value problem for n pairs of functions with measurable coefficients and its application to the study of singular integrals in L_p spaces with weights. *Izv. Akad. Nauk SSSR Ser. Mat.*, 28:277–306, 1964.

[213] I. B. Simonenko. A new general method of investigating linear operator equations of singular integral equation type. I. *Izv. Akad. Nauk SSSR Ser. Mat.*, 29:567–586, 1965.

[214] I. B. Simonenko and C. N. Min′. *Local method in the theory of one-dimensional singular integral equations with piecewise continuous coefficients. Noethericity.* Rostov. Gos. Univ., Rostov, 1986.

[215] I. H. Sloan. A quadrature-based approach to improving the collocation method. *Numer. Math.*, 54:41–56, 1988.

[216] I. H. Sloan and W. L. Wendland. A quadrature-based approach to improving the collocation method for splines of even degree. *Z. Anal. Anwendungen*, 8:361–376, 1989.

[217] I. S. Sokolnikoff. *Mathematical theory of elasticity.* Robert E. Krieger Publishing Co. Inc., Melbourne, Fla., second edition, 1983.

[218] A. P. Soldatov. *Kraevye zadachi teorii funktsii v oblastyakh s kusochno-gladkoi granitsei. Chast I, II.* Tbilis. Gos. Univ., Tbilisi, 1991.

[219] S. Timoshenko and S. Woinowsky-Krieger. *Theory of plates and shells.* McGraw-Hill, New York, 1959.

[220] M. S. Titcombe, M. J. Ward, and M. C. Kropinski. A hybrid method for low Reynolds number flow past an asymmetric cylindrical body. *Stud. Appl. Math.*, 105(2):165–190, 2000.

[221] I. N. Vekua. *Generalized analytic functions.* Nauka, Moscow, second edition, 1988.

[222] N. P. Vekua. *Systems of singular integral equations.* P. Noordhoff Ltd., Groningen, 1967.

[223] E. E. Vitrichenko and V. D. Didenko. The collocation method for the approximate solution of a singular integro-differential equation. In *Boundary value problems and their applications*, pages 24–31. Chuvash. Gos. Univ., Cheboksary, 1989.

[224] N. E. Wegge-Olsen. *K-theory and C^*-algebras. A friendly approach.* Univ. Press, Oxford, 1993.

[225] N. N. Yukhanonov. Exact and approximate solution of a boundary value problem of conjugation of analytic functions with derivatives on the circle. *Dokl. Akad. Nauk Tadzhik. SSR*, 29(2):78–81, 1986.

[226] N. N. Yukhanonov. On the determination of the exact and approximate solutions of a generalized characteristic singular integral equation. *Dokl. Akad. Nauk Tadzhik. SSR*, 33(11):720–724, 1990.

[227] V. A. Zolotarevskiĭ and V. I. Njaga. Approximate solution of a certain class of singular integral equations with shift. *Izv. Vysš. Učebn. Zaved. Matematika*, (11(174)):105–108, 1976.

[228] A. Zygmund. *Trigonometric series. Vol. I, II.* Cambridge Mathematical Library. Cambridge University Press, Cambridge, third edition, 2002.

Index

Operator Theory: Advances and Applications (OT)

Edited by
Israel Gohberg, Tel Aviv University, Israel

This series is devoted to the publication of current research in operator theory, with particular emphasis on applications to classical analysis and the theory of integral equations, as well as to numerical analysis, mathematical physics and mathematical methods in electrical engineering.

BIRKHÄUSER

OT 183: Nazaikinskii, V.E.. / Savin, A.Yu. / Sternin, B.Yu., Elliptic Theory and Noncommutative Geometry (2008). Subseries **A**dvances in **P**artial **D**ifferential **E**quations. ISBN 978-3-7643-8774-7

OT 182: van der Mee, C., Exponentially Dichotomous Operators and Applications (2008). Subseries **L**inear **O**perators and **L**inear **S**ystems. ISBN 978-3-7643-8732-7

OT 181: Bastos, A. / Gohberg, I. / Lebre, A.B. / Speck, F.-O. (eds.), Operator Algebras, Operator Theory and Applications (2008). ISBN 978-3-7643-8683-2

OT 180: Okada, S. / Ricker, W.J. / Sánchez Pérez, E.A., Optimal Domain and Integral Extension of Operators Acting in Function Spaces (2008). ISBN 978-3-7643-8647-4

OT 179: Ball, J.A. / Eidelman, Y. / Helton, J.W. / Olshevsky, V. / Rovnyak, J. (eds.), Recent Advances in Matrix and Operator Theory (2007). ISBN 978-3-7643-8538-5

OT 178: Bart, H. / Gohberg, I. / Kaashoek, M.A. / Ran, A.C.M., Factorization of Matrix and Operator Functions: The State Space Method (2007). ISBN 978-3-7643-8267-4

OT 177: López-Gómez, J. / Mora-Corral, C., Algebraic Multiplicity of Eigenvalues of Linear Operators (2007). ISBN 978-3-7643-8400-5

OT 176: Alpay, D. / Vinnikov, V. (Eds.), System Theory, the Schur Algorithm and Multidimensional Analysis (2007). ISBN 978-3-7643-8136-3

OT 175: Förster, K.-H. / Jonas, P. / Langer, H. / Trunk, C. (Eds.), Operator Theory in Inner Product Spaces (2007). ISBN 978-3-7643-8269-8

OT 174: Janas, J. / Kurasov, P. / Laptev, A. / Naboko, S. / Stolz, G. (Eds.), Operator Theory, Analysis and Mathematical Physics (2007). ISBN 978-3-7643-8134-9

OT 173: Emel'yanov, E.Yu., Non-spectral Asymptotic Analysis of One-Parameter Operator Semigroups (2006). ISBN 978-3-7643-8095-3

OT 172: Toft, J. / Wong, M.W. / Zhu, H. (Eds.), Modern Trends in Pseudo-Differential Operators (2007). ISBN 3-7643-8097-7

OT 171: Dritschel, M. (Ed.), The Extended Field of Operator Theory (2007). ISBN 3-7643-7979-0

OT 170: Erusalimsky, Ya.M. / Gohberg, I. / Grudsky, S.M. / Rabinovich, V. / Vasilevski, N. (Eds.), Modern Operator Theory and Applications (2007). ISBN 3-7643-7736-4

OT 169: Haase, M., The Functional Calculus for Sectorial Operators (2006). ISBN 3-7643-7697-X

OT 168: Koelink, E. / van Neerven, J. / de Pagter, B. / Sweers, G. (Eds.), Partial Differential Equations and Functional Analysis. ISBN 3-7643-7600-7

OT 167: Alpay, D. (Ed.), Wavelets, Multiscale Systems and Hypercomplex Analysis (2006). ISBN 3-7643-7587-6

OT 166: De Gosson, M., Symplectic Geometry and Quantum Mechanics (2006). Subseries **A**dvances in **P**artial **D**ifferential **E**quations ISBN 3-7643-7574-4

OT 165: Alpay, D. / Gohberg, I. (Eds.), Interpolation, Schur Functions and Moment Problems (2006). Subseries **L**inear **O**perators and **L**inear **S**ystems. ISBN 3-7643-7546-9

Frontiers in Mathematics

This series is designed to be a repository for up-to-date research results which have been prepared for a wider audience. Graduates and post-graduates as well as scientists will benefit from the latest developments at the research frontiers in mathematics and at the "frontiers" between mathematics and other fields like computer science, physics, biology, economics, finance, etc.

BIRKHÄUSER

Advisory Board

Leonid Bunimovich (Atlanta), Benoît Perthame (Paris), Laurent Saloff-Coste (Rhodes Hall), Igor Shparlinski (Sydney), Wolfgang Sprössig (Freiberg), Cédric Villani (Lyon)

■ **Bouchut, F.**, CNRS & Ecole Normale Sup., Paris, France

Nonlinear Stability of Finite Volume Methods for Hyperbolic Conservation Laws and Well-Balanced Schemes for Sources

2004. 142 pages. Softcover.
ISBN 978-3-7643-6665-0

■ **Catoni, F. / Boccaletti, D. / Cannata, R. / Catoni, V. / Nichelatti, E. / Zampetti, P.**

The Mathematics of Minkowski Space-Time. With an Introduction to Commutative Hypercomplex Numbers

2008. 250 pages. Softcover.
ISBN 978-3-7643-8613-9

■ **Brešar, M.**, University of Maribor, Slovenia / **Chebotar, M.A.**, Kent State University, USA / **Martindale III, W.S.**, University of Massachusetts, Amherst, MA, USA

Functional Identities

2007. 272 pages. Softcover.
ISBN 978-3-7643-7795-3

■ **Cardinali, I.**, Università di Siena, Italy / **Payne, S.E.**, University of Colorado, Denver, CO, USA

q-Clan Geometries in Characteristic 2

2007. 180 pages. Softcover.
ISBN 978-3-7643-8507-1

■ **Clark, J.**, Otago Univ., New Zealand / **Lomp, C.**, Univ. di Porto, Portugal / **Vanaja, N.**, Mumbai Univ., India / **Wisbauer, R.**, Univ. Düsseldorf, Germany

Lifting Modules

2006. 408 pages. Softcover.
ISBN 978-3-7643-7572-0

■ **De Bruyn, B.**, Ghent University, Belgium

Near Polygons

2006. 276 pages. Softcover.
ISBN 978-3-7643-7552-2

■ **Henrot, A.**, Université Henri Poincaré, Vandoeuvre-les-Nancy, France

Extremum Problems for Eigenvalues of Elliptic Operators

2006. 216 pages. Softcover.
ISBN 978-3-7643-7705-2

■ **Kasch, F.**, Universität München, Germany / **Mader, A.**, Hawaii University

Rings, Modules, and the Total

2004. 148 pages. Softcover.
ISBN 978-3-7643-7125-8

■ **Krausshar, R.S.**, Ghent University, Belgium

Generalized Analytic Automorphic Forms in Hypercomplex Spaces

2004. 182 pages. Softcover.
ISBN 978-3-7643-7059-6

■ **Kurdachenko, L.**, Dnipropetrovsk National University, Ukraine / **Otal, J.**, University of Zaragoza, Spain / **Subbotin, I.Ya.**, National University, Los Angeles, CA, USA

Artinian Modules over Group Rings

2006. 259 pages. Softcover.
ISBN 978-3-7643-7764-9

■ **Lindner, M.**, University of Reading, UK

Infinite Matrices and their Finite Sections. An Introduction to the Limit Operator Method

2006. 208 pages. Softcover.
ISBN 978-3-7643-7766-3

■ **Perthame, B.**, Ecole Normale Supérieure, Paris, France

Transport Equations in Biology

2006. 206 pages. Softcover.
ISBN 978-3-7643-7841-7

■ **Thas, K.**, Ghent University, Belgium

Symmetry in Finite Generalized Quadrangles

2004. 240 pages. Softcover.
ISBN 978-3-7643-6158-7

■ **Xiao, J.**, Memorial University of Newfoundland, St. John's, NF, Canada

Geometric Q_p Functions

2006. 250 pages. Softcover.
ISBN 978-3-7643-7762-5

■ **Zaharopol, R.**, Math. Reviews, Ann Arbor, USA

Invariant Probabilities of Markov-Feller Operators and Their Supports

2005. 120 pages. Softcover.
ISBN 978-3-7643-7134-0